CHECKLIST FOR USABILITY

(Numbers in parentheses refer to the first page of major discussion in the text.)

CONTENT

☐ Is all the material relevant to this user for this task?   (30)

☐ Has all the material been checked for accuracy?   (150)

☐ Is the level of technicality appropriate for this audience?   (24)

☐ Are warnings and cautions inserted where needed?   (470)

☐ Are claims, conclusions, and recommendations supported by evidence?   (526)

☐ Is the material free of gaps, foggy areas, or needless details?   (31)

☐ Are all key terms defined?   (424)

☐ Are all sources documented?   (578)

ORGANIZATION

☐ Is the structure of the document visible at a glance?   (193)

☐ Is there a clear line of reasoning that emphasizes what is important?   (195)

☐ Is material organized in the sequence users are expected to follow?   (208)

☐ Is everything easy to locate?   (193)

☐ Is the material "chunked" into easily digestible parts?   (193)

STYLE

☐ Is each sentence understandable the first time it is read?   (216)

☐ Is rich information expressed in the fewest words possible?   (224)

☐ Are sentences put together with enough variety?   (231)

☐ Are words chosen for exactness?   (235)

☐ Is the tone appropriate?   (244)

DESIGN

☐ Is page design inviting, accessible, and appropriate for the user's needs?   (305)

☐ Are there adequate aids to navigation (heads, lists, typestyles)?   (308)

☐ Are adequate visuals used to clarify, emphasize, or summarize?   (257)

☐ Do supplements accommodate the needs of a diverse audience?   (324)

ETHICAL, LEGAL, AND CULTURAL CONSIDERATIONS

☐ Does the document indicate sound ethical judgment?   (61)

☐ Does the document respect copyright law and other legal standards?   (73)

☐ Does the document respect users' cultural diversity?   (48)

The cover illustration, Book of Knowledge of Mechanical Processes, *by Al Djazari (c. 1300), represents the fusion of old and new that embodies technical communication. While technical communicators use classical rhetorical principles, they do so with the most modern tools and in the service of up-to-the-moment technology.*

Technical Communication

EIGHTH EDITION

JOHN M. LANNON

University of Massachusetts, Dartmouth

 LONGMAN

An imprint of Addison Wesley Longman, Inc.

New York • Reading, Massachusetts • Menlo Park, California • Harlow, England
Don Mills, Ontario • Sydney • Mexico City • Madrid • Amsterdam

Sponsoring Development Manager: Arlene Bessenoff
Development Editor: David Munger
Marketing Manager: Renée Ortbals
Supplements Editor: Donna Campion
Full Service Production Manager: Patti Brecht
Project Coordination, Text Design, and Electronic Page Makeup: Nesbitt Graphics, Inc.
Cover Designer/Manager: Nancy Danahy
Cover Illustration: "Livre de la Connaissance des Procèdes Mécaniques," Book of Knowledge of Mechanical Processes, Al Djazari, c. 1300, Turkey. Topkapi Serail-Museum, Istanbul, Turkey/Giraudon, Paris/Superstock.
Photo Researcher: Julie Tesser
Senior Print Buyer: Hugh Crawford
Printer and Binder: World Color Book Services
Cover Printer: The Lehigh Press, Inc.

Library of Congress Cataloging-in-Publication Data
Lannon, John M.
    Technical communication/John M. Lannon. — 8th ed.
        p.   cm.
    Rev. ed. of: Technical writing. 7th ed. 1997.
    Includes bibliographical references and index.
    ISBN 0-321-02395-1
        1. Technical writing.   2. Communication of technical information.
    I. Lannon, John M.   Technical writing.   II. Title.
    T11.L24 2000
    808'.0666—dc21   99-23623
                                                                                 CIP

Please visit our website at http://www.awlonline.com/lannontech

ISBN 0-321-02395-1

345678910—WCT—02010099

Brief Contents

Detailed Contents

Preface

Whether handwritten, electronically mediated, or face-to-face, workplace communication is more than a value-neutral exercise in "information transfer"; it is a complex social transaction. Each rhetorical situation has its own specific interpersonal, ethical, legal, and cultural demands. Moreover, today's workplace professional is not only a fluent *communicator,* but also a discriminating *consumer* of information, skilled in the methods of inquiry, retrieval, evaluation, and interpretation essential to informed decision making.

Designed in response to these issues, *Technical Communication,* eighth edition, addresses a wide range of interests for classes in which students from a variety of majors are enrolled. The text explains, illustrates, and applies rhetorical principles to an array of assignments, from brief memos and summaries to formal reports and proposals. To help students develop awareness of audience and accountability, exercises incorporate the problem-solving demands typical in college and on the job. Self-contained chapters allow for various course plans and customized assignments.

__Note:__ While "writing" remains the central component in this edition, the title has been changed (from Technical Writing *to* Technical Communication*) to reflect the focus on the whole range of communication tasks in today's workplace—from managing collaborative groups to designing pages for the World Wide Web.*

HALLMARKS OF THE EIGHTH EDITION

The hallmarks of the eighth edition of *Technical Communication* include:

- **Increased coverage of computers and the Internet as used in technical communication.** Fully integrated computing advice is supplemented by "In Brief" discussions of technology and interpersonal issues that are shaping workplace communication.
- **A full chapter dedicated to collaboration.** In acknowledgment of the increasing importance of this topic, the eighth edition of *Technical Communication* includes a new chapter on collaboration (Chapter 6). The chapter provides an emphasis on computer-mediated and Internet collaboration. Collaborative projects are featured throughout the text.
- **Increased emphasis on information literacy.** *Information literate people* are those who "know how knowledge is organized, how to find information, and how to use information in such a way that others can learn from them."* Crit-

---

American Library Association Presidential Committee on Information Literacy: Final Report. Chicago: ALA, 1989.

ical thinking—the basis of information literacy—is covered intensively in Part II and integrated throughout the text.

◆ **Increased coverage of usability testing.** Usability receives consistent and explicit emphasis throughout, with a new chapter on usability (Chapter 17) and usability checklists at the end of relevant chapters in Parts IV and V.

◆ **Expanded treatment of ethical and legal issues.** The book now features integrated coverage of legal and ethical considerations in word choice, product descriptions, instructions, and other forms of hard copy and electronic communication.

◆ **Stronger coverage of international issues in technical communication.** Advice on connecting with global audiences includes analysis of the specific cultural context for various types of communication.

ADDITIONAL FEATURES OF THE EIGHTH EDITION

Technical Communication, eighth edition, also includes the following:

◆ **A new four-color design throughout** allows the book to make clearer distinctions between different types of documents and underscores the trend in four-color workplace documents.

◆ **"In Brief" boxes** provide interesting and topical applications of the important issues discussed in various chapters, such as collaborating, technology, and ethics.

◆ **Guidelines** help students apply and synthesize the information in the chapter and offer practical suggestions for real workplace situations.

◆ **Exercises/Collaborative activities** at the end of each chapter help students apply what they've learned. More collaborative activities have been added to reflect the presence of collaboration in all aspects of technical communication.

◆ **Usability Checklists** help students polish their writing by giving them points to consider and page cross-references so that they can refer to specific passages in the text to find more information on each point.

◆ **Marginal annotations** highlight important concepts in the text. The new **Note** annotation adds clarification or points out up-to-the-minute business and technological advances.

ORGANIZATION OF *TECHNICAL COMMUNICATION,* EIGHTH EDITION

The text begins with a brief overview of workplace communication in Chapter 1, followed by five major sections:

Part I: Communicating in the Workplace treats job-related communication as a problem-solving process. Students learn to think critically about the informative, persuasive, and ethical dimensions of their communications. They also learn how to adapt to the interpersonal challenges of collaborative work and to the various needs and expectations of global audiences.

Part II: Retrieving, Analyzing, and Synthesizing Information treats research as a deliberate inquiry process. Students learn to formulate significant research questions; to explore primary and secondary sources in hard copy and electronic form; to evaluate and interpret their findings; and to summarize for economy, accuracy, and emphasis.

Part III: Structural and Style Elements offers strategies for organizing and conveying messages that users can follow and understand. Students learn to control their material and to develop a readable style.

Part IV: Visual, Design, and Usability Elements treats the rhetorical implications of graphics, page design, and document supplements. Students learn to enhance a document's access, appeal, and visual impact for audiences who need to locate, understand, and use the information successfully.

Part V: Specific Documents and Applications applies earlier concepts and strategies to the preparation of print and electronic documents and oral presentations. Various letters, memos, reports, and proposals offer a balance of examples from the workplace and from student writing. Each sample document has been chosen so that students can emulate it easily.

Finally, the appendixes contain instructions for recording research findings and for documenting them in MLA or APA style; a demonstration of the writing process in a workplace setting; and a brief handbook of grammar, usage, and mechanics.

NEW TO THIS EDITION: A CHAPTER-BY-CHAPTER GUIDE

In developing *Technical Communication,* eighth edition, I have condensed and reorganized all chapters for greater clarity, conciseness, and emphasis. I have included new and updated examples throughout the text, and provided more annotated writing samples. I have expanded the number of collaborative projects and exercises in critical thinking. See especially Chapters 2, 6, 8, 10, 12, 17, 20, and 23. New material includes:

- A new Chapter 6, on working collaboratively. The chapter covers guidelines and strategies for organizing a team, managing conflict, listening, thinking creatively, and editing the work of others.
- A fully revised Chapter 7, on critical thinking in the research process. New sections discuss how expert opinion influences decision making and offer guidelines for evaluating expert opinion.
- A fully revised Chapter 8, on exploring hard copy, online, and Internet sources. New sections discuss the benefits of hard copy versus electronic sources, an-

swer frequently asked questions about copyright, and present guidelines for Internet research.

- ◆ A fully revised Chapter 9, on exploring primary sources. The chapter presents guidelines for interviews and questionnaires and discusses public records that are available through the Freedom of Information Act.
- ◆ An expanded Chapter 10, on critically evaluating and interpreting information. New sections cover guidelines for evaluating sources (including Web sources) and for deciphering evidence, discuss fallacies inherent in causal claims and statistical data, and present the types of potential errors in any legitimate research.
- ◆ Chapter 12 includes a new section on storyboards as an organizing tool.
- ◆ Chapter 14 now includes guidelines for fitting visuals with printed text.
- ◆ Chapter 15 now provides a section on designing on-screen pages.
- ◆ A new Chapter 17, on usability testing, covers criteria and guidelines for testing a document's usability and discusses usability issues in online or multimedia documents.
- ◆ A fully revised Chapter 18, on memo reports and electronic mail. New sections cover interpersonal considerations in writing a memo and guidelines for choosing email versus paper or the phone.
- ◆ A fully revised Chapter 19, on workplace letters. New sections cover electronic job hunting (including computer-scannable and hyperlinked résumés) and guidelines for successful job interviews.
- ◆ A new Chapter 20, on creating Web pages and other electronic documents. The chapter presents introductions to hypertext and HTML, usability criteria, and guidelines for creating a Web site.
- ◆ Chapter 23 now includes a section on faulty instructions and legal liability.
- ◆ A fully revised Chapter 25, on analytical reports. New sections illustrate the reasoning processes in causal analysis, comparative analysis, and feasibility analysis; the perils of excessive information; and the importance of self-assessment during the analytical process.
- ◆ A new appendix (Appendix A) covers recording and documenting research findings.
- ◆ A new appendix (Appendix B) illustrates the problem-solving process in an actual workplace situation.
- ◆ A revised and redesigned appendix (Appendix C) presents the essentials of grammar, usage, and mechanics.

INSTRUCTIONAL SUPPLEMENTS

These ancillary materials are available to accompany *Technical Communication,* eighth edition:

- ◆ An **Instructor's Manual** includes a test bank, chapter quizzes, and master sheets for overhead or opaque projection.

- ◆ A set of **transparencies** on disk supplements the master sheets in the Instructor's Manual.
- ◆ A **dedicated Web site** provides activities that take students and instructors beyond the textbook. The site includes chapter overviews, individual and collaborative projects, and links to resources for both students and instructors. For more information visit <http://www.awlonline.com/lannontech>.
- ◆ *Daedalus Online®* provides an Internet-based collaborative writing environment for students. The program offers prewriting strategies and prompts, computer-mediated conferencing, peer collaboration and review, comprehensive writing support, and secure, twenty-four-hour availability. For educators, *Daedalus Online* offers a comprehensive suite of online course management tools for managing an online class, dynamically linking assignments, and facilitating a heuristic approach to writing instruction. *Daedalus Online* is available at a discount when bundled with *Technical Communication,* eighth edition. For more information, visit <http://www.awlonline.com/daedalus>, or contact your Addison Wesley Longman sales representative.
- ◆ *The Longman English Pages,* our free content-rich Web site, includes additional reading selections and writing exercises. *The Longman English Pages* has a section dedicated to technical communication at <http://longman.awl.com/englishpages>.
- ◆ *Visual Communication: A Writer's Guide,* by Susan Hilligoss, introduces document design principles that writers can apply across different genres of writing, including academic papers, résumés, business letters, Web pages, brochures, newsletters, and proposals. Emphasizing audience and genre analysis, the guide shows how readers' expectations influence and shape a document's look. Practical discussions of space, type, organization, pattern, graphic elements, and visuals are featured, along with planning worksheets and design samples and exercises.

ACKNOWLEDGMENTS

Many of the refinements in this and earlier editions were inspired by generous and insightful suggestions from the following reviewers: Mary Beth Bamforth, Wake Technical Community College; Marian G. Barchilon, Arizona State University East; Christiana Birchak, University of Houston, Downtown; Gene Booth, Albuquerque Technical Vocational Institute; Alma Bryant, University of South Florida; Joanna B. Chrzanowski, Jefferson Community College; Jim Collier, Virginia Tech; Daryl Davis, Northern Michigan University; Charlie Dawkins, Virginia Polytechnical Institute and State University; Pat Dorazio, SUNY Institute of Technology; Julia Ferganchick-Neufang, University of Arkansas; Clint Gardner, Salt Lake Community College; Lucy Graca, Arapahoe Community College; Susan Guzman-Trevino, Temple College; Wade Harrell, Howard University; Linda Harris, University of Maryland, Baltimore County; Michael Joseph Hassett, Brigham Young University;

Cecilia Hawkins, Texas A & M University; Robert Hogge, Weber State University; Gloria Jaffe, University of Central Florida; Bruce L. Janoff, University of Pittsburgh; Jack Jobst, Michigan Technical University; Maxine Turner, Georgia Institute of Technology; JoAnn Kubala, Southwest Texas State University; Karen Kuralt, Louisiana Tech University; Elizabeth A. Latshaw, University of South Florida; Sherry Little, San Diego State University; Linda Loehr, Northeastern University; Devonee McDonald, Kirkwood Community College; James L. McKenna, San Jacinto College; Troy Meyers, California State University, Long Beach; Mohsen Mirshafiei, California State University, Fullerton; Thomas Murphy, Mansfield University; Thomas A. Murray, SUNY Institute of Technology; Shirley Nelson, Chattanooga State Technical Community College; Gerald Nix, San Juan College; Megan O'Neill, Creighton University; Don Pierstorff, Orange Coast College; Carol Clark Powell, University of Texas at El Paso; Mark Rollins, Ohio University, Athens; Beverly Sauer, Carnegie Mellon University; Sharla Shine, Terra Community College; Susan Simon, City College of the City University of New York; Tom Stuckert, University of Findlay; Anne Thomas, San Jacinto College; Jeff Wedge, Embry-Riddle University; Kristin Woolever, Northeastern University; Carolyn Young, University of Wyoming; Stephanee Zerkel, Westark College; Beverly Zimmerman, Brigham Young University; and Don Zimmerman, Colorado State University.

At the University of Massachusetts, Raymond Dumont was a constant source of help and ideas. Many other colleagues, graduate students, teaching assistants, and alumni offered countless suggestions. As always, students gave me feedback and inspiration.

This edition is the product of exceptional editorial guidance and production support from Anne Elizabeth Smith, Arlene Bessenoff, Tom Maeglin, David Munger, Patti Brecht, and Janet Nuciforo. Thank you all for your tolerance and generosity.

A special thank you to those I love: Chega, Daniel, Sarah, Patrick, and Max.

John M. Lannon

Introduction to Technical Communication

TECHNICAL COMMUNICATION SERVES
PRACTICAL NEEDS

WRITING IS PART OF MOST CAREERS

COMMUNICATION HAS AN ELECTRONIC
AND A HUMAN SIDE

COMMUNICATION REACHES A GLOBAL
AUDIENCE

In Brief Transferable Skills for the Twenty-first
Century

Industry, business, and government all rely on *technical communicators:* people who create, locate, analyze, and distribute information—within their organizations and around the world.

Information provides a basis for action: to perform tasks, answer questions, solve problems, or make decisions. A company, for example, might need information like the following (Davenport 39, 61, 146):

Typical information
needs in the
workplace

- ◆ What is our competition doing and how should we respond?
- ◆ Are customer preferences changing, and, if so, how?
- ◆ What new government regulations do we need to address?
- ◆ How can we cut production costs without sacrificing quality?
- ◆ Should we build, rent, or buy?
- ◆ What new technology should our company be thinking about?

Information is not
merely raw data

To help answer such questions, Web sites, intranets, and other resources—online or off—provide all sorts of *data* (measurements, observations, prices, statistics). But only when "endowed with relevance and purpose" does data become *information* (Davenport 9). In short, we create information by filtering, evaluating, and interpreting raw data and placing it in a usable context. And we usually shape our final product as a *document* (memo, letter, report, manual, email, or Web page).

Whenever you prepare a document that will serve as a basis for action, you work as a "technical communicator."

TECHNICAL COMMUNICATION SERVES PRACTICAL NEEDS

Unlike poetry or fiction, which appeal mainly to our *imagination,* technical documents appeal to our *understanding.* Technical communication therefore rarely seeks to entertain, create suspense, or invite differing interpretations. If you have written any type of lab or research report, you already know that a technical document leaves little room for ambiguity. To serve practical needs in the workplace, technical documents must be user oriented and efficient.

Technical Documents Are User Oriented

Instead of focusing on the writer's desire for self-expression, a technical document addresses the user's desire for information. This doesn't mean you should seem robotic, without any personality (or *voice*) at all. Your document may in fact reveal a lot about you (your competence, knowledge, integrity), but it rarely focuses on you personally. Readers are interested in *what you have done, what you recommend,* or *how you speak for your company;* they have only a professional interest in *who you are* (your feelings, hopes, dreams, visions). A personal essay, then, would not be technical communication. Consider this essay fragment:

What users expect

Focuses on the feelings

> Computers are not a particularly forgiving breed. The wrong key struck or the wrong command typed is almost sure to avenge itself on the inattentive user by banishing the document to some electronic trash can.

This personal view conveys a good deal about the writer's resentment and anxiety but very little about computers themselves.

The following example can be called technical communication because it focuses (see italics) on the subject, on what the writer has done, and on what the user should do:

Focuses on the subject, actions taken, and actions required

> On VR 320 terminals, *the BREAK key* is adjacent to keys used for text editing and special functions. Too often, users inadvertently strike the BREAK key, causing the program to quit prematurely. To prevent the problem, *we have* modified all database management terminals: to quit a program, *you must* now strike BREAK twice successively.

This next example can also be called technical communication because it focuses on what the writer recommends:

Focuses on the recommendation

> I recommend that our Web server be upgraded by a maximum addition to RAM, a new T–1 Internet connection, and a 40 GB RAID disk array. This expansion will (1) increase the system's responsiveness, (2) allow for real-time videoconferencing, and (3) allow us to move all third-party Web hosting to our own network.

And so your document never makes you "disappear," but it does focus on that which is most important to users.

Technical Documents Strive for Efficiency

How workplace and school writing differ

Professors read to *test* our knowledge; colleagues, customers, and supervisors read to *use* our knowledge. In the U.S. workplace, users of a document want only what they need; instead of reading from beginning to end, they are likely to use the document for reference. An efficient document saves time and energy.

NOTE

For any type of global communication, keep in mind that many cultures consider a direct, straightforward communication style offensive (page 8).

In any system, "efficiency" is the ratio of useful output to input. For the product that comes out, how much energy goes in?

When a system is efficient, the output nearly equals the input.

Similarly, a document's efficiency can be measured by how hard the user works to understand the message.

USER SPENDS ENERGY (input) → **DOCUMENT** → **USER GETS THE MESSAGE (output)**

No one should have to spend ten minutes deciphering a message worth only five minutes, as in this example:

An inefficient message

> At this point in time, we are presently awaiting an on-site inspection by vendor representatives relative to electrical utilization adaptations necessary for the new computer installation. Meanwhile, all staff are asked to respect the off-limits designation of said location, as requested, due to liability insurance provisions requiring the online status of the computer.

Notice how hard we had to work with the previous message to extract information that could be expressed this efficiently:

A more efficient version

> Hardware consultants soon will inspect our new computer room to recommend appropriate wiring. Because our insurance covers only an *operational* computer, this room must remain off limits until the computer is fully installed.

Inefficient documents have varied origins. Even when the information is accurate, errors like the following create needless labor:

Causes of inefficient documents

- ◆ more (or less) information than people need
- ◆ irrelevant or uninterpreted information
- ◆ confusing organization
- ◆ jargon or technical expressions people cannot understand
- ◆ more words than people need
- ◆ uninviting appearance or confusing layout
- ◆ no visual aids when people need or expect them

An efficient document sorts, organizes, and interprets information to suit the user's needs, abilities, and interests.

An efficient document is carefully designed to include these elements:

Elements of efficient documents

- ◆ *content* that makes the document worth reading
- ◆ *organization* that guides the user and emphasizes important material
- ◆ *style* that promotes rapid reading and accurate understanding
- ◆ *visuals* (graphs, diagrams, pictures) that clarify concepts and relationships, and that substitute for words whenever possible
- ◆ *format* (layout, typeface) that is accessible and appealing

> • *supplements* (abstracts, appendices) that allow users with different needs to read only those sections required for their work

A writer's legal accountability

User orientation and efficiency are more than abstract rules: In the event of a lawsuit, faulty writing is no different from any other faulty product. If your inaccurate or unclear or incomplete information leads to injury or damage or loss, *you* and your company can be held legally responsible.

WRITING IS PART OF MOST CAREERS

Although you might not anticipate a "writing" career, your skills as a professional who writes will be tested routinely in situations like these:

Ways in which your career may test your writing skills

- proposing various projects to management or to clients
- writing progress reports
- preparing company news releases for the public
- describing a new product to employees or customers
- writing procedures and instructions for employees or customers
- justifying to management a request for funding or personnel
- editing and reviewing documents written by colleagues
- designing material that will be read on a computer screen or transformed into sound and pictures

The role of collaboration

You might write alone or as part of a collaborative team, and you will face strict deadlines.

Working professionals often spend at least 40 percent of their time writing or dealing with someone else's writing (Barnum and Fisher 9–11). Here is a corporate executive's description of some audiences you can expect:

Workplace audiences for your writing

> The technical graduate entering industry today will, in all probability, spend a portion of his or her career explaining technology to lawyers—some friendly and some not—to consumers, to legislators or judges, to bureaucrats, to environmentalists and to representatives of the press. (Florman 23)

Here is what two top managers for an automaker say about the effect a document can have on the organization *and* on the writer:

Writing is an indicator of job performance

> A written report is often the only record which is made of results that have come out of years of thought and effort. It is used to judge the value of the *person's* work and serves as the foundation for all future action on the project. If it is written clearly and precisely, it is accepted as the result of sound reasoning and careful observation. If it is poorly written, the results presented in it are placed in a bad light

and are often dismissed as the work of a careless or incompetent worker. (Richards and Richards 6)

Good writing gives you and your ideas *visibility* and *authority* within your organization. Bad writing, on the other hand, is expensive: Written communication[1] in American business and industry costs billions of dollars yearly. More than 60 percent of that writing is inefficient: unclear, misleading, irrelevant, deceptive, or otherwise wasteful of time and money (Max 5–6).

Whatever your job description (engineer, technician, production manager), expect to be evaluated, at least in part, on your skills as a technical communicator. At one IBM™ subsidiary, for example, 25 percent of an employee's evaluation is based on how effectively the employee shares information (Davenport 99).

As you advance in your field, you increasingly share information and establish human contacts. Managers and executives, for example, spend much of their time negotiating, setting policies, and promoting their ideas—often among various cultures around the globe. In short, the higher your career goals, the better you need to communicate.

COMMUNICATION HAS AN ELECTRONIC AND A HUMAN SIDE

The rise of information technology

Electronic mail, voice mail, fax, teleconferencing, videoconferencing, Internet chat rooms, hypertext, multimedia—these and other resources, collectively known as *information technology* (IT)—enhance the speed, volume, and variety of ways of transmitting information.

Electronically mediated communication can reach a limitless audience instantly and globally, and can solicit immediate feedback from the audience. It is no wonder that recent investments in information technology total more than 3 trillion dollars—among U.S. companies alone (Davenport 6).

Limitations of information technology

Despite the tremendous advantages IT gives today's communicators, their information still needs to be *written*. Also, only humans can give *meaning* to all the information they convey and receive. Information technology, in short, is a tool, not a substitute for human interaction.

People add value to information by posing and answering questions no computer can answer.

Today, more than ever, people who communicate on the job need to sort, organize, and interpret their material so users can understand it and act on it. With so much information required, and so much available, no one can afford to "let the data speak for themselves."

---

[1]"Written communication" includes the full range of activities involved in preparing, producing, processing, storing, and retrieving documents for reuse.

QUESTIONS ONLY HUMANS CAN ANSWER

- *Which information is most relevant to this situation?*
- *Can I verify the accuracy of this source?*
- *What does this information mean?*
- *What action does it suggest?*
- *How does this information affect me or my colleagues?*
- *With whom should I share it?*

COMMUNICATION REACHES A GLOBAL AUDIENCE

Electronically linked, our global community shares social, political, and financial interests. Multinational corporations often use parts manufactured in one country and shipped to another for assembly into a product to be marketed elsewhere. Cars may be assembled in the United States for a German automaker, or farm equipment manufactured in East Asia for a U.S. company. Research crosses national boundaries, and professionals transact across cultures with documents like these (Weymouth 143):

Documents that address global audiences

- scientific reports and articles on AIDS and other diseases
- studies of global pollution and industrial emissions
- specifications for hydroelectric dams and other engineering projects
- operating instructions for appliances and electronic equipment
- catalogs, promotional literature, and repair manuals
- contracts and business agreements

In Brief TRANSFERABLE SKILLS FOR THE TWENTY-FIRST CENTURY

A truism for today's workplace might be "nothing lasts forever." High-tech start-up companies emerge and vanish overnight. Even large, established companies expanding at one moment may be "downsizing" at the next.

To lower their costs and remain flexible amidst rapidly shifting markets, employers increasingly offer jobs that are merely temporary: for contract workers, part-timers, consultants, and the like (Jones 51). Instead of joining a company, climbing up through the ranks, and retiring with a gold watch and a comfortable pension, today's college graduate can expect to have multiple employers—and careers.

Success in the twenty-first-century workplace will require *transferable skills*. And the ultimate transferable skill is that of *information worker*, a person who can locate information, verify its quality, figure out what it means, and shape it for the user's specific purposes.

To connect global communities, any document must respect not only language differences, but also cultural differences:

"Culture" defined

> Our accumulated knowledge and experiences, beliefs and values, attitudes and roles—in other words, our cultures—shape us as individuals and differentiate us as a people. Our cultures, inbred through family life, religious training, and educational and work experiences . . . manifest themselves . . . in our thoughts and feelings, our actions and reactions, and our views of the world.
>
> Most important for communicators, our cultures manifest themselves in our information needs and our styles of communication . . . our expectations as to how information should be organized, what should be included in its content, and how it should be expressed. (Hein 125)

Cultures differ over which behaviors seem appropriate for social interaction, business relationships, contract negotiation, and communication practices. An effective communication style in one culture may be offensive elsewhere. A recent survey of top international executives reveals the following attitudes toward U.S. communication style (Wandycz 22–23):

How various cultures view U.S. communication style

- *Latin America:* "Americans are too straightforward, too direct."
- *Eastern Europe:* "An imperial tone. . . . It's always about how [Americans] know best."
- *Southeast Asia:* "To get my respect, American business[people] should know something about [our culture]. But they don't."
- *Western Europe:* "Americans miss the small points."
- *Central Europe:* "Americans tend to oversell themselves."

In short, global communication requires documents that achieve "efficiency" without being offensive. For more discussion, see pages 30, 41, 252.

EXERCISES

1. Locate a brief example of a technical document or Web page (or a section of one). Make a photocopy, or printout, bring it to class, and explain why your selection can be called technical writing.
2. Research the kinds of writing you will do in your career. (Begin with the *Dictionary of Occupational Titles* in your library.) You might interview a member of your chosen profession. Why will you write on the job? For whom will you write? Explain in a memo to your instructor. (See pages 345, 346 for memo elements and format.)

3. Write a memo to your boss, justifying reimbursement for this course. Explain how the course will help you become more effective on the job.

COLLABORATIVE PROJECT

INTRODUCING A CLASSMATE

Class members will work together often this semester. So that everyone can become acquainted, your task is to introduce to the class the person seated next to you. (That person, in turn, will introduce you.) To prepare your introduction, follow this procedure:

a. Exchange with your neighbor whatever personal information you think the class needs: background, major, computer experience (email, Internet, other), career plans, communication needs of your intended profession, and so on. Each person gets five minutes to tell her or his story.

b. Take careful notes; ask questions if you need to.

c. Take your notes home and select only what you think will be useful to the class.

d. Prepare a one-page memo telling your classmates *who this person is.* (See pages 345, 346 for memo elements and format.)

e. Ask your neighbor to review the memo for accuracy; revise as needed.

f. Present the class with a two-minute oral paraphrase of your memo, and submit a copy of the memo to your instructor.

PART

I

Communicating
in the Workplace

Problem Solving in Workplace Communication

TECHNICAL COMMUNICATORS SOLVE
INTERRELATED PROBLEMS

PROBLEM SOLVING REQUIRES CREATIVE
AND CRITICAL THINKING

In Brief Communication in a Virtual Company

Guidelines for Writing with a Computer

All professionals specialize in solving problems (how to repair that equipment, how to improve this product, how to diagnose that ailment). Central to all specialized activity is the problem of *communicating*—a complex problem examined more easily by considering several subordinate problems.

TECHNICAL COMMUNICATORS SOLVE INTERRELATED PROBLEMS

No matter how sophisticated our information technology, computers cannot *think* for us or solve the problems we encounter (Figure 2.1) as technical communicators:

The communicator as problem solver

- the **information problem,** because different people doing different tasks have different information needs
- the **persuasion problem,** because people often disagree about what the information means and what action should be taken
- the **ethics problem,** because the interests of your employer may conflict with the interests of other people involved
- the **collaboration problem,** because an estimated 90 percent of U.S. workers "spend at least part of their day in a team situation" ("People" 57).

Figure 2.1
Problems in
Workplace
Communication

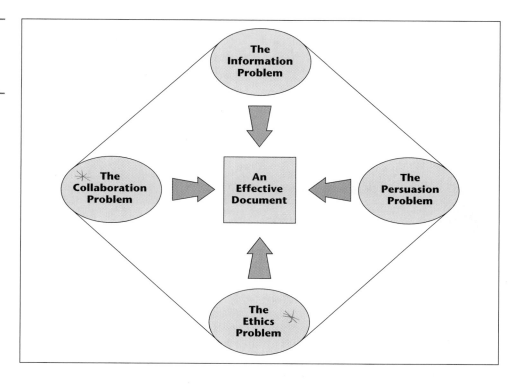

The scenarios that follow illustrate how a typical professional confronts these problems of communicating in the workplace.

The Information Problem

Sarah Burnes was hired two months ago as a chemical engineer for Millisun, a leading maker of cameras, multipurpose film, and photographic equipment. Sarah's first major assignment is to evaluate the plant's incoming and outgoing water. (Waterborne contaminants can taint film during production, and the production process itself can pollute outgoing water.) Management wants an answer to this question: How often should we change water filters? The filters are expensive and difficult to change, halting production for up to a day at a time. The company wants as much "mileage" as possible from these filters, without incurring government fines or tainting its film production.

Sarah will study endless printouts of chemical analysis, review current research and government regulations, do some testing of her own, and consult with her colleagues. When she finally decides on what all the data mean, Sarah will prepare a recommendation report for her bosses.

Later, she will collaborate with the company training manager and the maintenance supervisor to prepare a manual, instructing employees how to check and change the filters. Trying to cut secretarial and printing costs, the company has asked Sarah to design and produce this manual using its new desktop publishing system. ▲

Sarah's report, above all, needs to be accurate; otherwise, the company gets fined or lowers production. Once she has processed all the information, she faces the problem of giving users what they need: *How much explaining should I do? How will I organize? Do I need visuals?* And so on.

In other situations, Sarah will face a persuasion problem as well: for example, when decisions must be made or actions taken on the basis of incomplete or inconclusive facts or conflicting interpretations (Hauser 72). In these instances, Sarah will seek consensus for *her* view.

The Persuasion Problem

Millisun and other electronics producers are located on the shores of a small harbor, the port for a major fishing fleet. For twenty years, these companies had discharged effluents containing metal compounds, PCBs, and other toxins directly into the harbor. Sarah is on a multicompany team, assigned to work with the Environmental Protection Agency to clean up the harbor. Much of the team's collaboration occurs via email.

Enraged local citizens are demanding immediate action, and the companies themselves are anxious to end this public relations nightmare. But the team's analysis reveals that any type of cleanup would stir up harbor sediment, possibly dispersing the solution into surrounding waters and the atmosphere. (Many of the contaminants can be airborne.) Premature action might actually *increase* danger, but team members disagree on the degree of risk and on how to proceed.

Sarah's communication here takes on a persuasive dimension: She and her team members first have to resolve their own disagreements and produce an environmental-impact report that reflects the team's consensus. If the report recommends further study, Sarah will have to justify the delays to her bosses and the public relations office. She will have to make people understand the dangers as well as she understands them. ▲

In the above situation, the facts are neither complete nor conclusive, and views differ about what these facts mean. Sarah will have to balance the various political pressures and make a case for *her* interpretation. Also, as company *spokesperson*, Sarah will be expected to protect her company's interests. Some elements of Sarah's persuasion problem: *Are other interpretations possible? Is there a better way? Can I expect political or legal fallout?*

Sarah also will have to reckon with the ethical implications of her writing, with the question of "doing the right thing." For instance, Sarah might feel pressured to overlook or sugarcoat or suppress facts that would be costly or embarrassing to her company.

The Ethics Problem

"Can I be honest and still keep my job?"

To ensure compliance with OSHA[1] standards for worker safety, Sarah is assigned to test the air purification system in Millisun's chemical division. After finding the filters hopelessly clogged, she decides to test the air quality and discovers dangerous levels of benzene (a potent carcinogen). She reports these findings in a memo to the production manager, with an urgent recommendation that all employees be tested for benzene poisoning. The manager phones and tells Sarah to "have the filters replaced, and forget about it." Now Sarah has to decide what to do next: bury the memo in some file cabinet or defy her boss and jeopardize her job by sending copies to other people who might take action. ▲

Situations that compromise truth and fairness present the hardest choices of all: remain silent and look the other way or speak out and risk being fired. Some elements of Sarah's ethics problem: *Is this fair? Who might benefit or suffer? What other consequences could this have?*

In addition to solving these various problems, Sarah has to reckon with the implications of working in a team setting: Much of her writing, done at the computer, will be produced in collaboration with others (editors, managers, graphic artists), and her audience will extend beyond her own culture.

---

[1]Occupational Safety and Health Administration.

The Collaboration Problem

Recent mergers have transformed Millisun into a multinational corporation with branches in eleven countries, all connected by an intranet. Sarah can expect to collaborate with coworkers from diverse cultures on research and development, with government agencies of the host countries on safety issues, patents and licensing rights, product liability laws, and environmental concerns.

In order to standardize the sensitive management of the toxic, volatile, and even explosive chemicals used in film production, Millisun is developing automated procedures for quality control, troubleshooting, and emergency response to chemical leakage. Sarah has been assigned to a team that is preparing computer-based instructional packages for all personnel involved in Millisun's chemical management worldwide. ▲

As a further complication, Sarah will have to develop working relationships with people she has never met, people from other cultures, people she knows only via an electronic medium.

For Sarah Burnes, or any of us, writing is a process of *discovering* what we want to say, "a way to end up thinking something [we] couldn't have started out thinking" (Elbow 15). Throughout this process in the workplace, we rarely work alone, but instead collaborate with others for information, help in writing, and feedback (Grice 29–30). We must satisfy not only our audience, but also our employer, whose goals and values ultimately shape the document (Selzer 46–47). Almost any document for people outside our organization will be *reviewed* for accuracy, appropriateness, usefulness, and legality before it is finally approved (Kleimann 521).

PROBLEM SOLVING REQUIRES CREATIVE AND CRITICAL THINKING

In *creative thinking,* we explore new ideas; we build on information; we devise better ways of doing things. (For example, "How do we get as much mileage as possible from our water filters?")

In *critical thinking,* we test the strength of our ideas or the worth of our information. Instead of accepting the idea at face value, we examine, evaluate, verify, analyze, weigh alternatives, and consider consequences—at every stage of that idea's development. We employ critical thinking to examine our evidence and our reasoning, to discover new connections and new possibilities, and to test the effectiveness and the limits of our solutions.

We apply creative and critical thinking throughout the four stages in the *writing process:*

1. We work with our ideas and information.
2. We plan the document.

3. We draft the document.

4. We revise the document.

One engineering professional describes how creative and critical thinking enrich every stage of the writing process:

> Good writing is a process of thinking, writing, revising, thinking, and revising, until the idea is fully developed. An engineer can develop better perspectives and even new technical concepts when writing a report of a project. Many an engineer, at the completion of a laboratory project, senses a new interpretation or sees a defect in the results and goes back to the laboratory for additional data, a more thorough analysis, or a modified design. (Franke 13)

As the arrows in Figure 2.2 indicate, no one stage of the writing process is complete until all stages are complete. Figure 2.3 lists the kinds of questions we answer at various stages. On the job, we must often complete these stages under deadline pressure. Like the exposed tip of an iceberg, the finished document provides the only visible evidence of our labor.

Computers, of course, are essential tools in the writing process. The guidelines on page 20 will help you capitalize on all the benefits a computer has to offer.

Figure 2.2
The Writing Process for Technical Documents

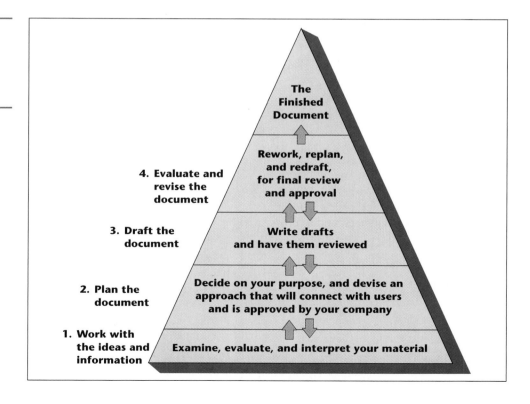

1. Work with the ideas and information:

- Have I defined the problem accurately?
- Is the information complete, accurate, reliable, and unbiased?
- Can it be verified?
- How much of it is useful?
- Do I need more information?
- What do these facts mean?
- What connections seem to emerge?
- Do the facts conflict?
- Are other interpretations or conclusions possible?
- Is a balance of viewpoints represented?
- What, if anything, should be done?
- Is it honest and fair?
- Is there a better way?
- What are the risks and benefits?
- What other consequences might this have?
- Should I reconsider?

2. Plan the document:

- When is it due?
- What do I want it to do?
- Who is my audience, and why will they use it?
- What do they need to know?
- What are the "political realities" (feelings, egos, cultural differences, and so on)?
- How will I organize?
- What format and visuals should I use?
- Whose help will I need?

3. Draft the document:

- How do I begin, and what comes next?
- How much is enough?
- What can I leave out?
- Am I forgetting anything?
- How will I end?
- Who needs to review my drafts?

4. Evaluate and revise the document:

- Is this draft usable?
- Does it do what I want it to do?
- Is the content worthwhile?
- Is the organization sensible?
- Is the style readable?
- Is everything easy to find?
- Is the format appealing?
- Is everything accurate, complete, appropriate, and correct?
- Who needs to review and approve the final version?
- Does it advance my organization's goals?
- Does it advance my audience's goals?

Figure 2.3   Creative and Critical Thinking in the Writing Process

In Brief COMMUNICATION IN A VIRTUAL COMPANY

Office communication has evolved dramatically, as illustrated in the practices below.

- Instead of being housed in one location, the virtual company may have branches across the state, the nation, or the world, to which many employees "commute" electronically. These telecommuters include freelance workers who are employed by other companies as well.
- Instead of relying on secretaries, managers compose their own letters and memoranda for distribution via email to readers across the building or across the globe.
- On desktop publishing (DTP) networks, the composition, layout, graphics design, typesetting, and printing of external documents and Web pages are done in-house.
- Optical scanners take an electronic snapshot of any paper document produced or received, including incoming mail. Stored online, this image can be retrieved and edited, printed out, faxed or emailed, or posted on an electronic bulletin board or Web page. Company forms (requisitions, accident reports, etc.) can be produced, filed, updated, distributed, filled out, and delivered electronically.
- Workplace discussions and document sharing occur via email, voice mail, or videoconferencing networks. Email listservs announce daily developments for employees or readers worldwide such as price and inventory lists, changes or updates in policies or procedures, and press releases.
- Employees work and write collaboratively, sharing resources in a joint effort (as in developing a proposal or a marketing plan). Drafts circulated electronically enable colleagues to add comments directly on the manuscript. Multimedia systems present text, graphics, sound, and animated material retrieved from a computer file. People networked from various locations work on the electronic document and view and comment on one another's "work in progress."
- Online databases store information from books, magazines, newspapers, journals, and so forth, and are searched via the World Wide Web for the latest stock market quotations, trends in global weather patterns, sites of recent disease outbreaks, and so on. A single compact disk can store an entire encyclopedia, a medical dictionary, or interactive manuals and lessons.
- Unlike printed texts, which tend to be read front to back, electronic texts often are read nonsequentially (Grice and Ridgway 37). Readers navigate their own paths and choose various routes to explore.

GUIDELINES

for Writing
with a
Computer

1. *Beware of computer junk.* The ease of cranking out words on a computer can produce long, windy pieces that say nothing. Cut anything that fails to advance your meaning. (See pages 224–31 for ways to achieve conciseness.)

2. *Never confuse style with substance.* Laser printers and choices of typefaces, type sizes, and other design options can produce attractive documents. But not even the most attractive design redeems a document whose content is worthless or inaccessible.

3. *Save and print your work often.* Save each paragraph as you write it; print out each page as you complete it; and keep a copy of your document on a backup disk.

4. *Consider the benefits of revising from hard copy.* Nothing beats scribbling on the printed page. The hard copy provides the whole text, right in front of you.

5. *Never depend only on automated "checkers."* Not even the most sophisticated writing aids can replace careful proofreading. A synonym found in an electronic thesaurus may distort your meaning. The spell checker cannot differentiate among correctly spelled words such as "their," "they're" or "there" or "it's" versus "its." And neither spell nor grammar checkers can evaluate stylistic *appropriateness* (those subtle choices of phrasing that determine tone and emphasis). Page 253 summarizes the limitations of computerized aids.

6. *Select a design and a transmission medium that your audience favors.* Should the document be primarily verbal, visual, or some combination? Should it travel by conventional mail, interoffice mail, email? Who are the users and what would they prefer in this situation—the solid feel of paper or the "hi-tech" lure of a computer screen? Research suggests that younger audiences prefer flashy graphics; older audiences prefer traditional text; and people in general trust text more than visual images (Horton, "Mix Media" 781).

EXERCISE
(INDIVIDUAL OR COLLABORATIVE)

1. As you respond to the following scenario,[2] carefully consider the information, persuasion, and ethical problems involved (and be prepared to discuss them in class).

> You are Manager of Product Development at High-Tech Toys, Inc. You need to send a memo to the Vice President of Information Services, explaining the following:
>
> a. The laser printer in your department is often out of order.

[2]My thanks to Teresa Pawelcyzk for the original version of this exercise.

b. The laser printer is seldom repaired satisfactorily.

c. Either the machine is faulty or the repairperson is incompetent (but this person always appears promptly and cheerfully when summoned from Corporate Maintenance—and is a single parent raising three young children).

d. It is difficult to get things done in your department without being able to use the laser printer.

e. The members of your department share ideas and plans daily.

f. You want the problem solved—but without getting the repairperson fired.

In your memo, recommend a solution, and justify briefly your recommendation.

COLLABORATIVE PROJECT

AN ISSUE OF ETHICS

Working in small groups, analyze Sarah Burnes's "Ethics Problem" (page 15). What could happen if Sarah follows her boss's orders? What could happen if she doesn't? After discussing the issues involved and the possible consequences, try to reach a consensus about what action Sarah should take in this situation. Appoint one member to present your group's conclusion to the class.

Solving the Information Problem

ASSESS THE AUDIENCE'S INFORMATION NEEDS

IDENTIFY LEVELS OF TECHNICALITY

DEVELOP AN AUDIENCE AND USE PROFILE

In Brief Human Factors in Communication Failure

All technical communication is for people who will use and react to the information. Your task might be to *define* something—as in explaining what "variable annuity" means for insurance clients; to *describe* something—as in showing an architectural client what a new office building will look like; to *explain* something—as in telling a stereo technician how to eliminate bass flutter in your company's new line of speakers. As depicted in Figure 3.1, the information problem calls for **audience analysis,** in which you learn all you can about those who will use your document.

People are not interested in how smart or eloquent you are, but in how to find what they need, quickly and easily.

ASSESS THE AUDIENCE'S INFORMATION NEEDS

Good writing connects with its audience by recognizing its unique background, needs, and preferences. The same basic message can be conveyed in different ways for different audiences. For instance, an article describing a new cancer treatment might appear in a medical journal for doctors and nurses. A less technical version might appear in a textbook for medical and nursing students. A more simplified version might appear in *Reader's Digest.* All three versions treat the same topic, but each meets the needs of a different audience.

Because your audience knows less than you, it will have questions.

TYPICAL AUDIENCE QUESTIONS ABOUT WORKPLACE DOCUMENTS
- What is the purpose of this document?
- Why should I read it?
- What happened, and why?
- How should I perform this task?
- What action should be taken?
- How much will it cost?
- What are the risks?

IDENTIFY LEVELS OF TECHNICALITY

When you write for a close acquaintance (coworker, engineering colleague, chemistry professor who reads your lab reports, or supervisor), you adapt your report to that person's knowledge, interests, and needs. But some audiences are larger and less defined (say, for a journal article, a computer manual, a set of first-aid procedures, or an accident report). When you have only a general notion about your audience's background, decide whether your document should be *highly technical, semitechnical,* or *nontechnical,* as depicted in Figure 3.2.

Figure 3.1
One Problem
Confronted by
Communicators

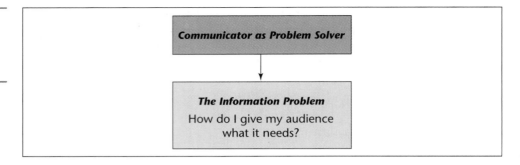

The Highly Technical Document

Users at a specialized level expect the facts and figures they need—without long explanations. In the following description of emergency treatment, an emergency-room physician reports to the patient's doctor, who needs an exact record of symptoms, treatment, and results.

A TECHNICAL VERSION

Expert users need just the facts and figures, which they can interpret for themselves

The patient was brought to the ER by ambulance at 1:00 A.M., September 27, 2000. The patient complained of severe chest pains, dyspnea, and vertigo. Auscultation and EKG revealed a massive cardiac infarction and pulmonary edema marked by pronounced cyanosis. Vital signs: blood pressure, 80/40; pulse, 140/min; respiration, 35/min. Lab: wbc, 20,000; elevated serum transaminase; urea nitrogen, 60 mg%. Urinalysis showed 4+ protein and 4+ granular casts/field, indicating acute renal failure secondary to the hypotension.

The patient received 10 mg of morphine stat, subcutaneously, followed by nasal oxygen and 5% D & W intravenously. At 1:25 A.M. the cardiac monitor recorded an irregular sinus rhythm, indicating left ventricular fibrillation. The patient was defibrillated

Figure 3.2
Deciding on a
Document's
Level of
Technicality

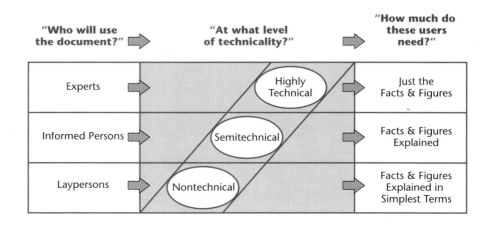

stat and given a 50 mg bolus of Xylocaine intravenously. A Xylocaine drip was started, and sodium bicarbonate administered until a normal heartbeat was established. By 3:00 A.M., the oscilloscope was recording a normal sinus rhythm.

As the heartbeat stabilized and cyanosis diminished, the patient received 5 cc of Heparin intravenously, to be repeated every six hours. By 5:00 A.M. the BUN had fallen to 20 mg% and vital signs had stabilized: blood pressure, 110/60; pulse, 105/min; respiration, 22/min. The patient was now conscious and responsive.

For her expert colleague, this physician defines no technical terms (*pulmonary edema, sinus rhythm*). Nor does she interpret lab findings (4+ *protein, elevated serum transaminase*). She uses abbreviations her colleague understands (*wbc, BUN, 5% D & W*). Because her colleague knows all about specific treatments and medications (*defibrillation, Xylocaine drip*), she doesn't explain their scientific bases. Her report answers concisely the main user questions she anticipates: *What was the problem? What was the treatment? What were the results?*

The Semitechnical Document

One broad class of users has some technical background, but less than the experts. For instance, first-year medical students have specialized knowledge, but less than advanced students. Yet all medical students could be considered semitechnical. Therefore, when you write for a semitechnical audience, identify the *lowest* level of understanding in the group, and write to that level. Too much explanation is better than too little.

The following partial version of the medical report might appear in a textbook for medical or nursing students, in a report for a medical social worker, or in a monthly report for the hospital administration.

A SEMITECHNICAL VERSION

Informed but nonexpert users need enough explanation to understand what the data mean

Examination by stethoscope and electrocardiogram revealed a massive failure of the heart muscle along with fluid buildup in the lungs, which produced a cyanotic **discoloration of the lips and fingertips from lack of oxygen.**

The patient's blood pressure at 80 mm Hg (systolic)/40 mm Hg (diastolic) was **dangerously below its normal measure of 130/70.** A pulse rate of 140/minute was **almost twice the normal rate of 60–80.** Respiration at 35/minute was more than **twice the normal rate of 12–16.**

Laboratory blood tests yielded a white blood cell count of 20,000/cu mm (normal value: 5,000–10,000), **indicating a severe inflammatory response by the heart muscle.** The elevated serum transaminase enzymes (**produced in quantity only when the heart muscle fails**) confirmed the earlier diagnosis. A blood urea nitrogen level of 60 mg% (normal value: 12–16 mg%) **indicated that the kidneys had ceased to filter out metabolic waste products.** The 4+ protein and casts reported from the urinalysis (normal value: 0) **revealed that the kidney tubules were degenerating as a result of the lowered blood pressure.**

> The patient immediately received **morphine to ease the chest pain,** followed by **oxygen to relieve strain on the cardiopulmonary system,** and an intravenous solution of **dextrose and water to prevent shock.**

This version explains the raw data (in boldface). Exact dosages are omitted because no one in this audience actually will be treating this patient. Normal values of lab tests and vital signs, however, help readers interpret. (Experts know these values.) Knowing what medications the patient received would be especially important to answering this audience's central question: *How is a typical heart attack treated?*

The Nontechnical Document

People with no training look for the big picture instead of complex details. They expect technical data to be translated into words most people understand. Laypersons are impatient with abstract theories but want enough background to help them make the right decision or take the right action. They are bored or confused by excessive detail, but frustrated by raw facts left unexplained or uninterpreted. They expect to understand the document after reading it only once.

The following nontechnical version of the medical report might be written for the patient's spouse who is overseas on business, or as part of a script for a documentary about emergency room treatment.

A NONTECHNICAL VERSION

Heart sounds and electrical impulses were both abnormal, **indicating a massive heart attack caused by failure of a large part of the heart muscle.** The lungs were swollen with fluid and the lips and fingertips showed **a bluish discoloration from lack of oxygen.**

Blood pressure was **dangerously low, creating the risk of shock.** Pulse and respiration were **almost twice the normal rate, indicating that the heart and lungs were being overworked** in keeping oxygenated blood circulating freely.

Blood tests confirmed the heart attack diagnosis and **indicated that waste products usually filtered out by the kidneys were building up in the bloodstream. Urine tests showed that the kidneys were failing as a result of the lowered blood pressure.**

The patient was given **medication to ease the chest pain, oxygen to ease the strain on the heart and lungs, and intravenous solution to prevent the blood vessels from collapsing and causing irreversible shock.**

Laypersons need everything translated into terms they understand

Nearly all interpretation (in boldface), this version mentions no specific medications, lab tests, or normal values. It merely summarizes events and briefly explains them.

In a different situation, however (say, a malpractice trial), the nontechnical audience might need detailed information about medication and treatment. Such a report would, of course, be much longer—a short course in emergency coronary treatment.

Primary and Secondary Audiences

Whenever you prepare a single document for multiple users, classify your audience as *primary* or *secondary*. Primary users usually are those who requested the document as a basis for decisions or actions. Secondary users are those who will carry out the project, who will advise the decision makers or who will be affected by this decision in some way.

Primary and secondary audiences often differ in technical background. When you must write for audiences at different levels, follow these guidelines:

How to tailor a single document for multiple users

- ◆ If the document is short (a letter, memo, or anything less than two pages), rewrite it at various levels.
- ◆ If the document exceeds two pages, address the primary users. Then provide appendixes for secondary users. Transmittal letters, informative abstracts, and glossaries can also help nonexperts understand a highly technical report. (See pages 323–34 for use and preparation of appendices and other supplements.)

The document in this next scenario must be tailored for both primary and secondary users.

Tailoring a Single Document for Different Users

Different users have different information needs

You are a metallurgical engineer in an automotive consulting firm. Your supervisor has asked that you test the fractured rear axle of a 1998 Delphi pickup truck recently involved in a fatal accident. Your assignment is to determine whether the fractured axle *caused* or *resulted from* the accident.

After testing the hardness and chemical composition of the metal and examining microscopic photographs of the fractured surfaces (fractographs), you conclude that the fracture resulted from stress that developed *during* the accident. Now you must report your procedure and your findings to a variety of readers.

"What do these findings mean?"

Because your report may serve as courtroom evidence, you must explain your findings in meticulous detail. But your primary users (the decision makers) will be nonspecialists (the attorneys who have requested the report, insurance representatives, possibly a judge and a jury), so you must translate your report, explaining the principles behind the various tests, defining specialized terms such as "chevron marks," "shrinkage cavities," and "dimpled core," and showing the significance of these features as evidence.

"How did you arrive at these conclusions?"

Secondary users will include your supervisor and outside consulting engineers who will be evaluating your test procedures and assessing the validity of your findings. Consultants will be focusing on various parts of your report, to verify that your procedure has been exact and faultless. For this group, you will have to include appendixes spelling out the technical details of your analysis: *how* hardness testing of the axle's case and core indicated that the axle had been properly carburized; *how* chemical analysis ruled out the possibility that the manufacturer had used inferior alloys; *how* light-microscopic fractographs revealed that the origin of the fracture, its direction of propagation, and the point of final rupture indicated a ductile fast fracture, not one caused by torsional fatigue. ▲

In the previous situation, primary users need to know *what your findings mean*, whereas secondary users need to know *how you arrived at your conclusions*. Unless it serves the needs of each group independently, your information will be worthless.

DEVELOP AN AUDIENCE AND USE PROFILE

Focus sharply on your audience by asking the questions below.

QUESTIONS ABOUT A DOCUMENT'S INTENDED AUDIENCE AND USE

- ◆ Who wants the document? Who else will read it?
- ◆ Why do they want the document? How will they use it?
- ◆ What is the technical background of the primary audience? Of the secondary audience?
- ◆ How might cultural differences shape readers' expectations and interpretations?
- ◆ How much does the audience already know? What material will have informative value?
- ◆ What exactly does the audience need to know, and in what format? How much is enough?
- ◆ When is the document due?

To answer these questions, consider the suggestions that follow, and use a version of the Audience and Use Profile Sheet shown in Figure 3.3 for all your writing.

Audience Characteristics

Identify the primary audience by name, job title, and specialty (Martha Jones, Director of Quality Control, B.S. and M.S. in mechanical engineering). Are they superiors, colleagues, or subordinates? Are they inside or outside your organization? What is their attitude toward this topic likely to be? Are they apt to accept or reject your conclusions and recommendations? Will your report convey good or bad news? How might cultural differences affect audience expectations and interpretations?

Identify others who might be interested in or affected by the document, or who will advise the primary audience.

Audience Identity and Needs

Primary audience: _____ *(name, title)*

Secondary audience: _____

Relationship: _____ *(client, employer, other)*

Intended use of document: _____ *(perform a task, solve a problem, other)*

Prior knowledge about this topic: _____ *(knows nothing, a few details, other)*

Additional information needed: _____ *(background, only bare facts, other)*

Probable questions: _____ ?

_____ ?

_____ ?

_____ ?

_____ ?

Audience's Probable Attitude and Personality

Attitude toward topic: _____ *(indifferent, skeptical, other)*

Probable objections: _____ *(cost, time, none, other)*

Probable attitude toward this writer: _____ *(intimidated, hostile, receptive, other)*

Persons most affected by this document: _____

Temperament: _____ *(cautious, impatient, other)*

Probable reaction to document: _____ *(resistance, approval, anger, guilt, other)*

Risk of alienating anyone: _____

Audience Expectations about the Document

Reason document originated: _____ *(audience request, my idea, other)*

Acceptable length: _____ *(comprehensive, concise, other)*

Material important to this audience: _____ *(interpretations, costs,*

_____ *conclusions, other)*

Most useful arrangement: _____ *(problem-causes-solutions, other)*

Tone: _____ *(businesslike, apologetic, enthusiastic, other)*

Cultural considerations: _____ *(level of detail or directness, other)*

Intended effect on this audience: _____ *(win support, change behavior, other)*

Due date: _____

Figure 3.3 Audience and Use Profile Sheet

Purpose of the Document

Learn why people want the document and how they will use it. Do they simply want a record of activities or progress? Do they expect only raw data, or conclusions and recommendations as well? Will people act immediately on the information? Do they need step-by-step instructions? Will the document be read and discarded, or filed, published, or distributed electronically? In your audience's view, what is most important? Try asking them directly.

Audience's Technical Background

Colleagues who speak your technical language will understand raw data. Supervisors responsible for several technical areas may want interpretations and recommendations. Managers who have limited technical knowledge expect definitions and explanations. Clients with no technical background expect versions that spell out what the facts mean to them (to their health, pocketbook, financial prospects). However, none of these generalizations might apply to your situation. When in doubt, aim for low technicality.

Audience's Cultural Background

Some information needs are culturally determined. Germans, for example, tend to value thoroughness and complexity, with every detail included and explained in a businesslike tone. Japanese prefer multiple perspectives on the material, lots of graphics and a friendly, encouraging tone (Hein 125–26).

Anglo-American business culture generally values plain talk that gets right to the point, but Asian cultures consider this rude, preferring indirect and somewhat ambiguous messages, which leave interpretation up to the reader (Leki 151; Martin and Chaney 276–77). To avoid seeming impolite, some readers might hesitate to ask for clarification or additional information. In Asian cultures even disagreement or refusal might be expressed as "We will do our best" or "This is very difficult," instead of "No"—to avoid offending and to preserve harmony (Rowland 47).

Also, many U.S. idioms ("breaking the bank," "cutthroat competition," "sticking your neck out") and cultural references ("the crash of '29," "Beantown") make no sense outside of U.S. culture (Coe "Writing" 97).[1]

Audience's Knowledge of the Subject

People expect to find something new and useful in your document. Writing has informative value[2] when it does at least one of these things:

---

[1]For more information on how to analyze business practices in different cultures, consult one of the many available texts on this subject, for example, David A. Victor's *International Business Communication*.

[2]Adapted from James L. Kinneavy's assertion that discourse ought to be unpredictable, in *A Theory of Discourse* (Englewood Cliffs, NJ: Prentice, 1971).

Elements of
"informative value"
- shares something new and relevant
- reminds us about something we know but ignore
- offers fresh insight or perspective on something we already know

In short, informative writing gives people exactly what they need.

People approach most topics with some prior knowledge (or old information). They might need reminding, but they don't need a rehash of old information; they can "fill in the blanks" for themselves. Nor does anyone need every bit of new information you can think of.

The more nonessential information people receive, the more likely they are to overlook or misinterpret the important material.

As you read this book, you expect an introduction to technical communication, and my purpose is to provide that. Which of these statements would you find useful?

Not all facts are
equally useful
1. Technical communication is hard work.
2. "The computerized version of the *Oxford English Dictionary* incorporates a modified SGML syntax" (Fawcett 379).
3. "Technical communication is a process of making deliberate decisions in response to a specific situation. In this process, you discover important meanings in your topic, and give your audience the information they need to understand your meanings" (Hogge 3).

Statement 1 is old news to anyone who has ever picked up a pencil. You probably find Statement 2's information new but not relevant. Statement 3 offers new insight into a familiar process, so we can say it has informative value. On the other hand, a technical communication professional may find Statement 2 more useful than 3.

Appropriate Details, Format, and Design

The amount of detail in your document (*How much is enough?*) depends on what you can learn about your audience's needs. Were you asked to "keep it short" or to "be comprehensive"? Can you summarize, or does everything need spelling out? Are people more interested in conclusions and recommendations, or do they want all the details? Have they requested a letter, a memo, a short report, or a long, formal report with supplements (title page, table of contents, appendixes, and so on)? Can visuals and page design (charts, graphs, drawings, headings, lists) make the material more accessible?

Due Date and Timing

Does your document have a deadline? Workplace documents almost always do. Or is there an optimum time for submission? In the workplace, due dates and timing are vital for competing effectively (for example, in submitting bids for a project). When possible, ask users to review an early draft and to suggest improvements.

In Brief HUMAN FACTORS IN COMMUNICATION FAILURE

Below is a sampling of "honest mistakes" in communication caused by human limitations:

Neglecting to Convey Vital Information

November, 1973: The Vermont Yankee Nuclear Power Plant experienced near-meltdown "after day workers installing a closed-circuit television cut off power to a primary safety system and failed to inform the night shift" (Monmonier 209–10).

Not Being Assertive about Vital Information

January, 1982: A Boeing 737 crashes on takeoff from Washington International Airport, killing 78 people. "The copilot had warned the captain of possible trouble several times—icy conditions were causing false readings on an engine-thrust gauge—but the copilot had not spoken forcefully enough, and the pilot ignored him" (Pool 44).

Underestimating Vital Information

December, 1942: "Some people within the U.S. Army knew that a large group of airplanes was headed toward Pearl Harbor; others knew that six Japanese aircraft carriers weren't where they were supposed to be, yet nobody acted on that information" (Davenport 7).

Overlooking Vital Errors

August, 1996: A 2-month-old boy with a heart defect is admitted to a major hospital in apparent heart failure. He is treated with the drug Digoxin and expected to recover fully. Within hours, his heart stops. All revival efforts fail.

A medical resident had mistakenly ordered a dose of 0.9 milligrams—instead of the correct .09 mgs. The error was then overlooked by the attending doctor, the pharmacy technician who filled the order, and a second resident physician who was asked by a nurse to check the dosage. Because of this error of one decimal point, the patient received 10 times the correct dose of Digoxin (Belkin 28+).

EXERCISES

1. Locate a short article from your field (or part of a long article or a selection from your textbook for an advanced course). Choose a piece written at the highest level of technicality you understand and then translate the piece for a layperson, as in the example on page 26. Exchange translations with a classmate from a different major. Read your neighbor's translation and write a paragraph evaluating its level of technicality. Submit to your instructor a copy of the original, your translated version, and your evaluation of your neighbor's translation.

2. Assume that a new employee is taking over your job because you have been promoted. Identify a specific problem in the job that could cause difficulty for the new employee. Write for the employee instructions for avoiding or dealing with the problem. Before writing, create an audience and use profile by answering (on paper) the questions on page 28.

COLLABORATIVE PROJECTS

Form teams according to major (electrical engineering, biology, etc.), and respond to the following situation.

Assume that your team has received the following assignment from your major department's chairperson: An increasing number of first-year students are dropping out of the major because of low grades, stress, or inability to keep up with the work. Your task is to prepare a "Survival Guide," for distribution to incoming students. This one- or two-page memo should focus on the challenges and the pitfalls of the first year in a given major and should include a brief motivational section, along with whatever else team members think readers need.

ANALYZE YOUR AUDIENCE

Use Figure 3.3 as a guide for developing your audience and use profile.

DEVISE A PLAN FOR ACHIEVING YOUR GOAL

From the audience traits you have identified, develop a plan for communicating your information. Express your goal and plan in a statement of purpose. For example:

> The purpose of this document is to explain the challenges and pitfalls of the first year in our major. We will show how dropouts have increased, discuss what seems to go wrong, give advice on avoiding some common mistakes, and emphasize the benefits of remaining in the program.

PLAN, DRAFT, AND REVISE YOUR DOCUMENT

Brainstorm for worthwhile content (for this exercise, make up some reasons for the dropout rate if you need to), do any research that may be needed,

write a workable draft, and revise until it represents your team's best work.

Appoint a team member to present the finished document (along with a complete audience and use analysis) for class evaluation, comparison, and response.

ALTERNATIVE PROJECTS

a. In a one- or two-page memo to *all* incoming students develop a "First-Year Survival Guide." Spell out the least information anyone should know in order to get through the first year of college.

b. In one or two pages, describe the job outlook in your field (prospects for the coming decade, salaries, subspecialties, promotional opportunities, etc.). Write for high school seniors interested in your major. Your team's description will be included in the career handbook published by your college.

c. Identify an area or situation on campus that is dangerous or inconvenient or in need of improvement (endless cafeteria lines, poorly lit intersections or parking lots, noisy library, speeding drivers, inadequate dorm security, etc.). Observe the situation as a group during a peak-use period. Spell out the problem in a letter to a specified decision maker (dean, campus police chief, head of food service) who will presumably use your information as a basis for action.

Solving the
Persuasion Problem

Persuasion means trying to influence someone's actions, thinking, or decision making (Figure 4.1). People in the workplace use persuasion daily: to win support or cooperation from coworkers, to attract clients and customers, to request a raise. The size of your persuasion problem depends on who you are trying to persuade, how you want them to respond, and how firmly they are committed to their own viewpoints.

You face a persuasion problem whenever you take a stance on an issue about which people disagree. A persuasive stance ordinarily takes the form of a thesis or a *claim* (a statement of the point you are trying to prove). For instance, you might want people to recognize facts they've ignored:

A claim about what the facts are

> The O-rings in the space shuttle's booster rockets have a serious defect that could have disastrous consequences.

Or you might want to influence their interpretation of the facts:

A claim about what the facts mean

> Taking time to redesign and test the O-rings is better than taking unacceptable risks to keep the shuttle program on schedule.

Or you might want someone to take immediate action:

A claim about what should be done

> We should call for a delay of tomorrow's shuttle launch because the risks are simply too great.

Even when a fact is obvious, people may disagree about what it means or what action it indicates.

Your own letters, memos, and reports will ask people to accept and act on claims like these (Gilsdorf, "Executives,' and Academics' Perception" 59–62):

Claims require support

- ◆ We cannot meet this production deadline without sacrificing quality.
- ◆ We're doing all we can to correct your software problem.
- ◆ This hiring policy is discriminatory.

Figure 4.1
Two Problems
Confronted by
Communicators

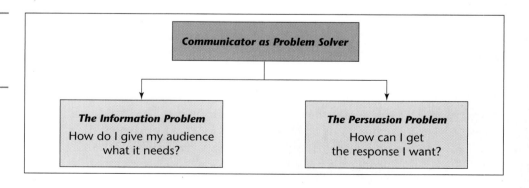

- ◆ Our software is superior to the competing brand.
- ◆ I deserve a raise.

Without support that people find convincing, none of these claims are likely to be accepted. Before changing their minds, people need to see "what's in it for them" (Hahn 19).

ASSESS THE POLITICAL REALITIES

Any document can evoke different reactions depending on a user's temperament, interests, fears, biases, ambitions, or assumptions. Whenever peoples' views are challenged, they react with questions like these:

TYPICAL AUDIENCE QUESTIONS ABOUT ANY ATTEMPT TO PERSUADE

- ◆ Says who?
- ◆ So what?
- ◆ Why should I?
- ◆ Why rock the boat?
- ◆ What's in it for me?
- ◆ What will it cost?

- ◆ What are the risks?
- ◆ What are you up to?
- ◆ What's in it for you?
- ◆ What does this really mean?
- ◆ Will it mean more work for me?
- ◆ Will it make me look bad?

People read between the lines. Some might be impressed and pleased by your suggestions for increasing productivity or cutting expenses; some might feel offended or threatened. Some might suspect you of trying to undermine your boss. Furthermore, no one wants bad news; some people prefer to ignore it (as the O-ring flaw on the shuttle *Challenger's* booster rockets made tragically clear). Such are the political realities in any organization (Hays 19).

EXPECT AUDIENCE RESISTANCE

People who haven't made up their minds are more receptive to persuasion than those who have:

We rely on persuasion to help us make up our minds

> We need others' arguments and evidence. We're busy. We can't and don't want to discover and reason out everything for ourselves. We look for help, for short cuts, in making up our minds. (Gilsdorf, "Write Me" 12)

People who *have* decided what to think, however, naturally assume they're right, and they often refuse to budge. The O-ring claims cited on page 35 were made—and well supported—by engineers before the shuttle *Challenger* exploded on January 28, 1986. That NASA management ignored these claims illustrates how people can resist the most compelling arguments.

Whenever you question people's stance on an issue or try to change their behavior, expect resistance:

Why persuasion is so difficult

> By its nature, informing "works" more often than persuading does. While most people do not mind taking in some new facts, many people do resist efforts to change their opinions, attitudes, or behaviors. (Gilsdorf, "Executives' and Academics' Perception" 61)

The bigger the stake in the issue—the more personal the involvement—the more resistance you can expect.

When people do yield to persuasion, they may respond grudgingly, willingly, or enthusiastically (Figure 4.2). Researchers categorize these responses as compliance, identification, or internalization (Kelman 51–60):

Some ways of yielding to persuasion are more productive than others

- ◆ *Compliance:* "I'm giving in to your demand in order to get a reward or to avoid punishment. I really don't accept it, but I feel pressured so I'll go along to get along."
- ◆ *Identification:* "I'm going along with your appeal because I like and believe you, I want you to like me, and I believe we have something in common."
- ◆ *Internalization:* "I'm going along because what you're saying makes good sense and it fits my goals and values."

Although mere compliance is sometimes necessary (as in military orders or workplace safety regulations), nobody likes to be coerced. If people merely comply because they have no choice, you can expect to lose their loyalty and goodwill— and as soon as the threat or reward disappears, so will their compliance. Persuasion relies on identification and, especially, internalization.

Figure 4.2
The Levels of Response to Persuasion

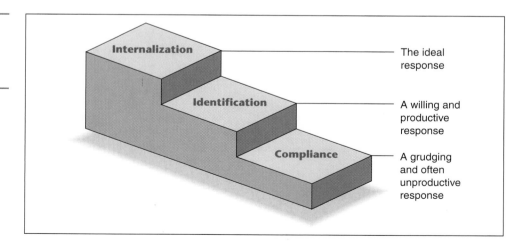

KNOW HOW TO CONNECT WITH THE AUDIENCE

Persuasive people know when to simply declare what they want, when to reach out and create a relationship, or when to appeal to reason (Kipnis and Schmidt 40–46). These strategies can be labeled the *power connection,* the *relationship connection,* and the *rational connection,* respectively (Figure 4.3).

For an illustration of these different connections, picture this situation: Your Company, XYZ Engineering, has just developed a fitness program, based on findings that healthy employees work better, take fewer sick days, and cost less to insure. This program offers clinics for smoking, stress reduction, and weight loss, along with group exercise. In your second month on the job you read this notice in your email:

POWER CONNECTION

To: Employees@XYZ.com
FROM: GMaximus@XYZ.com
RE: Physical Fitness

Orders readers to show up

On Monday, June 10, all employees will report to the company gymnasium at 8:00 a.m. for the purpose of choosing a walking or jogging group. Each group will meet for 30 minutes three times weekly during lunch time.

How would you react? Despite the reference to "choosing," you are given no real choice but simply ordered to show up. This kind of power connection is typically used by bosses and others in power. Although it often achieves compliance, the power connection almost always alienates the audience.

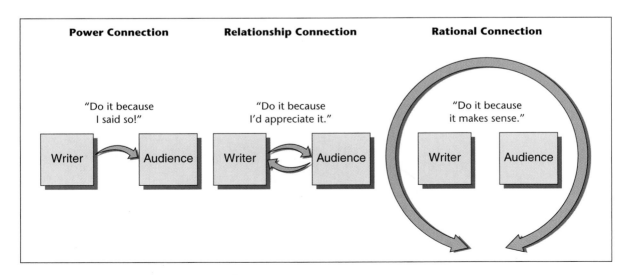

Figure 4.3　Three Strategies for Connecting with an Audience

Suppose, instead, that you receive the following version of the memo. How would you react?

RELATIONSHIP CONNECTION

To: Employees@XYZ.com
FROM: GMaximus@XYZ.com
RE: An Invitation to Physical Fitness

Invites readers to participate

I realize most of you spend lunch hour playing cards, reading, or just enjoying a bit of well-earned relaxation in the middle of a hectic day. But I'd like to invite you to join our lunchtime walking/jogging club.

Leaves the choice to the reader

We're starting this club in hopes that it will be a great way for us all to feel better. Why not give it a try?

This second version evokes a sense of identification, of shared feelings and goals. Instead of being commanded, readers are invited—given a real choice.

Often, the biggest factor in persuasion is an audience's perception of the writer. People are more receptive to someone they like, trust, and respect. The relationship connection is especially vital in cross-cultural communication—as long as it is not too "chummy" and informal. (see pages 48–49).

NOTE

Of course, you would be unethical in appealing to—or faking—the relationship merely to hide the fact that you have no evidence to support your claim (Ross 28). People need to find the claim believable ("Exercise will help me feel better") and relevant ("I, personally, need this kind of exercise").

Here is a third version of the memo. As you read, think about the ways it makes a persuasive case.

RATIONAL CONNECTION

To: Employees@XYZ.com
FROM: GMaximus@XYZ.com
RE: Invitation to Join One of Our Jogging or Walking Groups

Presents authoritative evidence

I want to share a recent study from the *New England Journal of Medicine*, which reports that adults who walk two miles a day could increase their life expectancy by three years.

Other research shows that 30 minutes of moderate aerobic exercise, at least three times weekly, has a significant and long-term effect in reducing stress, lowering blood pressure, and improving job performance.

Offers alternatives

As a first step in our exercise program, XYZ Engineering is offering a variety of daily jogging groups: The One-Milers, Three-Milers, and Five-Milers. All groups will meet at designated times on our brand new, quarter-mile, rubberized clay track.

For beginners or skeptics, we're offering daily two-mile walking groups. For the truly resistant, we offer the option of a Monday–Wednesday–Friday two-mile walk.

Offers a compromise

Coffee and lunch breaks can be rearranged to accommodate whichever group you select.

Leaves the choice
to the reader

Offers incentives

> Why not take advantage of our hot new track? As small incentives, XYZ will reimburse anyone who signs up as much as $100 for running or walking shoes, and will even throw in an extra fifteen minutes for lunch breaks. And with a consistent turnout of 90 percent or better, our company insurer may be able to eliminate everyone's $200 yearly deductible in medical costs.

This version expresses respect for the reader's intelligence *and* for the relationship. With any reasonable audience, the rational connection stands the best chance of succeeding.

NOTE   *Keep in mind that no cookbook formula exists, and in many situations, even the best persuasive attempts may be rejected.*

ALLOW FOR GIVE-AND-TAKE

Persuasion requires flexibility on your part. Instead of just pushing your own case forward, consider other people's viewpoints. In advocating your position, for example, you need to (Senge 8):

How to promote
your view

- explain the reasoning and evidence behind it
- invite people to find weak spots in your case, and to improve on it
- invite people to challenge your ideas (say, with alternative reasoning or data)

When others offer an opposing view, you need to:

How to respond to
opposing views

- try to see things their way, instead of insisting on your way
- rephrase their position in your own words, to be sure you understand it accurately
- try reaching agreement on what to do next, to resolve any insurmountable differences
- explore possible compromises others might accept

Perhaps some XYZ employees, for example, have better ideas for making the exercise program work for everyone.

ASK FOR A SPECIFIC RESPONSE

Unless you are giving an order, diplomacy is essential in persuasion. But don't be afraid to ask for what you want:

Spell out what you
want

> The moment of decision is made easier for people when we show them what the desired action is, rather than leaving it up to them. . . . No one likes to make deci-

sions: there is always a risk involved. But if the writer asks for the action, and makes it look easy and urgent, the decision itself looks less risky. (Cross 3)

Tell people exactly what you want them to do or think.

NOTE   *Keep in mind that overly direct communication can offend audiences from other cultures. Don't mistake bluntness for clarity.*

NEVER ASK FOR TOO MUCH

Stick with what is achievable

People never accept anything they consider unreasonable. And the definition of "reasonable" depends on the individual. Employees at XYZ, for example, differ as to which walking/jogging option they might accept.

To the jock writing the memo, a daily five-mile jog might seem perfectly reasonable, but to most people this would seem outrageous. XYZ's program, therefore, has to offer something most people (except, say, couch potatoes and those in poor health) accept as reasonable. Any request that exceeds its audience's "latitude of acceptance" (Sherif 39–59) is doomed.

Identify your audience's latitude of acceptance

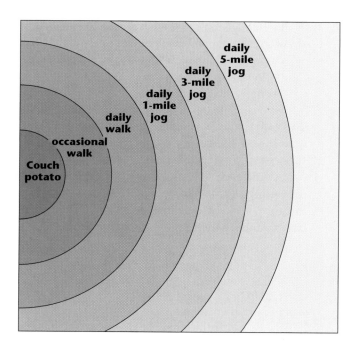

Options offered by XYZ

To get a clear sense of exactly how much is *achievable* in the given situation, ask around beforehand and feel out the people involved.

RECOGNIZE ALL COMMUNICATION CONSTRAINTS

Constraints are limits or restrictions imposed by the situation—who can say what to whom, and so on. The workplace presents constraints like these:

Communication constraints in the workplace

- ◆ What can I say around here, to whom and how?
- ◆ Should I say it in person, by phone, in print, online?
- ◆ Could I be creating any ethical or legal problems?
- ◆ Is this the best time to say it?
- ◆ What's my relationship with the audience?
- ◆ Who are the personalities involved?
- ◆ Any peer pressure to overcome?
- ◆ How big an issue is this?

Organizational Constraints

Organizations announce their own official constraints: deadlines; budgets; guidelines for organizing, formatting, and distributing documents; and so on. But communicators also face *unofficial* constraints:

Decide carefully when to say what to whom

> Most organizations have clear rules for interpreting and acting on statements made by colleagues. Even if the rules are unstated, we know who can initiate interaction, who can be approached, who can propose a delay, what topics can or cannot be discussed, who can interrupt or be interrupted, who can order or be ordered, who can terminate interaction, and how long interaction should last. (Littlejohn and Jabusch 143)

The exact rules vary among organizations, and most are unspoken, but anyone who breaks the rules (say, by going over a supervisor's head with a complaint or suggestion) invites disaster.

Airing even a legitimate gripe in the wrong way through the wrong medium to the wrong person can be fatal to your work relationships. The following email, for instance, is likely to be interpreted by the executive officer as petty and whining behavior, and by the maintenance director as a public attack.

Wrong way to the wrong person

To: CEO@XYZ.com
CC: MaintenanceDirector@XYZ.com
FROM: Middle Manager@XYZ.com
RE: Trash Problem
Please ask the Maintenance Director to get his people to do their job for a change. I realize we're all understaffed, but I've gotten dozens of complaints this week about the filthy restrooms and overflowing wastebaskets in my department. If he wants us to empty our own wastebaskets, why doesn't he let us know?

Instead, why not address the message directly to the key person—or better yet, phone the person?

A better way to the right person

> TO: MaintenanceDirector@XYZ.com
> FROM: MiddleManager@XYZ.com
> RE: Staffing Shortage
> I wonder if we could meet to exchange some ideas about how our departments might be able to support one another during these staffing shortages.

Can you identify the unspoken rules in companies where you have worked? What happens when such rules are ignored?

Legal Constraints

Sometimes what you can say is limited by contract or by laws protecting confidentiality or customers' rights, or laws affecting product liability. These constraints include:

Major legal constraints on communication

- ◆ In a collection letter for nonpayment, you can threaten legal action but you cannot threaten any kind of violence or to publicize the refusal to pay, nor can you pretend to be an attorney (Varner and Varner 31–40).
- ◆ If someone requests information on one of your employees, you can "respond only to specific requests that have been approved by the employee. Further, your comments should relate only to job performance which is documented" (Harcourt 64).
- ◆ When writing sales literature or manuals, you and your company are liable for faulty information that leads to injury or damage.

As you prepare any document, try to anticipate the possible legal problems that could arise. For instance, suppose that an employee of XYZ Corporation drops dead during the new exercise program you've marketed so persuasively. Could you and your company be liable? Should you require physical exams and stress tests (at company expense) for participants?

Ethical Constraints

Ethical constraints are defined not by law, but by honesty and fair play. For example, it may be perfectly legal to promote a new pesticide by emphasizing its effectiveness, while downplaying its carcinogenic effects; whether such action is ethical, however, is another issue. To earn people's trust, you will find that "saying the right thing" involves more than legal considerations.

NOTE   *Persuasive skills carry tremendous potential for abuse. "Presenting your best case" does not mean deceiving others—even if the dishonest answer is the one people want to hear.*

Time Constraints

Persuasion is often a matter of good timing. Should you wait for an opening, release the message immediately, or what? Let's assume that you're trying to "bring out the vote" among members of your professional society on some hotly debated issue, say, whether to refuse to work on any project related to biological warfare. You might prefer to wait until you have all the information you need or until you've analyzed the situation and planned a strategy. But if you delay too long, rumors or paranoia could cause people to harden their positions—and their resistance to your appeals.

Social and Psychological Constraints

Too often, the human side of communication leads to misunderstanding, due to constraints like those on page 45.

In Brief HOW YOU SPEAK SHOWS WHERE YOU RANK*

Popular discussion of communication style in recent years has centered on differences between the sexes. The subject has been fodder for TV talk shows, corporate seminars, and bestsellers, notably Deborah Tannen's You Just Don't Understand *and John Gray's* Men Are From Mars, Women Are From Venus. *But Sarah McGinty, a teaching supervisor at Harvard University's school of education, believes language style is based more on power than on gender—and that marked differences distinguish the powerful from the powerless loud and clear. As a consultant, she is often called on to help clients develop more effective communication styles. FORTUNE's Justin Martin spoke with McGinty about her ideas:*

What style of speaking indicates that someone possesses power?

A person who feels confident and in control will speak at length, set the agenda for a conversation, stave off interruptions, argue openly, make jokes, and laugh. Such a person is more inclined to make statements, less inclined to ask questions. They are more likely to offer solutions or a program or a plan. All this creates a sense of confidence in listeners.

What about people who lack power? How do they speak?

The power deficient drop into conversations, encourage other speakers, ask numerous questions, avoid argument, and rely on gestures such as nodding and smiling that suggest agreement. They tend to offer empathy rather than solutions. They often use unfinished sentences. Unfinished sentences are a language staple for those who lack power.

How do you figure out what style of communication you lean toward?

It's quite hard to do. We're often quite ignorant about our own way of communicating. Everyone comes home at night occasionally and says, "I had that idea, but no one heard me, and everyone thinks it's Harry's idea." People like to pin that on gender and a lot of other things as well. But it's important to find out what really did happen. Maybe it was the volume of your voice, and you weren't heard. Maybe you overexplained, and the person who followed up pulled out the nugget of your thought.

But it's important to try to get some insight into what your own language habits are so that

"What's our
relationship?"
- ◆ *Relationship with the audience:* Is your reader a superior, a subordinate, a peer? (Try not to dictate to subordinates or shield superiors from bad news.) How well do you know each other? Can you joke around or should you be serious? Do you get along or have a history of conflict or mistrust? What you say and how you say it—and how it is interpreted—will be influenced by the relationship.

"How receptive is
this audience?"
- ◆ *Audience's personality:* A person's willingness to be persuaded depends largely on personality: self-esteem, assertiveness, conformity, open- or closed-mindedness, and so on (Stonecipher 188–89). The less persuadable your audience, the harder you must work. If your audience is totally resistant, you may want to back off—or give up altogether.

"How unified is this
audience?"
- ◆ *Audience's sense of identity and affiliation as a group:* Does the group have a strong identity (union members, conservationists, engineers)? Will group loy-

---

you can be analytical about whether you're shooting yourself in the foot. You can tape your side of phone calls, make a tape of a meeting, or sign up for a communications workshop. That's a great way to examine how you conduct yourself in conversations and in meetings.

Does power language differ from company to company?

Certainly. The key is figuring out who gets listened to within your corporate culture. That can make you a more savvy user of language. Try to sit in on a meeting as a kind of researcher, observing conversational patterns. Watch who talks, who changes the course of the discussion, who sort of drops in and out of the conversation. Then try to determine who gets noticed and why.

One very effective technique is to approach the person who ran the meeting a couple of days after the fact and ask for an overall impression. What ideas were useful? What ideas might have a shot at being implemented?

How can you get more language savvy?

You can start by avoiding bad habits, such as always seeking collaboration in the statements

you make. Try to avoid "as Bob said" and "I pretty much agree with Sheila." Steer clear of disclaimers such as "I may be way off base here, but . . . " All these serve to undermine the impact of your statements.

The amount of space you take up can play a big part in how powerful and knowledgeable you appear. People speaking before a group, for instance, should stand with their feet a little bit apart and try to occupy as much space as possible. Another public-speaking tip: Glancing around constantly creates a situation in which nobody really feels connected to what you're saying.

Strive to be bolder. Everyone tends to worry that they will offend someone by stating a strong opinion. Be bold about ideas, tentative about people. Saying "I think you're completely wrong" is not a wise strategy. Saying "I have a plan that I think will solve these problems" is perfectly reasonable. You're not attacking people. You're being bold with an idea.

*Reprinted with permission from *Fortune* 2 Feb. 1998: 15.

alty or pressure prevent certain appeals from working? Address the group's collective concerns.

"Where are people coming from on this?"

◆ *Perceived size and urgency of the issue:* Does the audience see this as a threat or an opportunity? Has the issue been understated or overstated? Big problems often cause people to exaggerate their fears, loyalties, and resistance to change—or to seek quick solutions. Assess the problem realistically. Don't downplay a serious problem, but don't cause panic, either.

SUPPORT YOUR CLAIMS CONVINCINGLY

The persuasive argument is the one that makes the best case—in the audience's view:

A persuasive case offers reasons that matter to the audience

> When we seek a project extension, argue for a raise, interview for a job . . . we are involved in acts that require good reasons. Good reasons allow our audience and ourselves to find a shared basis for cooperating. . . . [Y]ou can use marvelous language, tell great stories, provide exciting metaphors, speak in enthralling tones, and even use your reputation to advantage, but what it comes down to is that you must speak to your audience with reasons they understand. (Hauser 71)

Imagine the following situation: As documentation manager for Bemis Software, a rapidly growing company, you supervise preparation and production of all user manuals. The present system for producing manuals is inefficient because three different departments are involved in (1) assembling the material, (2) word processing and designing, and (3) publishing the manuals. Much time and energy are wasted as a manual goes back and forth among software specialists, communication specialists, and the art and printing department. After studying the problem and calling in a consultant, you decide that greater efficiency would result if desktop publishing software were installed in all computer terminals. Everyone involved could then contribute during all three phases. To sell this plan to bosses and coworkers you need reasons based on *evidence* and *appeals to everyone's needs and values* (Rottenberg 104–06).

Offer Convincing Evidence

Evidence (factual support from an outside source) is a powerful persuader—as long as it meets an audience's standards. A discerning audience evaluates evidence using these criteria (Perloff 157–58):

Criteria for worthwhile evidence

◆ *The evidence has quality.* People expect evidence that is strong, specific, new, or different.

◆ *The sources are credible.* People want to know where the evidence comes from, how it was collected, and who collected it.

◆ *The evidence is considered reasonable.* The evidence falls within the audience's "latitude of acceptance" (discussed on page 41).

Common types of evidence include factual statements, statistics, examples, and expert testimony.

FACTUAL STATEMENTS. A *fact* is something that can be demonstrated by observation, experience, research, or measurement—and that your audience is willing to recognize.

Offer the facts

> Most of our competitors already have desktop publishing networks in place.

Be selective. Decide which facts best support your case.

STATISTICS. Numbers can be highly convincing. Workplace audiences are usually interested in the "bottom line": costs, savings, losses, profits.

Cite the numbers

> After a cost/benefit analysis, our accounting office estimates that an integrated desktop publishing network will save Bemis 30 percent in production costs and 25 percent in production time—savings that will enable the system to pay for itself within one year.

But numbers can mislead. Your statistics must be accurate, trustworthy, and easy to understand and verify (see pages 157–64.) Always cite your source.

EXAMPLES. Examples help people visualize and remember the point. For example, the best way to explain what you mean by "inefficiency" in your company is to show "inefficiency" occurring:

Show what you mean

> The figure illustrates the inefficiency of Bemis's present system for producing manuals:

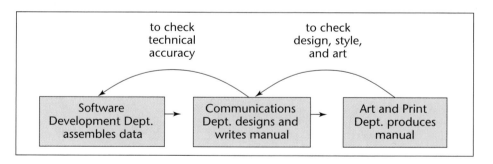

> A manual typically goes back and forth through this cycle three or four times, wasting time and effort in all three departments.

Always explain how the example fits the point it is designed to illustrate.

EXPERT TESTIMONY. Expert opinion—if it is unbiased and the expert is recognized—lends authority and credibility to any claim.

Cite the experts

> Ron Catabia, nationally recognized networking consultant, has studied our needs and strongly recommends we move ahead with the integrated network.

See page 104 for the limits of expert testimony.

Evidence alone isn't always enough to change a person's mind. At Bemis, for example, the bottom line might be very persuasive for company executives but managers and employees might be asking: Does this threaten my authority? Will I have to work harder? Will I fall behind? Is my job in danger? This group expects some benefit beyond company profit.

Appeal to Common Goals and Values

Identify at least one goal you and your audience have in common: "What do we all want most?"

We can assume that everyone at Bemis, for example, wants job security and some control over their career. A persuasive recommendation will therefore take these goals into account:

> I'd like to show how desktop publishing skills, instead of threatening anyone's job, would only increase career mobility for all of us.

People's goals are shaped by their values (qualities they believe in, ideals they stand for): friendship, loyalty, honesty, equality, fairness, and so on (Rokeach 57–58).

At Bemis, you might appeal to the commitment to quality and achievement shared by the company and individual employees:

> None of us needs reminding of the fierce competition in the software industry. The improved collaboration among networking departments will result in better manuals, keeping us on the front line of quality and achievement.

Give people reasons that have real meaning for them personally. For example, in a recent study of teenage attitudes about the hazards of smoking, respondents listed these reasons for not smoking: bad breath, difficulty concentrating, loss of friends, and trouble with adults. No respondents listed dying of cancer—presumably because this reason carries little meaning for young people personally (Baumann et al. 510–30).

CONSIDER THE CULTURAL CONTEXT

An estimated 60 percent of business ventures between the United States and other countries fail (Isaacs 43). Many of these failures result from cultural differences that prevent people from responding to persuasive appeals and reaching agreement.

Depending on social customs and values, people from different cultures react differently to persuasive appeals[1]:

Margin notes:

"What makes these people tick?"

Appeal to shared goals

Appeal to shared values

---

[1]Adapted from Beamer 293–95; Gestelend 24; Hulbert, "Overcoming" 42; Jameson 9–11; Kohl et al. 65; Martin and Chaney 271–77; Nydell 61; Thrush 276–77; Victor 159–66.

How cultural
differences influence
a persuasive
situation

- Some cultures hesitate to debate, criticize, or express disagreement.
- Some cultures observe special formalities in communicating (say, expressions of concern for one's family).
- Some cultures consider the source of a message as important as its content. Establishing trust and building a relationship are essential preludes to doing business.
- Cultures might differ in their attitudes toward big business, technology, competition, or women in the workplace. They might value delayed gratification more than immediate reward, stability more than progress, time more than profit, politeness more than candor, age more than youth.

One especially volatile cause of clashes among different cultures is related to *face saving:*

Face saving is every
person's bottom
line

Face saving [is] the act of preserving one's prestige or outward dignity. People of all cultures, to a greater or lesser degree, are concerned with face saving. (Victor 159–61)

People lose face in situations like these:

How people lose
face

- *When they are offended or embarrassed by blatant criticism:* A U.S. businessperson in China criticizes the Tiananmen Square massacre or illegal contributions to American political parties (Stepanek 4).
- *When their customs are ignored:* An American female arrives to negotiate with older, Japanese males; Silicon Valley businesspeople show up in T-shirts and baseball caps to meet with hosts wearing suits.
- *When their values are trivialized:* An American businessperson in Paris greets the French host as "Pierre," slaps him on the back, and jokes that the "rich French food" on the Concorde flight had him "throwing up all the way over" (Isaacs 43).

When someone feels insulted, meaningful interaction ceases. Effective communication in any cultural context enables people to save face.

Show respect for a country's cultural heritage by learning all you can about its history, landmarks, famous people, and especially its customs and values (Isaacs 43). The questions on page 50 can get you started.

NOTE   *Violating a person's cultural frame of reference is offensive, but so is reducing individual complexity to a laundry list of cultural stereotypes. Any generalization about any group presents a limited picture and in no way accurately characterizes any or all members of the group.*

QUESTIONS FOR ANALYZING CULTURAL DIFFERENCES*

What behavior does the culture consider acceptable?

- Casual versus formal interaction
- Directness and plain talk versus indirectness and ambiguity
- Rapid decision making versus extensive analysis and discussion
- Willingness to request clarification
- Willingness to argue, criticize, or disagree
- Willingness to be contradicted
- Willingness to express emotion

What values does the culture consider most important?

- Attitude toward big business, competition, and U.S. culture
- Youth versus age
- Rugged individualism versus group loyalty
- Status of women in the workplace
- Feelings versus logic
- Candor versus face saving
- Progress and risk taking versus stability
- Importance of trust and relationship building
- Importance of time ("Time is money!" or "Never rush!")

*Adapted from Beamer 293–95; Gesteland 24; Hulbert, "Overcoming" 42; Jameson 9–11; Kohl et al. 65; Martin and Chaney 271–77; Nydell 61; Thrush 276–77; Victor 159–66.

GUIDELINES

for Making Your Case

Any attempt at persuasion must focus on the audience:

> No matter how brilliant, any argument rejected by its audience is a failure.

If people dislike you or decide that what you say has no meaning for them personally, they reject your argument. Begin by trying to see things the audience's way, instead of insisting that they see things your way.

Planning Your Case

1. *Assess the political climate.* Whom will this affect? How will people react? How will they interpret your motives? Can you be outspoken? Could the document cause legal problems? To avoid backlash:

   - Be aware of your status in the group; don't overstep.
   - Don't expect perfection from anyone—including yourself.
   - Don't make anyone look bad, lose face, or feel coerced.
   - Don't make promises you can't keep.
   - Ask directly for support: "Is this idea worthy of your commitment?" (Senge 7).
   - Invite intended readers—especially the group opinion-makers—to review early drafts.

When reporting something that others do not want to hear, expect fallout. Decide beforehand whether you want to keep your job or your dignity (see Chapter 5 for more on ethics).

2. *Learn the unspoken rules.* Know the constraints (especially the legal ones) on what you can say to whom and how and when. Consider the cultural context.

3. *Decide on a connection (or combination of connections).* Should you *require* compliance, or appeal to a relationship or to common sense and reason?

4. *Anticipate reactions.* Will people be defensive, shocked, annoyed, angry? Try to neutralize big objections beforehand. Express your judgments ("We could do better") without blaming people ("It's all your fault").

Preparing Your Case

1. *Be clear about what you want.* Diplomacy is important—especially in cross-cultural communication—but don't leave people guessing about your purpose.

2. *Never assert or request something you know people will reject outright.* Can people live with what you're requesting or proposing? Get a sense of what is achievable by asking what people are thinking. Invite them to share in decision-making. Offer real choices.

3. *Project a likable and reasonable persona.* Persona is the image or impression the writer projects in her or his tone. Resist the urge to "sound off." Audiences tune out aggressive people—no matter how sensible the argument. Admit the imperfections in your case. Pat deserving people on the back. Invite people to respond.

4. *Find points of agreement with your audience.* What does everyone involved want? To reduce conflict, focus early on a shared value, goal, or concern. Emphasize your similarities.

5. *Never distort the opposing position.* A sure way to alienate people is to cast the opponent in a more negative light than the facts warrant.

6. *Concede* **something** *to the opposing position.* Admit the merits of the opposing case before arguing the merits of your own. Show empathy and willingness to compromise.

7. *Stick to claims or assertions you can support.* Show people exactly what's in it for them—but never distort the facts just to please the audience.

8. *Stick to your best material.* Some points are stronger than others. Decide which material—from your audience's view—best advances your case.

Presenting Your Case

1. *Get a second opinion before releasing the document.* Ask someone you trust and who has no stake in the issue to read the document. If needed, have your company's legal department approve the document.

2. *Get the timing right.* When will your case most likely fly—or crash and burn? Look for a good opening.

3. *Get the medium right.* Should this message travel in person, via hard copy, phone, email, fax, newsletter, bulletin board, or what? (See also page 92). Should everyone receive it via the same medium? If your written message will surprise readers, try to warn them face-to-face or by phone.

4. *Don't be defensive about negative reactions.* Instead, admit mistakes, invite people to improve on your ideas (Bashein and Marcus 43).

5. *Know when to back off.* People who feel bullied, manipulated, or deceived often become your enemies.

NOTE

People rarely change their minds quickly or without good reason. A truly resistant audience will dismiss even the best arguments and may end up feeling threatened and resentful. Even with a receptive audience, attempts at persuasion can fail. Often, the best you can do is avoid disaster and create an opportunity for people to appreciate the merits of the case.

Figure 4.4 illustrates the guidelines in an actual persuasive situation. This letter is from a company that distributes systems for generating electrical power from recycled steam (cogeneration). President Tom Ewing writes a persuasive answer to a potential customer's question: "Why should I invest in the system you are proposing for my plant?" As you read the letter, notice the kinds of evidence and appeals that support the opening claim. Notice also the focus on reasons important to the reader.

Figure 4.5 shows the audience and use profile for the above writing situation.

July 20, 20XX

Mr. Richard White, President
Southern Wood Products
Box 84
Memphis, TN 37162

Dear Mr. White:

The writer states his claim

In our meeting last week, you asked me to explain why we have such confidence in the project we are proposing. Let me outline what I think are excellent reasons.

Offers first reason

Gives examples

First, you and Don Smith have given us a clear idea of your needs, and our recent discussions confirm that we fully understand these needs. For instance, our proposal specifies an air-cooled condenser rather than a water-cooled condenser for your project because water in Memphis is expensive. And besides saving money, an air-cooled condenser will be easier to operate and maintain.

Offers second reason
Appeals to shared value (quality)

Gives example

Further examples

Appeals to reader's goal (security)

Second, we have confidence in our component suppliers and they have confidence in this project. We don't manufacture the equipment; instead, we integrate and package cogeneration systems by selecting for each application the best components from leading manufacturers. For example, Alias Engineering, the turbine manufacturer, is the world's leading producer of single-stage turbines, having built more than 40,000 turbines in 70 years. Likewise, each component manufacturer leads the field and has a proven track record. We have reviewed your project with each major component supplier, and each guarantees the equipment. This guarantee is of course transferable to you and is supplemented by our own performance guarantee.

Offers third reason

Cites experts

Third, we have confidence in the system design. We developed the CX Series specifically for applications like yours, in which there is a need for both a condensing and a backpressure turbine. In our last meeting, I pointed out the cost, maintenance, and performance benefits of the CX Series. And although the CX Series is an innovative design, all components are fully proven in many other applications, and our suppliers fully endorse this design.

Figure 4.4 Supporting a Claim with Good Reasons   *Reprinted with permission of Thomas S. Ewing, President, Ewing Power Systems, So. Deerfield, MA 01373.*

Richard White, July 20, 20XX, p. 2

Closes with best reason

Appeals to shared value (trust) and shared goal (success)

Finally, and perhaps most important, you should have confidence in this project because we will stand behind it. As you know, we are eager to establish ourselves in Memphis-area industries. If we plan to succeed, this project must succeed. We have a tremendous amount at stake in keeping you happy.

If I can answer any questions, please phone me. We look forward to working with you.

Sincerely,

Tom Ewing

Thomas S. Ewing
President
EWING POWER SYSTEMS, INC.

Figure 4.4   Supporting a Claim with Good Reasons   *Continued*

Audience Identity and Needs

Primary audience: _Richard White, President of Southern Wood Products_

Secondary audience: _Don Smith, Plant Engineer; several plant managers_

Relationship: _A possible customer for a customized cogeneration system_

Intended use of document: _To provide information for a purchasing decision_

Prior knowledge about this topic: _Has compared various power-generation systems_

Additional information needed: _Seems to doubt the reliability of our system_

Probable questions: _How much money will your proposed system really save?_
How reliable is the equipment?
Can we depend on the innovative design you are proposing?
What quality of service can we expect?

Audience's Probable Attitude and Personality

Attitude toward topic: _Highly interested, but somewhat skeptical_

Probable objections: _This technology is too recent to have a solid track record_

Probable attitude toward this writer: _Receptive but cautious_

Organizational climate: _Open and flexible; lots of collaboration_

Persons most affected by this document: _White and other decision makers_

Temperament: _White takes a deliberate and conservative approach to decision making_

Probable reaction to document: _Readers should feel somewhat reassured_

Risk of alienating anyone: _No apparent risks_

Audience Expectations about the Document

Reason document originated: _Response to management concerns about reliability_

Cultural considerations: _None in particular_

Acceptable length: _Average business letter that gets right to the point_

Material important to this audience: _Solid evidence to back up our claim of "confidence" in this project_

Most useful arrangement: _A convincing list of reasons and appeals_

Tone: _Encouraging, friendly, and confident_

Intended effect on this audience: _To pave the way for a decision to purchase_

Due date: _ASAP—to illustrate our responsiveness to customer concerns_

Figure 4.5   Audience and Use Profile Sheet for Figure 4.4

Checklist FOR CROSS-CULTURAL DOCUMENTS

Use this checklist* to verify that your documents respect audience diversity. (Page numbers in parentheses refer to the first page of discussion.)

☐ Does the document allow everyone to save face? (49)

☐ Is the document sensitive to the culture's customs and values? (48)

☐ Does the document conform to the safety and regulatory standards of the country? (75)

☐ Does the document provide the expected level of detail? (30)

☐ Does the document avoid possible misinterpretation? (30)

☐ Is the document organized in a way that readers will consider appropriate? (193)

☐ Does the document observe interpersonal conventions important to the culture (accepted forms of greeting or introduction, politeness requirements, first names, titles, and so on)? (49)

☐ Does the document's tone reflect the appropriate level of formality or casualness? (248)

☐ Is the document's style appropriately direct or indirect? (49)

☐ Is the document's format consistent with the culture's expectations? (320)

☐ Does the document embody universal standards for ethical communication? (69)

*This list was largely adapted from Caswell-Coward 265; Weymouth 144; Beamer 293–95; Martin and Chaney 271–77; Victor 159–61.

EXERCISES

1. You work for a technical marketing firm proud of its reputation for honesty and fair dealing. A handbook being prepared for new personnel includes a section titled "How to Avoid Abusing Your Persuasive Skills." All employees have been asked to contribute to this section by preparing a written response to the following:

   Share a personal experience in which you or a friend were the victim of persuasive abuse in a business transaction. In a one- or two-page memo, describe the situation and explain exactly how the intimidation, manipulation, or deception occurred.

   Write the memo and be prepared to discuss it in class.

2. Find an example of an effective persuasive letter. In a memo to your instructor, explain why and how the message succeeds. Base your evaluation on the persuasion guidelines, pages 50–52. Attach a copy of the letter to your evaluation memo. Be prepared to discuss your evaluation in class.

   Now, evaluate a poorly written document, explaining how and why it fails.

3. Think about some change you would like to see on your campus or at work. Perhaps you would like to make something happen, such as a campus-wide policy on plagiarism, changes in course offerings or requirements, more access to computers, a policy on sexist language, or a day care center. Or perhaps you would like to improve something, such as the grading system, campus lighting, the system for student evaluation of teachers, or the promotion system at work. Or perhaps you would like to stop something from happening, such as noise in the library or sexual harassment at work.

   Decide whom you want to persuade, and write a memo to that audience. Anticipate carefully your audience's implied questions, such as:

- *Do we really have a problem or need?*
- *If so, should we care enough about it to do anything?*
- *Can the problem be solved?*
- *What are some possible solutions?*
- *What benefits can we anticipate? What liabilities?*

Can you envision additional audience questions? Do an audience and use analysis based on the profile sheet, page 55.

   Don't think of this memo as the final word but as a consciousness-raising introduction that gets the reader to acknowledge that the issue deserves attention. At this early stage, highly specific recommendations would be premature and inappropriate.

4. Challenge an attitude or viewpoint that is widely held by your audience. Maybe you want to persuade your classmates that the time required to earn a bachelor's degree should be extended to five years or that grade inflation is watering down your school's reputation. Maybe you want to claim that the campus police should (or should not) wear guns. Or maybe you want to ask students to support a 10 percent tuition increase in order to make more computers and software available.

   Do an audience and use analysis based on the profile sheet, page 55. Write specific answers to the following questions:

- What are the political realities?
- What kind of resistance could you anticipate?
- How would you connect with readers?
- What about their latitude of acceptance?
- Any other constraints?
- What reasons could you offer to support your claim?

   In a memo to your instructor, submit your plan for presenting your case. Be prepared to discuss your plan in class.

5. Use the questions on page 50 as a basis for interviewing a student from another country. Be prepared to share your findings with the class.

COLLABORATIVE PROJECTS

1. You work for an environmental consulting firm that is under contract with various countries for a range of projects, including these:

- A plan for rain forest regeneration in Latin America and Sub-Saharan Africa
- A plan to decrease industrial pollution in Eastern and Western Europe
- A plan for "clean" industries in developing countries
- A plan for organic agricultural development in Africa and India
- A joint American/Canadian plan to decrease acid rain
- A plan for developing alternative energy sources in Southeast Asia

Each project will require environmental impact statements, feasibility studies, grant proposals, and a legion of other documents, often prepared in collaboration with members of the host country, and in some cases prepared by your company for audiences in the host country—from political, social, and industrial leaders to technical experts and so on.

   For such projects to succeed, people from different cultures have to communicate effectively and sensitively, creating goodwill and cooperation.

   Before your company begins work in earnest with a particular country, your coworkers will need to develop a degree of cultural awareness. Your assignment is to select a country and to research that culture's behaviors, attitudes, values, and social system in terms of how these variables influence the culture's communication preferences and expectations. What should your colleagues know about this culture in order to communicate effectively and diplomatically? Do the necessary research using the questions from page 50 as a guide.

Prepare a recommendation report in memo form. Be prepared to present your findings in class.

2. Often, workplace readers need to be *persuaded* to accept recommendations that are controversial or unpopular. This project offers practice in dealing with the persuasion problems of communicating within organizations.

Divide into teams. Assume that your team agrees strongly about one of these recommendations and is seeking support from classmates and instructors (and administrators, as potential readers) for implementing the recommendations.

CHOOSE ONE GOAL

a. Your campus Writing Center always needs qualified tutors to help first-year composition students with writing problems. On the other hand, students of professional writing need to sharpen their own skill in editing, writing, motivation, and diplomacy. All students in your class, therefore, should be assigned to the Writing Center during the semester's final half, to serve as tutors for twenty hours (beyond normal course time).

b. To prepare students for communicating in an automated work environment, at least one course assignment (preferably the long report) should be composed, critiqued, and revised online. Students not yet skilled in HTML will be required to develop the skill by midsemester.

c. This course should help individuals improve at their own level, instead of forcing them to compete with stronger or weaker writers. All grades, therefore, should be Pass/Fail.

d. To prepare for the world of work, students need practice in peer evaluation as well as

self-evaluation. Because this textbook provides definite criteria and checklists for evaluating various documents, students should be allowed to grade each other and to grade themselves. These grades should count as heavily as the instructor's grades.

e. In preparation for writing in the workplace, no one should be allowed to limp along, just getting by with minimal performance. This course, therefore, should carry only three possible grades: A, B, or F. Those whose work would otherwise merit a C or D would instead receive an Incomplete, and be allowed to repeat the course as often as needed to achieve a B grade.

f. To ensure that all graduates have adequate communication skills for survival in a world in which information is the ultimate product, each student in the college should pass a writing proficiency examination as a graduation requirement.

ANALYZE YOUR AUDIENCE

Your audience here consists of classmates and instructors (and possibly administrators). From your recent observation of this audience, what reader characteristics can you deduce?

Follow the model in Figure 4.5 for designing a profile sheet to record your audience and use analysis, and to duplicate for use throughout the semester. (Feel free to improve on the design and content of this model.)

Following is one possible set of responses to questions about audience identity and needs for Goal *e* from the previous list.

- *Who is my audience?* Classmates and instructors (and possibly some college administrators).
- *How will readers use my information?* Readers will decide whether to support our recommendation for limiting possible grades in this course to three: A, B, or F.
- *How much is the audience likely to know already about this topic?* Everyone here is already a

grade expert, and will need no explanation of the present grading system.

- *What else does the audience need to know?* Instructors should need no persuading; they know all about the quality of writing expected in the workplace. But some of our classmates probably will have questions like these: Why should we have to meet such high expectations? How can this grading be fair to the marginal writers? How will I benefit from these tougher requirements? Don't we already have enough work here?

We will have to answer questions by explaining how the issue boils down to "suffering now" or "suffering later," and that one's skill in communication will determine one's career advancement.

DEVISE A PLAN FOR ACHIEVING YOUR GOAL

From the audience traits you have identified, develop a plan for justifying your recommendation.

Express your goal and plan in a statement of purpose. Here is an example for Goal *e*:

> The purpose of this document is to convince classmates that our recommendation for an A/B/F grading system deserves your support. We will explain how skill in workplace writing affects career advancement, how higher standards for grading would help motivate students, and how our recommendation could be implemented realistically and fairly.

PLAN, DRAFT, AND REVISE YOUR DOCUMENT

Brainstorm (page 89) for worthwhile content, do any research that may be needed, write a draft, and revise as often as needed to produce a document that stands the best chance of connecting with your audience.

Appoint a member of your team to present the finished document (along with a complete audience and use analysis) for class evaluation and response.

Solving the
Ethics Problem

An effective message (one that achieves its purpose) isn't necessarily an ethical message. Think of examples from advertising: "Our artificial sweetener is composed of proteins that occur naturally in the human body (amino acids)" or "Our potato chips contain no cholesterol." Such claims are technically accurate but misleading: amino acids in certain sweeteners can alter body chemistry to cause headaches, seizures, and possibly brain tumors; potato chips are loaded with saturated fat—which produces cholesterol.

Communication is unethical when it leaves readers at a disadvantage or prevents them from making their best decisions (Figure 5.1). Ethical communication is measured by standards of honesty, fairness, and concern for everyone involved (Johannesen 1).

RECOGNIZE UNETHICAL COMMUNICATION

Thousands of people are injured or killed yearly in avoidable accidents—the result of faulty communication. Following is a notable example.

The Challenger Accident

Unethical communication has consequences

On January 28, 1986, the space shuttle *Challenger* exploded 43 seconds after launch, killing all seven crew members.

> *Immediate cause:* Two rubber O-ring seals in a booster rocket permitted hot exhaust gases to escape, igniting the adjacent fuel tank (Figure 5.2).

> *Ultimate cause:* The O-ring hazard had been recognized since 1977 and documented by engineers but largely ignored by management. (Managers had claimed that the O-ring system was safe because it was "redundant": each primary O-ring was backed up by a secondary O-ring.)

> Moreover, in the final hours, engineers argued against launching because that day's low temperature would drastically increase the danger of both primary and secondary

Figure 5.1
Three Problems Confronted by Communicators

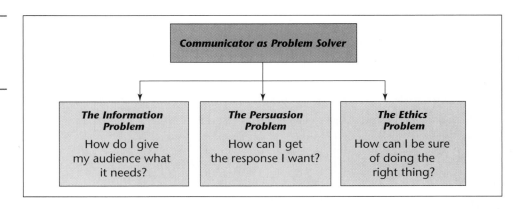

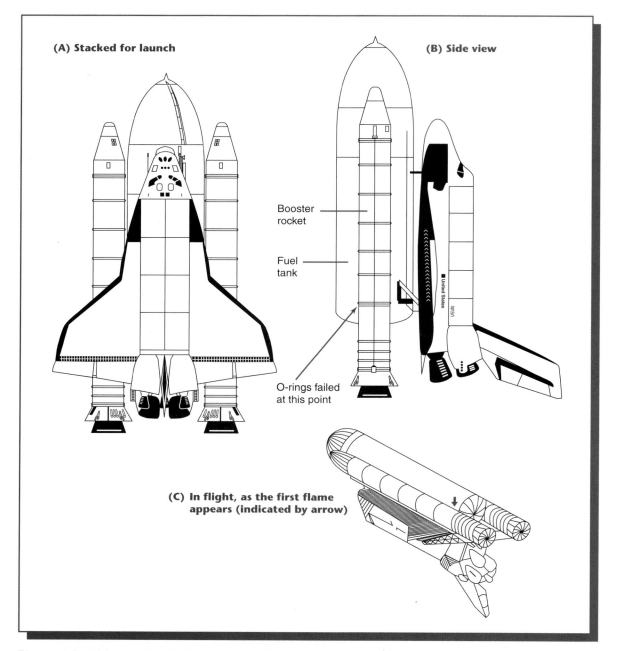

Figure 5.2 Views of the Challenger   *Source: Report to the Presidential Commission on the Space Shuttle* Challenger *Accident. Washington, D.C.: U.S. Government Printing Office, 1986: 3, 26.*

O-rings failing. But, under pressure of deadlines, managers chose to relay only a down-played version of these warnings to the NASA decision makers who were to make *Challenger's* fatal launch decision.

Here is how miscommunication contributed to the Challenger accident:[1]

1. More than six months before the explosion, officials at Morton Thiokol, Inc., manufacturer of the booster rockets, received a memo from engineer R. M. Boisjoly (Presidential Commission 49). Boisjoly described how exhaust gas leakage on an earlier flight had eroded O-rings in certain noncritical joints. The memo emphatically warned of a possible "catastrophe" in some future flight if O-rings should fail to seal a critical joint. Marked COMPANY PRIVATE, Boisjoly's urgent message never reached top-level decision makers at NASA.

2. Boisjoly and other company engineers continued to complain, in writing, about the lack of management attention or support on the O-ring issue. These complaints were largely ignored within the company—and they never reached the customer, NASA decision makers.

3. On the evening of January 27, managers and engineers at Morton Thiokol debated whether to recommend the January 28 launch. Besides the problem of O-ring erosion from exhaust gases, engineers worried that cold weather could harden the rubber O-rings, preventing them from sealing the joint at all. No prior flight had launched below 53 degrees Fahrenheit. At this temperature—and even up to 75 degrees—some blow-by (exhaust leakage) had occurred. The temperature on January 28 would reach barely 30 degrees.

   Arguing against the launch, Roger Boisjoly and his supporters persuaded Thiokol management to recommend no launch until the temperature reached at least 53 degrees (Presidential Commission 89).

4. In a teleconference with NASA's Marshall Space Center, Thiokol relayed its no-launch recommendation to the next decision maker. One Thiokol manager's testimony:

   > Mr. Mulloy [a NASA official] said he did not accept that recommendation, and Mr. Hardy said he was appalled that we would make such a recommendation. (Presidential Commission 94)

   Rejecting the facts and the engineers' risk assessment, NASA asked Thiokol to reconsider its recommendation.

5. The Thiokol staff met again, the engineers and one manager still opposing the launch. The managers then met *without* the engineers, and the one reluctant manager was told to "take off your engineering hat and put on your management hat" (Presidential Commission 93).

6. After deciding that the O-ring system had an acceptable margin of safety, Thiokol reversed its recommendation (Presidential Commission 108). Top NASA decision makers never were informed about engineers' objections to the low-temperature launch or about their concern over O-ring erosion in prior shuttle flights. The launch took place on schedule.

---

[1]Adapted from Winsor; Pace; R. Rowland; and Gouran et al.

7. The Presidential Commission investigating the fatal launch reached these conclusions (104):

- ◆ The Commission was troubled by what appears to be a propensity of management at Marshall to contain potentially serious problems and to attempt to resolve them internally rather than communicate them forward.
- ◆ The Commission concluded that the Thiokol management reversed its position and recommended the launch . . . at the urging of Marshall and contrary to the views of its engineers in order to accommodate a major customer. ▲

Along with failures in leadership, management, and mechanical systems, failed communication played a key role in the *Challenger* disaster.

The *Challenger* catastrophe made for dramatic headlines, as did the recent revelation that a government-operated nuclear facility in Hanford, Washington had knowingly leaked radiation for years without local residents being informed. But more routine examples of deliberate miscommunication are rarely publicized:

Routine instances of unethical communication

- ◆ A person lands a great job by exaggerating his credentials, experience, or expertise.
- ◆ A marketing specialist for a chemical company negotiates a huge bulk sale of its powerful new pesticide by downplaying the carcinogenic hazards.
- ◆ A manager writes a strong recommendation to get a friend promoted, while overlooking someone more deserving.

Other instances of unethical communication are not simply black and white. Here is one engineer's description of the gray area in which debates over product safety versus quality often occur:

Ethical decisions are not always "black and white"

> The company must be able to produce its products at a cost low enough to be competitive. . . . To design a product that is of the highest quality and consequently has a high and uncompetitive price may mean that the company will not be able to remain profitable, and be forced out of business. (Burghardt 92)

Do you emphasize to a customer the need for scrupulous maintenance of a highly sensitive computer—and risk losing the sale? Or do you focus instead on the computer's positive features? The decisions we make in these situations are often influenced by the pressures we feel.

KNOW THE MAJOR CAUSES OF UNETHICAL COMMUNICATION

To save face, escape blame, or get ahead, anyone might be tempted to say what people want to hear or to suppress bad news. But normally honest people usually break the rules only when compelled by employer, coworkers, or their own bad judgment.

Yielding to Social Pressure

You might have to choose between doing what you know is right and doing what your employer expects ("just follow orders" or "look the other way"):

Pressure to "look the other way"

> Just as your automobile company is about to unveil its hot, new pickup truck, your safety engineering team discovers that the reserve gas tanks (installed beneath the truck but outside the frame) may explode on impact in a side collision. The company has spent a small fortune developing and producing this new model and doesn't want to hear about this problem.

Companies often face the contradictory goals of *production* (which means *making* money on the product) and *safety* (which means *spending* money to avoid accidents that may or may not happen). When productivity receives exclusive priority, safety concerns may suffer (Wickens 434–36). Thus it seems no surprise that well over 50 percent of managers surveyed nationwide feel "pressure to compromise personal ethics for company goals" (Golen et al. 75). These pressures take varied forms (Lewis and Reinsch 31):

Typical workplace pressures

- the drive for profit
- the need to beat the competition (other organizations or coworkers)
- the need to succeed at any cost, as when superiors demand more productivity or savings without questioning the methods
- an appeal to loyalty—to the organization and to its way of doing things

Figure 5.3 depicts the assault on ethical values by ordinary, common pressures in organizations.

Mistaking Groupthink for Teamwork

Organizations rely on teamwork and collaboration to get a job done. But teamwork should not be confused with *groupthink* (Janis 9).

Groupthink occurs when group pressure prevents individuals from questioning, criticizing, or "making waves." Group members feel a greater need for acceptance and a sense of belonging than for critically examining the issues. Anyone who has endured adolescent peer pressure has experienced groupthink.

In a large company or on a complex project, individual responsibility is easily camouflaged:

How some corporations evade responsibility for their actions

> Lack of accountability is deeply embedded in the concept of the corporation. Shareholders' liability is limited to the amount of money they invest. Managers' liability is limited to what they choose to know about the operation of the company. The corporation's liability is limited by Congress (the Price-Anderson Act, for example, caps the liability of nuclear power companies in the aftermath of a nuclear disaster), by insurance, and by laws allowing corporations to duck liability by altering their . . . structure. (Mokhiber 16)

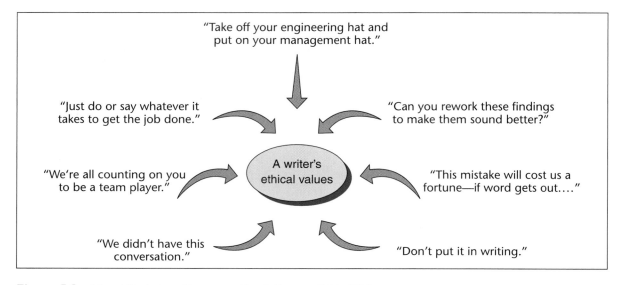

Figure 5.3 How Workplace Pressures Can Influence Ethical Values

Countless people may work on a major project (say, a new airplane). Decisions at any level can affect the whole project. But with so many people collaborating, identifying those responsible for an error is often impossible—especially in errors of *omission,* that is, of overlooking something that should have been done (Unger 137).

"I was only following orders!"

People commit unethical acts inside corporations that they never would commit as individuals representing only themselves. (Bryan 86)

Figure 5.4 depicts the kind of thinking that allows individuals to deny responsibility.[2]

Groupthink was exemplified by the decision to relay to top NASA officials only downplayed warnings about *Challenger*'s launch. In such cases, the group suffers an "illusion of invulnerability," which creates "excessive optimism and encourages taking extreme risks" (Janis 197).

Groupthink encourages subordinates to suppress bad news in their reports to superiors, instead "stressing what they think the superior wants to hear" (Littlejohn and Jabusch 159). Morton Thiokol's final launch recommendation to NASA serves as a memorable example.

---

[2]My thanks to Judith Kaufman for this idea.

Figure 5.4
Hiding behind
Groupthink

My job is to put together a persuasive ad for this brand of diet pill. It is someone else's job to make sure the claims are accurate and cause the customer no harm. It was someone else's job to make sure the stuff was safe. My job is only to promote the product. If someone does get hurt, it won't be my problem!

Allowing Personal Bias to Influence Judgment

All information is a matter of personal or social interpretation (Dombrowski 97); therefore, "objective reporting," practically speaking, is impossible. Even the most "impartial judgment" is unconsciously biased by a person's self-interest—what psychologists call the **self-serving bias:**

We process information on the basis of our own self-interest

> When presented with identical information . . . people first determine their preference for a certain outcome on the basis of self-interest and then justify this preference on the basis of fairness by changing the importance of attributes affecting what is fair. Thus the problem lies not in our desire to be unfair, but in our inability to interpret information in an unbiased manner. . . . People tend to confuse what is personally beneficial with what is fair or moral. (Bazerman, Morgan, and Loewenstein 91).

How we interpret and share information is influenced by our employer's expectations and our own self-interest. Consider, for example, this claim by Thiokol engineers before *Challenger*'s launch (Clark 194): "Low temperatures will harden the O-rings and will compromise their ability to seal the rocket joint." The meaning of this claim depended on its interpretation as a *technical fact* ("Exhaust leakage means an explosion") or a *social fact* ("Another launch delay means our company looks bad") (Ornatowski 98).

Any fact is subject to social interpretation

- ♦ To the engineers—technical experts, but not decision makers—the *technical fact* meant that the risk was serious enough to warrant a launch delay.
- ♦ To lower-level decision makers, the *social fact* meant that the risk was acceptable because of the social pressure to launch on schedule.
- ♦ To the ultimate decision makers at NASA, the data meant little, because lower-level decision makers provided information that was neither sufficient nor clear enough for the level of risk to be understood or recognized.

In the best circumstances, NASA estimates a risk of catastrophic failure during the 8.5 minutes after shuttle launch at about 1 in 75 flights; private experts, at about 1 in 60 (Broad B5). In *Challenger*'s case, the influence of self-serving bias on human judgment decreased these odds dramatically.

UNDERSTAND THE POTENTIAL FOR COMMUNICATION ABUSE

On the job, your effectiveness is judged by how well your documents speak for the company and advance its interests and agendas (Ornatowski 100–01). You walk the proverbial line between telling the truth and doing what your employer expects (Dombrowski 97).

Workplace communication influences the thinking, actions, and welfare of numerous people: customers, investors, coworkers, the public, policymakers—to name a few. These people are victims of communication abuse whenever we give them information that is less than the truth as we know it, as in the following examples.

Suppressing Knowledge the Public Deserves

Pressure to downplay the dangers of technology can result in censorship. For instance, some prestigious science journals have refused to publish legitimate studies linking chlorine and fluoride in drinking water with cancer risk, and fluorescent lights with childhood leukemia (Begley 63).

Threat of a lawsuit might cause a journal editor to suppress an article about, say, adverse effects of a popular over-the-counter medication.

MIT professor Arnold Barnett has found that information about airline safety is often suppressed by air traffic controllers because of "a natural tendency not to call attention to events in which their own performance was not exemplary" or their hesitation to "squeal" about pilot error (qtd. in Ball 13).

Conversely, breakthroughs, say, in cancer treatment, may be kept secret for months until lucrative patents can be obtained. One survey revealed this practice among 82 percent of biomedical companies and 50 percent of university researchers (Gibbs 116).

Exaggerating Claims about Technology

Organizations that have a stake in a particular technology may be especially tempted to exaggerate its benefits, potential, or safety. If your organization depends on outside funding (as in the defense or space industry), you might find yourself pressured to make unrealistic promises.

Falsifying or Fabricating Data

Research data might be manipulated or invented to support specific agendas (say, by a tobacco company facing lawsuits or by a scientist seeking grant money). Developments in fields such as biotechnology often occur too rapidly to allow for adequate peer review of articles before they are published ("Misconduct Scandal" 2).

Stealing or Divulging Proprietary Information

Information that originates in a specific company is the exclusive intellectual property of that company. *Proprietary* information includes insider financial information, employee records, product formulas, test and experiment results, surveys financed by clients, market research, plans, specifications, and minutes of meetings (Lavin 5). In theory, such information is legally protected, but it remains vulnerable to sabotage, theft, or exposure to the press. Fierce competition for the latest intelligence among rival companies leads to measures like these:

Examples of
corporate
espionage

> Companies have been known to use business school students to garner information on competitors under the guise of conducting "research." Even more commonplace is interviewing employees for slots that don't exist and wringing them dry about their current employer. (Gilbert 24)

The law prohibits employees who switch jobs from revealing proprietary information about a previous employer, or from soliciting its clients. A court recently ruled that even a collection of customer business cards can be a trade secret. The "doctrine of inevitable disclosure" asserts that an employee in a new job, sooner or later and deliberately or not, will disclose sensitive information. Acting on this doctrine, courts require some employees to remain idle between jobs until the information they possess has become outdated (Lenzer and Shook 100–02).

Misusing Electronic Information

With so much information stored in databases (by schools, employers, government, mail order retailers, credit bureaus, banks, credit card companies, insurance companies, pharmacies), how we combine, use, and disseminate the information is vital (Finkelstein 471). Also, compared with hard copy, a database is easier to alter; one simple command can wipe out or transform the facts. Inaccurate information can be sent from one database to others, and "correcting information in one database does not guarantee that it will be corrected in others" (Turner 5).

Other misuses include plagiarizing or republishing electronic information without obtaining authorization and without paying royalties to the intellectual property owner (Berry 1).

Withholding Information People Need for Their Jobs

Nowhere is the adage that "information is power" truer than among coworkers. One sure way to sabotage a colleague is to deprive that person of information about the task at hand. Studies show that people withhold information from coworkers for more benign reasons as well, such as fear that someone else might take credit for their work or might "shoot them down" (Davenport 90).

Exploiting Cultural Differences

Cross-cultural communication carries great potential for abuse. Based on its level of business experience, technological development, or social values, a particular culture might be especially vulnerable to manipulation or deception. Some countries, for example, place greater reliance on interpersonal trust than on lawyers, and a handshake can be worth more than the fine print of a legal contract. Those who are more trusting are more easily exploited.

RELY ON CRITICAL THINKING FOR ETHICAL DECISIONS

Because of their effects on people and on your career, ethical decisions challenge your critical thinking skills:

Ethical decisions require critical thinking

- ◆ How can I know the "right action" in this situation?
- ◆ What are my obligations, and to whom, in this situation?
- ◆ What values or ideals do I want to stand for in this situation?
- ◆ What is likely to happen if I do X—or Y?

Ethical issues resist simple formulas, but the following criteria offer some guidance.

Reasonable Criteria for Ethical Judgment

Reasonable criteria (standards that most people consider acceptable) take the form of *obligations, ideals,* and *consequences* (Ruggiero 33–34; Christians et al 17–18).

Obligations are the responsibilities we have to everyone involved:

Our obligations are varied and often conflicting

- ◆ *Obligation to ourselves,* to act in our own self-interest and according to good conscience
- ◆ *Obligation to clients and customers,* to stand by the people to whom we are bound by contract—and who pay the bills
- ◆ *Obligation to our company,* to advance its goals, respect its policies, protect confidential information, and expose misconduct that would harm the organization
- ◆ *Obligation to coworkers,* to promote their safety, and well-being
- ◆ *Obligation to the community,* to preserve the local economy, welfare, and quality of life
- ◆ *Obligation to society,* to consider the national and global impact of our actions

When the interests of these parties conflict—as they often do—we have to decide where our primary obligations lie.

Ideals are the values that we believe in or stand for: loyalty, friendship, compassion, dignity, fairness, and whatever qualities make us who we are (Ruggiero 33).

Consequences are the beneficial, or harmful, results of our actions. Consequences may be immediate or delayed, intentional or unintentional, obvious or subtle (Ruggiero 33). Some consequences are easy to predict; some are difficult; some are impossible.

Figure 5.5 depicts the relationship among these three criteria.

The above criteria help us understand why even good intentions can produce bad judgments, as in the situation on page 71.

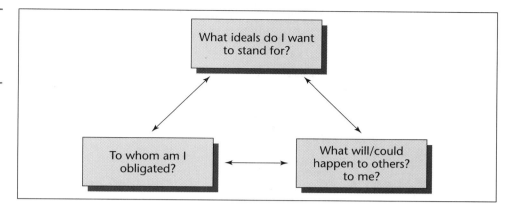

Figure 5.5
Reasonable
Criteria for
Ethical Judgment

What seems like
the "right action"
might be the wrong
one

Someone observes . . . that waste from the local mill is seeping into the water table and polluting the water supply. This is a serious situation and requires a remedy. But before one can be found, extremists condemn the mill for lack of conscience and for exploiting the community. People get upset and clamor for the mill to be shut down and its management tried on criminal charges. The next thing you know, the plant does close, 500 workers are without jobs, and no solution has been found for the pollution problem. (Hauser 96)

Because of their zealous dedication to the *ideal* of a pollution-free environment, the extremists failed to anticipate the *consequences* of their protest or to respect their *obligation* to the community's economic welfare.

In the *Challenger* disaster, Thiokol's sense of *obligation* to its client (i.e., NASA) caused it to ignore the *ideal* of product safety and the potential *consequences* of its recommendation to launch.

Ethical Dilemmas
Ethical decisions are especially frustrating when no single choice seems acceptable (Ruggiero 35).

Ethical questions
often resist easy
answers

In private and public ways, such dilemmas are inescapable. For example, the announced intention of "welfare reform" includes freeing people from lifelong dependence. One could argue that dedication to this *consequence* would violate our *obligations* (to the poor and the sick) and our *ideals* (of compassion or fairness). On the basis of our three criteria, how else might the welfare-reform issue be considered?

Ethical dilemmas also confront the medical community. For instance, in late 1992, a brain-dead, pregnant woman in Western Europe was kept alive for several weeks until her child could be delivered. Also in 1992, a California court debated the ethics of using the medical findings of Nazi doctors who had experimented on prisoners in concentration camps. In terms of our three criteria, how might these dilemmas be considered?

ANTICIPATE SOME HARD CHOICES

Communicators' ethical choices basically involve revealing or concealing information:

Ethical decisions about communication

- ◆ What exactly do I report, and to whom?
- ◆ How much do I reveal or conceal?
- ◆ How do I say what I have to say?
- ◆ Could misplaced obligation to one party be causing me to deceive others?

For illustration of a hard choice in workplace communication, consider the following scenario:

A Hard Choice

You are a structural engineer working on the construction of a nuclear power plant in a developing country. After years of construction delays and cost overruns, the plant finally has received its limited operating license from the country's Nuclear Regulatory Commission.

During your final inspection of the nuclear core containment unit, on February 15, you discover a ten-foot-long, hairline crack in a section of the reinforced concrete floor, twenty feet from where the cooling pipes enter the containment unit. (The especially cold and snowless winter likely has caused a frost heave beneath the foundation.) The crack has either just appeared or was overlooked by NRC inspectors on February 10.

The crack might be perfectly harmless, caused by normal settling of the structure; and this is, after all, a "redundant" containment system (a shell within a shell). But the crack might also signal some kind of serious stress on the entire containment unit, which could damage the entry and exit cooling pipes or other vital structures.

You phone your boss, who is just about to leave on vacation. He tells you, "Forget it; no problem," and hangs up.

You know that if the crack is reported, the whole start-up process scheduled for February 16 will be delayed indefinitely. More money will be lost; excavation, reinforcement, and further testing will be required—and many people with a stake in this project (from company executives to construction officials to shareholders) will be furious—especially if your report turns out to be a false alarm. Media coverage will be widespread. As the bearer of bad news—and bad publicity—you suspect that, even if you turn out to be right, your own career could be damaged by your apparent overreaction.

On the other hand, ignoring the crack could have unforeseeable consequences. Of course, no one would ever be able to implicate you. The NRC has already inspected and approved the containment unit—leaving you, your boss, and your company in the clear. You have very little time to decide. Start-up is scheduled for tomorrow, at which time the containment system will become intensely radioactive.

What would you do? Come to class prepared to justify your decision on the basis of the obligations, ideals, and consequences involved. ▲

You may have to choose between the goals of your organization and what you know is right

Working professionals commonly face similar choices, which they must often make alone or on the spur of the moment, without the luxury of meditation or consultation.

NEVER DEPEND ONLY ON LEGAL GUIDELINES

Can the law tell you how to communicate ethically? Sometimes. If you stay within the law, are you being ethical? Not always. Legal standards "sometimes do no more than delineate minimally acceptable behavior." In contrast, ethical standards "often attempt to describe ideal behavior, to define the best possible practices for corporations" (Porter 183).

Except for the instances listed below, lying is rarely illegal. Common types of legal lies are depicted in Figure 5.6. Later chapters cover other kinds of legal lying, such as page design that distorts the real emphasis or words that are deliberately unclear, misleading, or ambiguous.

What, then, are a communicator's legal guidelines? Among the laws regulating workplace communication are these:

Laws that govern workplace communication

- *Laws against deception* prohibit lying under oath, lying to a federal agent, lying about a product so as to cause injury, or breaking a contractual promise.
- *Libel law* prohibits any false written statement that maliciously attacks or ridicules anyone. A statement is considered libelous when it damages someone's reputation, character, career, or livelihood or when it causes humiliation or mental suffering. Material that is damaging but truthful would not be considered libelous unless it were used intentionally to cause harm. In the event of a libel suit, a writer's ignorance is no defense; even when the damaging material has been obtained from a presumably reliable source, the writer (and publisher) are legally accountable.[3]
- *Copyright law* (pages 116, 125) protects the ownership rights of authors—or of their employers, in cases where the writing was done as part of one's employment.
- *Law against software theft* prohibits unauthorized duplication of copyrighted software. A first offense carries up to five years in prison and fines up to $250,000. The Software Publisher's Association estimates that software piracy costs the industry more than $2 billion yearly ("On Line" 1).
- *Law against electronic theft* prohibits unauthorized distribution of copyrighted material via the internet as well as possession of ten or more electronic copies of any material worth $2,500 or more (Evans 22).
- *Laws against deceptive or fraudulent advertising* prohibit false claims or suggestions that a product or treatment will cure disease, or representation of a used

---

[3]Thanks to my colleague Peter Owens for the material on libel.

In Brief ETHICAL STANDARDS ARE GOOD FOR BUSINESS

In business transactions, people look for companies they can trust. And, as these examples illustrate, trustworthiness pays:

A Company's Truthfulness to Investors May Enhance Its Stock Value

Research indicates that publicly traded companies that disclose the truth about profits and losses are tracked by more securities analysts than companies that use "accounting smoke screens" to suppress bad economic news. "All it takes is the inferential leap that more analysts touting your stock means a higher stock price" (Fox 303).

High Standards Earn Customer Trust

Wetherill Associates, Inc., a car-parts supply company, was founded on the principle of honesty and "taking the right action," instead of trying to maximize profits and minimize losses. Among Wetherill's policies:

◆ Employees are given no sales quotas, so that no one will be tempted to camouflage disappointing sales figures.
◆ Employees are required to be honest in all business practices.
◆ Lies (including "legal lies") to colleagues or customers are grounds for being fired.
◆ Employees who gossip or backbite are penalized.

From a $50,000 start-up budget and 45 people who shared this ethical philosophy in 1978, the company has grown to 480 employees and $160 million in yearly sales and $16 million in profit—and continues growing at 25 percent annually (Burger 200–01).

Socially Responsible Action Earns International Goodwill

John Brown, CEO of British Petroleum Co., is committed to the economic and social prosperity of all locations in which his global corporation operates. Some of BP's efforts to earn employee loyalty and community goodwill:

◆ Building schools, providing job training for local employees, and supporting small business development;
◆ Providing medical equipment and assorted technology;
◆ Repairing environmental damage from forest fires and other natural disasters;
◆ Keeping detailed, open records of workplace and environmental accidents, and working constantly to eliminate such accidents;
◆ Listening to the concerns of local residents and seeking their feedback.

Rather than occurring at shareholder expense, BP's investment in social responsibility apparently improves its bottom line: Annual returns of 33 percent have outperformed the Dow-Jones industrial average by more than 50 percent since 1992 (Garten 26).

---

product as new. Fraud is defined as "lying that causes another person monetary damage" (Harcourt 64).

◆ *Liability laws* define the responsibilities of authors, editors, and publishers for damages resulting from incomplete, unclear, misleading, or otherwise defective information. The misinformation might be about a product (such as failure to warn about the toxic fumes from a spray-on oven cleaner) or a procedure (such as misleading instructions in an owner's manual for using a tire jack). Even if misinformation is given out of ignorance, the writer is liable (Walter and Marsteller 164–65).

Figure 5.6
Some Legal Lies in the Workplace

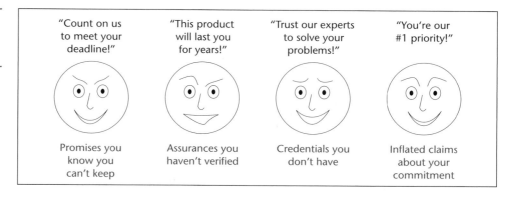

"Count on us to meet your deadline!"

Promises you know you can't keep

"This product will last you for years!"

Assurances you haven't verified

"Trust our experts to solve your problems!"

Credentials you don't have

"You're our #1 priority!"

Inflated claims about your commitment

Legal standards for product literature vary from country to country. A document must satisfy the legal standards for safety, health, accuracy, or language for the country in which it is distributed. For example, instructions for assembly or operation must carry warnings stipulated by the country in which the product is sold. Inadequate documentation, as judged by that country's standards, can result in a lawsuit (Caswell-Coward 264–66; Weymouth 145).

> NOTE
>
> *Laws regulating communication practices are few because such laws have traditionally been seen as threats to freedom of speech (Johannesen 86). Large companies typically have legal departments to advise you about a document's legal aspects. Also, most professions have their own ethics guidelines. If your field has its own formal code, obtain a copy.*

DECIDE WHERE AND HOW TO DRAW THE LINE

Suppose your employer asks you to cover up fraudulent Medicare charges or a violation of federal pollution standards. If you decide to resist, your choices seem limited: resign or go public (i.e., blow the whistle).

Walking away from a job isn't easy, and whistle-blowing can cause career disaster (Rubens 330). Many organizations refuse to hire anyone blacklisted as a whistle-blower (Wicclair and Farkas 19). Even if you aren't fired, expect animosity on the job. Consider this excerpt from a 1995 study of sixty-eight whistle-blowers by the Research Triangle Institute:

Consequences of whistle-blowing

"More than two-thirds of all whistle-blowers reported experiencing at least one negative outcome. . . ." Those most likely to experience adverse consequences were "lower ranking [personnel]." Negative consequences included pressure to drop their allegations, [ostracism] by colleagues, reduced research support, and threatened or actual legal action. Interestingly, . . . three-fourths of these whistle blowers experiencing "severe negative consequences" said they would definitely or probably blow the whistle again. (qtd. in "Consequences of Whistle Blowing" 2)

Despite the negative consequences, few people surveyed regretted their decision to go public.

Employers are generally immune from lawsuits by employees who have been dismissed unfairly but who have no contract or union agreement specifying length of employment (Unger 94). Current law, however, offers some protection for whistle-blowers.

Limited legal protections for whistle-blowers

- ◆ The Federal False Claims Act allows an employee to sue, in the government's name, a contractor who defrauds the government (say, by overcharging for military parts). The employee receives up to 25 percent of money recovered by the suit. Also, this law allows employees of government contractors to sue when they are punished for whistle-blowing (Stevenson 1).
- ◆ Anyone who is punished for reporting employer violations to a regulatory agency (Federal Aviation Administration, Nuclear Regulatory Commission, Occupational Health and Safety Administration, and so on) can request a Labor Department investigation. A claim ruled valid leads to reinstatement and reimbursement for back pay and legal expenses.[4]
- ◆ Laws in several states protect employees who report discrimination or harassment on the basis of sexual orientation (Fisher, "Can I Stop" 205).
- ◆ Laws protecting whistle-blowers from employer retaliation are on the books in roughly one-half of the states.

Despite the best legal protections, an employee who takes on a company without the backing of a labor union or professional group can expect lengthy court battles and disruption of life and career.

Before accepting a job, do some discreet research about the company's reputation. (Of course you can learn only so much before working there.) Learn whether the company has *ombudspersons* (who help employees file complaints) or hotlines for advice on ethics problems or for reporting violations. Ask whether the company has a formal code for personal and organizational behavior. Finally, assume that no employer, no matter how ethical, will tolerate any statement that makes the company look bad.

NOTE   *Sometimes the right ethical choice is obvious, but often it is not so obvious. No one can always know without a doubt what to do. This chapter is only an introduction to the inevitable hard choices that will be yours to make and to live with throughout your career.*

---

[4]Although employees are legally entitled to speak confidentially with OSHA inspectors about health and safety violations, a recent survey reveals that inspectors themselves believe such laws offer little protection against company retribution (Kraft 5).

GUIDELINES

for Ethical Communication*

How do we balance self-interest with the interests of others—our employers, the public, our customers? Listed below are guidelines:

Satisfying the Audience's Information Needs

1. *Give the audience everything it needs to know.* To see things as accurately as you do, people need more than just a partial view. Don't bury readers in needless details, but do make sure they get all the facts straight. If you're at fault, admit it and apologize.

2. *Give the audience a clear understanding of what the information means.* Facts can be misinterpreted. Ensure that readers understand the facts as you do.

Taking a Stand versus the Company

1. *Get your facts straight, and get them on paper.* Don't blow matters out of proportion, but do keep a paper (and an electronic) trail in case of legal proceedings.

2. *Appeal your case in terms of the company's interests.* Instead of being pious and judgmental ("This is a racist and sexist policy, and you'd better get your act together"), focus on what the company stands to gain or lose ("Promoting too few women and minorities makes us vulnerable to legal action").

3. *Aim your appeal toward the right person.* If you have to go to the top, find someone who knows enough to appreciate the problem and who has enough clout to make something happen.

4. *Get professional advice.* Contact an attorney and your professional society for advice about your legal rights.

Leaving the Job

1. *Make no waves before departure.* Discuss your departure only with people "who absolutely need to know." Say nothing negative about your employer to clients, coworkers, or anyone else.

2. *Leave all proprietary information behind.* Take no hard copy documents or computer disks prepared on the job—except for those records tracing the process of your resignation or termination.

The Ethics Checklist (page 78) incorporates additional guidelines from other chapters.

---

* Adapted from Clark 194; Unger 127–30; Lenzer and Shook 102.

Checklist FOR ETHICAL COMMUNICATION

Use this checklist to help your documents reflect reasonable, ethical judgment. (Numbers in parentheses refer to the first page of discussion.)*

☐ Do I avoid exaggeration, understatement, sugarcoating, or any distortion or omission that leaves readers at a disadvantage? (61)

☐ Do I make a clear distinction between "certainty" and "probability"? (73)

☐ Am I being honest and fair with everyone involved? (61)

☐ Have I explored all sides of the issue and all possible alternatives? (99)

☐ Are my information sources valid, reliable, and relatively unbiased? (145)

☐ Do I actually believe what I'm saying, instead of being a mouthpiece for groupthink or advancing some hidden agenda? (65)

☐ Would I still advocate this position if I were held publicly accountable for it? (66)

☐ Do I provide enough information and interpretation for readers to understand the facts as I know them? (77)

☐ Am I reasonably sure this document will harm no innocent persons or damage their reputations? (74)

☐ Am I respecting all legitimate rights to privacy and confidentiality? (68)

☐ Do I inform readers of the consequences or risks (as I am able to predict) of what I am advocating? (68)

☐ Do I state the case clearly, instead of hiding behind jargon and generalities? (235)

☐ Do I give candid feedback or criticism, if it is warranted? (67)

☐ Am I distributing copies of this document to every person who has the right to know about it? (69)

☐ Do I credit all contributors and sources of ideas and information? (578)

*Adapted from Brownell and Fitzgerald 18; Bryan 87; Johannesen 21–22; Larson 39; Unger 39–46; Yoos 50–55.

EXERCISES

1. Prepare a memo (one or two pages) for distribution to first-year students in which you introduce the ethical dilemmas they will face in college. For instance:

   - If you received a final grade of A by mistake, would you inform your professor?
   - If the library lost the record of books you've signed out, would you return them anyway?
   - Would you plagiarize—and would that change in your professional life?
   - Do you support lowering standards for student athletes if the team's success was important for the school's funding and status?
   - Would you allow a friend to submit a paper you've written for some other course?

   What other ethical dilemmas can you envision? Tell your audience what to expect, and give them some *realistic* advice for coping. No sermons, please.

2. In your workplace communications, you may face hard choices concerning what to say, how much to say, how to say it, and to whom. Whatever your choice, it will have definite consequences. Be prepared to discuss the following cases in terms of the obligations, ideals, and consequences involved. Can you think of similar choices you or someone you know has already faced? What happened? How might the problems have been avoided?

   - While traveling on an assignment that is being paid for by your employer, you visit an area in which you would really like to live and work, an area in which you have lots of contacts but never can find time to visit on your own. You have five days to complete your assignment, and then you must report on your activities. You complete the assignment in three days. Should you spend the remaining two days checking out other job possibilities, without reporting this activity?

- As a marketing specialist, you are offered a lucrative account from a cigarette manufacturer; you are expected to promote the product. Should you accept the account? Suppose instead the account were for beer, junk food, suntanning parlors, or ice cream. Would your choice be different? Why, or why not?

- You have been authorized to hire a technical assistant, and so you are about to prepare an advertisement. This is a time of threatened cutbacks for your company. People hired as "temporary," however, have never seemed to work out well. Should your ad include the warning that this position could be only temporary?

- You are one of three employees being considered for a yearly production bonus, which will be awarded in six weeks. You've just accepted a better job, at which you can start any time in the next two months. Should you wait until the bonus decision is made before announcing your plans to leave?

- You are marketing director for a major importer of coffee beans. Your testing labs report that certain African beans contain roughly twice the caffeine of South American varieties. Many of these African varieties are big sellers, from countries whose coffee bean production helps prop otherwise desperate economies. Should your advertising of these varieties inform the public about the high caffeine content? If so, how much emphasis should this fact be given?

- You are research director for a biotechnology company working on an AIDS vaccine. At a national conference, a researcher from a competing company secretly offers to sell your company crucial data that could speed discovery of an effective vaccine. Should you accept the offer?

COLLABORATIVE PROJECT

After dividing into groups, study the following scenario and complete the assignment: You belong to the Forestry Management Division in a state whose year-round economy depends almost totally on forest products (lumber, paper, etc.) but whose summer economy is greatly enriched by tourism, especially from fishing, canoeing, and other outdoor activities. The state's poorest area is also its most scenic, largely because of the virgin stands of hardwoods. Your division has been facing growing political pressure from this area to allow logging companies to harvest the trees. Logging here would have positive and negative consequences: for the foreseeable future, the area's economy would benefit greatly from the jobs created; but traditional logging practices would erode the soil, pollute waterways, and decimate wildlife, including several endangered species—besides posing a serious threat to the area's tourist industry. Logging, in short, would give a desperately needed boost to the area's standard of living, but would put an end to many tourist-oriented businesses and would change the landscape forever.

Your group has been assigned to weigh the economic and environmental effects of logging, and prepare recommendations (to log or not to log) for your bosses, who will use your report in making their final decision. To whom do you owe the most loyalty here: the unemployed or underemployed residents, the tourist businesses (mostly owned by residents), the wildlife, the land, future generations? The choices are by no means simple. In cases like this, it isn't enough to say that we should "do the right thing," because we are sometimes unable to predict the consequences of a particular action—even when it seems the best thing to do. In a memo to your supervisor, tell what action you would recommend and explain why. Be prepared to defend your group's ethical choice in class on the basis of the obligations, ideals, and consequences involved.

Solving the Collaboration Problem

The power of
collaboration

Successful collaboration combines the best that each team member has to offer. It enhances creative thinking by providing new and different perspectives, innovative ideas, and solutions. It enhances critical thinking by providing feedback, group support, and the chance to test ideas in group discussion.

EXAMPLES OF SUCCESSFUL COLLABORATION

Our notion of a solitary engineer, scientist, or businessperson laboring in some quiet corner is rapidly disappearing. For inspiration and feedback, workplace professionals collaborate with peers and coworkers—across the company, the nation, or the globe.

Companywide
collaboration to
design a new
refrigerator

> Various components and aspects must be considered once the [refrigerator's] size is determined. These include the compressor, . . . the structure of the motor that drives the compressor, . . . the control system, . . . aesthetic considerations [and so on]. The point is that various individuals work on each of the components or subsystems, and then share information as they design the entire refrigerator system. (Burghardt 209)

Nationwide
collaboration to
restore the
Everglades

> A team of civil engineers, ecologists, and biologists is designing new flowways and levees to transport unpolluted water into the Everglades ecosystem. Team members might work from various sites, meeting electronically to share and refine design ideas, compare research findings, and edit reports and proposals. (Boucher 32–33)

Worldwide
collaboration to
build the
International Space
Station

> To study the effects of prolonged space travel on the human mind and body, as well as on plants and animals, American astronauts have been living with Russian cosmonauts on the Russian space station, Mir, since 1995. This is the first phase of the International Space Station, to be built with the collaboration of 15 countries. This $60 billion project will require experts worldwide to collaborate on problems in designing, building, staffing, and operating a multi-national space station. ("From Mir to Mars")

Uniting each of these projects are the countless documents that must be produced: proposals, specifications, progress reports, feasibility reports, operating manuals, and so on. And, like other project assignments, the related documents are completed jointly. The refrigerator "Owner's Manual," for instance, is produced through a joint effort of writers, engineers, graphic artists, editors, reviewers, marketing personnel, and lawyers (Debs, "Recent Research" 477).

How a collaborative
document is
produced

In collaborating to produce a document, not all members of the team do the actual "writing"; for example, some might research, edit, proofread, or test the document's usability (Chapter 17). Each participant on a well-focused collaborative team makes a specific contribution, as illustrated in the following guidelines.

GUIDELINES

for Organizing
a Collaborative
Team*

1. *Appoint a group manager.* The manager assigns tasks, enforces deadlines, conducts meetings and keeps them on track, consults with supervisors, and generally "runs the show."

2. *Define a clear and definite goal.* Compose a statement of purpose that spells out the project's goal and the group's plan for achieving it.

3. *Decide how the group will be organized.* Some possibilities:
   a. The group researches and plans together, but each person writes a different part of the document.
   b. Some members plan and research; one person writes a complete draft; others review, edit, revise, and produce the final version. Keep in mind that the final version should display one consistent style throughout—as if written by one person only.

4. *Divide the task.* Who will be responsible for which parts of the document or which phases of the project? Who is best at doing what (writing early or final drafts, editing, layout, design and graphics, oral presentation)?

5. *Establish specific completion dates for each phase.* This will keep everyone focused on what is due, and when.

6. *Decide on a meeting schedule and format.* How often will the group meet, and where and for how long? Who will take notes (or minutes)? Set a strict time limit for each meeting.

7. *Establish a procedure for responding to the work of other members.* Will reviewing and editing be done in writing, face-to-face, as a group, one-on-one, or online? Will this process be supervised by the project manager?

8. *Establish procedures for dealing with group problems.* How will gripes and disputes be aired and resolved (by vote, the manager, other)? How will irrelevant discussion be curtailed? Can inevitable conflict be used positively?

9. *Select a group decision-making style.* To focus group effort, Intel Corporation requires every group to decide on a specific decision-making style before each meeting. Some possible styles (Matson, "The Seven Sins" 30):
   a. *authoritative:* the group leader makes the decisions
   b. *consultative:* the leader makes decisions on the basis of group input
   c. *voting:* decisions are made by majority vote

10. *Appoint a different "observer" for each meeting.* At Charles Schwab & Co., the designated observer keeps a list of what worked well during the meeting, and what didn't. The list is then added to that meeting's minutes (Matson, "The Seven Sins" 31).

*Adapted from Debs, "Collaborative Writing" 38–41; Hill-Duin 45–50; Hulbert, "Developing" 53–54; McGuire 467–68; Morgan 540–41.

11. *Decide how to evaluate each member's contribution.* Will the manager assess each member's performance and, in turn, be evaluated by each member? Will members evaluate each other? What are the criteria? Figure 6.1 depicts one possible form for a manager's evaluation of members. Equivalent criteria for evaluating the manager include open-mindedness, ability to organize the team, fairness in assigning tasks, ability to resolve conflicts, or ability to motivate. (Members might keep a journal of personal observations, for overall evaluation of the project.)

> **NOTE** *Any evaluation of strengths and weaknesses should be backed up by comments that explain the ratings (as in Figure 6.1). A group needs to decide beforehand what constitutes "effort," "cooperation," and so on.*

12. *Prepare a project management plan.* Figure 6.2 shows a sample plan sheet. Distribute completed copies to members and the instructor.

13. *Submit progress reports regularly.* Progress reports (page 350) allow everyone to track activities, problems, and progress.

Figure 6.1
Sample Form for Evaluating Team Members

Performance Appraisal for J. Fishkill

(Rate each element as *[superior]*, *[acceptable]*, or *[unacceptable]* and use the "Comment" section to explain each rating briefly)

- *Cooperation:* [superior]
  Comment: works extremely well with others; always willing to help out; responds positively to constructive criticism

- *Dependability:* [acceptable]
  Comment: Arrives on time for meetings; completes all assigned work

- *Effort:* [acceptable]
  Comment: does fair share of work; needs no prodding

- *Quality of work produced:* [superior]
  Comment: produces work that is carefully researched, well documented and clearly written

- *Ability to meet deadlines:* [superior]
  Comment: delivers all assigned work on or before the deadline; helps other team members with last-minute tasks

R. P. Ketchum
Project manager's signature

Management Plan Sheet

Project title:
Audience:
Project manager:
Team members:
Purpose of the project:

Specific Assignments   **Due Dates**

Research: Research due:
Planning: Plan and outline due:
Drafting: First draft due:
Revising: Reviews due:
Preparing final document: Revision due:
Presenting oral briefing: Final document due:
                                                 Progress report(s) due:

Work Schedule

Group meetings: *Date*   *Place*   *Time*   *Note taker*
    #1
    #2
    #3
    etc.
Mtgs. w/instructor
    #1
    #2
    etc.

Miscellaneous

How will disputes and grievances be resolved?
How will performances be evaluated?
Other matters (Internet searches, email routing, computer conferences, etc.)?

Figure 6.2   Sample Plan Sheet for Managing a Collaborative Project

SOURCES OF CONFLICT IN COLLABORATIVE GROUPS

Workplace surveys show that people view meetings as "their biggest waste of time" (Schrage 232). This fact alone accounts for the boredom, impatience, or irritability that might crop up in any meeting. But even the most dynamic group setting can produce conflict because of individual differences like the following.

Interpersonal Differences

People might clash because of differences in personality, working style, commitment, standards, or ability to take criticism. Some might disagree about exactly what or how much the group should accomplish, who should do what, or who should have the final say. Some might feel intimidated or hesitant to speak out.[1] These interpersonal conflicts can actually worsen when the group transacts exclusively online (page 92).

Gender and Cultural Differences

Collaboration involves working with peers—those of equal status, rank, and expertise. But gender and cultural differences can cause some participants to feel less than equal.

GENDER CODES AND COMMUNICATION STYLE.   Research on ways women and men communicate in meetings indicates a gender gap. Communication specialist Kathleen Kelley-Reardon offers this assessment of gender differences in workplace communication:

How gender codes influence communication

> Women and men operate according to communication rules for their gender, what experts call "gender codes." They learn, for example, to show gratitude, ask for help, take control, and express emotion, deference, and commitment in different ways. (88–89)

Reardon explains how women tend to communicate during meetings: women are more likely to take as much time as needed to explore an issue, build consensus and relationship among members, use tact in expressing views, use care in choosing their words, consider the listener's feelings, speak softly, and allow interruptions. Women generally issue requests instead of commands (*Could I have the report by Friday?* versus *Have this ready by Friday.*) and qualify their assertions in ways that avoid offending (*I don't want to seem disagreeable here, but . . .*).

One study of mixed-gender peer interaction indicates that women, in contrast to men, tend to: be agreeable, solicit and admit the merits of other opinions, ask

---

[1]Adapted from Bogert and Butt 51; Burnett 533–34; Debs, "Collaborative Writing" 38; Hill Duin 45–46; Nelson and Smith 61.

questions, and express uncertainty (say, with qualifiers such as *maybe, probably, it seems as if*) (Wojahn 747).

None of these traits, of course, is gender specific. People of either gender can be soft-spoken and reflective. But such traits often attach to the "feminine" stereotype. Supreme Court Justice Sandra Day O'Connor, for example, recalls the difficulty of getting male colleagues to pay attention to what she had to say (Hugenberg, LaCivita, and Lubanovic 215).

Any woman who breaches the gender code, say, by being assertive, may be seen as "too controlling" (Kelley-Reardon 6). Studies suggest women have less freedom than male peers to alter their communication strategies: less assertive males often are still considered persuasive, whereas more assertive females often are not (Perloff 273). In the words of one researcher, fitting into the workplace culture requires that women decide "to be quiet and popular, or speak out, and not be accepted" (Jones 50).

As one consequence of these differences, males tend "to become leaders of task-oriented groups, whereas women emerge as social leaders more frequently than men" (Dillard, Solomon, and Samp 709).

CULTURAL CODES AND COMMUNICATION STYLE.   International business expert David A. Victor describes cultural codes that influence interaction in group settings:

How cultural codes influence communication

- ◆ Some cultures value silence more than speech, intuition and ambiguity more than hard evidence or data, politeness, and personal relationships more than business relationships (145–46).
- ◆ Cultures differ in their perceptions of time. Some are "all business" and in a big rush; others take as long as needed to weigh the issues, engage in small talk and digressions, chat about family, health, and other personal matters (233).
- ◆ Cultures differ in willingness to express disagreement, question or be questioned, leave things unstated, touch, shake hands, kiss, hug, or backslap (209–11).
- ◆ Direct eye contact is not always a good indicator of listening. In some cultures it is offensive. Other eye movements, such as squinting, closing the eyes, staring away, staring at legs or other body parts, are acceptable in some cultures but insulting in others (206).

MANAGING GROUP CONFLICT

No team can afford to assume that all members share one viewpoint, one communication style, one approach to problem solving. Before any group can reach final agreement, conflicts must be expressed and addressed openly. Pointing out that "conflict can be good for an organization—as long as it's resolved quickly," management expert David House offers these strategies for overcoming personal differences (Warshaw 48):

How to manage group conflict

- ◆ Give everyone a chance to be heard.
- ◆ Take everyone's feelings and opinions seriously.
- ◆ Don't be afraid to disagree.
- ◆ Offer and accept constructive criticism.
- ◆ Find points of agreement with others who hold different views.
- ◆ When the group does make a decision, support it fully.

Central to all these strategies is the ability to *listen* to what others have to say.

OVERCOMING DIFFERENCES BY ACTIVE LISTENING

Listening is key to learning, getting along, and building relationships. Nearly half our time communicating is spent listening (Pearce, Johnson, and Barker, 28). But poor listening behaviors cause us to retain only a fraction of what we hear.

How effective are your listening behaviors? Use the questions below for an assessment.

QUESTIONS FOR ASSESSING YOUR LISTENING BEHAVIORS

- ◆ Do I remember people's names after being introduced?
- ◆ Do I pay close attention to what is being said, or am I easily distracted?
- ◆ Do I make eye contact with the speaker, or stare off elsewhere?
- ◆ Do I actually appear interested and responsive, or bored and passive?
- ◆ Do I allow the speaker to finish, or do I interrupt?
- ◆ Do I tend to get the message straight, or misunderstand it?
- ◆ Do I remember important details from previous discussions, or forget who said what?
- ◆ Do I ask people to clarify complex ideas, or just stop listening?
- ◆ Do I know when to keep quiet, or do I insist on being heard?

When you communicate, are you "listening or just talking" (Bashein and Markus, 37)? Many of us seem more inclined to speak, to say what's on our minds, than to listen. We often hope someone else will do the the listening.

Effective listening requires *active* involvement—not merely passive reception. Stages in the process of active listening (Dumont and Lannon) 649–50:

Active listening is a complex process

a. *hearing* the speaker clearly
b. *focusing* on the message by tuning out distractions (often difficult during a busy workday)
c. *decoding* the message, deciding what it means

d. *accepting* the message as the speaker intended, without distorting it through one's own biases or preconceptions

e. storing the message in memory for later recall

The following guidelines will help you become an effective listener.

GUIDELINES

for Active Listening[*]

1. *Don't dictate.* If you are leading the group, don't express your view until everyone else has their chance.

2. *Be receptive.* Instead of resisting different views, develop a "learner's" mind-set: take it all in first, and evaluate it later.

3. *Keep an open mind.* Judgment stops thought (Hayakawa 42). Reserve judgment until everyone has had their say.

4. *Be courteous.* Don't smirk, roll your eyes, whisper, or wisecrack.

5. *Show genuine interest.* Eye contact is vital, and so is body language (nodding, smiling, leaning toward the speaker). Make it a point to remember everyone's name.

6. *Hear the speaker out.* Instead of "tuning out" a message you find disagreeable, allow the speaker to continue without interruption (except to ask for clarification). Delay your own questions, comments, and rebuttals until the speaker has finished.

7. *Focus on the message.* Instead of thinking about what you want to say next, try to get a clear understanding of the speaker's position.

8. *Ask for clarification.* If anything is unclear, say so: "Can you run that by me again?" To ensure accuracy, paraphrase the message: "So what you're saying is. . . . Is that right?"

9. *Be agreeable.* Don't turn the conversation into a contest, and don't insist on having the last word.

10. *Observe the 90/10 rule.* You rarely go wrong spending 90 percent of your time listening, and 10 percent speaking. President Calvin Coolidge claimed that "Nobody ever listened himself out of a job." Some historians would argue that "Silent Cal" listened himself right into the White House.

[*]Adapted from Armstrong 24+; Bashein and Markus 37; Cooper 78–84; Dumont and Lannon 648–51; Pearce, Johnson, and Barker 28–32.

THINKING CREATIVELY

Today's rapidly changing workplace demands new and better ways of doing things:

Creativity is a key asset

More than one-fourth of U.S. companies employing more than 100 people offer some kind of creativity training to employees. (Kiely 33)

Many ideas are
better than one

Creative thinking is especially effective in group settings, using the following techniques.

Brainstorming

When we begin working with a problem, we search for useful material: insights, facts, statistics, opinions, images—anything that sharpens our view of the problem and potential solutions ("How can we increase market share for Zappo software?"). *Brainstorming* is a technique for coming up with useful material. Its aim is to produce as many ideas as possible (on paper, screen, whiteboard, or the like). Although brainstorming can be done individually, it is especially effective in a group setting.

The procedure for
brainstorming

1. *Choose a quiet setting and agree on a time limit.*
2. *Decide on a clear and specific goal for the session.* For instance, "We need at least five good ideas about why we are losing top employees to other companies."
3. *Focus on the issue or problem.*
4. *As ideas begin to flow, record every one.* Don't stop to judge relevance or worth and don't worry about spelling or grammar.
5. *If ideas are still flowing at session's end, keep going.*
6. *Take a break.*
7. *Now confront your list.* Strike out what is useless and sort the remainder into categories. Include any new ideas that pop up.

Limitations of
brainstorming

Because of intimidation, groupthink (page 65), and other social pressures (page 66), group brainstorming often fails to achieve its "nonjudgmental ideal" (Kiely 34). Lower-status members, for instance, might feel reluctant to express their own ideas or criticize those of others.

Brainwriting

An alternative to brainstorming, *brainwriting*, enables group members to record their ideas—anonymously—on slips of paper or on a networked computer file. Ideas are then exchanged or posted on a large screen for comment and refinement by other members (Kiely 35).

Mind-Mapping

A more structured version of brainstorming, *mind-mapping* (Figure 6.3A) helps visualize relationships. Group members begin by drawing a circle around the main issue or concept, centered on the paper or whiteboard. Related ideas are then added, each in its own box, connected to the circle by a ruled line (or "branch"). Other branches are then added, as lines to some other distinct geometric shape containing supporting ideas. Unlike a traditional outline, a mind-map doesn't

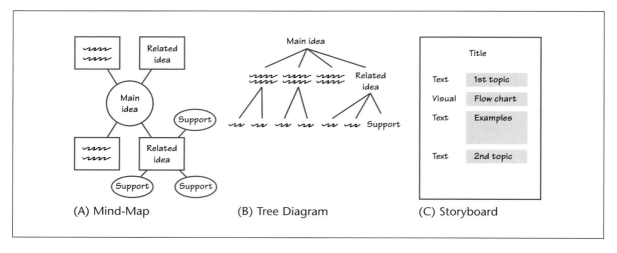

Figure 6.3   Techniques for Thinking Creatively

require sequential thinking: as each idea pops up, it is connected to related ideas by its own branch.

A simplified form of mind-mapping is the *tree diagram* (Figure 6.3B), in which major topic, minor topics, and subtopics are connected by branches that indicate their relationships. Page 98 shows a sample tree diagram for a research project.

Storyboarding

A technique for visualizing the shape of an entire process (or a document) is *storyboarding* (Figure 6.3C). Group members write each idea and sketch each visual on a large index card. Cards are then displayed on a wall or bulletin board so that others can comment, add, delete, refine, or reshuffle ideas, topics, and visuals (Kiely 35–36). Page 204 shows a final storyboard for a long report.

REVIEWING AND EDITING OTHERS' WORK

Documents produced collaboratively are reviewed and edited extensively. *Reviewing* means evaluating how well a document connects with its intended audience and meets its intended purpose. Reviewers typically examine a document for these specific qualities:

What reviewers look for

- ◆ accurate, appropriate, useful, and legal content
- ◆ material organized for the reader's understanding
- ◆ clear, easy to read, and engaging style
- ◆ effective visuals and page design
- ◆ a document that is safe, dependable, and easy to use

GUIDELINES

for Peer Reviewing and Editing

1. *Read the entire piece at least twice before you comment.* Develop a clear sense of the document's purpose and its intended audience. Try to visualize the document as a whole before you evaluate specific parts or features.

2. *Remember that correctness does not guarantee effectiveness.* Poor usage, punctuation, or mechanics do distract readers and harm the writer's credibility. However, a "correct" piece of writing might still contain faulty rhetorical elements (inappropriate content, confusing organization, wordy style, or the like).

3. *Understand the acceptable limits of editing.* In the workplace, editing can range from cleaning up and fine-tuning to an in-depth rewrite (in which case editors are cited prominently as consulting editors or coauthors). In school, however, rewriting someone else's piece to the extent that it ceases to belong to that writer may constitute plagiarism (page 575).

4. *Be honest but diplomatic.* Most of us are sensitive to criticism—even when it is constructive—and we all respond more favorably to encouragement. Begin with something positive before moving to material needing improvement. Be supportive instead of judgmental.

5. *Always explain why something doesn't work.* Instead of "this paragraph is confusing," say "because this paragraph lacks a clear topic sentence, I had trouble discovering the main idea." (See page 342 for sample criteria.) Help the writer identify the cause of the problem.

6. *Make specific recommendations for improvements.* Write out suggestions in enough detail for the writer to know what to do.

7. *Be aware that not all feedback has equal value.* Even professional reviewers and editors can disagree. Keep in mind that your job as a reviewer or editor is to help clarify and enhance a document—without altering its original meaning.

In reviewing, you explain to the writer how you respond as a reader. This commentary helps writers to think about ways of revising. Criteria for reviewing various documents appear in checklists throughout this book. (See also Chapter 17, on Usability).

Editing means actually "fixing" the piece by making it more precise and readable. Editors typically suggest improvements like these:

Ways in which editors "fix" writing

- ◆ rephrasing or reorganizing sentences
- ◆ clarifying a topic sentence

- ◆ choosing a better word or phrase
- ◆ correcting spelling, usage, or punctuation, and so on

Criteria for editing appear in Chapter 13, Appendix C, and inside the rear cover.

FACE-TO-FACE VERSUS ELECTRONICALLY MEDIATED COLLABORATION

When face-to-face meetings are preferable

Should participants meet in the same physical space or in virtual space? Face-to-face collaboration seems preferable when people don't know each other, when the issue is sensitive or controversial, or when people need to interact on a personal level (Munter 81).

When electronic meetings are preferable

Electronically mediated collaboration, using *groupware* (group authoring systems), is preferable when people are in different locations, have different schedules, or when it is important to neutralize personality clashes, to encourage shy people to participate, or to prevent intimidation by dominant participants (Munter 81, 83).

Here are some of the technologies that Facilitate electronic collaboration:

Tools for electronic collaboration

- ◆ *Basic email,* in which people exchange ideas and individual members contribute and receive messages as their schedules permit.
- ◆ *Electronic document sharing,* using file transfer software for sharing and editing drafts. The best software allows participants to comment on each other's work online so that each participant's comments are distinct from everyone else's and from the original text.
- ◆ *Teleconferencing,* using speakerphones; good for small rooms with fifteen or fewer people seated around a table.
- ◆ *Videoconferencing,* for live online meetings in which participants at different sites can edit and comment on each other's work while observing each other's body language.
- ◆ *Digital whiteboards,* a roughly 2 by 3 foot screen allowing participants from various sites to write, sketch, and erase in real time. Using individual computers, each person types in ideas which are then displayed on the large screen and on personal monitors.

Some experts argue that computer-based meeting tools eliminate equality issues that crop up in face-to-face meetings. As a result, some people feel more secure about saying "what they really think" (Matson, "The Seven Sins" 30). Moreover, "status cues" such as age, gender, appearance, or ethnicity virtually disappear online (Wojahn 747–48). Other experts disagree, arguing that equality issues require human solutions—namely, respect for other people's views and customs, and a willingness to listen.

NOTE   *One danger, of course, is that electronic media will encourage open hostility (as in email "flaming") among opposing participants.*

In Brief ETHICAL ISSUES IN WORKPLACE COLLABORATION

Our *lean* and *downsized* corporate world spells competition among workers—along with this dilemma:

> Many companies send mixed signals . . . saying they value teamwork while still rewarding individual stars, so that nobody has any real incentive to share the glory. (Fisher, "My Team Leader" 291)

The resulting mistrust interferes with fair and open teamwork and promotes unethical behavior.

Intimidating Coworkers

A dominant personality may intimidate peers into silence or agreement, or the group leaders may allow no one their say (Matson, "The Seven Sins" 30). Intimidated employees resort to "mimicking"—merely repeating what the boss says (Haskin, "Meetings without Walls" 55).

Claiming Credit for Others' Work

Workplace plagiarism occurs when the team or project leader claims all the credit. Even with good intentions, "the person who speaks for a team often gets the credit, not the people who had the ideas or did the work" (Nakache 287–88).

Team expert James Stern offers this solution:

> Some companies list 'core' and 'contributing' team members, to distinguish those who did most of the heavy lifting from those who were less involved. (qtd. in Fisher, "My Team Leader" 291)

Stern advises groups to decide beforehand—and in writing—what credit will be given for which contributions.

Hoarding Information

Surveys reveal that the biggest obstacle to workplace collaboration is employees' tendency "to hoard their own know-how" (Cole-Gomolski 6). Examples:

- ◆ *Whom do we contact for what?*
- ◆ *Where do we get the best price, the quickest repair, the most dependable service?*
- ◆ *What's the best way to do X?*

People might withhold exclusive information when they think it provides job security or to sabotage peers.

COLLABORATIVE PROJECTS

1. **Gender Differences:** Divide into small groups of mixed genders. Review pages 44–45 and 85–86. Then test the hypothesis that women and men communicate differently in the workplace.

    Each group member prepares the following brief messages—without consulting with other members:

    - A thank-you note to a coworker who has done you a favor.
    - A note asking a coworker for help with a problem or project.
    - A note asking a collaborative peer to be more cooperative or stop interrupting or complaining.
    - A note expressing impatience, frustration, confusion, or satisfaction to members of your group.
    - A recommendation for a friend who is applying for a position with your company.
    - A note offering support to a good friend and coworker.
    - A note to a new colleague, welcoming this person to the company.
    - A request for a raise, based on your hard work.

- The meeting is out of hand, so you decide to take control. Write what you would say.
- Some members of your group are dragging their feet on a project. Write what you would say.

As a group, compare messages, draw conclusions about the original hypothesis, and appoint one member to present findings to the class.

2. *A Matter of Ethics:* As an "observer" (page 82), keep a journal during a collaborative project, noting carefully what succeeded and what did not, what interpersonal conflicts developed and how they were resolved, what other issues contributed to progress or delay, the role and effectiveness of electronic tools, and so on. In a memo, report to your classmates and instructor, summarize the achievements and setbacks in your project and prepare recommendations for improving collaboration on future projects.

    In this evaluation/recommendation report, avoid attacking, blaming, or offending anyone. Offer constructive suggestions for improving collaborative work *in general.*

3. *Listening Competence:* Use the questions on page 87 to:

a. assess the listening behaviors of one member in your group during collaborative work,

b. have some other member assess your behaviors, and

c. do a self-assessment.

Record the findings and compare each self-assessment with the corresponding outside assessment. Discuss your findings with the class.

4. *Setting Up a Group* **Listserv:** If possible, have your teacher set up a *listserv* for your class. Or set up your own list with members of your collaborative group by listing their addresses and assigning the list a "nickname" or "alias": a single, short name, that, when typed into the address line of a message, will direct the message to all the addresses you have designated. Use your listserv (or your email network) to confer electronically on all phases of the collaborative project, including peer review (page 90) and usability testing (Chapter 17).

    When your collaborative project is complete, write an explanation telling how your group's use of a listserv eased or hampered your efforts and how it improved or detracted from the overall quality of your document.

PART

II

Retrieving, Analyzing, and Synthesizing Information

Thinking Critically about the Research Process

ASKING THE RIGHT QUESTIONS

EXPLORING A BALANCE OF VIEWS

ACHIEVING ADEQUATE DEPTH IN YOUR SEARCH

EVALUATING YOUR FINDINGS

INTERPRETING YOUR FINDINGS

In Brief **The Role of Expert Opinion in Decision Making**

Guidelines **for Evaluating Expert Information**

Major decisions in the workplace are typically based on careful research, with the findings recorded in a written report. Managers spend an estimated seventeen percent of their time searching for information (Davenport 157).

Research is a deliberate form of inquiry, a process of problem solving in which certain procedures follow a recognizable sequence (Figure 7.1). But research is never merely a by-the-numbers set of procedures ("First, do this; then, do that"). The stages depend on the many decisions that accompany any legitimate inquiry (Figure 7.2).[1]

Figure 7.1
Procedural
Stages of the
Research Process

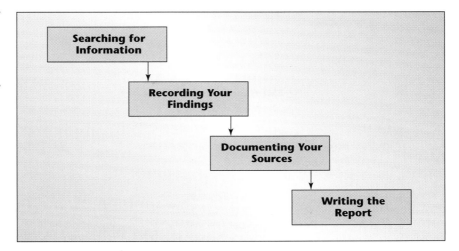

Figure 7.2
The Inquiry
Stages of the
Research Process

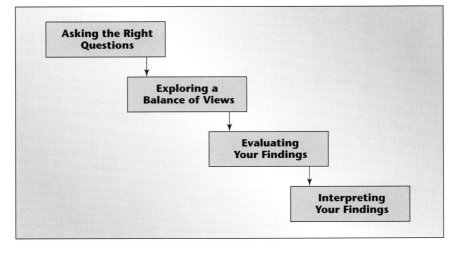

[1]My thanks to University of Massachusetts Dartmouth librarian Shaleen Barnes for inspiring this entire chapter.

ASKING THE RIGHT QUESTIONS

The answers you uncover will depend on the questions you ask. Assume, for instance, that you face the following scenario:

DEFINING AND REFINING A RESEARCH QUESTION

You are the public health manager for a small, New England town in which high-tension power lines run within one hundred feet of the elementary school. Parents are concerned about danger from electromagnetic radiation (EMR) emitted by these power lines in energy waves known as electromagnetic fields (EMFs). Town officials ask you to research the issue and prepare a report to be distributed at the next town meeting in six weeks.

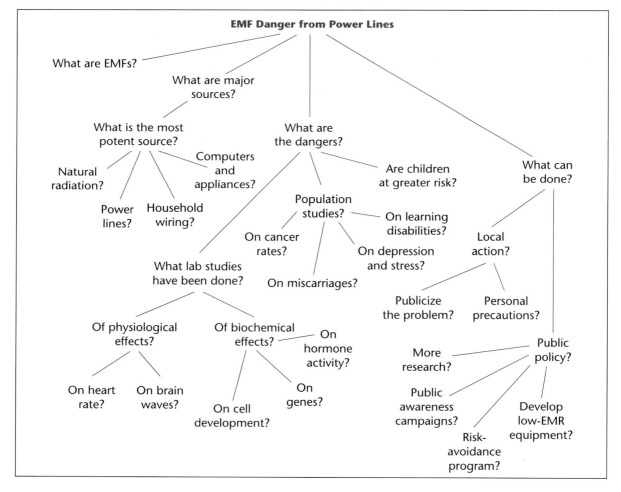

Figure 7.3   How the Right Questions Help Define a Research Problem

First, you need to identify the exact question or questions you want answered. Initially, the major question might be: *Do the power lines pose any real danger to our children?* After some phone calls around town and discussions at the coffee shop, you discover that townspeople actually have three major questions about electromagnetic fields: *What are they? Do they endanger our children? If so, then what can be done?*

Before settling on the right questions, you need to navigate a long list of possible questions, like those in the tree diagram in Figure 7.3. As research progresses, this diagram will grow. For instance, after some preliminary reading, you learn that electromagnetic fields radiate not only from power lines but from all electrical equipment, and even from the earth itself. So you face this additional question: *Do power lines present the greatest hazard as a source of EMFs?*

You now wonder whether the greater hazard comes from power lines or from other EMF sources. Critical thinking, in short, has helped you define and refine the essential questions.

EXPLORING A BALANCE OF VIEWS

Instead of settling for the first or most comforting or most convenient answer, pursue the *best* answer. Even "expert" testimony may not be the final word, because experts can disagree or be mistaken. To answer fairly and accurately, you are ethically obligated to consider a balance of perspectives from up-to-date and reputable sources (Figure 7.4).

Let's say you've chosen this question: *Do electromagnetic fields from various sources endanger our children?* Now you can consider sources to consult (journals,

Figure 7.4
Effective
Research
Considers
Multiple
Perspectives

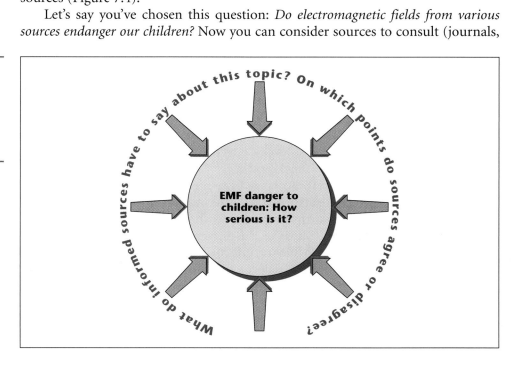

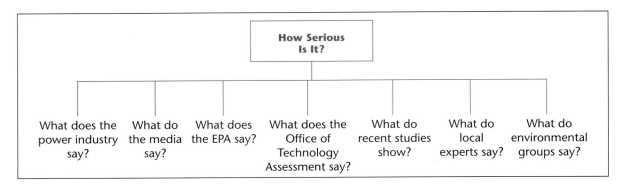

Figure 7.5   A Range of Essential Viewpoints

interviews, reports, database searches). Figure 7.5 illustrates some likely sources for information on the EMF topic.

NOTE

Recognize the difference between "balance" (sampling a full range of opinions) and "accuracy" (getting at the facts). Government or industry spokespersons, for example, might present a more positive view than the facts warrant. Not every source is equal, nor should we report points of view as if they were equal (Trafford 137).

ACHIEVING ADEQUATE DEPTH IN YOUR SEARCH[2]

Balanced research examines a broad *range* of evidence; thorough research, however, examines that evidence in sufficient *depth*. Different sources of information have different levels of detail and dependability (Figure 7.6).

The depth of a source often determines its quality

1. At the surface level are items from popular media (newspapers, radio, TV, magazines, certain Internet newsgroups and Web sites). Designed for general consumption, this layer of information often merely skims the surface of an issue.

2. At the next level are trade, business, and technical publications or Web sites (*Frozen Food World, Publisher's Weekly,* Internet listservs, and so on). Designed for users who range from moderately informed to highly specialized, this level of information focuses more on practice than on theory, on items considered newsworthy to group members, on issues affecting the field, and on public relations. Much of this material is prepared by science, business, or technical writers. While the information is usually accurate, viewpoints and interpretations tend to reflect a field's particular biases.

---

[2]My thanks to University of Massachusetts Dartmouth librarian Ross LaBaugh for inspiring this section.

Figure 7.6
Effective
Research
Achieves
Adequate Depth

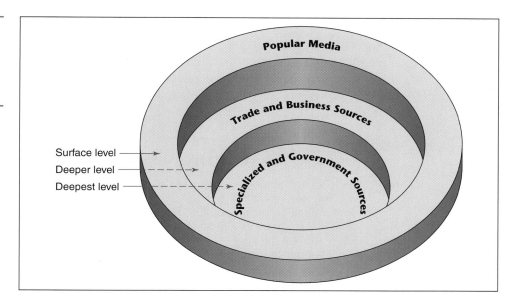

3. At a deeper level is the specialized literature (journals from professional associations—academic, medical, legal, engineering, and so on). Designed for practicing professionals, this level of information focuses on theory as well as practice, on descriptions of the latest studies—written by the researchers themselves and scrutinized by peers for accuracy and objectivity—on debates among scholars and researchers, on reviews and critiques of prior studies and publications.

Also at this deeper level are government sources (studies and reports by NASA, EPA, FAA, the Defense Department, Congress) and corporate documents available through the Freedom of Information Act (page 115). Designed for anyone willing to investigate its complex resources, this information layer offers hard facts and highly detailed and (in many instances) *relatively* impartial views.

NOTE *Web pages, of course, offer links to increasingly specific levels of detail. But the actual "depth" and quality of information from a particular Web site ultimately depend on the sponsorship and reliability of that site (page 147).*

How deep is deep enough? This depends on your purpose, your audience, and your topic. But the real story and the facts likely reside at deeper levels. Research on the EMF issue, for example, would need to look beneath popular "headlines" and biased special interests (say, electrical industry or environmental groups), focusing instead on studies done by experts.

EVALUATING YOUR FINDINGS

Not all findings have equal value. Some information might be distorted, incomplete, or misleading. Information might be tainted by *source bias,* in which a source might understate or overstate certain facts, depending on whose interests that source represents (say power company, government agency, or a reporter seeking headlines).

QUESTIONS FOR EVALUATING A PARTICULAR FINDING

- Is this information accurate, reliable, and relatively unbiased?
- Do the facts verify the claim?
- How much of the information is useful?
- Is this the whole, or the real, story?
- Does something seem to be missing?
- Do I need more information?

Remember, ethical researchers don't try to prove the "rightness" of their initial assumptions; instead, they research to find the *right* answers. Wait until the end of your inquiry to settle on a *definite* conclusion—based on what the facts suggest.

INTERPRETING YOUR FINDINGS

Once you have decided which findings seem legitimate, you must decide what they mean.

QUESTIONS FOR INTERPRETING YOUR FINDINGS

- What are my conclusions and do they address my original research question?
- Do any findings conflict?
- Are other interpretations possible?
- Should I reconsider the evidence?
- What, if anything, should be done?

Perhaps you will reach a definite conclusion. (For example, "The evidence about EMF dangers seems persuasive enough for us to be concerned and to take the following actions.") Perhaps you will not.

Even the best research can produce contradictory or indefinite conclusions. For instance, some scientists claim that studies linking electromagnetic radiation to health hazards are flawed. They point out that some EMF studies indicate increased cancer risk while others indicate beneficial health effects. Other scientists claim that stronger EMFs are emitted by natural sources, such as Earth's magnetic

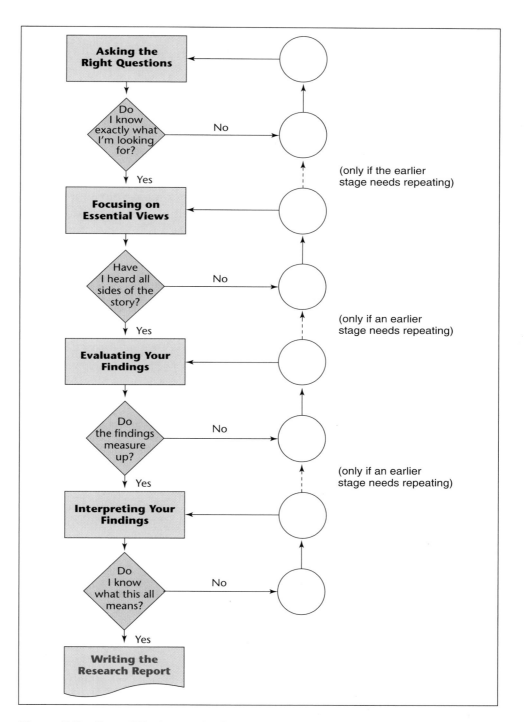

Decisions in the research process are recursive. No single stage is complete until all stages are complete. Stages are revisited as often as necessary.

Figure 7.7 Critical Thinking in the Research Process

field, than by electrical sources (McDonald, "Some Physicists" 5). An accurate conclusion would have to come from your analyzing all views and then deciding that one outweighs the others—or that only time will tell.

> **NOTE** *Never force a simplistic conclusion on a complex issue. Sometimes the best you can offer is an indefinite conclusion: "Although controversy continues over the extent of EMF hazards, we can take simple precautions to reduce our exposure." A wrong conclusion is far worse than no definite conclusion at all.*

Figure 7.7 shows the critical-thinking decisions crucial to worthwhile research: asking the right questions about your topic, your sources, your findings, and your conclusions. The quality of your entire research project will be determined by the quality of your *thinking* at each stage.

In Brief THE ROLE OF EXPERT OPINION IN DECISION MAKING

An expert is someone capable of doing the right thing at the right time.
(Holyoak, qtd. in Woodhouse and Nieusma 23).

What We Expect from Experts

Whenever we face uncertainty, we consult experts to help us make informed decisions about complex issues:

- *What should we do about global warming?*
- *Where and how should we store nuclear waste?*
- *How promising is the newly announced "cancer cure"?*
- *What is causing the massive die-off of frogs worldwide?*

How We Confer Expert Status

As information consumers, we confer expert status onto someone who seems to know more than we do—based on credentials, and relevant skills, experience, and knowledge. But researchers point out that "expert status is . . . in the eye of the beholder"—not necessarily based on a person's knowledge or *analytical skills,* but instead on that person's *linguistic skills:* use of technical language and persuasion strategies (Rifkin and Martin 31–36).

The Limits of Expert Opinion

Even though experts tend to consider themselves neutral, in controversial issues, outside influences often cause neutrality to disappear (Rifkin and Martin 31–33):

- The expert might have a financial stake in the issue—say an environmental researcher who receives financial support from nuclear power companies. So, even though we might recognize this person's knowledge and skill, we might have cause to mistrust her advice, recommendations, conclusions, or interpretations.
- The expert might have an extreme point of view, radically different from mainstream, accepted, scientific opinion—for example, about the risks or benefits of human cloning experiments.
- The expert might be venturing in areas beyond his expertise—for example, a real estate lawyer dabbling in copyright law.
- The expert can be mistaken—say, a meteorologist predicting the weather.

A Typical Case of Dueling Experts

For years, scientists and engineers have debated whether Yucca Mountain, Nevada, is an appropriate site for burying high-level nuclear waste deep underground. Some $3 billion worth of technical studies have assessed risks and benefits to health and safety.

Supporting Arguments

- Storing nuclear waste in one secure facility is safer and cheaper than storing it at the various power plants.
- A number of power plants are already running out of storage space.
- The Yucca Mountain site is remote, has a dry climate and stable geology, and abuts a desert already contaminated by nuclear-weapons testing over forty years ago.

Opposing Arguments

- Some scientists claim that earthquake possibilities have been greatly underestimated, pointing out that tremors occur periodically in this area.
- Leaking waste could contaminate ground water.
- Some of this material will remain highly dangerous for at least ten thousand years—a period longer than any written language (for warning) or "human-made edifice" has lasted.

Both sides of the argument are based on *expert* opinion or analysis. In short, "when it comes to Yucca Mountain, scientists do not have the answers" (Gross 134+).

GUIDELINES

for Evaluating Expert Information

To use expert information effectively, follow these suggestions:

1. *Look for common ground.* When opinions conflict, consult as many experts as possible and try to identify those areas in which they agree (Detjen 170).

2. *Consider all reasonable opinions.* Science writer Richard Harris notes that "Often [extreme views] are either ignored entirely or given equal weight in a story. Neither solution is satisfying. . . . Putting [the opinions] in balance means . . . telling . . . where an expert lies on the spectrum of opinion. . . . The minority opinion isn't necessarily wrong—just also Galileo" (170).

3. *Be sure the expert's knowledge is relevant in this context.* Don't seek advice about a brain tumor from a podiatrist.

4. *Don't expect certainty.* In complex issues, experts cannot *eliminate* uncertainty; they can merely help us cope with it.

5. *Expect special interests to produce their own experts to support their position.*

6. *Learn all you can about the issue before accepting anyone's final judgment.*

EXERCISES

1. Students in your major want a listing of one or two discipline-specific information sources from different depths of specialization:

   a. the popular press (newspaper, radio, TV, magazines)
   b. trade/business publications (newsletters and trade magazines)
   c. professional literature (journals)
   d. government sources (corporate data, technical reports, etc.)

   Prepare the list (in memo form) and include a one-paragraph description of each source.

2. Begin researching for the analytical report (Chapter 25) due at semester's end. Complete these steps. (Your instructor might establish a timetable.)

PHASE ONE: PRELIMINARY STEPS

   a. Choose a topic of immediate practical importance, something that affects you, your workplace, or your community directly.
   b. Develop a tree diagram (page 98) to help you ask the right questions.
   c. Identify a specific audience and its intended use of your information. Complete an audience and use profile (page 29).
   d. Narrow your topic, and check with your instructor for approval and advice.
   e. Make a working bibliography to ensure sufficient sources. Don't delay this step.
   f. List the information you already have about your topic.
   g. Write a clear statement of purpose and submit it in a proposal memo (page 497) to your instructor.
   h. Make a working outline

PHASE TWO: COLLECTING DATA (READ CHAPTERS 8–10 IN PREPARATION FOR THIS PHASE.)

   a. In your research, move from general to specific; begin with general works for an overview.
   b. Skim sources, looking for high points.
   c. Take selective notes. Use notecards or electronic file software.
   d. Plan and administer questionnaires, interviews, and inquiries.
   e. Whenever possible, conclude your research with direct observation.
   f. Evaluate each finding for accuracy, validity, reliability, and completeness.
   g. Decide what your findings mean.
   h. Use the checklist on page 170 to reassess your research methods and reasoning.

PHASE THREE: ORGANIZING YOUR DATA AND WRITING YOUR REPORT

   a. Revise your working outline as needed.
   b. Compose an audience and use analysis, like the sample on page 540.
   c. Fully document all sources of information.
   d. Write your final draft according to the checklist on page 550.
   e. Proofread and add all needed supplements (title page, letter of transmittal, abstract, summary, appendix, glossary).

DUE DATES: TO BE ASSIGNED BY YOUR INSTRUCTOR

List of possible topics due:

Final topic due:

Proposal memo due:

Working bibliography and working outline due:

Notecards due:

Copies of questionnaires, interview questions, and inquiry letters due:

Revised outline due:

First draft of report due:

Final draft with supplements and documentation due:

Exploring Hard Copy, Online, and Internet Sources

Although electronic searches for information are becoming the norm, a *thorough* search often requires careful examination of hard copy sources as well. Advantages and drawbacks of each search medium (Table 8.1) often provide good reason for exploring both.

HARD COPY VERSUS ELECTRONIC SOURCES

Benefits of hard copy sources

Hard copy libraries offer the judgment and expertise of librarians who organize and search for information. Compared with electronic files (on disks, tapes, hard drives), hard copy is easier to protect from tampering and to preserve from aging. (An electronic file's life span can be as brief as ten years).

Drawbacks of hard copy sources

Manual searches (flipping pages by hand), however, are time-consuming and inefficient: books can get lost; relevant information has to be pinpointed and retrieved, or "pulled," (page 119) by the user. Also, hard copy cannot be updated easily (Davenport 109–11).

Benefits of electronic sources

Compared with hard copy, electronic sources are more current, efficient, and accessible. Sources can be updated rapidly. Ten or fifteen years of an index can be reviewed in minutes. Searches can be customized: for example, narrowed to specific dates or topics. They can also be broadened: a keyword search (page 123) can uncover material that a hard copy search might have overlooked; Web pages can provide links to all sorts of material—much of which exists in no hard copy form.

Drawbacks of electronic sources

Drawbacks of electronic sources include the fact that databases rarely contain entries published before the mid-1960s and that material, especially on the Internet, can change or disappear overnight or be highly unreliable. Also, given the potential for getting lost in cyberspace, a thorough electronic search calls for a preliminary conference with a trained librarian.

TABLE 8.1

Hard Copy versus Electronic Sources: Benefits and Drawbacks		
	Benefits	**Drawbacks**
Hard Copy Sources	◆ organized and searched by librarians ◆ easier to preserve and keep secure	◆ time-consuming and inefficient to search ◆ hard to update
Electronic Sources	◆ more current, efficient, and and accessible ◆ searches can be narrowed or broadened ◆ can offer material that has no hard copy equivalent	◆ access to recent material only ◆ not always reliable ◆ user might get lost

For many automated searches, a manual search of hard copy is usually needed as well. One recent study found greater than fifty percent inconsistency among database indexers. Thus, even an electronic search by a trained librarian can miss improperly indexed material (Lang and Secic 174–75). In contrast, a manual search provides the whole "database" (the bound index or abstracts). As you browse, you often randomly discover something useful.

HARD COPY SOURCES

Where you begin your hard copy search depends on whether you are searching for background and basic facts or the latest information. Library sources appear in Figure 8.1.

If you are an expert in the field, you might simply do a computerized database search or browse through specialized journals and listservs. If you have limited knowledge or you need to focus your topic, you will probably want to begin with general reference sources.

University of Massachusetts Dartmouth librarian Ross LaBaugh suggests beginning with the popular, general literature, then working toward journals and other specialized sources: "The more accessible the source, the less valuable it is likely to be."

Reference Works

Reference sources provide background information

Reference works include encyclopedias, almanacs, handbooks, dictionaries, histories, and biographies. These provide background and bibliographies that can lead to more specific information. Make sure the work is current by checking the last copyright date.

BIBLIOGRAPHIES.   These comprehensive lists of publications about a subject are generally issued yearly, although some are published more frequently. However, they can quickly become dated.

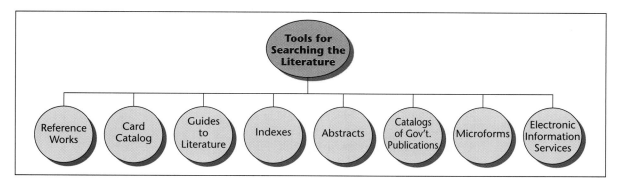

Figure 8.1 Library Sources

Annotated bibliographies that include an abstract for each entry are most helpful. Sample listings (with annotations):

Bibliographic Index. A list (by subject) of bibliographies that contain at least fifty citations. To see which bibliographies are published in your field, begin here.

A Guide to U.S. Government Scientific and Technical Resources. A list of everything published by the government in these broad fields.

Health Hazards of Video Display Terminals: An Annotated Bibliography. One of many bibliographies focused on a specific subject.

ENCYCLOPEDIAS. Encyclopedias provide basic information, some of which might be outdated. Sample listings:

Encyclopedia of Building and Construction Terms
Encyclopedia of Banking and Finance
Encyclopedia of Food Technology

The Encyclopedia of Associations lists over thirty thousand professional organizations worldwide (American Medical Association, Institute of Electrical and Electronics Engineers, and so on). Many organizations can be accessed via their Internet home pages.

DICTIONARIES. Dictionaries can be generalized or they can focus on specific disciplines or give biographical information. Sample listings:

Dictionary of Engineering and Technology
Dictionary of Telecommunications
Dictionary of Scientific Biography

HANDBOOKS. These research aids amass key facts (formulas, tables, advice, examples) about a field in condensed form. Sample listings:

Business Writer's Handbook
Civil Engineering Handbook
The McGraw-Hill Computer Handbook

ALMANACS. Almanacs contain factual and statistical data. Sample listings:

World Almanac and Book of Facts
Almanac for Computers
Almanac of Business and Industrial Financial Ratios

DIRECTORIES. Directories provide updated information about organizations, companies, people, products, services, or careers, often including addresses and phone numbers. Sample listings:

Directories

The Career Guide: Dun's Employment Opportunities Directory
Directory of American Firms Operating in Foreign Countries
The Internet Directory

Reference works are increasingly accessible by computer. Some, such as the *Free Online Dictionary of Computing,* are wholly electronic.

Card Catalog

All books, reference works, indexes, periodicals, and other materials held by a library are usually listed in its card catalog under three headings: *author, title,* and *subject.* Most libraries have automated their card catalogs. These electronic catalogs offer additional access points (beyond *author, title,* and *subject*) including:

Access points for
an electronic card
catalog

- ◆ *Descriptor:* for retrieving works on the basis of a key word or phrase (for example, "electromagnetic" or "power lines and health") in the subject heading, in the work's title, or in the full text of its bibliographic record (its catalog entry or abstract).
- ◆ *Document type:* for retrieving works in a specific format (videotape, audiotape, compact disc, motion picture).
- ◆ *Organizations and conferences:* for retrieving works produced under the name of an institution or professional association (Brookings Institution, American Heart Association).
- ◆ *Publisher:* for retrieving works produced by a particular publisher (for example, Little, Brown and Co.).

Figure 8.2 displays the first three screens you might encounter in an automated search using the descriptor ELECTROMAGNETIC. Through the Internet, a library's electronic catalog can be searched from anywhere in the world.

Guides to Literature

If you simply don't know which books, journals, indexes, and reference works are available for your topic, consult a guide to literature. For a general list of books in various disciplines, see Walford's *Guide to Reference Material* or Sheehy's *Guide to Reference Books.*

For scientific and technical literature, consult Malinowsky and Richardson's *Science and Engineering Literature: A Guide to Reference Sources.* Ask your librarian about literature guides for your discipline.

Indexes

Indexes offer
current information

Indexes are lists of books, newspaper articles, journal articles, or other works on a particular subject.

BOOK INDEXES.   A book index lists works by author, title, or subject. Sample indexes are listed on page 113.

You begin by pressing any key, and the computer responds with the screen:

This first screen lists your options for getting help or for searching the catalog from various access points

```
Type of searches:                          Press Help key for HELP

1  AU  =  Author                    8   PU  =  Publisher
2  OC  =  Organization or conference 9   SH  =  Subject heading
3  TI  =  Title                     10   DT  =  Document type
4  UT  =  Uniform or collective title 11        Combination
5  DE  =  Descriptor                12        ISBN
6  CN  =  Call number               13        ISBN
7  SE  =  Series                    14        Numeric

Enter the NUMBER of your search request and press RETURN:
```

After selecting the DE search mode, you type in your key word (ELECTROMAG-NETIC), and then press RETURN. This next screen appears (the first of several with all 108 entries):

A journal that might be of interest for this topic

Conference proceedings

A recent and relevant title for this topic

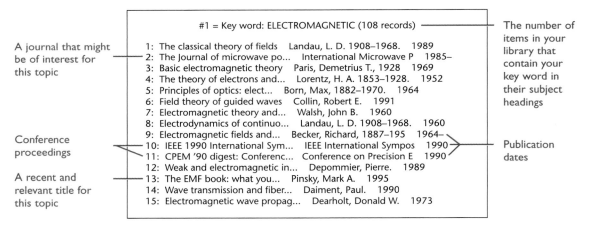

```
#1 = Key word: ELECTROMAGNETIC (108 records)

1:  The classical theory of fields   Landau, L. D. 1908–1968.   1989
2:  The Journal of microwave po...   International Microwave P   1985–
3:  Basic electromagnetic theory    Paris, Demetrius T., 1928   1969
4:  The theory of electrons and...   Lorentz, H. A. 1853–1928.   1952
5:  Principles of optics: elect...   Born, Max, 1882–1970.   1964
6:  Field theory of guided waves    Collin, Robert E.   1991
7:  Electromagnetic theory and...   Walsh, John B.   1960
8:  Electrodynamics of continuo...   Landau, L. D. 1908–1968.   1960
9:  Electromagnetic fields and...   Becker, Richard, 1887–195   1964–
10: IEEE 1990 International Sym...   IEEE International Sympos   1990
11: CPEM '90 digest: Conferenc...   Conference on Precision E   1990
12: Weak and electromagnetic in...   Depommier, Pierre.   1989
13: The EMF book: what you...   Pinsky, Mark A.   1995
14: Wave transmission and fiber...   Daiment, Paul.   1990
15: Electromagnetic wave propag...   Dearholt, Donald W.   1973
```

The number of items in your library that contain your key word in their subject headings

Publication dates

You select entry #13 and then press return. The computer responds with detailed bibliographical information on your selected item.

This screen shows the electronic equivalent of the printed catalog entry

```
Selection:
01-0211132

AUTHOR       Pinsky, Mark A.
TITLE        The EMF book: what you should know about electromagnetic
             fields, electromagnetic radiation, and your health
PUBLISHER    New York: Warner
DATE         1995
PHYS. FEAT.  246 p.; 20 cm.
SUBJECTS     Electromagnetic fields—Health aspects.
             ELF electromagnetic fields—Health aspects.
             Electric lines—Health aspects.
             Computer terminals—Health aspects.
```

Figure 8.2   Searching an Electronic Card Catalog

Book indexes

Scientific and Technical Books and Serials in Print. An annual listing of literature in science and technology.

New Technical Books: A Selective List with Descriptive Annotations. Issued ten times yearly.

Medical Books and Serials in Print. An annual listing of works from medicine and psychology.

NOTE   *Remember that no book is likely to offer the very latest information, because of the time required to publish a book manuscript (from several months to over one year).*

NEWSPAPER INDEXES.   These indexes list articles from major newspapers by subject. Sample titles:

Newspaper indexes

The New York Times Index
Christian Science Monitor Index
Wall Street Journal Index

PERIODICAL INDEXES.   A periodical index provides sources from magazines and journals. First, decide whether you seek general or specialized information. Two general indexes are the *Magazine Index,* a subject index on microfilm, and the *Readers' Guide to Periodical Literature,* which is updated every few weeks.

For specialized information, consult indexes that list journal articles by discipline, such as *Ulrich's International Periodicals Directory,* the *General Science Index,* the *Applied Science and Technology Index,* or the *Business Periodicals Index.* Specific disciplines have their own indexes:

Periodical indexes

Agricultural Index
Index to Legal Periodicals
International Nursing Index

Ask your librarian about the best indexes for your topic and about the many indexes that can be searched by computer.

CITATION INDEXES.   Citation indexes allow researchers to trace the development and refinement of a published idea. Using a citation index, you can track down the specific publications in which the original material has been cited, quoted, applied, critiqued, verified, or otherwise amplified (Garfield 200). In short, you can use them to answer this question: *Who else has said what about this idea?*

The Science Citation Index cross-references articles on science and technology worldwide. Both the *Science Citation Index* and its counterpart, the *Social Science Citation Index,* are searchable by computer.

TECHNICAL REPORT INDEXES.   Government and private sector reports prepared worldwide offer specialized and current information. (Proprietary or security restrictions limit public access to certain corporate or government documents.) Sample indexes:

Technical report indexes

Scientific and Technical Aerospace Reports
Government Reports Announcements and Index
Monthly Catalog of United States Government Publications

U.S. government report indexes are discussed on page 115.

PATENT INDEXES.   Countless patents are issued yearly to protect rights to new inventions, products, or processes. Information specialists Schenk and Webster point out that patents are often overlooked as sources of current information: "Since it is necessary that complete descriptions of the invention be included in patent applications, one can assume that almost everything that is new and original in technology can be found in patents" (121). Sample indexes:

Patent indexes

Index of Patents Issued from the United States Patent and Trademark Office
NASA Patent Abstracts Bibliography
World Patents Index

Patents in various technologies are searchable through databases such as *Hi Tech Patents, Data Communications,* and *WPI (World Patents Index).*

INDEX TO CONFERENCE PROCEEDINGS.   Schenk and Webster point out that many of the papers presented at the more than ten thousand yearly professional conferences are collected and then indexed in printed or computerized listings such as these:

Indexes to conference proceedings

Proceedings in Print
Index to Scientific and Technical Proceedings
Engineering Meetings (an Engineering Index database)

The very latest ideas, explorations, or advances in a field are often presented during such proceedings, before appearing as journal publications.

Abstracts
By indexing and summarizing each article, an abstract can save you from going all the way to the journal to decide whether to read the article or to skip it.
    Abstracts are usually titled by discipline:

Collections of abstracts

Biological Abstracts
Computer Abstracts
Forestry Abstracts

For some current research, you might consult abstracts of doctoral dissertations in *Dissertation Abstracts International.* Abstracts are increasingly searchable by computer.

Access Tools for U.S. Government Publications

The federal government publishes maps, periodicals, books, pamphlets, manuals, monographs, annual reports, research reports, and other information, often searchable by computer. A few of the countless titles available:

Government publications

Electromagnetic Fields in Your Environment
Major Oil and Gas Fields of the Free World
Journal of Research of the National Bureau of Standards

Your best bet for tapping this complex resource is to request assistance from the librarian in charge of government documents. Listed below are the basic access tools for documents issued or published at government expense, as well as for many privately sponsored documents.

Access tools

- *The Monthly Catalog of the United States Government.* The major pathway to government publications and reports.
- *Government Reports Announcements & Index.* A listing (with summaries) of more than one million federally sponsored research reports published and patents issued since 1964.
- *The Statistical Abstract of the United States.* Updated yearly, it offers an array of statistics on population, health, employment, and the like. It can be accessed via the World Wide Web. CD-ROM versions are available beginning with the 1997 edition.

Many unpublished documents are available under the *Freedom of Information Act.* The FOIA grants public access to all federal agency records except for classified documents, trade secrets, certain law enforcement files, records protected by personal privacy law, and the like.

Publicly accessible government records

Suppose you have heard that a certain toy has been recalled as a safety hazard and you want to know the details. In this case, the Consumer Product Safety Commission could help you. Perhaps you want to read the latest inspection report on conditions at a nursing home certified for Medicare. Your local Social Security office keeps such records on file. Or you might want to know if the Federal Bureau of Investigation has a file that includes you. In all these examples, you may use the FOIA to request information. (U.S. General Services Administration 1)

Contact the agency that would hold the records you seek: for workplace accident reports, the Department of Labor; for industrial pollution records, the Environmental Protection Agency; and so on.

Government information is increasingly posted to the Internet. Ask your reference librarian for electronic addresses of government agencies.

In Brief FREQUENTLY ASKED QUESTIONS ABOUT COPYRIGHT OF HARD COPY INFORMATION

Copyright laws ultimately have an ethical purpose: to balance the reward for intellectual labors with the public's right to use information freely.

1. *What is a copyright?*

   A copyright is the exclusive legal right to reproduce, publish, and sell a literary, dramatic, musical, or artistic work. Written permission must be obtained to use all copyrighted material.

2. *What are the limits of copyright protection?*

   Copyright protection covers the exact wording of the original, but not the ideas, concepts, theories, or factual information that it conveys. For example, Einstein's theory of relativity has no copyright protection but the exact wording in announcing or explaining the theory does (Abelman 33; Elias 3).

3. *How long does copyright protection last?*

   Works published before January 1, 1978 are protected for seventy-five years. Works published on or after January 1, 1978 are copyrighted for the author's life plus fifty years.

4. *Must a copyright be officially registered in order to protect a work?*

   No. Protection begins as soon as a work is created. But an author suing for copyright infringement would need to register the work before proceeding.

5. *Must a work be published in order to receive copyright protection?*

   No. In fact, a legal ruling of "fair use" for any unpublished work is less likely than for a published work because unauthorized use violates the author's right to decide if, when, and how to publish the work (Elias 108).

6. *What is "fair use"?*

   "Fair use" is the limited use of copyrighted material without permission. The use is considered "fair" if the material is not a large or representative portion of the whole. However, the source of any material that is not your own should be acknowledged. Fair use does not ordinarily apply to case studies, charts and graphs, author's notes, or private letters ("Copyright Protection" 30).

7. *What is the exact difference between copyright infringement and fair use?*

   Although using ideas from an original work is considered fair use, a paraphrase that incorporates too much of the original expression can be considered infringement—even when the source is cited (Abelman 41). The reproduction of a government document that *includes* material previously protected by copyright (graphs, images, company logos, slogans, or other material) is considered infringement. The United States Copyright Office offers this general caution:

   > The distinction between "fair use" and infringement may be unclear and not easily defined. There is no specific number of words, lines, or notes that may safely be taken without permission.
   >
   > Acknowledging the source of the copyrighted material does not substitute for obtaining permission. ("Fair Use" 1–2)

   When in doubt, obtain written permission from the copyright holder.

   For updated information about fair use issues, especially pertaining to works in electronic form, visit the Stanford University library Web site at: <http://www.fairuse.Stanford.edu>.

8. *What is material in the "public domain"?*

   "Public domain" refers to material not protected by copyright or material on which copyright has expired. Works published in the United States seventy-five years before the current year is in the public domain. Most government publications and commonplace information, such as height and weight charts or a [metric conversion] table are in the public domain. These works occasionally contain copyrighted material (used with permission and properly acknowledged). A new translation or version of a work in the public domain can be protected by copyright; if you are not sure whether something is in the public domain, request permission ("Copyright Protection" 31).

9. *What about international copyright protection?*

Copyright protection varies among individual countries, and some countries offer virtually no protection for foreign works (say, one produced in the United States):

There is no such thing as an "international copyright" that will automatically protect an author's writings throughout the world. . . . An author who wishes copyright protection . . . in a particular country should first determine the extent of protection available to works of foreign authors in that country . . . before the work is published anywhere, because protection may depend on the facts existing at the time of first publication. ("International Copyright" 1–2.)

In the United States, however, all foreign works that meet certain requirements are protected by copyright (Abelman 36).

10. *Who owns the copyright to a work prepared as part of one's employment?*

When a work is prepared in the service of one's employer or under contract for a client (by written agreement), that piece is a "work made for hire." The employer or client is legally considered the author and thus holds the copyright (Abelman 33–34). For example, a software manual researched, designed, and written as part of one's employment would be considered a work made for hire.

11. *What guidelines cover material downloaded on a computer?*

In recent rulings on suits for electronic copyright infringement, courts have followed the traditional copyright laws (Graybill 27). To keep up with new regulations and developments, visit the Web site of the United States Copyright Office at: <http://www.loc.gov/copyright>. For additional information on electronic copyright issues, see page 125, and visit *The Copyright Website* at <http://www.benedict.com> and the University of Texas legal site at <http://www.utsystem.edu/OGC>.

Microforms

Microform technology allows vast quantities of printed information to be stored on microfilm or microfiche. (This material is read on machines that magnify the reduced image.)

ONLINE SOURCES

Many hard copy sources discussed earlier are available in electronic form, accessible through your library or the World Wide Web.

Compact Discs

A single CD-Rom disc can store the equivalent of an entire encyclopedia and serves as a portable database, usually searchable via keyword.

One useful CD-ROM for business information is *ProQuest*™: its *ABI/IN-FORM* database indexes over eight hundred journals in management, marketing, and business published since 1989; its *UMI* database indexes major U.S. newspapers. A useful CD-ROM for information about psychology, nursing, education, and social policy is *SilverPlatter™*.

NOTE   *In many cases, CD-ROM access via the Internet is restricted to users who have their own passwords for entering a particular library's information system.*

Mainframe Databases

College libraries and corporations subscribe to online retrieval services that can access thousands of databases stored on centralized computers. Compared with CDs, mainframe databases are usually more specialized and more current, often updated daily (as opposed to weekly, monthly, or quarterly updating of CD databases).

Online retrieval services offer access to three types of databases: bibliographic, full-text, and factual (Lavin 14):

Types of online databases

- ◆ *Bibliographic databases* list publications in a particular field and sometimes include abstracts of each entry.
- ◆ *Full-text databases* display the entire article or document (usually excluding graphics) directly on the computer screen, and will print the article on command.
- ◆ *Factual databases* provide facts of all kinds: global and up-to-the-minute stock quotations, weather data, lists of new patents filed, and credit ratings of major companies, to name a few.

The following sections discuss four popular database services.

OCLC AND RLIN.   You easily can compile a comprehensive list of works on your subject at any library that belongs to the Online Computer Library Center (OCLC) or the Research Libraries Information Network (RLIN). OCLC and RLIN databases store millions of records of the same information in a printed card catalog. Using a networked terminal, you type in author or title. You will then get a listing of the publication you seek and information about where to find it.

DIALOG.   Many libraries subscribe to DIALOG, a network of more than one hundred fifty independent databases covering a range of subjects and searched by keywords. The system provides bibliographies and abstracts of the most recent medical articles on your topic.

Here are a few of DIALOG's databases:

DIALOG databases

Conference Papers Index
Electronic Yellow Pages (for Retailers, Services, Manufacturers)
ENVIROLINE

BRS.   Bibliographic Retrieval Services (BRS), another popular database network, provides bibliographies and abstracts from life sciences, physical sciences, business, or social sciences. On page 120 are a few of the more than fifty BRS databases.

In Brief "PUSH" VERSUS "PULL" STRATEGIES FOR DISTRIBUTING INFORMATION

The conventional strategy for distributing information requires recipients to *pull* the information—retrieve what they need when they need it, say, from the corporate library, the company files, or the Internet. An alternate strategy *pushes* hard copy or electronic information directly to the computers of selected recipients. "The best argument for [push] strategy is that people don't know what they don't know" (Davenport 148).

How Push Technology Works

Specialized software (such as *Headliner Professional* or *Pointcast Network*) allows users to stipulate the types of information they want. After searching, retrieving, and highlighting information tailored to a user's needs, push software automatically downloads the material to the client's desktop.

For Internet searches, users stipulate categories of information (say, municipal bond prices) or specific Web sites to be searched, with updates as requested. Intranet data or Internet material, links included, can be downloaded while the user works, and then browsed through at will.

A push-type search of the company intranet might combine field data from various salespeople (say, about orders or competition), or track the number of hits to the company Web site, or analyze email questions to the Web site about specific products. The user can then view search results (displayed in charts and graphs, spreadsheets, or reports organized by product category) via his or her email or browser (Baker 65; Cortese 152; Cronin 254; Haskin, "A Push" 75+).

Workplace Applications of Push Technology

The ultimate promise is that push technology will enable any company to become "event driven." "Every employee will know everything he or she needs to know as soon as the information becomes available" (Desmond 149). Examples:

- announcements of new developments, breaking news, weather conditions, and so on
- internal company information—say, new policies, sales updates, market events
- product updates, news about competing companies or research breakthroughs, conference proceedings
- electronic copies of relevant articles in journals or trade publications

Drawbacks of Push Technology

- No electronic search agent can discriminate like a human reviewer.
- Push software might introduce viruses, clutter your hard drive with worthless material, and breach the security of your system.

Examples of Push Technology

- Comprehensive retrieval systems, such as *SavvySearch,* that simultaneously employ dozens of Internet search engines (*Yahoo, InfoSeek, WebCrawler*), develop a customized search plan, and rank each source on the basis of its usefulness to the researcher.
- "Bozo filters" that sift through email messages, weeding out the nonessential and giving priority to others on the basis of particular names or other keywords.
- "Personalized newspaper" programs that monitor hundreds of news and information sources, select the news most relevant to the individual subscriber (e.g., about a particular company, industry, or medical treatment), and assemble and deliver the document via fax or email (Hafner 77).

A coming generation of electronic "mentors will search databases for information useful to a particular individual, and . . . spark the user's creativity with questions and facts like those from a human consultant" ("Electronic" 56).

BRS databases

Dissertation Abstracts International
Harvard Business Review
Pollution Abstracts

Comprehensive database networks such as DIALOG and BRS are accessible via the Internet, for a fee. Specialized databases, such as *MEDLINE* or *ENVIROLINE*, offer free bibliographies and abstracts, and copies of the full text can be ordered, for a fee. Ask your librarian for help searching online databases.

NOTE   *Never assume that computers yield the best material. Database specialist Charles McNeil points out that "the material in the computer is what is cheapest to put there." Reference librarian Ross LaBaugh warns of a built-in bias in databases: "The company that assembles the bibliographic or full-text database often includes a disproportionate number of its own publications." Like any collection of information, a database can reflect the biases of its assemblers.*

INTERNET SOURCES

Internet service providers (ISPs), including commercial services such as CompuServe, America Online, and Microsoft Network provide Internet access via "gateways," along with aids for navigating its many resources (Figure 8.3).

Usenet

Usenet denotes a worldwide system for online discussions via newsgroups, a type of electronic bulletin board at which users post and share information and discuss topics of common interest via email.

Newsgroups are either *moderated* or *unmoderated*. In a moderated group, all contributions are reviewed by a moderator who must approve the material before it can be posted. In an unmoderated group, all contributions are posted. Most newsgroups are unmoderated. Also available are *newsfeed* newsgroups, which gather and post news items from wire services such as the Associated Press.

Newsgroups typically publish answers to "frequently asked questions" (*FAQ lists*) about their particular topic of interest (acupuncture, AIDS research, sexual harassment, etc.). While these can be a good source of any group's distilled wisdom, FAQs reflect the biases of those who contribute to and edit them (Maeglin 5). A group's particular convictions might politicize information and produce all sorts of inaccuracies (Snyder 90).

Listservs

Like newsgroups, listservs are special-interest groups for email discussion and information sharing. In contrast to newsgroups, listserv discussions usually focus on specialized topics, with discussions usually among experts (say, cancer re-

Figure 8.3
Various Parts of
the Internet

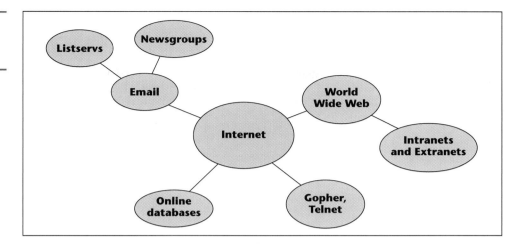

searchers), often before their findings or opinions appear in published form. Many listservs include a FAQ listing.

Listserv access is available to subscribers who receive mailings automatically via email. Like a newsgroup, a listserv may be moderated or unmoderated, but subscribers/contributors are expected to observe "netiquette" (proper Internet etiquette) and to stick to the topic without digressions, "flaming" (attacking someone), or "spamming" (posting irrelevant messages).

Electronic Magazines (Zines)

Zines offer information available only in electronic form. Despite the broad differences in quality among zines, this online medium offers certain benefits over hard copy magazines (Brody 40):

Benefits of online
magazines

- links to related information
- immediate access to earlier magazine issues
- interactive forums for discussions among readers, writers, and editors
- rapid updating and error correction.

Major news publications such as *The Wall Street Journal* offer interactive editions online.

Email Inquiries

The global email network is excellent for contacting knowledgeable people in any field. Email addresses are increasingly accessible via locator programs that search various local directories listed on the Internet (Steinberg 27). But unsolicited and indiscriminate email inquiries might offend the recipient.

World Wide Web

The Web is a global network of databases, documents, images, and sounds. All types of information from anywhere in the Web network can be accessed and explored through navigation programs such as *Lynx, Netscape Navigator,* or *Microsoft Internet Explorer,* known as "browsers." Hypertext links among Web resources allow users to explore information along different paths by clicking on key words or icons.

> **NOTE**
>
> *Telnet* and *Gopher* are older portions of the Internet that have largely been supplanted by the World Wide Web. These networks can still be explored by means of patched together search engines such as WAIS (Wide Area Information Service), Archie, and Veronica (the Very Easy Rodent-Oriented Net-Wide Index to Computerized Archives).

Each Web site has own *home page* that serves as an introduction to the site and is linked to additional "pages" that individual users can explore according to their information needs.

> **NOTE**
>
> Assume that any material obtained from the Internet is protected by copyright. Before using this material anywhere other than in a college paper (properly documented), obtain written permission from the owner.

Intranets and Extranets

An *intranet* is an in-house network that uses Internet technology for information access within a company. Examples may include the company's document library, price lists, online discussions, and progress reports—even the hard drives of colleagues. Fast-food chains and other franchises use intranets to respond to franchise questions and to distribute advice, industry/company news, and sales figures (Wallace 12). An intranet puts the company's knowledge and expertise at every employee's fingertips. Some organizations post their company "yellow pages," listing the expertise possessed by each employee.

An *extranet* integrates a company intranet with the Internet. External users (customers, subcontractors, outside vendors) with a password can browse nonrestricted areas of a company's Web site and download information. Extranets eliminate the need for printing and mailing information to clients or suppliers (Stedman 49). They also facilitate collaboration among organizations. At Caterpillar Tractors, Inc., for example, when a customer's equipment breaks down, the entire company can mobilize immediately, contact experts inside and outside the organization, access records of previous solutions to similar problems, and collaborate on a solution—for example, designing an improved mechanical part for the equipment (Haskin 57–60).

The extranet blend of in-house information and Internet access raises security concerns for any organization, leaving its network vulnerable to hackers, spies, or saboteurs (Myerson 35). Therefore, each extranet site has its own *firewall* (software that keeps out uninvited users and controls access to data) that includes password protection and *encryption* (coding) of sensitive information (Haskin 59).

KEYWORD SEARCHES USING BOOLEAN OPERATORS

Most engines that search by key word allow the use of Boolean[1] operators (commands such as "AND," "OR," "NOT," and so on) to define relationships among various key words. Table 8.2 shows how these commands can expand a search or narrow it by generating more or fewer "hits."

Boolean commands also can be combined, as in

(electromagnetic *OR* radiation) *AND* (fields *OR* tumors)

The hits produced here would contain any of these combinations:

electromagnetic fields, electromagnetic and tumors, radiation and fields, radiation and tumors

Using *truncation* (cropping a word to its root and adding an asterisk), as in *electromag**, would produce a broad array of hits, including these:

electromagnet, electromagnetic energy, electromagnetic impulse, electromagnetic wave

Different search engines use Boolean operators in slightly different ways; many include additional options (such as NEAR, to search for entries that contain search terms within ten or twenty words of each other). Click on the HELP option of your particular search engine to see which strategies it supports.

TABLE 8.2

Using Boolean Operators to Expand or Limit a Search	
If you enter these terms . . .	*The computer searches for . . .*
electromagnetism *AND* health	only entries that contain both terms
electromagnetism *OR* health	all entries that contain either term
electromagnetism *NOT* health	only entries that contain term 1 and do not contain term 2
electromag*	all entries that contain this root within other words

[1]British mathematician and logician George Boole (1815–1864) developed the system of symbolic logic (Boolean logic) now widely used in electronic information retrieval.

GUIDELINES

for Researching
on the
Internet*

1. *Focus your search beforehand.* The more precisely you identify the information you seek, the less your chance of wandering through cyberspace.

2. *Select key words or search phrases that are varied and technical, rather than general.* Some search terms generate better hits than others. In addition to "electromagnetic radiation," for example, try "electromagnetic fields," "power lines and health," or "electrical fields." Specialized terms (for example, *vertigo,* versus *dizziness*), offer the best access to sites that are reliable, professional, and specific. Always check your spelling.

3. *Look for Websites that are specific.* Specialized newsletters and trade publications offer good site listings.

4. *Set a time limit for searching.* Surveys indicate that employees waste a lot of time surfing for personal instead of business reasons. Set a ten to fifteen minute time limit, and try to avoid tangents.

5. *Expect limited results from any search engine.* Each search engine (*AltaVista, Excite, HotBot, Infoseek, WebCrawler, Yahoo!,* and so on) has strengths and weaknesses. Some are faster and more thorough while others yield more targeted and updated hits. Some search titles only—instead of the full text—for key words. Studies show that "Web content is increasing so rapidly that no single search engine indexes more than about one-third of it" (Peterson 286). Broaden your coverage by using multiple search engines.

6. *Use bookmarks and hotlists for quick access to favorite Web sites.* Mark a useful site with a bookmark and add it to your hotlist.

7. *Expect material on the Internet to have a brief life span.* Site addresses can change overnight; material is rapidly updated or discarded. If you find something of value, save or print it before it changes or disappears.

8. *Be selective about what you download.* Download only what you need. Unless they are crucial to your research, omit graphics, sound, and video files because these consume time and disk space. Focus on text files only.

9. *Never download copyrighted material without written authorization from the copyright holder (page 116).* The 1997 No Electronic Theft Act (NET) makes it a federal crime to possess or distribute "unauthorized electronic copies of copyrighted material valued over $1,000, even when no profit is involved" (Grossman 37). Only material in the public domain (page 116) is exempted.

   Before downloading *anything* from the Internet, ask yourself: Am I violating someone's privacy (as in forwarding an email or a newsgroup entry)? or Am I decreasing the value of this material for its owner in any way?" Obtain permission beforehand, and cite the source.

10. *Consider using information retrieval services.* An electronic service such as Inquisit or DIALOG protects copyright holders by selling access to all materials in its database. For a fee, users can download full texts of articles.

One drawback is that these retrieval services do not catalog material that exists only in electronic form (zines, newsgroup and listserv entries, and so on). Therefore, these databases exclude potentially valuable material (such as research studies not yet available in hard copy) accessible only through a general Web search.

*Guidelines adapted from Baker 57+; Branscum 78; Busiel and Maeglin 39–40, 76; Fugate 40–41; Kawasaki 156; Matson, "(Search) Engines" 249–52.

In Brief COPYRIGHT PROTECTION AND FAIR USE OF ELECTRONIC INFORMATION

Unanswered Questions

Copyright and fair use law is quite specific for printed works or works in other tangible form (paintings, photographs, music). But how do we define "fair use" (page 116) of intellectual property in electronic form? How does copyright protection apply (Dyson 137)? How do fair use restrictions apply to material used in multimedia presentations or to text or images that have been altered or reshaped to suit the user's specific needs (Steinberg 30)?

Information obtained via email or discussion groups presents additional problems: Sources often do not wish to be quoted or named or to have early drafts made public. How do we protect source confidentiality? How do we avoid infringing on works in progress that have not yet been published? How do we quote and cite this material without violating ownership and privacy rights (Howard 40–41)?

Present Status of Electronic Copyright Law

Subscribers to commercial online databases such as DIALOG pay fees, and copyholders in turn receive royalties (Communication Concepts, Inc. 13). But few specific legal protections exist for noncommercial types of electronic information.

Since April 1989, however, most works are considered copyrighted as soon as they are produced in *any* tangible form—even if they carry no copyright notice. Fair use of electronic information is generally limited to brief excerpts that serve as a basis for response—for example, in a discussion group. Except for certain government documents, no Internet posting is in the "public domain" (page 116) unless it is expressly designated as such by its author (Templeton).

Until specific laws are enacted, the following examples can be considered violations of copyrighted material in electronic form (Communication Concepts, Inc. 13; Elias 85, 86; Templeton):

- Downloading a work from the Internet and forwarding copies to other readers.
- Editing, altering, or incorporating an original work as part of your own document or multimedia presentation.
- Placing someone else's printed work online without the author's written permission.
- Reproducing and distributing original software or material from a privately owned database.

◆ Copying and forwarding an email message without the sender's authorization. The exact wording of an email message is copyrighted, but its content may legally be revealed—except for *proprietary information* (page 68).

Some copyright violations (say, reproducing and distributing trade secrets) may exceed the boundaries of civil law and be prosecuted as felonies (Templeton).

When in doubt, assume the work is copyrighted, and obtain written permission from the owner.

EXERCISES

1. Using the printed or electronic card catalog, locate and record the full bibliographic data for five books in your field or on your semester report topic, all published within the past year.

2. Consult the *Library of Congress Subject Headings* for alternative headings under which you might find information in the card catalog for your semester report topic.

3. List five major reference works in your field or on your topic by consulting Sheehy, Walford, or a more specific guide to literature.

4. List the titles of each of these specialized reference works in your field or on your topic: a bibliography, an encyclopedia, a dictionary, a handbook, an almanac (if available), and a directory.

5. Identify the major periodical index in your field or on your topic. Locate a recent article on a specific topic (e.g., use of artificial intelligence in medical diagnosis). Photocopy the article and write an informative abstract.

6. Consult the appropriate librarian and identify two databases you would search for information on the topic in Exercise 2.

7. Identify the major abstract collection in your field or on your topic. Using the abstracts, locate a recent article. Photocopy the abstract and the article.

8. Using technical report indexes, locate abstracts of three recent reports on one specific topic in your field. Provide complete bibliographic information.

9. Using patent indexes, locate and describe three recently patented inventions in your field, and provide complete bibliographic information.

10. Using indexes of conference proceedings, locate abstracts of three recent conference papers on *one* specific topic in your field. Provide complete bibliographic information.

11. Using the *Monthly Catalog* or *Government Reports Announcements and Index,* locate and photocopy (or download and print) a recent government publication in your field or on your topic.

12. Using OCLC, RLIN, ProQuest, SilverPlatter, or similar online or CD-ROM services, locate and copy the bibliographic record (including abstracts, if available) of four current books and four current articles in your field or on your topic.

13. If your library offers students a free search of commercial database networks such as DIALOG, ask your librarian for help in preparing an electronic search for your semester report.

14. Explore Internet databases via Prodigy, Pathfinder, the Microsoft Network or a similar access provider or "on-ramp." Prepare a list of promising database resources for your report topic.

15. Locate an Internet discussion list related to your report topic. Download and print out the group's FAQ list. For a directory of listservs, consult <*http://www.liszt.com*>.

16. Using Netscape Navigator, Internet Explorer, or a similar browsing program, search Web sites to locate resources for your report topic.

17. If your library belongs to a consortium of electronically networked libraries, search the holdings of other libraries on the network for topic

resources not available in your library. Prepare a list of promising possibilities.

18. Students in your major want a listing of at least *two* of each of the following discipline-specific sources: the main reference books; indexes; periodicals; government publications; commercial, Internet, and CD-ROM databases; online newsgroups and discussion groups. Prepare the list (in memo form) and include a one-paragraph description of each source. Be prepared to discuss your list in class.

19. Most Web browsers allow you to do keyword searches (page 123) by using a search engine such as *Yahoo!, Lycos, AltaVista,* or *InfoSeek.* Each engine has its own guidelines and peculiarities; these are usually explained in a "help" file or user's guide. Learn to use at least one search engine; for your classmates, write instructions for designing and conducting a Web search using that engine.

> URLs: *http://www.yahoo.com*
> *http://www.lycos.com*
> *http://www.altavista.com*
> *http://www.infoseek.com*

COLLABORATIVE PROJECTS

1. Group yourselves according to major. For other students in your major, prepare a guide, in the form of a brochure, to your library's electronic resources (CD-ROM services and commercial database services, electronic catalogs, network consortium, Internet gateways, World Wide Web access, and so on). Describe discipline-specific types of resources available via each electronic medium. Early in this project, arrange for a group tour and demonstration of your library's resources by a trained librarian. (In conjunction with this project, your instructor may assign Chapters 15 and 23.)

2. Divide into small groups and prepare a comparative evaluation of literature search media. Each group member will select *one* of the resources listed below and create an individual bibliography (listing at least twelve recent and relevant works on a specific topic of interest selected by the group):

   - conventional print media
   - electronic catalogs
   - CD-ROM services
   - a commercial database service such as DIALOG
   - the Internet and World Wide Web
   - an electronic consortium of local libraries, if applicable

   After carefully recording the findings and keeping track of the time spent in each search, compare the ease of searching and quality of results obtained from each type of search on your group's selected topic. Which medium yielded the most current sources (page 145)? Which provided abstracts and full texts as well as bibliographic data? Which consumed the most time? Which provided the most dependable sources (page 146)? The most diverse or varied sources (page 99)? Which cost the most to use? Finally, which yielded the greatest *depth* of resources (page 100).

   Prepare a report and present your findings to the class. (In conjunction with this project, your instructor may assign Chapter 25.)

3. Divide into small groups and decide on a campus or community issue or some other topic worthy of research. Elect a group manager to assign and coordinate tasks. At project's end, the manager will provide a performance appraisal by summarizing, in writing, the contribution of each team member. Assigned tasks will include planning, information gathering from primary and secondary sources, document preparation (including visuals) and revision, and classroom presentation. (See page 82 for collaboration guidelines.)

   Do the research, write the report, and present your findings to the class. (In conjunction with this project, your instructor may assign Chapter 25.)

4. Group yourselves according to major. Assume that several major employers in your field are

holding a job fair on campus next month and will be interviewing entry-level candidates. Each member of your group is assigned to develop a profile of *one* of these companies or organizations by researching its history, record of mergers and stock value, management style, financial condition, price/earnings ratio of its stock, growth prospects, products and services, multinational affiliations, ethical record, environmental record, employee relations, pension plan, employee stock options or profit-sharing plans, commitment to affirmative action, number of women in upper management, or any other features important to a prospective employee. The entire group will then edit each profile and assemble them in one single document to be used as a reference for students in your major.

Exploring
Primary Sources

**Figure 9.1
Sources for
Primary Research**

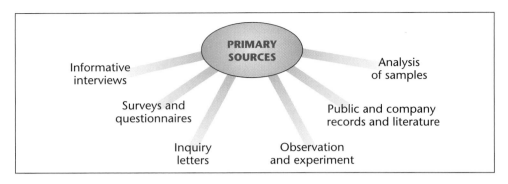

Workplace decisions often rely on primary research—an original, firsthand study of the topic, involving sources like those in Figure 9.1.

INFORMATIVE INTERVIEWS

An excellent primary source for information unavailable in any publication is the personal interview. Much of what an expert knows may never be published (Pugliano 6). Also, a respondent might refer you to other experts or sources of information.

Of course, an expert's opinion can be just as mistaken or biased as anyone else's (page 104). Like patients who seek second opinions about serious medical conditions, researchers seek a balanced range of expert opinions about a complex problem or controversial issue—not only from a company engineer and environmentalist, for example, but also from independent and presumably more objective third parties such as a professor or journalist who has studied the issue.

Selecting the Best Medium

Once you decide whom to interview about what, select your medium carefully:

How interview
media compare

- ◆ *In-person interviews* are most productive because they allow human contact (Hopkins-Tanne 24).
- ◆ *Phone interviews* are convenient but they lack the human contact of in-person interviews—especially when the interviewer and respondent have not met.
- ◆ *Email interviews* are convenient and inexpensive, and they allow time for respondents to consider their answers.
- ◆ Fax interviews are highly impersonal, and generally a bad idea.

Whatever your medium, obtain a respondent's approval *beforehand* instead of waylaying this person with an unwanted surprise.

GUIDELINES
for Informative
Interviews*

Purpose statement

Planning the Interview

1. *Focus on your purpose.* Determine exactly what you hope to learn from this interview. Write out your purpose.

> I will interview Anne Hector, Chief Engineer at Northport Electric, to ask her about the company's approaches to EMF risk avoidance—within the company as well as in the community.

2. *Do your homework.* Learn all you can about the topic beforehand. If the respondent has published anything relevant, read it before the interview. Be sure the information this person might provide is unavailable in print.

3. *Contact the intended respondent.* Do this by phone, letter, or email, and be sure to introduce yourself and your purpose. (See Karen Granger's letter on page 376.)

4. *Request the interview at your respondent's convenience.* Give the respondent ample notice and time to prepare, and ask whether she/he objects to being quoted or taped. If you use a tape recorder, insert fresh batteries and a new tape and set the recording volume loud enough.

Preparing the Questions

1. *Make each question clear and specific.* Vague, unspecific questions elicit vague, unspecific answers.

A vague question

> How is this utility company dealing with the problem of electromagnetic fields?

Which problem—public relations, potential liability, danger to electrical workers, to the community, or what?

A clear and specific question

> What safety procedures have you developed for risk avoidance on electrical work crews?

2. *Avoid questions that can be answered with "yes" or "no."*

An unproductive question

> In your opinion, can technology find ways to decrease EMF hazards?

Instead, phrase your question to elicit a detailed response:

*Several guidelines are adapted from Blum 88; Dowd, 13–14; Hopkins-Tanne 23, 26; Kotulak 147; McDonald 190; Rensberger 15; Young 114, 115, 116.

A productive
question

> Of the various technological solutions being proposed or considered, which do you consider most promising?

This is one instance in which your earlier homework pays off.

3. *Avoid loaded questions.* A loaded question invites or promotes a particular bias:

A loaded question

> Wouldn't you agree that EMF hazards have been overstated?

An impartial question does not lead the interviewee to respond in a certain way.

An impartial
question

> In your opinion, have EMF hazards been accurately stated, overstated, or understated?

4. *Save the most difficult, complex, or sensitive questions for last.* Leading off with your toughest questions might annoy respondents, making them uncooperative.

5. *Write out each question on a separate blank page.* Use a three-ring binder with $8\frac{1}{2}$" x 11" pages to arrange your questions in logical order. You can then flip to a new page for each question and record responses easily.

Conducting the Interview

1. *Make a good start.* Dress appropriately and arrive on time. Thank your respondent; restate your purpose; explain why you believe he/she can be helpful; explain exactly how you will use the information.

2. *Be sensitive to cultural differences.* If the respondent belongs to a different culture from your own, then consider the level of formality, politeness, directness, relationship building and other behaviors seen as appropriate in that culture. (See also pages 48–50.)

3. *Let the respondent do most of the talking.* Keep your opinions to yourself.

4. *Be a good listener.* Don't doodle or let your eyes wander. People reveal more when their listener seems genuinely interested.

5. *Stick to your interview plan.* If the respondent wanders, politely nudge the conversation back on track (unless the added information is useful).

6. *Ask for clarification or explanation whenever necessary.* If you don't understand an answer, say so. Request an example, an analogy, or a simplified version—and keep asking until you understand.

Clarifying questions

- ◆ Could you go over that again?
- ◆ Is there a simpler explanation?

Science writer Ronald Kotulak argues that, "No question is dumb if the answer is necessary to help you understand something. . . . Don't pretend to know more than you do" (144).

7. *Keep checking on your understanding.* Repeat major points in your own words and ask if the technical details are accurate and if your interpretation is correct.

8. *Be ready with follow-up questions.* Some answers may lead to additional questions.

Follow-up questions

- ◆ Why is it like that?
- ◆ Could you say something more about that?
- ◆ What more needs to be done?
- ◆ What happened next?

9. *Keep note taking to a minimum.* Record statistics, dates, names, and other precise data, but not every word. Jot key terms or phrases that later can refresh your memory.

Concluding the Interview

1. *Ask for closing comments.* Perhaps the respondent can lead you to additional information.

Concluding questions

- ◆ Would you care to add anything?
- ◆ Is there anything else I should talk to?
- ◆ Is there anyone who has a different point of view?
- ◆ Are there any other sources you are aware of that might help me better understand this issue?

2. *Invite the respondent to review your version.* If the interview will be published, ask the respondent to check your final draft (for misspelled names, inaccurate details, misquotations, and so on) and to approve it. Offer to provide copies of any document in which this information appears.

3. *Thank your respondent and leave promptly.*

4. *As soon as you leave the interview, write a complete summary (or record one verbally).* Do this while responses are fresh in your memory.

Q. *Could you please summarize your communication responsibilities?*

A. The corporate relations office oversees three departments: customer service (which handles claims, adjustments, and queries), public relations, and employee relations. My job is to supervise the production of all documents generated by this office.

Probing and following up

Q. *Isn't that a lot of responsibility?*

A. It is, considering we're trying to keep some people happy, getting others to cooperate, and trying to get everyone to change their thinking and see things in a positive light. Just about every document we write has to be persuasive.

Seeking clarification

Q. *What exactly do you mean by "persuasive"?*

A. The best way to explain is through examples of what we do. The customer service department responds to problems like these: Some users are unhappy with our software because it won't work for a particular application, or they find a glitch in one of our programs, or they're confused by the documentation, or someone wants the software modified to meet a specific need. For each of these complaints or requests we have to persuade our audience that we've resolved the problem or that we're making a genuine effort to resolve it quickly.

The public relations department works to keep up our reputation through links outside the company. For instance, we keep in touch with this community, with consumers, the general public, government and educational agencies. . . .

Seeking clarification

Q. *Can you be more specific? "Keeping in touch" doesn't sound much like persuasion.*

A. Okay, right now we're developing programs with colleges and universities, in which we offer heavily discounted software, backed up by an extensive support network (regional consultants, an 800 phone hotline, and workshops). We're hoping to persuade them that our software is superior to our well-entrenched competitor's. And locally we're offering the same kind of service and support to business clients.

Following up

Q. *What about employee relations?*

A. Day to day we face the usual kinds of problems: trying to get 100 percent employee contributions to the United Way, or persuading employees to help out in the community, or getting them to abide by new company regulations restricting smoking or to limit personal phone calls. Right now, we're facing a real persuasive challenge. Because of market saturation, software sales have flattened across the board. This means temporary layoffs for roughly 28 percent of our employees. Our only alternative is to persuade *all* employees to accept a 10-percent salary and benefit cut until the market improves.

Probing

Q. *How, exactly, do you persuade employees to accept a cut in pay and benefits?*

A. Basically, we have to make them see that by taking the cut, they're really investing in the company's future—and, of course, in their own.

(The interview continues.)

Figure 9.2   Partial Text of an Informative Interview

A Sample Interview

Figure 9.2 shows the partial text of an interview on persuasive challenges in the workplace. Notice how the interviewer probes, seeks clarification, and follows up on responses from XYZ's Director of Corporate Relations.

SURVEYS AND QUESTIONNAIRES

Surveys help us to develop profiles and estimates about the concerns, preferences, attitudes, beliefs, or perceptions of a large, identifiable group (a *target population*) by studying representatives of that group (a *sample group*).

Surveys help us make assessments like these

- ◆ *Do consumers prefer brand A or brand B?*
- ◆ *What percentage of students feel safe on our campus?*
- ◆ *Is public confidence in technology increasing or decreasing?*

The questionnaire is the tool for conducting surveys. While interviews allow for greater clarity and depth, questionnaires offer an inexpensive way to survey a large group. Respondents can answer privately and anonymously—and often more candidly than in an interview.

Questionnaires do have certain limitations, though:

Limitations of survey research

- ◆ *A low rate of response (often less than 30 percent).* People refuse to respond to a questionnaire that seems too long, too complicated, or threatening. They might be embarrassed by the topic or afraid of how their answers could be used.
- ◆ *Responses that might be nonrepresentative.* A survey will get responses from the people who want to respond, but you will know nothing about the people who don't respond. Those who responded might have extreme views, a particular stake in the outcome, or some other motive that inaccurately represents the population being surveyed (Plumb and Spyridakis 625–26).
- ◆ *Lack of follow-up.* Survey questions do not allow for the kind of follow-up and clarification possible with interview questions.

Even surveys designed by professionals carry potential for error. As consumers of survey research, we need to understand how surveys are designed, administered, and interpreted, and what can go wrong. Following is an introduction to creating surveys and to avoiding pitfalls along the way.

Defining the Survey's Purpose and Target Population

Why is this survey being done? What, exactly, is it measuring? How much background research is needed? How will the survey findings be used?

Who is the exact population being studied (the chronically unemployed, part-time students, computer users)? For example, in its research on science and technology activity, the 1997 *Statistical Abstract of the United States* differentiates "scientists and engineers" from "technicians":

Target populations clearly defined

> Scientists and engineers are defined as persons engaged in scientific and engineering work at a level requiring a knowledge of sciences equivalent at least to that acquired through completion of a 4-year college course. Technicians are defined as persons engaged in technical work at a level requiring knowledge acquired through a technical institute, junior college, or other type of training less extensive than 4-year college training. Craftspersons and skilled workers are excluded. (603)

Identifying the Sample Group

How will intended respondents be selected? How many respondents will there be? Generally, the larger the sample surveyed, the more dependable the results (assuming a well-chosen and representative sample). Will the sample be randomly chosen? In the statistical sense, "random" does not mean "chosen haphazardly": a random sample means that each member of the target population stands an equal chance of being in the sample group.

Even a sample that is highly representative of the target population carries a measure of *sampling error.*

A type of survey error

> The particular sample used in a survey is only one of a large number of possible samples of the same size which could have been selected using the same sampling procedures. Estimates derived from the different samples would, in general, differ from each other. (*Statistical Abstract* 949)

The larger the sampling error (usually expressed as the *margin of error,* page 161), the less dependable the survey findings.

Defining the Survey Method

What type of data (opinions, ideas, facts, figures) will be collected? Is timing important? How will the survey be administered—in person, by mail, by phone? How will the data be collected, recorded, analyzed, and reported (Lavin 277)?

Phone, email, and in-person surveys yield fast results and high response rates, but respondents consider phone surveys annoying and, without anonymity, people tend to be less candid. Mail surveys are less expensive and more confidential. Computerized surveys create the sense of a video game: the program analyzes each response and then automatically designs the next question. Respondents who dislike being quizzed by a human researcher seem more comfortable with this automated format (Perelman 89–90).

GUIDELINES

for Developing
a Questionnaire

Open-ended
questions

1. *Decide on the types of questions* (Adams and Schvaneveldt 202–12; Velotta 390). Questions can be *open-ended* or *closed-ended*. Open-ended questions allow respondents to express exactly what they're thinking or feeling in a word, phrase, sentence, or short essay:

- How much do you know about electromagnetic radiation at our school?
- What do you think should be done about electromagnetic fields (EMFs) at our school?

Since you never know what people will say, open-ended questions are a good way to uncover attitudes and obtain unexpected information. But essay-type questions are hard to answer and tabulate.

When you want to measure exactly where people stand, choose closed-ended questions:

Closed-ended
questions

Are you interested in joining a group of concerned parents?
YES _____ NO _____
Rate your degree of concern about EMFs at our school.
HIGH _____ MODERATE _____ LOW _____ NO CONCERN _____
Circle the number that indicates your view about the town's proposal to spend $20,000 to hire its own EMF consultant.
     1 . . . . 2 . . . . 3 . . . . 4 . . . . 5 . . . . 6 7
 Strongly   No   Strongly
 Disapprove   Opinion   Approve

Respondents may be asked to *rate* one item on a scale (from high to low, best to worst), to *rank* two or more items (by importance, desirability), or to select items from a list. Other questions measure percentages or frequency:

How often do you . . . ?
ALWAYS _____ OFTEN _____ SOMETIMES _____ RARELY _____ NEVER _____

Although closed-ended questions are easy to answer, tabulate, and analyze, they might elicit biased responses. Some people, for instance, automatically prefer items near the top of a list or the left side of a rating scale (Plumb and Spyridakis 633). Also, people are prone to agree rather than disagree with assertions in a questionnaire (Sherblom, Sullivan, and Sherblom 61).

2. *Design an engaging introduction and opening questions.* Persuade respondents that the survey relates to their concerns, that their answers matter, and that their anonymity is assured. Explain how respondents will benefit from your findings, or offer an incentive (say, a copy of your final report).

A survey
introduction

Your answers will help our school board to speak accurately for your views at our next town meeting. Results of this survey will appear in our campus newspaper. Thank you.

Researchers often include a cover letter with the questionnaire.

Begin with the easiest questions. Once respondents commit to these, they are likely to complete later, more difficult questions.

3. *Make each question unambiguous.* All respondents should be able to interpret identical questions identically. An ambiguous question leaves room for different interpretations than the one intended.

> Do you favor weapons for campus police?   YES _____ NO _____

An ambiguous question

"Weapons" might mean tear gas, clubs, handguns, all three, or two out of three. Consequently, responses to the above question would produce a misleading statistic, such as "Over 95 percent of students favor handguns for campus police" when the accurate conclusion might be "Over 95 percent of students favor some form of weapon." Moreover, the limited "yes/no" format reduces an array of possible opinions to an either/or choice.

A clear, incisive question

> Do you favor (check all that apply):
> _____ Having campus police carry mace and a club?
> _____ Having campus police carry nonlethal "stun guns"?
> _____ Having campus police store handguns in their cruisers?
> _____ Having campus police carry small-caliber handguns?
> _____ Having campus police carry large-caliber handguns?
> _____ Having campus police carry no weapons?
> _____ Don't know

To ensure a full range of possible responses, include options such as "Other _____," "Don't know," "Not Applicable," or an "Additional Comments" section.

4. *Make each question unbiased.* Avoid *loaded questions* that invite or advocate a particular viewpoint or bias:

A loaded question

> Should our campus tolerate the needless endangerment of innocent students by lethal weapons?
> YES _____ NO _____

Using emotionally loaded and judgmental words ("endangerment," "innocent," "tolerate," "needless," "lethal") in a survey is unethical because their built-in judgments manipulate people's responses (Hayakawa 40).

5. *Make it brief, simple, and inviting.* Try to limit questions and their response spaces to two sides of a single page. Include a stamped, self-addressed return envelope, and specify a return date. Address each respondent by name, sign your letter or your introduction, and give your title.

Figure 9.3
A Questionnaire
Cover Letter

April 5, 20xx

House 10
University of Massachusetts, Dartmouth
North Dartmouth, MA 02747

Name, Title
Company Name
Address

Dear _____:

I am exploring ways to enhance relationships between UMD's Professional
Communication Program and the local business community.

Specific areas of inquiry:
1. the communication needs of local companies and industries
2. the feasibility of on-campus and in-house seminars for
   employees
3. the feasibilty of expanding communication course offerings
   at UMD

Please take a few minutes to complete the attached survey. Your response
will provide an important contribution to my study.

All respondents will receive a copy of my report, scheduled to appear in
the fall issue of *The Business and Industry Newsletter.* Thank you.

Sincerely,

L.S. Taylor
Technical Communication Student

A Sample Questionnaire

The student-written letter and questionnaire in Figures 9.3 and 9.4, sent to presidents of local companies, is designed to elicit responses that can be tabulated easily. (For a sample usability questionnaire, see page 341.)

Written reports of survey findings often include an appendix (page 332) that contains a copy of the questionnaire as well as the tabulated responses.

Communication Questionnaire

1. Describe your type of company (e.g., manufacturing, high tech)

   _____

2. Number of employees, (Please check one.)

   _____ 5–25   _____ 50–100   _____ 150–300
   _____ 25–50   _____ 100–150   _____ 300–450

3. What types of written communication occur in your company? (Label by frequency: never, rarely, sometimes, often.)

   _____ memos   _____ letters   _____ advertising
   _____ manuals   _____ reports   _____ newsletters
   _____ procedures   _____ proposals   _____ other (Specify.)
   _____ email   _____ catalogs   _____

4. Who does most of the writing? (Pls. give titles.)   _____

5. Please characterize your employees' writing effectiveness.

   _____ good   _____ fair   _____ poor

6. Does your company have formal guidelines for writing?

   _____ no   _____ yes   (Pls. describe briefly.) _____

7. Do you offer in-house communication training?

   _____ no   _____ yes   (Pls. describe briefly.) _____

8. Please rank the usefulness of the following areas in communication training (from 1–10).

   _____ organizing information   _____ audience awareness
   _____ summarizing information   _____ persuasive writing
   _____ editing for style   _____ grammar
   _____ document design   _____ researching
   _____ email etiquette   _____ Web page design
   _____ other (Pls. specify.)_____

9. Please rank these skills in order of importance (from 1–6).

   _____ reading   _____ listening   _____ speaking to groups
   _____ writing   _____ collaborating   _____ speaking face-to-face

10. Do you provide tuition reimbursement for employees?

    _____ no   _____ yes

11. Would you consider having UMD communication interns work for you part-time?

    _____ no   _____ yes

12. Should UMD offer Saturday seminars in communication?

    _____ no   _____ yes

    Additional comments/suggestions: _____
    _____

Figure 9.4   A Sample Questionnaire

INQUIRY LETTERS, PHONE CALLS, AND EMAIL INQUIRIES

Letters, phone calls, or email inquiries to experts listed in Web pages are handy for obtaining specific information from government agencies, legislators, private companies, university research centers, trade associations, and research foundations such as the Brookings Institution and the Rand Corporation (Lavin 9).

NOTE *Keep in mind that unsolicited inquiries, especially by phone or email, can be intrusive and offensive.*

PUBLIC RECORDS AND ORGANIZATIONAL PUBLICATIONS

The *Freedom of Information Act* and state public record laws grant access to an array of government, corporate, and organizational documents. Obtaining these documents (from state or federal agencies) takes time, but in them you can find answers to questions like these (Blum 90–92):

Public records may hold answers to tough questions

- ◆ Which universities are being investigated by the USDA (Dept. of Agriculture) for mistreating laboratory animals?
- ◆ Are IRS auditors required to meet quotas?
- ◆ What are the results of state and federal water-quality inspections in this region?

Organization records (reports, memos, computer printouts, and so on) are good primary sources. Most organizations also publish pamphlets, brochures, annual reports, or prospectuses for consumers, employees, investors, or voters.

NOTE *Be alert for bias in company literature. In evaluating the safety measures at a local nuclear power plant, you would want the complete picture. Along with the company's literature, you would want studies and reports from government agencies and publications from environmental groups.*

PERSONAL OBSERVATION AND EXPERIMENT

Observation should be your final step because you now know what to look for. Know how, where, and when to look, and jot down observations immediately. You might even take photos or make drawings.

Informed observations can pinpoint real problems. Here is an excerpt from a report investigating low morale at an electronics firm. This researcher's observations and interpretation are crucial in defining the problem:

Direct observation is often essential

Our on-site communications audit revealed that employees were unaware of any major barriers to communication. More than seventy-five percent of employees claimed they felt free to talk to their managers, but the managers, in turn, estimated that fewer than fifty percent of employees felt free to talk to them.

The problem involves misinterpretation. Because managers don't ask for complaints, employees are afraid to make them, and because employees never ask for an evaluation, they never get one. Each side has inaccurate perceptions of what the other side expects, and because of ineffective communications, each side fails to realize that its perceptions are wrong.

NOTE   *Even direct observation is not foolproof: for instance, you might be biased about what you see (focusing on the wrong events or ignoring something important), or, instead of behaving normally, people being observed might behave in ways they think you expect (Adams and Schvaneveldt 244).*

An experiment is a controlled form of observation designed to verify an assumption (e.g., the role of fish oil in preventing heart disease) or to test something untried (the relationship between background music and worker productivity). Each field has its own guidelines for experiment design.

ANALYSIS OF SAMPLES

Workplace research can involve collecting and analyzing samples: water or soil or air, for contamination and pollution; foods, for nutritional value; ore, for mineral value; or plants, for medicinal value. Investigators analyze material samples to find the cause of an airline accident. Engineers analyze samples of steel, concrete, or other building materials to determine their load-bearing capacity. Medical specialists analyze tissue samples for disease.

EXERCISES

1. Revise these questions to make them appropriate for inclusion in a questionnaire:

   a. Would a female president do the job as well as a male?
   b. Don't you think that euthanasia is a crime?
   c. Do you oppose increased government spending?
   d. Do you think welfare recipients are too lazy to support themselves?
   e. Are teachers responsible for the decline in literacy among students?
   f. Aren't humanities studies a waste of time?
   g. Do you prefer Rocket Cola to other leading brands?
   h. In meetings, do you think men are more interruptive than women?

2. Identify and illustrate at least six features that enhance the effectiveness of the questionnaire in Figure 9.4. (Review pages 137–38 for criteria.) Be prepared to discuss your evaluation in class.

3. Arrange an interview with someone in your field. Decide on general areas for questioning: job opportunities, chances for promotion, salary range, requirements, outlook for the next decade, working conditions, job satisfaction, and so on. Compose specific interview questions; conduct the interview, and summarize your findings in a memo to your instructor.

COLLABORATIVE PROJECT

Divide into small groups and decide on a survey of views, attitudes, preferences, or concerns in regard to some issue affecting your campus or the surrounding community. Expand on this short list of possible survey topics:

- campus codes prohibiting hate speech or offensive language in general
- campus alcohol policy
- campus safety
- facilities for disabled students
- campus racial or gender issues
- access to computers

Once you have identified your survey's exact purpose and your target population, follow these steps:

a. Decide on the size and makeup of a randomly selected sample group.
b. Try to identify all sources of potential error.
c. Develop a questionnaire that will measure accurately what your survey intends to measure. Design questions that are engaging, unambiguous, unbiased, and easy to answer and tabulate.
d. Administer the survey to a representative sample group.
e. Tabulate, analyze, and interpret the responses.
f. Compose a written report summarizing your survey purpose, process, findings, and conclusions. Discuss any survey limitations. Include a copy of the questionnaire as well as the tabulated responses.
g. Appoint one group member to present your findings to the class.

In addition to reviewing pages 135–40, look over Chapter 10, especially the section on validity and reliability (page 165).

Evaluating and Interpreting Information

Not all information is equally valuable. Not all interpretations are equally valid. For instance, if you really want to know how well the latest innovation in robotic surgery works, you need to check with other sources besides, say, its designer (from whom you could expect an overly optimistic or insufficiently critical assessment).

Whether you work with your own findings or those of other researchers, you need to decide if the information is any good for your purposes. Then you need to decide what this information means. Figure 10.1 outlines your challenge.

For producers as well as consumers of research, evaluating and interpreting information is a process that allows much room for error. The critical-thinking strategies in this section will help you to distinguish legitimate inquiry from simple information gathering.

EVALUATE THE SOURCES

Not all sources are equally dependable. A source might offer information that is out of date, inaccurate, incomplete, mistaken, or biased.

Is the Source Up-to-Date?

Even newly published books contain information that can be more than a year old, and journal articles often undergo a lengthy process of peer review.

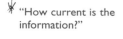
"How current is the information?"

Certain types of information become outdated more quickly than others. For topics that focus on technology (multimedia law, superconductivity, alternative cancer treatments), information more than a few months old may be outdated. But

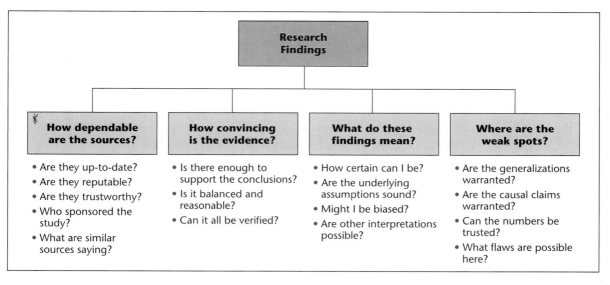

Figure 10.1   Decisions in Evaluating and Interpreting Information

for topics that focus on people (business ethics, management practices, workplace gender issues), historical perspectives often help.

The most recent information is not always the most reliable—especially in scientific research, a process of ongoing inquiry in which what seems true today may be proven false tomorrow: "Physicist and philosopher of science John Ziman has suggested that the physics in undergraduate textbooks is 90 percent true, while that in the primary research journals is 90 percent false" (Taubes 76).

Is the Printed Source Dependable?

"What is the source's reputation?"

Assess a publication's reputation by checking its copyright page. Is the work published by a university, professional society, museum, or respected news organization? Do members of the editorial and advisory board have distinguished titles and degrees? Is the publication refereed (all submissions reviewed by experts before acceptance)?

Assess the document's thoroughness by checking its bibliography or list of references to see how extensively the author has researched the issue (Barnes).

Assess an author's reputation by checking citation indexes (page 113) to see what other experts have said about this research. Many periodicals also provide brief biographies or descriptions of authors' earlier publications and other achievements.

Is the Electronic Source Dependable?

"Can the source be trusted?"

The Internet offers information that never appears in other sources, for example from listservs and newsgroups. But much of this material may reflect the bias of the special-interest groups that provide it. Moreover, anyone can publish almost anything on the Internet—including misinformation—without having it verified, edited, or reviewed for accuracy. Don't expect to find everything you need on the Internet. (Pages 147–48 offer suggestions for evaluating Web sites.)

Is the Information Relatively Unbiased?

"Who sponsored the study, and why?"

Much of today's research is funded by private companies or special-interest groups, which have their own social, political, or economic agendas (Crossen 14, 19). Medical research may be sponsored by drug or tobacco companies; nutritional research, by food manufacturers; environmental research, by oil or chemical companies. Even public policy research (on gun control, school prayer, endangered species) sponsored by opposing groups (say, environmentalists versus the logging industry) produces contradictory findings.

Instead of a neutral and balanced inquiry, this kind of *strategic research* is designed to support one special interest or another (Crossen 132–34). Furthermore, those who pay for strategic research are not likely to publicize findings that contradict their original claims or opinions (lower profits than expected, greater losses

GUIDELINES

for Evaluating
Sources on
the Web*

1. *Consider the site's domain type and sponsor.* In this typical address, <**http://www.umass.edu**>, the site or *domain* information follows the *www*. The *.edu* signifies the type of organization from which the site originates. Standard domain types in the United States:

   .com = business/commercial organization
   .edu = educational institution
   .gov = government organization
   .mil = military organization
   .net = any group or individual with simple software and Internet access
   .org = nonprofit organization

   The domain type might signal a certain bias or agenda that could skew the data. For example, at a *.com* site, you might find accurate information, but also a sales pitch. At an *.org* site, you might find a political or ideological bias (say, The Heritage Foundation's conservative ideology versus the Brookings Institution's more liberal slant). Knowing a site's sponsor can help you evaluate the credibility of its postings.

2. *Identify the purpose of the page or message.* Decide whether the message is intended simply to relay information, to sell something, or to promote a particular ideology or agenda.

3. *Look beyond the style of a site.* Fancy graphics, video, and sound do not always translate into dependable information. Sometimes the most reliable material resides in the less attractive, text-only sites.

4. *Assess the site's/material's currency.* An up-to-date site should indicate when the material was created or published, and when it was posted and updated.

5. *Assess the author's credentials.* Learn all you can about the author's reputation, expertise on this topic, institutional affiliation (a university, Fortune 500 Company, reputable environmental group). Do this by following links to other sites that mention the author or by using search engines to track the author's name. Newsgroup postings often contain "a *signature file* that includes the author's name, location, institutional or organizational affiliation, and often a quote that suggests something of the writer's personality, political leanings, or sense of humor" (Goubil-Gambrell 229–30).

6. *Compare the site with other sources.* Check related sites and publications to compare the quality of information and to discover what others might have said about this site or author. Comparing many similar sites

*Guidelines adapted from Barnes; Busiel and Maeglin 39; Elliot; Fackelmann 397; Grassian; Hall 60–61; Hammett; Harris, Robert; Stemmer).

helps you create a *benchmark,* a standard for evaluating any particular site. Ask a librarian for help.

7. *Decide whether the assertions/claims make sense.* Decide where, on the spectrum of informed opinion and accepted theory, this author's position resides. Is each assertion supported by convincing evidence? Never accept any claim that seems extreme without verifying it through other sources, such as a professor, a librarian, or a specialist in the field.

8. *Look for other indicators of quality.*
   - *Worthwhile content:* The material is technically accurate. All sources of data presented as "factual" are fully documented. (Appendix A)
   - *Sensible organization:* The material is organized for the user's understanding, with a clear line of reasoning.
   - *Readable style:* The material is well written (clear, concise, easy to understand) and free of typos, misspellings, and other mechanical errors.
   - *Objective coverage:* Debatable topics are addressed in a balanced and impartial way, with fair, accurate representation of opposing views. The tone is reasonable, with no "sounding off."
   - *Expertise:* The author refers to related theory and other work in the field and uses specialized terminology accurately and appropriately.
   - *Peer review:* The material has been evaluated and verified by related experts.
   - *Links to reputable sites:* The site offers a gateway to related sites that meet quality criteria.
   - *Follow-up option:* The material includes a signature block or a link for contacting the author or the organization.

or risks than expected). Research consumers need to know exactly what the sponsors of a particular study stand to gain or lose from the results (234).

NOTE   *Keep in mind that any research ultimately stands on its own merits. Thus, funding by a special interest should not automatically discredit an otherwise valid and reliable study.*

How Does This Source Measure Up to Others?

"What are similar sources saying?"

Most studies have some type of flaw (page 166). Therefore, instead of relying on a single source or study, seek a consensus among various respected sources (Cohn 106).

EVALUATE THE EVIDENCE

Evidence is any finding used to support or refute a particular claim. While evidence can serve the truth, it also can create distortion, misinformation, and deception. For example:

Questions that invite distorted evidence

- ◆ How much money, material, or energy does recycling really save?
- ◆ How well are public schools educating children?
- ◆ Which investments or automobiles are safest?

Competing answers to such questions often rest on evidence that has been stacked to support a particular view or agenda.

Is the Evidence Sufficient?

"Is there enough evidence?"

Evidence is sufficient when nothing more is needed to reach an accurate judgment or conclusion. A study of the stress-reducing benefits of low-impact aerobics, for example, would require a broad survey sample: people who have practiced aerobics for a long time; people of both genders, different ages, different occupations, and different lifestyles before they began aerobics; and so on. Even responses from hundreds of practitioners might be insufficient unless those responses were supported by laboratory measurements of metabolic and heart rates, blood pressure, and so on.

NOTE

Although anecdotal evidence ("This worked great for me!") might offer a good starting point for investigation, your personal experience rarely provides enough evidence from which to generalize. No matter how long you might have practiced aerobics, for instance, you need to determine whether your experience is representative.

Is the Presentation of Evidence Balanced and Reasonable?

How evidence can be misused

Misuse of evidence in a courtroom often makes headlines. But evidence routinely is misused beyond the courtroom as well, as in the following instances.

"Is this claim too good to be true?"

OVERSTATEMENT.   Consumers are offered a daily menu of cures for ailments ranging from insomnia to cancer. Overzealous researchers might exaggerate their achievements, without mentioning the limitations of their study.

"Is there a downside?"

OMISSION OF VITAL FACTS.   Aspirin is widely promoted for *decreasing* heart attack and stroke risk caused by clotting, but far less emphasized is its role in *increasing* stroke risk caused by brain bleeding (Lewis 222). Acetaminophen products are advertised as a "safe" alternative pain reliever, without aspirin's side effects (stomach irritation, Reye's syndrome), but even small overdoses of acetaminophen have caused liver failure. Moreover, acetaminophen is the leading cause of U.S. drug fatalities (Easton and Herrara 42–44).

DECEPTIVE FRAMING OF THE FACTS.   A *frame of reference* is a set of ideas, beliefs, or views that influences our interpretation of other ideas. For example, people with a fundamentalist view of the Bible might reject the concept of evolution. Or consider the well-known optimist/pessimist test: Is the glass half full (a positive frame of reference) or half empty (a negative frame of reference)? In medical terms, is a "90-percent survival rate" more acceptable than a "10-percent mortality rate"? *Framing* sways our perception (Lang and Secic 239–40).

Framing can be a powerful—and often unethical—persuasive tool, as in this example: In June 1998 the McCain Tobacco Bill was defeated by a $40 million advertising campaign (financed by tobacco companies) that framed the bill as "pro-tax" instead of "anti-tobacco" by focusing on the proposed $1.10 cigarette tax (Goodman 6).

Whether the language used is provocative ("rape of the environment," "bureaucrats," "tree huggers") or euphemistic ("teachable moment" versus "mistake"), deceptive framing is designed to obscure the real issues.

Can the Evidence Be Verified?

Hard evidence consists of factual statements, expert opinion, or verifiable statistics. Soft evidence consists of uninformed opinion or speculation, data obtained or analyzed unscientifically, and findings not replicated or reviewed by experts. Reputable news organizations employ fact-checkers to verify information before it's printed.

Is the glass "half-full" or "half-empty"?

How framing swayed voters

"Is the evidence hard or soft?"

"One question: If this is the Information Age, how come nobody knows anything?"

Source: © The New Yorker Collection 1998 Robert Mankoff from cartoonbank.com. All Rights Reserved.

INTERPRET YOUR FINDINGS

Raw information does not equal understanding. To *interpret* information you must reach an overall judgment about what the findings mean and what conclusion or action they suggest.

Unfortunately, research does not always yield clear or conclusive answers. Instead of settling for the most *convenient* answer, pursue the most *reasonable* answer by critically examining a full range of possible meanings.

What Level of Certainty Is Warranted?

Research can yield three distinctly different levels of certainty:

1. The ultimate truth—the *conclusive answer:*

A practical definition of "truth"

> Truth is what is so about something, the reality of the matter, as distinguished from what people wish were so, believe to be so, or assert to be so. . . . In the words of Harvard philosopher Israel Scheffler, truth is the view "which is fated to be ultimately agreed to by all who investigate." The word *ultimately* is important. Investigation may produce a wrong answer for years, even for centuries. . . . Does the truth ever change? No. . . .
>
> One easy way to spare yourself any further confusion about truth is to reserve the word *truth* for the final answer to an issue. Get in the habit of using the words *belief, theory,* and *present understanding* more often. (Ruggiero 21–22)

We all seek conclusive interpretations, but often we must settle for uncertain answers.

2. *The probable answer:* the answer that stands the best chance of being true or accurate—given the most we can know at this particular time. Probable answers are subject to revision in light of new information.

3. *The inconclusive answer:* the realization that the truth of the matter is more elusive, ambiguous, or complex than we expected.

"Exactly how certain are we?"

We need to decide what level of certainty our findings warrant. For example, we are *highly certain* about the perils of smoking or sunburn, *reasonably certain* about the health benefits of fruits and vegetables and moderate exercise, but *less certain* about the perils of coffee drinking or the benefits of vitamin supplements.

The "truth" never changes; however, our notions about "truthfulness" do, as in these examples:

Some changing notions about "Truth"

- ◆ *"The earth is the center of the universe."* Though dead wrong, Ptolemy's cosmology was based on the best information available in the second century.

And this certainty survived thirteen centuries—even after new information had discredited Ptolemy's theory: When Copernicus and Galileo proposed more truthful views in the fifteenth century, they were labeled heretics.

◆ *"Brush your teeth and blow your nose with asbestos."* Considered a "miracle fiber" for two thousand years, asbestos—soft, flexible, and fire resistant—was used in countless products—ranging from asbestos handkerchiefs in the first century to tablecloths, toothpaste, and cigarette filters in the twentieth century. Not until the 1970s was the truth about the long-suspected role of asbestos in lung disease publicized (Alleman and Mossman 70–74).

◆ *"The vast majority of Asian American students excel in school."* This popular, high-achiever stereotype ignores all diversity among major Asian American ethnic groups (South Asian, Korean, Filipino, Chinese, Japanese, and Southeast Asian). A recent study by the Educational Testing Service reveals the truth: "Many Asian Americans are undereducated and have low socioeconomic status. . . . The myth of them all being educational high achievers has kept many students from needed services and support" (Kim).

Most scientific, technical, and social controversies are open-ended. Therefore, no irrefutable presentation of "the facts" will likely settle the controversy over questions like these:

Questions that invite controversy

◆ *How rapidly is global warming progressing?*
◆ *What is causing the death and disfigurement of frogs worldwide?*
◆ *Can vitamins prevent cancer or heart disease?*
◆ *Should Affirmative Action programs be expanded or discontinued?*

Does this mean that such questions should be ignored? Of course not. Even though some claims cannot be proven, we can still reach reasonable conclusions on the basis of the evidence.

Are the Underlying Assumptions Sound?

Assumptions are notions we take for granted, things we accept without proof. The research process rests on assumptions like these: that a sample group accurately represents a larger target group, that survey respondents remember certain facts accurately, that mice and humans share enough biological similarities for meaningful generalization of research results. For a particular study to be valid (page 165), its underlying assumptions must be accurate.

How underlying assumptions affect research validity

Consider this example: You are an education consultant evaluating the accuracy of IQ testing as a predictor of academic performance. Reviewing the evidence, you perceive an association between low IQ scores and low achievers. You then verify your statistics by examining a cross-section of reliable sources. Can you then conclude that IQ tests *do* predict performance accurately? This conclusion might be invalid unless you could verify the following assumptions:

1. That neither parents, teachers, nor children had seen individual test scores, which could produce biased expectations.
2. That, regardless of score, each child had completed an identical curriculum at an identical pace, instead of being "tracked" on the basis of his or her score.

NOTE

Assumptions are often easier to identify in someone else's thinking and writing than in our own. During collaborative discussions, request colleagues' help in identifying your own assumptions (Maeglin).

To What Extent Has Bias Influenced the Interpretation?
When the issue is controversial, our own biased assumptions might cause us to overestimate (or deny) the certainty of findings.

Personal bias is a fact of life

Unless you are perfectly neutral about the issue, an unlikely circumstance, at the very outset . . . you will believe one side of the issue to be right, and that belief will incline you to . . . present more and better arguments for the side of the issue you prefer. (Ruggiero 134)

How bias can outweigh evidence

The following example illustrates how *cognitive bias* (seeing what we expect to see) can blind us to the most compelling evidence:

The 1989 spill from the oil tanker *Exxon Valdez* polluted more than 1,000 miles of Alaskan shoreline and led to a massive recovery effort that included using high-pressure hot water to clean oil from the beaches. But a respected study shows that the *uncleaned* beaches are now healthier than those sterilized by the hot water: "Whatever [beach life] survived the oiling did not survive the cure" (Holloway 109). But this finding—that cleaning up is more harmful than helpful—remains highly unpopular with the Alaskan public, who continue to insist on the removal of virtually every drop of oil (109–12).

Because personal bias is hard to transcend, rationalizing often becomes a substitute for reasoning:

Rationalizing versus reasoning

You are reasoning if your belief follows the evidence—that is, if you examine the evidence first and then make up your mind. You are rationalizing if the evidence follows your belief—if you first decide what you'll believe and then select and interpret evidence to justify it. (Ruggiero 44)

Personal bias is often subconscious until we examine our own value systems: attitudes long held but never analyzed, assumptions we've inherited from our backgrounds, and so on. Recognizing our biases is the crucial first step toward managing them.

Are Other Interpretations Possible?

"What else could this mean?"

Perhaps other researchers disagree with the meaning of these findings. Settling on a final meaning can be hard. For instance, how should we interpret the reported increase in violent crime on U.S. college campuses—especially in light of statistics that show violent crime decreasing in general (Lederman 5)? Should we conclude that (a) the college population is becoming more violent, (b) some drugs and guns from high schools end up on campuses, (c) off-campus criminals see students as easy targets, or (d) all of these?

Or could these findings mean something else entirely? For example, that (a) increased law enforcement has produced more campus arrests—and thus, greater awareness of crime or (b) campus crimes really haven't increased, but fewer now go unreported? Depending on our interpretation, we might conclude that the situation is worsening—or improving!

NOTE

Not all interpretations are equally valid. Never assume that any interpretation that is possible is also allowable—especially in terms of its ethical consequences. Certain interpretations in the college crime example, for instance, might justify an overly casual or overly vigilant response—either of which could have disastrous consequences.

AVOID ERRORS IN REASONING

In the process of reasoning, we make *inferences:* We derive conclusions about what we don't know by reasoning from what we do know (Hayakawa 37). For example, we might infer that a drug that boosts immunity in laboratory mice will boost immunity in humans, or we might attribute a reported rise in campus crime to the fact that young people have become more violent. Whether a particular inference is on target or dead wrong depends largely on our answers to these questions:

Questions for testing inferences

- ◆ *To what extent can these findings be generalized?*
- ◆ *Is Y really caused by X?*
- ◆ *How much can the numbers be trusted, and what do they mean?*

Following are three major reasoning errors that can distort our interpretations.

Faulty Generalization

We engage in faulty generalization when we jump from a limited observation to a sweeping conclusion. Even "proven" facts can invite mistaken conclusions.

Factual observations

1. "Some studies have shown that gingko [an herb] improves mental functioning in people with dementia [mental deterioration caused by maladies such as Alzheimer's Disease]" (Stix 29).
2. "For the period 1992–2005, two thirds of the fastest-growing occupations will call for no more than a high-school degree" (Harrison 62).
3. "Adult female brains are significantly smaller than male brains—about 8% smaller, on average" (Seligman 74).

In Brief HOW STANDARDS OF PROOF VARY FOR DIFFERENT AUDIENCES AND CULTURAL SETTINGS

How much evidence is enough to "prove" a particular claim? The answer often depends on whether the inquiry occurs in the science lab, the courtroom, or the boardroom, as well as the specific cultural setting:

◆ *The scientist demands evidence that indicates at least ninety-five percent certainty.* A scientific finding must be evaluated and replicated by other experts. Good science looks at the entire picture. Findings are reviewed before they are reported. Inquiries and answers in science are never "final," but open-ended and ongoing: what seems probable today may be shown improbable by tomorrow's research.

◆ *The juror demands evidence that indicates only fifty-one percent certainty (a "preponderance of the evidence").* Jurors are not scientists. Instead of the entire picture, jurors get only the information made available by lawyers and witnesses. A jury bases its opinion on evidence that exceeds "reasonable

doubt" (Monastersky, "Courting" 249; Powell 32+). Based on such evidence, courts have to make decisions that are final.

◆ *The corporate executive demands immediate (even if insufficient) evidence.* In a global business climate of overnight developments (in world markets, political strife, military conflicts, natural disasters) important business decisions are often made on the spur of the moment. On the basis of incomplete or unverified information—or even hunches—executives must make quick decisions to react to crises and capitalize on opportunities (Seglin 54).

◆ *Specific cultures may have their own standards for authentic, reliable, and persuasive evidence.* "For example, African cultures rely on storytelling for authenticity. Arabic persuasion is dependent on universally accepted truths. And Chinese value ancient authorities over recent empiricism." (Byrd and Reid 109–11).

Invalid conclusions

1. Gingko is food for the brain!
2. Higher education . . . Who needs it?!
3. Women are the less intelligent gender.

"How much can we generalize from these findings?"

When we accept findings uncritically and jump to conclusions about their meaning (as in 1 and 2, above) we commit the error of *hasty generalization*. When we overestimate the extent to which the findings reveal some larger truth (as in 3, above) we commit the error of *overstated generalization*.

NOTE

We often need to generalize, and we should. For example, countless studies support the generalization that fruits and vegetables help lower cancer risk. But we ordinarily limit general claims by inserting qualifiers such as "usually," "often," "sometimes," "probably," "possibly," or "some."

Faulty Causal Reasoning

Causal reasoning explains why something happened or what happens as a result of something. *A definite cause is usually obvious* ("The engine's overheating is caused by a faulty radiator cap"). We reason about definite causes when we explain why the combustion in a car engine causes the wheels to move, or why the moon's orbit makes the tides rise and fall.

But causal reasoning often explores *causes that are not so obvious, but only possible or probable.* In these cases, much searching, thought, and effort are usually needed to argue for a specific cause.

"Did X probably, possibly, or definitely cause Y?"

Suppose you set out to answer this question: "Why does our college campus lack day care facilities for children?" Brainstorming produces these possible causes:

- ◆ lack of need among students
- ◆ lack of interest among students, faculty, and staff
- ◆ high cost of liability insurance
- ◆ lack of space and facilities on campus
- ◆ lack of trained personnel
- ◆ prohibition by state law
- ◆ lack of legislative funding for such a project

Suppose you proceed with interviews, questionnaires, and research into state laws, insurance rates, and availability of personnel. First, you rule out some causes: specifically, you find a need among students, high campus interest, plenty of qualified staff candidates, and no legal prohibitions. Three probable causes remain: lack of funding, high insurance rates, and lack of space. Further inquiry reveals that lack of funding and high insurance rates are indeed issues. But you conclude that these obstacles could be eliminated through new sources of revenue: charging a modest fee per child, soliciting donations, diverting funds from other campus organizations, and so on. Finally, after examining available campus space and interviewing school officials, you settle on one *definite* cause: lack of space and facilities.

When you report on your research, be sure readers can draw conclusions identical to your own on the basis of the evidence. The process might be diagrammed like this:

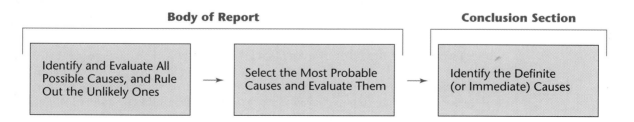

The persuasiveness of your causal argument will depend on the quality of evidence you offer, as well as on your ability to clearly explain the links in the chain. Also, you must convince your audience that you haven't overlooked important alternative causes.

NOTE

Any complex effect likely has more than one cause; therefore, you need to make the case that the cause you have isolated is the real one. In the day care scenario, for example, one might argue that the lack of space and facilities is somehow related to funding. And the college's inability to find funds might be related to student need or interest, which is not high enough to exert real pressure. Lack of space and facilities, however, does seem to be the "immediate" cause.

Here are common errors that distort or oversimplify cause-effect relationships:

Ignoring other causes

Investment builds wealth. *[Ignores the roles of knowledge, wisdom, timing, and luck in successful investing.]*

Ignoring other effects

Running improves health. *[Ignores the fact that many runners get injured, and that some even drop dead while running.]*

Inventing a causal sequence

Right after buying a rabbit's foot, Felix won the state lottery. *[Posits an unwarranted causal relationship merely because one event follows another.]*

Confusing correlation with causation

Women in Scandinavian countries drink a lot of milk. Women in Scandinavian countries have a high incidence of breast cancer. Therefore, milk must be a cause of breast cancer. *[The association between these two variables might be mere coincidence and might obscure other possible causes, such as environment, fish diet, and genetic predisposition (Lemonick 85).]*

Rationalizing

My grades were poor because my exams were unfair. *[Denies the real causes of one's failures.]*

AVOID STATISTICAL FALLACIES

How numbers can mislead

Numbers seem more precise, more objective, more scientific and less ambiguous than words. They are easier to summarize, measure, compare, and analyze. But numbers also can mislead. For example, radio or television phone-in surveys often produce distorted data: Although "90 percent of callers" might express support for a particular viewpoint, people who actually call tend to be those with the greatest anger or most extreme feelings—representing only a fraction of overall attitudes (Fineman 24). Mail-in surveys can produce similar distortion.

Before relying on any set of numbers, we need to know exactly where they come from, how they were collected, and how they were analyzed.

Common Statistical Fallacies

Faulty statistical reasoning produces conclusions that are unwarranted, inaccurate, or downright deceptive. Some typical errors in reasoning are discussed below.

"Exactly how well are we doing?"

THE SANITIZED STATISTIC.   Numbers can be manipulated (or "cleaned up") to create a rosy picture or obscure the facts. For example, the College Board's 1996 re-centering of SAT scores raised the "average" math score from 478 to 500 and the average verbal score from 424 to 500 (a boost of almost 5 and 18 percent, respectively) although actual student performance remains unchanged (Samuelson 44).

In Brief CORRELATION VERSUS CAUSATION

Causal reasoning often relies on statistical analysis. The following definitions help us evaluate a statistical analysis of causal relationships.

◆ *Correlation:* a numerical measure of the strength of the relationship between two variables—say, smoking and lung cancer risk, or education and income (Black 513).

We depict the strength of a correlation by plotting data in *scatter diagrams* (Lang and Secic 94):

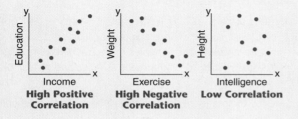

| High Positive Correlation | High Negative Correlation | Low Correlation |

As correlation increases, all data points approach the pattern of a perfect slanting line. In a *positive correlation,* as one variable (say, education) increases, so does the other (say, income). In a *negative correlation,* one variable (say, exercise) increases, while the other (say, weight) decreases.

But instead of "proving" that one thing "causes" another, correlation merely signals a possible relationship.

◆ *Causation:* the demonstrable production of a specific effect (smoking causes lung cancer). Correlations between smoking and lung cancer or between education and income signal a causal relationship that has been amply demonstrated.

Causation Actually Demonstrated

Multiple studies (especially in labs) have demonstrated how specific causes produce specific effects: say how asbestos fibers invade lung tissue to cause a cancer known as *mesothelioma,* or how ionizing radiation alters cellular DNA (Harris, Richard 172).

Causation Strongly Indicated

Consider this: "7 percent of Americans eat at McDonald's on any given day, and the average child watches 10,000 food commercials on television a year" ("In Brief" 28). Can we speculate about the relationship between these eating and viewing habits and the fact that the U.S. obesity rate has risen by one-third since 1980 ("Fat Chance" 3)?

Improbable Causation

A recently discovered correlation between moderate alcohol consumption and decreased heart disease risk offers insufficient proof that moderate drinking "causes" less heart disease. Only detailed research might answer this question.

No Demonstrable Causation

A strong correlation has been found between sharp drops in income for middle-class people and a doubling of their risk of death within five years (Koretz 32). But is income instability a "cause" of death? Of course not.

◆ *Coincidence:* the random but simultaneous occurrence of two or more events.
◆ *Confounding (or confusing) factors:* other reasons or explanations for a particular outcome. For instance, studies indicating that regular exercise improves health might be overlooking this confounding factor: healthy people tend to exercise more than those who are unhealthy ("Walking" 3–4).

THE OVERLY PRECISE STATISTIC.   Precise numbers can be used to quantify something so inexact, vaguely defined, or difficult to count that it should only be approximated (Huff 247; Lavin 278):

"How many rats was that?"

- ◆ Boston has 3,247,561 rats.
- ◆ Zappo detergent makes laundry 10 percent brighter.

An exact number looks impressive, but certain subjects (child abuse, cheating in college, virginity, drug and alcohol abuse on the job, eating habits) cannot be quantified exactly because respondents don't always tell the truth (because of denial, embarrassment, or merely guessing). Or they respond in ways they think the researcher expects.

Overly precise numbers are also harder to read:

Needlessly precise and hard to read

Computer engineers' salaries increased from $34,717 to $41,346, while programmers' salaries increased from $26,807 to $32,112.

Easier to decipher and compare

Computer engineers' salaries increased roughly from $35,000 to $41,000, while programmers' salaries increased from $27,000 to $32,000.

Unless greater precision is required, round your numbers to a maximum of two significant digits ("0.00") (Lang and Secic 40).

THE UNDEFINED AVERAGE.   The mean, median, and mode are three measures of *central tendency* (the intermediate, or middle, value in a set of numbers) (Black 53–56; Huff 244; Lavin 279).

Three ways of reporting an "average"

- ◆ The *mean* is the result of adding up the value of each item in a given set, and then dividing by the number of items.
- ◆ The *median* is the result of ranking all the values from high to low, and then identifying the middle value (or the 50th percentile, as in calculating SAT scores). In an even-numbered set, the median would be a number midway between the two middle values.
- ◆ The *mode* is the value that occurs most often in a given set of numbers.

Each of these three measurements represents some kind of middle value, or average, but to interpret a set of numbers accurately, we need to know which "average" is being reported.

"Why is everyone griping?"

Assume we are calculating the "average" salary among female vice presidents at XYZ Corporation (ranked from high to low):

Vice President	Salary
A	$90,000
B	90,000
C	80,000
D	65,000
E	65,000
F	55,000
G	50,000

From this list, the "mean salary" (total salaries divided by number of salaries) equals $70,000; the "median salary" (middle value) equals $65,000; the "mode" (most frequent—or "typical"—value) equals $90,000. Each is, technically, a middle—or "average"—and each could be used to support or refute a particular assertion (say, "Women vice presidents receive too little" or "Women vice presidents receive too much").

Research expert Michael Lavin sums up the potential for bias in the reporting of averages:

Unethical uses of "averages"

> Depending on the circumstances, any one of these measurements [*mean, median,* or *mode*] may describe a group of numbers better than the other two. . . . [But] people typically choose the value which best presents their case, whether or not it is the most appropriate. (279)

What the "mean" doesn't tell us

Although it is the most commonly computed average, the mean ignores the data's *variability* (how values in a data set differ from each other and how they deviate from the mean figure). Reporting the "mean" figure alone can mislead when *outliers* (extreme values on either end) deviate from the normal range of values. For instance, if Vice President A's salary (above) were $200,000, this outlier would distort the group average—increasing the "mean salary" by over 20 percent (Plumb and Spyridakis 636). In that instance, the "median" figure—unaffected by outlying values—would present a more realistic picture.

What the "median" and the "mode" don't tell us

A drawback of the "median" figure is that it ignores extreme values. But the "mode" focuses only on the "typical" values, ignoring the data's *range* (distance between highest and lowest values), as well as their variability. In short, the clearest picture of a data set may require two of these measures—if not all three.

NOTE

> *Failure to report "outliers" (the small percentage of the results occurring at a great distance from the mean) in calculating the mean or secretly leaving the outliers out of the calculation is deceptive. An ethical approach is to calculate the results with and without the outliers and to report both figures (Lang and Secic 31).*

THE DISTORTED PERCENTAGE FIGURE.   Percentages are often reported without explanation of the original numbers used in the calculation (Adams and Schvaneveldt 359; Lavin 280): "Seventy-five percent of respondents prefer our brand over the competing brand"—without mention that only four people were surveyed. Three out of four respondents is a far less credible number than, say, 3,000 out of 4,000—yet each of these ratios equals "75 percent."

NOTE

In small samples, percentages can mislead because the percentage size can dwarf the number it represents: "In this experiment, 33% of the rats lived, 33% died, and the third rat got away." (Lang and Secic 41). When your sample is small, report the actual numbers: "Five out of ten respondents agreed. . . ."

"How much of a majority is 51 percent?"

Another distortion occurs when a percentage report omits the *margin of error* (the margin within which the true figure lies, based on estimated sampling errors in a survey). For example, a claim that "the majority of people surveyed prefer Brand X" might be based on 51 percent of respondents expressing this preference; but if the survey carried a "2 percent margin of error," the true figure could be as low as 49 or as high as 53 percent. In a survey with a high margin of error, the true figure might be so uncertain that no definite conclusion may be drawn.

"Which car should we buy?"

THE BOGUS RANKING. Items might be compared on the basis of ill-defined criteria (Adams and Schvaneveldt 212; Lavin 284): "Last year, the Batmobile was the number-one selling car in Gotham City"—without mentioning that some competing car makers actually sold *more* cars to private individuals, and that the Batmobile figures were inflated by hefty sales—at huge discounts—to rental car companies and corporate fleets. Unless we know how the ranked items were chosen and how they were compared (the *criteria*), a ranking can produce a scientific-seeming number based on a wholly unscientific method.

The Limitations of Number Crunching

Computers are great for comparing, synthesizing, and predicting—because of their speed, processing power, and "what if" capabilities. But limitless capacity to crunch numbers cannot guarantee accurate results.

DRAWBACKS OF DATA MINING. Many highly publicized correlations are the product of *data mining,* in which a software program searches databases and randomly compares one set of variables (say, buying habits) with another set (age, income, or gender). From these countless comparisons, certain patterns, or associations, appear (perhaps between coffee drinking and pancreatic cancer risk). At one retail company, data mining revealed a correlation between diaper sales and beer sales (presumably because young fathers go out at night to buy diapers). The retailer then displayed the diapers next to the beer and reportedly sold more of both (Rao 128).

Because it detects hidden relationships and has predictive power, data mining is a popular business tool. Companies assemble their own *data warehouses* (databases of information about customers, research and development, market conditions, legal matters). Much of what is "mined," however, can turn out to be trivial or absurd:

Bizarre correlations uncovered by data mining

- Venereal disease rates correlate with air pollution levels (Stedman, "Data Mining" 28).
- The hourly wages of self-proclaimed "natural blondes" (both male and female) averaged 75 cents more than the wages of nonblonde people in 1993 (Sklaroff and Ash 85).

NOTE   *Despite its limitations, data mining is potentially invaluable for "uncovering correlations that require computers to perceive but that thinking humans can evaluate and research further" (Maeglin).*

THE BIASED META-ANALYSIS.   Meta-analysis is a broad overview of related studies, conducted to synthesize the larger meaning of these collective findings and to identify some common pattern (for example, the role of high-fat diets in cancer risk).

As "objective" as it may seem, meta-analysis does carry potential for error (Lang and Secic):

Potential errors in meta-analysis

- *Selection bias.* Because results ultimately depend on which studies have been included and which omitted, a meta-analysis can reflect the biases of the researchers who select the material (174).
- *Publication bias.* Small studies have less chance of being published than large ones (175–76).
- "*Head counting.*" This questionable method tallies the studies that have positive, negative, or insignificant results, and announces the winning category based on sheer numbers—without accounting for a study's relative size, design, or other differences.

THE FALLIBLE COMPUTER MODEL.   Computer models process complex assumptions to produce impressive but often inaccurate estimates about costs, benefits, risks, or probable outcomes.

How assumptions influence a computer model

For instance, models to predict global warming levels are based on differing assumptions about wind and weather patterns, ozone levels, carbon dioxide concentrations, airborne sediment from volcanic eruptions, and so on. Different global warming models predict sea level rises ranging from a few inches to several feet (Barbour 121) or global temperature rises ranging from 1.8° F. to 8.1° degrees (Begley, "Odds" 72). Other models suggest that warming effects could be offset by evaporation of ocean water and by clouds reflecting sunlight back to outer space (Monatersky 69). Still other models suggest that the warming over the last 100 years might not be caused by the greenhouse effect at all, but by "random fluctuations in global temperatures" (Stone 38).

Estimates produced by any model depend on the assumptions (and data) programmed in. Assumptions might be influenced by researcher bias or the sponsors' agenda. For example, a prediction of human fatalities from a nuclear reactor meltdown might rest on assumptions about availability of safe shelter, evacuation routes, time of day, season, wind direction, and the structural integrity of the containment unit. But these assumptions could be manipulated to overstate or understate the risk (Barbour 228). For computer-modeled estimates of accident risk (oil spill, plane crash) or of the costs and benefits of a proposed project or policy (International Space Station, welfare reform), consumers rarely know the assumptions behind the numbers. We wonder, for example, about the assumptions underlying NASA's pre-*Challenger* risk assessment, in which a 1985 computer model reportedly showed an accident risk of less than 1 in 100,000 shuttle flights (Crossen 54).

Misleading Terminology

The language used to interpret statistics sometimes hides their real meaning.

"Do we all agree on what these terms mean?"

- *People treated for cancer have a "50 percent survival rate."* This claim is doubly misleading: (1) "survival," to laypersons, means "staying alive," but to medical experts, staying alive for only five years after diagnosis qualifies as survival; (2) the "50 percent" survival figure covers all cancers, including certain skin or thyroid cancers that have extremely high cure rates, as well as other cancers (such as lung or ovarian) that are rarely curable and have extremely low survival rates ("Are We" 6; *Facts and Figures* 2).

- *"The BODCARE Health Plan is proud of achieving 99 percent customer satisfaction."* The rate of customer "satisfaction," in this context, provides little real information about the quality of the health plan itself. Because most HMO customers are reasonably healthy, they rarely visit a doctor and so have little cause for dissatisfaction. Only those in poor health would be able to evaluate "quality of service." Also, customer service is not the same as "quality of treatment"—which is much harder for laypersons to evaluate. Finally, are we talking about customers who are "satisfied," "somewhat satisfied," or "very satisfied"—all combined into one result (Spragins 77)?

INTERPRET THE REALITY BEHIND THE NUMBERS

Even the most accurate statistics require that we interpret the reality behind the numbers:

"Is this news good, bad, or insignificant?"

- *"Rates for certain cancers double after prolonged exposure to electromagnetic radiation."* Does the cancer rate actually increase from "1 in 10,000" to "2 in 10,000," or from "1 in 50" to "2 in 50"? Without knowing the *base rate* (original rate) we cannot possibly decide how alarming this "doubling" of risk actually is.

- *"Saabs and Volvos are involved in 75 percent fewer fatal accidents than average."* Is this only because of superior engineering or also because Saab and Volvo owners tend to drive very carefully ("The Safest" 72)?
- *"Eating charbroiled meat may triple the risk of stomach cancer."* The actual added risk of dying from a weekly charbroiled steak is 1 in 4 million—lower than the risk of drowning in the bathtub (Lee 280).

The above numbers may be "technically accurate" and may seem highly persuasive in the interpretations they suggest. But the actual "truth" behind these numbers is far more elusive. Any interpretation of statistical data carries the possibility that other, more accurate interpretations have been overlooked or deliberately excluded (Barnett 45).

ACKNOWLEDGE THE LIMITS OF RESEARCH

Legitimate researchers live with uncertainty. They expect to be wrong far more often than right. Experimentation and exploration often produce confusion, mistakes, and dead ends. Following is a brief list of things that can go wrong with research and interpretation.

"Scientists confirmed today that everything we know about the structure of the universe is wrongedy-wrong-wrong."

Source: © The New Yorker Collection 1998 Jack Ziegler from cartoonbank.com. All Rights Reserved.

Obstacles to Validity and Reliability

What makes a survey valid

Validity and *reliability* determine the dependability of any research (Adams and Schvaneveldt 79–97; Burghardt 174–75; Crossen 22–24; Lang and Secic 154–55; Velotta 391). *Valid research* produces correct findings. A survey, for example, is valid when (1) it measures what you want it to measure, (2) it measures accurately and precisely, and (3) its findings can be generalized to the target population. A valid survey question helps each respondent to interpret the question exactly as the researcher intended, and it asks for information respondents are qualified to provide.

Survey validity depends largely on trustworthy responses. Even clear, precise, and neutral questions can produce mistaken, inaccurate, or dishonest answers for these reasons:

Why survey responses can't always be trusted

- People often see themselves as more informed, responsible, or competent than they really are. For example, in surveys of job skills, respondents consistently rank themselves in the top twenty-fifth percentile (Fisher 206). Also, ninety percent of adults rate their driving ability as "above average." And a mere two percent of high school students rate their leadership ability "below average," while twenty-five percent rate themselves in the top one percent (Baumeister 21).
- Respondents might suppress information that reflects poorly on their behavior, attitudes, or will power when asked questions like: "How often do you take needless sick days?" "Would you lie to get ahead?" "How much TV do you watch?"
- Respondents might exaggerate or invent facts or opinions that reveal a more admirable picture when asked questions like: "How much do you give to charity?" "How many books do you read?" "How often do you hug your children?"
- Even when respondents don't know, don't remember, or have no opinion, they tend to guess in ways designed to win the researcher's approval.

What makes a survey reliable

Reliable research produces findings that can be replicated. A survey is reliable when its results are consistent, for instance, when a respondent gives identical answers to the same survey given twice or to different versions of the same questions. A reliable survey question can be interpreted identically by all respondents. Factors that compromise survey reliability include the following:

Why surveys aren't always reliable

- Each sample group has its own peculiarities (in distribution of age, gender, religious background, educational level, and so on).
- Various observers might interpret identical results differently.
- A single observer might interpret the same results differently at different times.

Much of your communication will be based on the findings of other researchers, so you will need to assess the validity and reliability of their research as well as your own.

Flaws in Study Design

While some types of studies are more reliable than others, each type has limitations (Cohn 106; Harris, Richard 170–72; Lang and Secic 8–9; Murphy 143).

EPIDEMIOLOGIC STUDIES.   Observing various populations (human, animal, or plant), epidemiologists search for correlations (say, between computer use and cataracts). Conducted via observations, interviews, surveys, or review of records, these studies involve no controlled experiments. Their major limitations:

Common flaws in epidemiologic studies

- Faulty sampling techniques (page 136) may distort results.
- Observation bias (seeing what one wants to see) may occur.
- Coincidence can easily be mistaken for correlation.
- Confounding factors (other explanations) often affect results.

Even with a correlation that is ninety-nine percent certain, an epidemiological study alone doesn't "prove" anything. (The larger the study, however, the more credible.)

LABORATORY STUDIES.   Although laboratories offer controlled conditions, these studies also have certain limitations:

Common flaws in laboratory studies

- The reaction of an isolated group of cells does not always predict the reaction of the entire organism.
- The reactions of experimental animals to a treatment or toxin often are not generalizable to humans. For example, massive, short-term doses given to animals differ vastly from the lower, long-term doses taken by humans.
- Faulty lab technique may distort results. For example, a recent study linking Vitamin C pills to genetic damage created panic. Experts have since concluded that "researchers themselves may have created 90 to 99% of the genetic damage when they ground up the cells to examine the DNA" ("Vitamin C" 1).

HUMAN EXPOSURE STUDIES (CLINICAL TRIALS).   These studies compare one group of people receiving medication or treatment with an untreated group, the *control group.* Major limitations of human exposure studies:

Common flaws in clinical trials

- The study group may be nonrepresentative or too different from the general population in overall health, age, or ethnic background. (For example, the fact that gingko might slow memory loss in sick people doesn't mean it will boost the memory of healthy people.)
- Anecdotal reports are unreliable. Respondents often invent answers to questions like: "How often do you eat ice cream?"
- Lack of objectivity may distort results. A drug's effectiveness is tested in a *randomized trial,* which compares a treated group of patients with a control group, who are given *placebos* (substitutes containing no actual medicine).

These studies are *masked (or blind):* specific group assignment is concealed from patients. But doctors sometimes sneak their sickest patients into the treatment group for their possible benefit from new treatments. Despite good intentions, this practice subverts the random selection vital to such trials, making precise assessment of treatment impossible (Wallich 20+).

It is best to look for some consensus among a combination of studies.

Sources of Measurement Error

How scientific
measurement can
go wrong

All measurements (of length, time, temperature, weight, population characteristics) are prone to error (Taylor 3–4). The two basic types of measurement error are discussed below.

RANDOM ERROR.   Anything that causes variations in values from measure to measure is a random error (Taylor 46, 94):

Causes of random
error in
measurement

- Technique can vary from measure to measure: in running a stopwatch, reading markings on a scale, locating two points for measuring distance, changing body position in reading an instrument (say, a thermometer).
- Each observation differs with different observers or with the same observer in repeated observations.
- Each sample in a survey differs. Because of this sampling error, the same computation applied to multiple random samples drawn from the same given population produces varying results.

Researchers help neutralize random errors by repeating the measurements and averaging the range of values.

SYSTEMATIC ERROR.   Any consistent bias that causes researchers to overestimate or underestimate a measurement's true value is a systematic error (Taylor 94, 97):

Causes of
systematic error in
measurement

- A measuring device can be faulty: for example, a watch that runs slow or an improperly calibrated instrument.
- The measuring device (for example, a scale) might be positioned improperly.

Systematic errors are harder to neutralize than random errors because *all* systematically flawed measurements are inaccurate.

Sources of Deception
One problem in reviewing scientific findings is "getting the story straight." Deliberately or not, consumers are often given a distorted picture.

"Has bad or
embarrassing news
been suppressed?"

UNDERREPORTED HAZARDS.   Although twice as many people in the United States are killed by medications as in auto accidents—and countless others harmed—doctors rarely report adverse drug reactions. One Rhode Island study showed some 26,000 such reactions noted in doctors' files, of which only 11 had been reported to the Food and Drug Administration (Freundlich 14).

"Is the topic 'too
weird' for
researchers?"

THE UNTOUCHABLE RESEARCH TOPIC.   Paranormal phenomena and extraterrestrials are rarely the topics of respectable research. Neither is alternative medicine—despite the fact that one U.S. citizen in three uses some type of alternative treatment (chiropractic, acupuncture, acupressure, herbal therapy) at least once a year. Little or no scientific evidence suggests that any or all of these therapies have actual medical benefits. Why? "Partly because few 'respectable' scientists are willing to risk their reputations to do the testing required, and partly because few firms would be willing to pay for it if they were." Drug companies have little interest because "herbal medicines, not being new inventions, cannot be patented" ("Any Alternative?" 83).

"Does this report
get the story
straight?"

A "GOOD STORY" BUT BAD SCIENCE.   Spectacular claims that are even remotely possible are more appealing than spectacular claims that have been disproven. This is why we hear plenty about claims like the following—but little about their refutation:

Even bad science
makes good news

- Giant comet headed for earth!
- Tabletop device achieves cold fusion—producing more energy than it consumes!
- Insects may carry the AIDS virus!

Does all this mean we can't believe anything? Of course not. But we should be very selective about what we do choose to believe.

GUIDELINES

for Evaluating
and Interpreting
Information

Evaluate the Sources

1. *Check the source's date of posting or publication.* Although the latest information is not always the best, it's important to keep up with recent developments.
2. *Assess the reputation of each printed source.* Check the copyright page, for background on the publisher; the bibliography, for the quality and extent of research; and (if available) the author's brief biography, for credentials.
3. *Assess the quality of each electronic source.* See page 147 for evaluating Internet sources. Don't expect comprehensive sources on any single database.

4. *Identify the study's sponsor.* If the study acclaims the crash-worthiness of the Batmobile, but is sponsored by the Batmobile Auto Company, be skeptical.

5. *Look for corroborating sources.* Usually, no single study produces dependable findings. Learn what other sources say.

Evaluate the Evidence

1. *Decide whether the evidence is sufficient.* Evidence should surpass mere personal experience, anecdote, or news reports. It should be substantial enough for reasonable and informed observers to agree on its value, relevance, and accuracy.

2. *Look for a reasonable and balanced presentation of evidence.* Suspect any claims about "breakthroughs," or "miracle cures," as well as loaded words that invite emotional response or anything beyond accepted views on a topic. Expect a discussion of drawbacks as well as benefits.

3. *Do your best to verify the evidence.* Examine the facts that support the claims. Look for replication of findings. Go beyond the study to determine the direction in which the collective evidence seems to be leaning.

Interpret Your Findings

1. *Don't expect "certainty."* Most complex questions are open-ended and a mere accumulation of "facts" doesn't "prove" anything. Even so, the weight of solid evidence usually points toward some reasonable conclusion.

2. *Examine the underlying assumptions.* As opinions taken for granted, assumptions are easily mistaken for facts.

3. *Identify your personal biases.* Examine your own assumptions. Don't ignore evidence simply because it contradicts your way of seeing, and don't focus only on evidence that supports your assumptions.

4. *Consider alternate interpretations.* Consider what else this evidence might mean.

Check for Weak Spots

1. *Scrutinize all generalizations.* Decide whether the "facts" are indeed facts or assumptions, and whether the evidence supports the generalization. Suspect any general claim not limited by some qualifier ("often," "sometimes," "rarely," or the like).

2. *Treat causal claims skeptically.* Differentiate correlation from causation, as well as possible from probable or definite causes. Consider confounding factors (other explanations for the reported outcome).

3. *Look for statistical fallacies.* Determine where the numbers come from, and how they were collected and analyzed—information that legitimate researchers routinely provide. Note the margin of error.

4. *Consider the limits of computer analysis.* Data mining often produces intriguing but random correlations; meta-analysis might oversimplify relationships among various studies; a computer model is only as accurate as the assumptions and data programmed into it.

5. *Look for misleading terminology.* Examine terms that beg for precise definition in their specific context: "survival rate," "success rate," "customer satisfaction," "average increase," "risk factor," and so on.

6. *Interpret the reality behind the numbers.* Consider the possibility of alternative, more accurate, interpretations of these numbers.

7. *Consider the study's possible limitations.* Small, brief studies are less reliable than large, extended ones; epidemiologic studies are less reliable than laboratory studies (which also carry flaws); animal or human exposure studies are often not generalizable to larger human populations; "masked" (or blind) studies are not always as objective as they seem; measurements are prone to error.

8. *Look for the whole story.* Consider whether bad news may be underreported; good news, exaggerated; bad science, camouflaged and sensationalized; or research on promising but unconventional topics (say, alternative energy sources) ignored.

Checklist FOR THE RESEARCH PROCESS

(Use this checklist to assess your research process. Numbers in parentheses refer to the first page of discussion.)

Methods

- [ ] Did I ask the right questions? (98)
- [ ] Are the sources appropriately up-to-date? (145)
- [ ] Is each source reputable, trustworthy, relatively unbiased, and borne out by other, similar sources? (146)
- [ ] Does the evidence clearly support the conclusions? (149)
- [ ] Is the evidence balanced and reasonable? (149)

- [ ] Can all evidence be verified? (150)
- [ ] Is a fair range of viewpoints presented? (99)
- [ ] Has my research achieved adequate depth? (100)
- [ ] Has my entire research process been valid and reliable? (165)

Reasoning

- [ ] Can I discern assumption from fact? (152)
- [ ] Am I reasoning instead of rationalizing? (153)
- [ ] Can I discern correlation from causation? (158)

- ☐ Is this the most reasonable conclusion (or simply the most convenient)? (99)
- ☐ Can I rule out other possible interpretations or conclusions? (153)
- ☐ Have I accounted for all sources of bias, including my own? (153)
- ☐ Are my generalizations warranted by the evidence? (154)
- ☐ Am I confident that my causal reasoning is accurate? (155)
- ☐ Can I rule out confounding factors? (158)
- ☐ Can all numbers, statistics, and interpretations be trusted? (157)
- ☐ Have I resolved (or at least acknowledged) any conflicts among my findings? (102)
- ☐ Can I rule out any possible error or distortion? (166)

- ☐ Am I getting the whole story, and getting it straight? (167)

Documentation

- ☐ Is the documentation consistent, complete, and correct? (Appendix A)
- ☐ Is all quoted material clearly marked throughout the text? (575)
- ☐ Are direct quotations used sparingly and appropriately? (575)
- ☐ Are all quotations accurate and integrated grammatically? (576)
- ☐ Are all paraphrases accurate and clear? (577)
- ☐ Have I documented all sources not considered common knowledge? (578)
- ☐ Are electronic sources cited clearly and appropriately? (587)

EXERCISES

1. Assume that you are an assistant communications manager for a new organization that prepares research reports for decision makers worldwide. (A sample topic: "What effect has the North American Free Trade Agreement had on the U.S. computer industry?") These clients expect answers based on the best available evidence and reasoning.

   Although your recently hired coworkers are technical specialists, few have experience in the kind of wide-ranging research your clients require. Training programs in the research process are being developed by your communications division but will not be ready for several weeks.

   Meanwhile, your boss directs you to prepare a one- or two-page memo that introduces employees to major procedural and reasoning errors that affect validity and reliability in the research process. Your boss wants this memo to be comprehensive but not vague.

2. From print or broadcast media or from personal experience, identify an example of each of the following sources of distortion or of interpretive error:

   - a study with questionable sponsorship or motives
   - reliance on insufficient evidence
   - unbalanced presentation
   - overestimating the level of certainty
   - biased interpretation
   - rationalizing
   - faulty causal reasoning
   - hasty generalization
   - overstated generalization
   - sanitized statistic
   - meaningless statistic
   - undefined average
   - distorted percentage figure
   - bogus ranking
   - fallible computer model
   - misinterpreted statistic
   - deceptive reporting

Submit your examples to your instructor along with a memo explaining each error, and be prepared to discuss your material in class.

3. Referring to the list in Exercise 2, identify the specific distortion or interpretive error in the following examples:

a. *The federal government excludes from unemployment figures an estimated 5 million people who remain unemployed after one year* (Morgenson 54).

b. *Only 38.268 percent of college graduates end up working in their specialty.*

c. *Sixty-six percent of employees we hired this year are women and minorities, compared to the national average of 40 percent.* No mention is made of the fact that only three people have been hired this year, by a company that employs 300 (mostly white males).

d. *Are you pro-life (or pro-choice)?*

4. Identify confounding factors (page 158) that might have been overlooked in the following interpretations and conclusions:

a. *The overall cancer rate today is higher than in 1910* ("Are We" 4). Does this mean the actual incidence of disease has increased or are other explanations for this finding possible?

b. *One out of every five patients admitted to Central Hospital dies* (Sowell 120). Does this mean that the hospital is bad?

c. *In a recent survey, rates of emotional depression differed widely among different countries—far lower in Asian than in western countries* (Horgan 24+). Are these differences due to culturally specific genetic factors, as many scientists might conclude? Or is this conclusion *confounded* by other variables?

d. *"Among 20-year-olds in 1979, those who said that they smoked marijuana 11 to 50 times in the past year had an average IQ 15 percentile points higher than those who said they'd only smoked once"* (Sklarof and Ash 85). Does

this indicate that pot increases brain power or could it mean something else?

e. *Teachers are mostly to blame for low test scores and poor discipline in public schools.* How is our assessment of this claim affected by the following information? *From age 2 to 17, Children in the U.S. average 12,000 hours in school, and 15,000 to 18,000 hours watching TV* ("Wellness Facts" 1).

COLLABORATIVE PROJECTS

1. Exercises from the previous section may be done as collaborative projects.

2. *Evaluating sources.* Figure A.4 (pages 592–94) lists the final sources for the research project discussed early in Chapter 7 (page 98). Review pages 145–48 and then turn to Figure A.4 and evaluate the sources on the basis of these criteria: currency, range, balance, relative objectivity, and reputability. Here are more specific questions:

- Should the sources generally have been more current? Why, or why not?
- Which types of sources are represented here: business, science, trade, or general interest publications; newspapers or newsmagazines; scholarly journals; government reports; or primary sources? Does the range of sources seem adequate (from general to specialized)? Explain.
- How many different viewpoints are represented here (representatives of the company, consumer advocates, independent researchers, print journalists, the media, people in the industry, investors, others)? Do the sources represent a fair balance of views on this controversial issue? Explain.
- Which sources seem most likely to be objective or impartial? Explain.
- Which seem most likely to be biased? Explain.
- Which seem most expert or authoritative? Explain.

- Which seem most comprehensive? Explain.
- Overall, do the sources seem adequate for the topic and situation described in Chapter 7? Explain.

Make the explanations brief but informative enough to justify your evaluation.

Appoint a group manager who will lead the discussion and assign the following tasks: taking notes, reporting the group's evaluation in a memo, editing and revising the memo, orally reporting the evaluation to the class.

3. Assess the findings below and then describe how you would rework the comparison to arrive at a meaningful conclusion.

A 1998 article used colorful graphics to underscore the "gap" between teachers' wages and the "average wages" of other workers. State-by-state, teachers' wages exceeded the "average wage" by a figure ranging from 2.9 percent (in the District of Columbia) to 65.2 percent (in Pennsylvania)—the gap in most states ranging from 20 to 60 percent. These figures were offered as evidence for the claim that teachers are indeed handsomely paid. (Brimelow 51)

Summarizing and Abstracting Information

PURPOSE OF SUMMARIES

Guidelines for Summarizing Information

WHAT USERS EXPECT FROM A SUMMARY

A SITUATION REQUIRING A SUMMARY

FORMS OF SUMMARIZED INFORMATION

PLACEMENT OF SUMMARIZED INFORMATION

Usability Checklist for Summaries

A summary is a concise statement of the main points in a longer document.

PURPOSE OF SUMMARIES

On the job, you have to write concisely about your work. You might report on meetings or conferences, describe your progress on a project, or propose a money-saving idea. A routine assignment for many new employees is to provide decision makers with summaries of the latest developments in their field.

Researchers and users who must act on information need to identify quickly what is most important in a long document. An abstract is a type of summary that does three things: (1) shows what the document is all about; (2) helps users decide whether to read all of it, parts of it, or none of it; and (3) gives users a framework for understanding what follows.

An effective summary communicates the *essential message* accurately and in the fewest words. Consider the following passage:

The original passage

> The lack of technical knowledge among owners of television sets leads to their suspicion about the honesty of television repair technicians. Although television owners might be fairly knowledgeable about most repairs made to their automobiles, they rarely understand the nature and extent of specialized electronic repairs. For instance, the function and importance of an automatic transmission in an automobile are generally well known; however, the average television owner knows nothing about the flyback transformer in a television set. The repair charge for a flyback transformer failure is roughly $150—a large amount to a consumer who lacks even a simple understanding of what the repairs accomplished. In contrast, a $450 repair charge for the transmission on the family car, though distressing, is more readily understood and accepted.

Three ideas make up the essential message: (1) television owners lack technical knowledge and are suspicious of repair technicians; (2) an owner usually understands even the most expensive automobile repairs; and (3) owners do not understand or accept expenses for television repairs. A summary of the above passage might read like this:

A summarized version

> Because television owners lack technical knowledge about their sets, they are often suspicious of repair technicians. Although consumers may understand expensive automobile repairs, they rarely understand or accept repair and parts expenses for their television sets.

NOTE *For letters, memos, or other short documents that can be read quickly, the only summary needed is usually an opening thesis or topic sentence that previews the contents.*

Summaries are vital to people who have no time to read in detail everything that crosses their desks. One recent U.S. president required that all significant world news for the last twenty-four hours be condensed to one page and placed on

GUIDELINES

for
Summarizing
Information

1. *Be considerate of later readers.* Unless you own the book, journal, or magazine, work from a photocopy.

2. *Read the entire original.* When summarizing someone else's work, grasp the complete picture before writing.

3. *Reread and underline.* Identify the issue that led to the article or report. Focus on the essential message: thesis and topic sentences, findings, conclusions, and recommendations.

4. *Pare down your underlined material.* Omit lengthy background, technical details, examples, explanations, and anything not essential to the overall meaning. In summarizing the work of others, avoid quotations; if you must quote some crucial word or phrase directly, use quotation marks.

5. *Rewrite in your own words.* Even if this draft is too long, include everything that seems essential for this version to stand alone; you can trim later.

6. *Edit for conciseness.* Once your draft contains all the material users need, find ways to trim the word count (page 224).

   a. Cross out all needless words—but keep sentences clear and grammatical:

Needless words
omitted

> As far as artificial intelligence is concerned, the technology is only in its infancy.

   b. Cross out needless prefaces:

Needless prefaces
omitted

> The writer argues . . . .
> Also discussed is . . . .

   c. Combine related ideas (page 232) and rephrase to emphasize relationships:

Disconnected and
rambling

> A recent study emphasized job opportunities in the computer field. Fewer of tomorrow's jobs will be for programmers and other people who know how to create technology. More jobs will be for people who can use technology—as in marketing and finance (Ross 206).

Compare this connected and more concise version:

Connected and
concise

> A recent study predicts fewer jobs for programmers and other creators of technology, and more jobs for users of technology—as in marketing and finance (Ross 206).

7. *Check your version against the original.* Verify this version's accuracy and completeness. Add no personal comments—unless you are preparing an executive abstract (page 184)

8. *Rewrite your edited version.* In this final version, strive for readability and conciseness. Add transitional expressions (page 646) to emphasize connections. Respect any stipulated word limit.

9. *Document your source.* Cite the full source below any summary not accompanied by its original (Appendix A).

his desk each morning. Another president employed a writer who summarized articles from more than two dozen major magazines.

WHAT USERS EXPECT FROM A SUMMARY

Whether you summarize your own documents (like the sample on page 331) or someone else's, users will have these expectations:

Elements of a
usable summary

- ◆ *Accuracy:* Users expect a summary to precisely sketch the content, emphasis, and line of reasoning from the original.
- ◆ *Completeness:* Users expect to consult the original document only for more detail—but not to make sense of the main ideas and their relationships.
- ◆ *Readability:* Users expect a summary to be clear and straightforward—easy to follow and understand.
- ◆ *Conciseness:* Users expect a summary to be informative yet brief, and they may stipulate a word limit (say, two hundred words).
- ◆ *Nontechnical style:* Unless they are all experts, users expect plain English.

Although the summary is written last, it is read first. Take the time to do a good job.

A SITUATION REQUIRING A SUMMARY

Assume that you work in the information office of your state's Department of Environmental Management (DEM). In the coming election, citizens will vote on a referendum proposal for constructing the state's first nuclear power plant. Referendum supporters argue that nuclear power would help solve the growing problem of acid rain and global warming from burning fossil fuels. Opponents argue that nuclear power is expensive and unsafe.

To clarify the issues for voters, the DEM is preparing a newsletter to be mailed to each registered voter. You have been assigned the task of researching the recent data on nuclear power and summarizing them for newsletter readers. Here is one of the articles marked up and then summarized according to our guidelines.

U.S. Nuclear Power Industry

Background and Current Status

Combine as
orienting statement
(controlling idea)

Omit background
details
Include causes of
problem

The U.S. nuclear power industry, while currently generating more than 20 percent of the Nation's electricity, faces an uncertain future. No nuclear power plants have been ordered since 1978, and more than 100 reactors have been cancelled, including all ordered after 1973. No units are currently under active construction; the Tennessee Valley Authority's Watts Bar 1 reactor, ordered in 1970 and licensed to operate in 1996, was the last U.S. nuclear unit to be completed. The nuclear power industry's troubles include a slowdown in the rate of growth of electricity demand, high nuclear power plant construction <u>costs, public concern about nuclear safety</u> and waste disposal, and a <u>changing regulatory environment.</u>

Obstacles to Expansion

Include major cause
Omit nonvital details

<u>High construction costs</u> are <u>perhaps</u> the <u>most serious obstacle</u> to nuclear power expansion. Construction costs for reactors completed within the last decade have ranged from $2 billion to $6 billion, averaging about $3,000 per kilowatt of electric generating capacity (in 1995 dollars). The nuclear industry predicts that new plant designs could be

Include key comparison

built for about half that amount, but construction costs would still <u>substantially exceed</u> the projected <u>costs of coal- and gas-fired plants.</u>

Omit speculation

Of more immediate concern to the nuclear power industry is the outlook for existing nuclear reactors in a deregulated electricity market. Electric utility restructuring, which is currently under way in several States, could increase the competition faced by

Include key facts

existing nuclear plants. <u>High operating costs</u> and the need for <u>costly improvements and equipment replacements have resulted in the permanent shutdown</u> during the past decade <u>of 10 U.S.</u> commercial <u>reactors</u> before completion of their 40-year licensed operating periods. <u>Several more</u> reactors are <u>currently</u> being <u>considered for early shutdown.</u>

Include key facts and comparisons

Nevertheless, <u>all is not bleak for the U.S.</u> nuclear power <u>industry,</u> which currently comprises <u>109 licensed reactors</u> at 68 plant sites <u>in 38 States. Electricity production</u> from U.S. nuclear power plants <u>is greater than that from oil, natural gas, and hydropower,</u> <u>and behind only coal, which accounts for approximately 55 percent</u> of U.S. electricity generation. Nuclear plants generate more than half the electricity in six States.

Omit visual

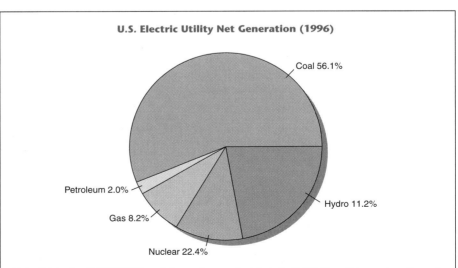

U.S. Electric Utility Net Generation (1996)

Coal 56.1%

Petroleum 2.0%

Gas 8.2%

Nuclear 22.4%

Hydro 11.2%

Note: Total value (2,087,977 kilowatt hours of generation) includes 3,213 kilowatt hours of geothermal generation and 1,245 kilowatt hours of other generation, which represent less than 1 percent of total generation.

Source: *U.S. Department of Energy, Energy Information Administration*

Average operating costs of U.S. nuclear plants have dropped during the 1990s, and costly downtime has been steadily reduced. Licensed commercial reactors generated electricity at an average of 75 percent of their total capacity in 1996, slightly below the previous year's record.

Global warming that may be caused by fossil fuels—the "greenhouse effect"—is cited by nuclear power supporters as an important reason to develop a new generation of reactors. But the large obstacles noted above must still be overcome before electric utilities will order new nuclear units.

Reactor manufacturers are working on designs for safer, less expensive nuclear plants, and the Nuclear Regulatory Commission (NRC) has approved new regulations to speed up the nuclear licensing process, consistent with the Energy Policy Act of 1992. Even so, the Energy Information Administration forecasts that no new U.S. reactors will become operational before 2010, if any are ordered at all.

Safety Concerns

Controversy over safety has dogged nuclear power throughout its development, particularly following the Three Mile Island accident in Pennsylvania and the April 1986 Chernobyl disaster in the former Soviet Union. In the United States, safety-related shortcomings have been identified in the construction quality of some plants, plant operation and maintenance, equipment reliability, emergency planning, and other areas. In addition, mishaps have occurred in which key safety systems have been disabled. NRC's oversight of the nuclear industry is an ongoing issue; nuclear utilities often complain that they are subject to overly rigorous and inflexible regulation, but nuclear critics charge that NRC frequently relaxes safety standards when compliance may prove difficult or costly to the industry.

In terms of public health consequences, the safety record of the U.S. nuclear power industry has been excellent. In more than 2,000 reactor-years of operation in the United States, the only incident at a commercial power plant that might lead to any deaths or injuries to the public has been the Three Mile Island accident, in which more than half the core melted. Public exposure to radioactive materials released during that accident is expected to cause fewer than five deaths (and perhaps none) from cancer over the following 30 years. An independent study released in September 1990 found no "convincing evidence" that the Three Mile Island accident had affected cancer rates in the area around the plant. However, a study released in February 1997 concluded that much higher levels of radiation may have been released during the accident than was previously believed.

The relatively small amounts of radioactivity released by nuclear plants during normal operation are not generally believed to pose significant hazards. Documented public exposure to radioactivity from nuclear power plant waste has also been minimal, although the potential long-term hazard of waste disposal remains controversial. There is substantial scientific uncertainty about the level of risk posed by low levels of radiation exposure; as with many carcinogens and other hazardous substances, health effects can be clearly measured only at relatively high exposure levels. In the case of radiation, the assumed risk of low-level exposure has been extrapolated mostly from health effects documented among persons exposed to high levels of radiation, particularly Japanese survivors of nuclear bombing.

Margin annotations (left column):

Include key fact
Omit nonvital details

Include key claim

Omit explanation

Include key fact

Omit nonvital fact
Include key fact

Include key facts
Omit examples

Include key claims

Include key fact

Include striking exception

Omit long explanation

Omit speculation

Include key issue

Omit explanation

Include key claim

Omit nonvital details

Include key claim

The <u>consensus among most safety experts is that a severe nuclear power plant accident in the United States is likely to occur less frequently than one every 10,000 reactor-years</u> of operation. These experts believe that most severe accidents would have small public health impacts, and that accidents causing as many as 100 deaths would be much rarer than once every 10,000 reactor-years. On the other hand, <u>some experts challenge the complex calculations that go into predicting such accident frequencies,</u> contending that accidents with serious public health consequences may be more frequent.

Regulation

Omit long explanation

For many years, a top priority of the nuclear industry was to modify the process for licensing new nuclear plants. No electric utility would consider ordering a nuclear power plant, according to the industry, unless licensing became quicker and more predictable, and designs were less subject to mid-construction safety-related changes ordered by NRC. The Energy Policy Act of 1992 largely implemented the industry's goals.

Nuclear plant licensing under the Atomic Energy Act of 1954 had historically been a two-stage process. NRC first issued a construction permit to build a plant and then, after construction was finished, an operating permit to run it. Each stage of the licensing process involved complicated proceedings. Environmental impact statements also are required under the National Environmental Policy Act.

Over the vehement objections of nuclear opponents, the Energy Policy Act provides a clear statutory basis for one-step nuclear licenses, allowing completed plants to operate without delay if construction criteria are met. NRC would hold preoperational hearings on the adequacy of plant construction only in specified circumstances.

Include key claims

<u>A fundamental concern</u> in the nuclear regulatory debate <u>is</u> the <u>performance of NRC in issuing and enforcing</u> nuclear <u>safety regulations. The</u> nuclear <u>industry and</u> its <u>supporters</u> have regularly <u>complained</u> that unnecessarily <u>stringent and inflexibly enforced</u> nuclear safety <u>regulations have burdened</u> nuclear <u>utilities</u> and their <u>customers with excessive costs. But many</u> environmentalists, nuclear <u>opponents,</u> and other groups <u>charge NRC with</u> being too close to the nuclear industry, a situation that they say has resulted in <u>lax oversight</u> of nuclear power plants and routine exemptions from safety requirements.

Omit explanation

Primary responsibility for nuclear safety compliance lies with nuclear utilities, which are required to find any problems with their plants and report them to NRC. Compliance is monitored directly by NRC, which maintains at least two resident inspectors at each nuclear power plant. The resident inspectors routinely examine plant systems, observe the performance of reactor personnel, and prepare regular inspection reports. For serious safety violations, NRC often dispatches special inspection teams to plant sites.

Decommissioning and Life Extension

Include key fact
Omit nonvital details

Include key fact

<u>When</u> nuclear power <u>plants end their useful lives, they must be safely removed from service,</u> a process called <u>decommissioning.</u> NRC requires nuclear utilities to make regular contributions to special trust funds to ensure that money is available to remove all radioactive material from reactors after they closed. Because <u>no full-sized U.S.</u> commercial <u>reactor has</u> yet <u>been completely decommissioned,</u> which <u>can take</u> several <u>decades,</u> the cost of the process can only be estimated. Decommissioning <u>cost estimates</u>

Include striking cost
figure
Omit speculation

cited by a 1996 Department of Energy report, for one full-sized commercial reactor, ranged from about $150 million to $600 million in 1995 dollars.

It is assumed that U.S. commercial reactors could be decommissioned at the end of their 40-year operating licenses, although several plants have been retired before their licenses expired and others could seek license renewals to operate longer. NRC rules allow plants to apply for a 20-year license extension, for a total operating time of 60 years. Assuming a 40-year lifespan, more than half of today's 109 licensed reactors could be decommissioned by the year 2016.

Source: *Congressional Digest* Jan. 1998: 7+.

Assume that in two early drafts of your summary, you rewrote and edited; for coherence and emphasis, you inserted transitions and combined related ideas. Here is your final draft.

A SUMMARY

U.S. Nuclear Power Industry: Background and Current Status

Although nuclear power generates more than 20 percent of U.S. electricity, no plants have been ordered since 1978, orders dating to 1973 are cancelled, and no units are now being built. Cost, safety, and regulatory concerns have led to zero growth in the industry.

Nuclear plant construction costs far exceed those for coal- and gas-fired plants. Also, high operating and equipment costs have forced permanent, early shutdown of 10 reactors, and the anticipated shutdown of several more.

On the positive side, the 109 licensed reactors in 38 states produce roughly 22 percent of the nation's electricity—more than oil, natural gas, and hydropower combined, and second only to coal, which produces roughly 55 percent. Moreover, nuclear power is cleaner than fossil fuels. Yet, despite declining costs and safer, less expensive designs, no new reactors could come online earlier than 2010—if any were ordered.

Safety concerns persist about plant construction, operation, and maintenance, as well as equipment reliability, emergency planning, and NRC's (Nuclear Regulatory Commission) oversight of the industry. Scientists disagree over the extent of long-term hazards from low-level emissions during plant operation and from waste disposal.

Except for the 1979 partial meltdown at Three Mile Island, however, the U.S. nuclear power industry has an excellent safety record for more than 2,000 reactor-years of operation. Most experts estimate that a severe nuclear accident in the United States will occur less than once every 10,000 reactor-years, but other experts are less optimistic.

Central to the nuclear power controversy is the NRC's role in policing the industry and enforcing safety regulations. Industry supporters claim that overregulation has created excessive costs. But opponents charge the NRC with lax oversight and enforcement.

One final unknown involves "decommissioning": safely closing down an aging power plant at the end of its 40-year operating life, a lengthy process expected to cost $150 million to $600 million per reactor.

Source: *Congressional Digest* Jan. 1998: 7+.

The version above is trimmed, tightened, and edited: word count is reduced to less than twenty percent of original length. A summary this long serves well in many situations, but other audiences might want a briefer and more compressed summary—say, roughly fifteen percent of the original:

A MORE COMPRESSED SUMMARY

Although nuclear power generates more than 20 percent of U.S. electricity, cost, safety, and regulatory concerns have led to zero growth in the industry. Moreover, operating and equipment costs are forcing many permanent, early shutdowns.

On the positive side, nuclear reactors generate more of the nation's electricity than all other fossil fuels except coal—and with far less pollution. Yet, despite declining operating costs and safer, less expensive designs, no new reactors could come online earlier than 2010—if any were ordered.

Safety concerns persist about plant construction, operation, and maintenance as well as equipment reliability, emergency planning, and NRC (Nuclear Regulatory Commission) oversight. Scientists disagree over the probability of a severe accident and the long-term hazards from normal, low-level emissions or from waste disposal. Except for the 1979 partial meltdown at Three Mile Island, however, the U.S. industry's safety record remains excellent.

Also controversial is the NRC's role in policing and enforcement. Industry supporters claim that excessive regulation has created excessive costs. But opponents charge the NRC with lax oversight and enforcement.

Finally, "decommissioning," safely closing down an aging power plant at the end of its operating life, is a lengthy and costly process.

Notice that the essential message remains intact; related ideas are again combined and fewer supporting details are included. Clearly, length is adjustable according to your audience and purpose.

FORMS OF SUMMARIZED INFORMATION

In preparing a report, proposal, or other document, you might summarize works of others as part of your presentation. But you often summarize your own presentations as well. For instance, if your document extends to several pages, you usually include, near the end or the beginning, a summary. Depending on its location and level of detail, this summarized information takes one of these four forms: *closing summary, informative abstract, descriptive abstract,* or *executive abstract*[1] (Figure 11.1).

The Closing Summary

Summarized information at the end of a document's body helps users review and remember the main points or major findings from the presentation. This look

---

[1]Adapted from Vaughan. Although I take liberties with his classification, Vaughan helped clarify my thinking about the overlapping terminology that blurs these distinctions.

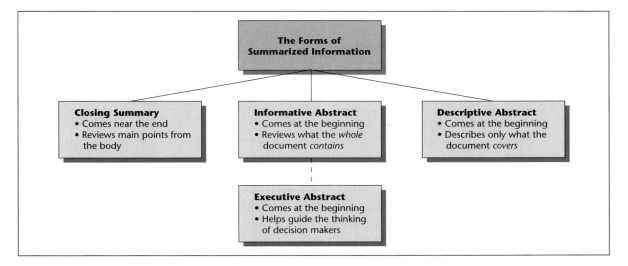

Figure 11.1   Summarized Information Takes Various Forms

back at "the big picture" helps users appreciate and understand any conclusions and recommendations that follow. (See pages 538 and 549 for examples.)

The Informative Abstract

In addition to a closing summary, you might include an opening summary to show users what the document is all about, to help them decide whether to read it, and to give them an orientation—a framework for understanding what follows. This opening summary is called an informative abstract, partly to differentiate it from its closing counterpart and partly because it is different: besides simply reviewing main points or major findings, the informative abstract identifies the issue or need that led to the report and includes condensed conclusions and recommendations. (See page 331 for an example.)

The Descriptive Abstract

As noted above, the informative abstract *explains what the original document contains* (its origins, findings, and conclusions). But another, more compressed form of summarized information can precede a document: the descriptive abstract, which simply *describes what the original is about* (its subject).

A descriptive abstract, then, conveys only the nature and extent of the document. It presents the broadest view and offers no major facts from the original.

Compare, for example, the abstract that follows with the article summary on page 181:

A DESCRIPTIVE ABSTRACT

U.S. Nuclear Power Industry: Background and Current Status

The track record of the U.S. nuclear power industry is examined and reasons for its lack of growth are identified and assessed.

Because they tend to focus on methodology rather than results, descriptive abstracts are used most often in the sciences and social sciences.[2]

On the job, you might prepare informative abstracts for a boss who needs the information but who has no time to read the original. Or you might write descriptive abstracts to accompany a bibliography of works you are recommending to colleagues or clients (an annotated bibliography).

The Executive Abstract

A special type of informative abstract, the executive abstract (or "executive summary"), generally has more of a persuasive emphasis: namely, to convince the reader to act. Executive abstracts are crucial to decision making—in cases when readers have no time to read the entire original document and when they expect the writer to help guide their thinking. ("Tell me how to think about this," instead of, "Help me understand this.") Unless the user stipulates a specific format, organize your executive abstract to answer these questions:

Users of an executive abstract have these questions

- ◆ *What did you find?*
- ◆ *What does it mean?*
- ◆ *What should be done?*

The following executive abstract addresses the problem of falling sales for a leading company in the breakfast cereal industry (Grant 223+).

Status Report: Market Share for Goldilocks Breakfast Cereals, Inc. (GBC)

Executive Abstract

In response to a request from GBC's Board of Directors, the accounting division analyzed recent trends in the company's sales volume and profitability.

---

[2]My thanks to Daryl Davis for this clarifying distinction.

Findings

"What did you
find?"

- Even though GBC is the cereal industry leader, its sales from 1995 to 1999 increased at a mere average of 2.5 percent annually, to $5.2 billion, and net income decreased 12 percent overall, to $459 million.
- This weak sales growth apparently results from consumer resistance to retail price increases for cereal, totalling 91 percent from 1985 to 1998, the highest increase of any processed-food product.
- GBC traditionally offers discount coupons to offset price increases, but consumers seem to prefer a lower everyday price.
- GBC introduces an average of two new cereal products annually (most recently, "Coconut Whammos" and "Spinach Crunchies"), but such innovations do little to increase consumer interest.
- A growing array of generic cereal brands have been underselling GBC's products by more than $1 per box, especially in giant retail outlets.
- In June 1998, GBC dropped its cereal prices by roughly 20 percent, but by this time, the brand had lost substantial market share to generic cereal brands.

Conclusions

"What does it
mean?"

- Slow but progressive loss of market share threatens GBC's dominance as industry leader.
- GBC must regain consumer loyalty to reinvigorate its market base.
- Not only have discount coupon promotions proven ineffective, but the manufacturer's cost for such promotions can total as much as 20 percent of sales revenue.
- New cereal products have done more to erode than to enhance GBC's brand image.

Recommendations

"What should be
done?"

To regain lost market share and ensure continued dominance, GBC should implement the following recommendations:

1. Eliminate coupon promotions immediately.
2. Curtail development of new cereal products, and invest in improving the taste and nutritional value of GBC's traditional products.
3. Capitalize on GBC's brand recognition with an advertising campaign to promote GBC's "best-sellers" as an "all-day" food (say, as a healthful snack or lunch or an easy and inexpensive alternative to microwave dinners).
4. Examine the possibility of high-volume sales at discounted prices through giant retail chains.

PLACEMENT OF SUMMARIZED INFORMATION

Different companies have different report design requirements, but these general guidelines serve in many situations:

Placing summarized
information in a
long report

- Place the closing summary in the concluding section of the report, usually preceding any conclusions and recommendations.
- Place a descriptive abstract single-spaced on the lower third of the title page.

Usability Checklist FOR SUMMARIES

Use this checklist to refine your summaries. (Page numbers in parentheses refer to first page of discussion.)

Content

☐ Does the summary contain only the essential message? (175)
☐ Does the summary make sense as an independent piece? (177)
☐ Is the summary accurate when checked against the original? (176)
☐ Is the summary free of any additions to the original? (176)
☐ Is the summary free of needless details? (176)
☐ Is the summary economical, yet clear and comprehensive? (177)
☐ Is the source documented? (176)

☐ Does the descriptive abstract tell what the original is about? (183)

Organization

☐ Is the summary coherent? (176)
☐ Are there enough transitions to reveal the line of thought? (176)

Style

☐ Is the summary's level of technicality appropriate for users? (177)
☐ Is the summary free of needless words? (176)
☐ Are all sentences clear, concise, and fluent? (215)
☐ Is the summary written in correct English? (Appendix C)

• Place an informative or executive abstract on its own separate page, immediately following the table of contents.

Reports and proposals in later chapters illustrate these various placement options.

EXERCISES

1. Read each of these two paragraphs, and then list the significant ideas that make up each essential message. Write a summary of each paragraph.

   In recent years, ski-binding manufacturers, in line with consumer demand, have redesigned their bindings several times in an effort to achieve a noncompromising synthesis between performance and safety. Such a synthesis depends on what appear to be divergent goals. Performance, in essence, is a function of the binding's ability to hold the boot firmly to the ski, thus enabling the skier to rapidly change the position of his or her skis without being hampered by a loose or wobbling connection. Safety, on the other

hand, is a function of the binding's ability both to release the boot when the skier falls, and to retain the boot when subjected to the normal shocks of skiing. If achieved, this synthesis of performance and safety will greatly increase skiing pleasure while decreasing accidents.

Contrary to public belief, sewage treatment plants do not fully purify sewage. The product that leaves the plant to be dumped into the leaching (sievelike drainage) fields is secondary sewage containing toxic contaminants such as phosphates, nitrates, chloride, and heavy metals. As the secondary sewage filters into the ground, this conglomeration is carried along. Under the leaching area develops a contaminated mound through which groundwater flows, spreading the waste products over great distances. If this leachate reaches the outer limits of a well's drawing radius, the water supply becomes polluted. And because all water flows essentially toward the sea, more pollution is added to the coastal regions by this secondary sewage.

2. Attend a campus lecture on a topic of interest and take notes on the significant points. Write a summary of the lecture's essential message.

3. Find an article about your major field or area of interest and write both an informative abstract and a descriptive abstract of the article.

4. Select a long paper you have written for one of your courses; write an informative abstract and a descriptive abstract of the paper.

5. After reading the article in Figure 11.2 prepare a descriptive abstract and an informative abstract, using the guidelines on page 176. Identify a specific audience and use for your material.

   A possible scenario: You are assistant communications manager for a leading software development company. Part of your job involves publishing a monthly newsletter for employees. After coming across this article, you decide to summarize it for the upcoming issue. (Aspirin is a popular item in this company, given the headaches, stiff necks, and other medical problems that often result from prolonged computer work.) You have 350–375 words of newsletter space to fill. Consider carefully what this audience does and doesn't need. In this situation, what information is most important?

   Bring your abstracts to class and exchange them with a classmate for editing according to the revision checklist. Revise your edited copies before submitting them to your instructor.

COLLABORATIVE PROJECT

Organize into small groups and choose a topic for discussion: an employment problem, a campus problem, plans for an event, suggestions for energy conservation, or the like. (A possible topic: Should employers have the right to require lie detector tests, drug tests, or AIDS tests for their employees?)

Discuss the topic for one class period, taking notes on significant points and conclusions. Afterward, organize and edit your notes in line with the directions for writing summaries. Next, write a summary of the group discussion in no more than 200 words. Finally, as a group, compare your individual summaries for accuracy, emphasis, conciseness, and clarity.

ASPIRIN
A New Look at an Old Drug
by Ken Flieger

Americans consume an estimated 80 billion aspirin tablets a year. The *Physician's Desk Reference* lists more than 50 over-the-counter drugs in which aspirin is the principal active ingredient. Yet, despite aspirin's having been in routine use for nearly a century, both scientific journals and the popular media are full of reports and speculation about new uses for this old remedy.

Almost a century after its development aspirin is the focus of extensive laboratory research and some of the largest clinical trials ever carried out in conditions ranging from cardiovascular disease and cancer to migraine headache and high blood pressure in pregnancy.

How Does It Work?

The mushrooming interest in aspirin has come about largely because of fairly recent advances in understanding how it works. What is it about this drug that, at small doses, interferes with blood clotting, at somewhat higher doses reduces fever and eases minor aches and pains, and at comparatively large doses combats pain and inflammation in rheumatoid arthritis and several other related diseases?

The answer is not yet fully known, but most authorities agree that aspirin achieves some of its effects by inhibiting the production of prostaglandins. Prostaglandins are hormone-like substances that influence the elasticity of blood vessels, control uterine contractions, direct the functioning of blood platelets that help stop bleeding, and regulate numerous other activities in the body.

In the 1970s, a British pharmacologist, John Vane, Ph.D., noted that many forms of tissue injury were followed by the release of prostaglandins. In laboratory studies, he found that two groups of prostaglandins caused redness and fever, common signs of inflammation. Vane and his co-workers also showed that by blocking the synthesis of prostaglandins, aspirin prevented blood platelets from aggregating, one of the initial steps in the formation of blood clots.

This explanation of how aspirin and other nonsteroidal anti-inflammatory drugs (NSAIDs) produce their intriguing array of effects prompted laboratory and clinical scientists to form and test new ideas about aspirin's possible value in treating or preventing conditions in which prostaglandins play a role. Interest quickly focused on learning whether aspirin might prevent the blood clots responsible for heart attacks.

A heart attack or myocardial infarction (MI) results from the blockage of blood flow not *through* the heart, but *to* heart muscle. Without an adequate blood supply, the affected area of muscle dies and the heart's pumping action is either impaired or stopped altogether.

The most common sequence of events leading to an MI begins with the gradual build-up of plaque (atherosclerosis) in the coronary arteries. Circulation through these narrowed arteries is restricted, often causing the chest pain known as angina pectoris.

An acute heart attack is believed to happen when a tear in plaque inside a narrowed coronary artery causes platelets to aggregate, forming a clot that blocks the flow of blood. About 1,250,000 persons suffer heart attacks each year in the United States, and some 500,000 of them die. Those who survive a first heart attack are at greatly increased risk of having another.

Could Aspirin Help?

To learn whether aspirin could be helpful in preventing or treating cardiovascular disease, scientists have carried out numerous large randomized controlled clinical trials. In these studies, similar groups of hundreds or thousands of people are randomly assigned to receive either aspirin or a placebo, an inactive, look-alike tablet. The participants— and in double-blind trials the investigators, as well—do not know who is taking aspirin and who is swallowing a placebo.

Over the last two decades, aspirin studies have been conducted in three kinds of individuals: persons with a history of coronary artery or cerebral vascular disease, patients in the immediate, acute phases of a heart attack, and healthy men with no indication of current or previous cardiovascular illness.

The results of studies of people with a history of coronary artery disease and those in the immediate phases of a heart attack have proven to be of tremendous importance in the prevention and treatment of cardiovascular disease. The studies showed that aspirin substantially reduces the risk of death and/or non-fatal heart attacks in patients with a previous MI or unstable angina pectoris, which often occurs before a heart attack.

On the basis of such studies, these uses for aspirin (unstable angina, acute MI, and survivors of an MI) are described in the professional labeling of aspirin products, information provided to physicians and other

Figure 11.2   An Article to Be Summarized   *Source: Excerpt from* FDA Consumer *Jan./Feb. 1994: 19–21.*

health professionals. Aspirin labeling intended for the general public does not discuss its use in arthritis or cardiovascular disease because treatment of these serious conditions—even with a common over-the-counter drug—has to be medically supervised. The consumer labeling contains a general warning about excessive or inappropriate use of aspirin, and specifically warns against using aspirin to treat children and teenagers who have chickenpox or the flu because of the risk of Reye syndrome, a rare but sometimes fatal condition.

Aspirin for Healthy People?
Once aspirin's benefits for patients with cardiovascular disease were established, scientists sought to learn whether regular aspirin use would prevent a first heart attack in healthy individuals. The findings regarding that critical question have thus far been equivocal. The major American study designed to find out if aspirin can prevent cardiovascular deaths in healthy individuals was a randomized, placebo-controlled trial involving just over 22,000 male physicians between 40 and 84 with no prior history of heart disease. Half took one 325-milligram aspirin tablet every other day, and half took a placebo.

The trial was halted early, after about four-and-a-half years, and the findings quickly made public in 1988 when investigators found that the group taking aspirin had a substantial reduction in the rate of fatal and non-fatal heart attacks compared with the placebo group. There was, however, no significant difference between the aspirin and placebo groups in number of strokes (aspirin-treated patients did slightly worse) or in overall deaths from cardiovascular disease.

A similar study in British male physicians with no previous heart disease found no significant effect nor even a favorable trend for aspirin on cardiovascular disease rates. The British study of 5,100 physicians, while considerably smaller than the American study, reported three-quarters as many vascular "events." FDA scientists believe the results of the two studies are inconsistent.

The U.S. Preventive Services Task Force, a panel of medical-scientific authorities in health promotion and disease prevention, is one of many groups looking at new information on the role of aspirin in cardiovascular disease. In its *Guide to Clinical Preventive Services,* issued in 1989, the task force recommended that low-dose aspirin therapy "should be considered for men aged 40 and over who are at significantly increased risk for myocardial infarction and who lack contraindications" to aspirin use. A revised *Guide,* scheduled for publication in the fall of 1994, is expected to include a slightly revised recommendation concerning aspirin and cardiovascular disease but no major change in advice to physicians about aspirin's possible role in preventing heart attacks.

Better understanding of aspirin's myriad effects in the body has led to clinical trials and other studies to assess a variety of possible uses: preventing the severity of migraine headaches, improving circulation to the gums thereby arresting periodontal disease, preventing certain types of cataracts, lowering the risk of recurrence of colorectal cancer, and controlling the dangerously high blood pressure (called preeclampsia) that occurs in 5 to 15 percent of pregnancies.

None of these uses for aspirin has been shown conclusively to be safe and effective, and there is concern that people may be misusing aspirin on the basis of unproven notions about its effectiveness. Last October, FDA proposed a new labeling statement for aspirin products advising consumers to consult a doctor before taking aspirin for new and long-term uses. The proposed statement would read. "IMPORTANT: See your doctor before taking this product for your heart or for other new uses of aspirin because serious side effects could occur with self-treatment."

The Other Side of the Coin
While examining new possibilities for aspirin in disease treatment and prevention, scientists do not lose sight of the fact that even at low doses aspirin is not harmless. A small subset of the population is hypersensitive to aspirin and cannot tolerate even small amounts of the drug. Gastrointestinal distress—nausea, heartburn, pain—is a well-recognized adverse effect and is related to dosage. Persons being treated for rheumatoid arthritis who take large daily doses of aspirin are especially likely to experience gastrointestinal side effects.

Aspirin's antiplatelet activity apparently accounts for hemorrhagic strokes, caused by bleeding into the brain, in a small but significant percentage of persons who use the drug regularly. For the great majority of occasional aspirin users, internal bleeding is not a problem. But aspirin may be unsuitable for people with uncontrolled high blood pressure, liver or kidney disease, pep-tic ulcer, or other conditions that might increase the risk of cerebral hemorrhage or other internal bleeding.

New understanding of how aspirin works and what it can do leaves no doubt that the drug has a far broader range of uses than imagined [nearly a century ago]. The jury is still out, however, on a number of key questions about the best and safest ways to use aspirin. And until some critical verdicts are handed down, consumers are well-advised to regard aspirin with appropriate caution.

Figure 11.2   An Article to Be Summarized *Continued*

PART

III

Structural and Style Elements

Organizing
for Users

PARTITIONING AND CLASSIFYING

OUTLINING

STORYBOARDING

PARAGRAPHING

SEQUENCING

Our thinking rarely occurs in a neat, predictable sequence, but we cannot report our ideas in the same random order in which they occur. Instead of forcing users to organize unstructured information themselves, we shape this material for their understanding. In setting out to organize, we face questions like these:

TYPICAL QUESTIONS IN ORGANIZING FOR USERS

- *What relationships do these data suggest?*
- *What do I want to emphasize?*
- *In which sequence will users approach this material?*

- *What belongs where?*
- *What do I say first?*
- *What comes next?*
- *How do I end?*

To answer these questions, we rely on the organizing strategies discussed below.

PARTITIONING AND CLASSIFYING

Partition and classification are both strategies for sorting things out. *Partition* deals with *one thing only*. It divides that thing into parts, chunks, sections, or categories for closer examination (say, a report divided into introduction, body, and conclusion).

Partition answers these user questions

- *What are its parts?*
- *What is it made of?*

Classification deals with *an assortment of things* that share certain similarities. It groups these things systematically (for example, grouping electronic documents into categories—reports, memos, Web pages).

Classification answers these user questions

- *What relates to what?*
- *What belongs where?*

Whether you choose to apply partition or classification depends on your purpose. For example, to describe a personal computer system to a novice, you might partition the system into *CPU, keyboard, printer, power cord,* and so on; for a seasoned user who wants to install an expansion card, you might partition the system into *processor-direct slot, video-in slot, communication slot,* and so on. On the other hand, if you have twenty-five software programs to arrange so you can easily locate the one you want, you will need to group them into smaller categories. You might want to classify programs according to function (*word processing, graphics, database management*) or according to expected frequency of use or relative ease of use—or some other basis.

Close examination of any complex problem usually requires both partition and classification, as in this example:

DATA IN RANDOM FORM

While researching the health effects of electromagnetic fields (EMFs), you encounter information about various radiation sources; ratio of risk to level of exposure; workplace studies; lab studies of cell physiology, biochemistry, and behavior; statistical studies of diseases in certain populations; conflicting expert views; views from local authorities, and so on. ▲

Figure 12.1 shows how classification might organize this random collection of EMF data into manageable categories. (Note that many of the Figure 12.1 categories might be divided further into subcategories, such as *kitchen sources, work-*

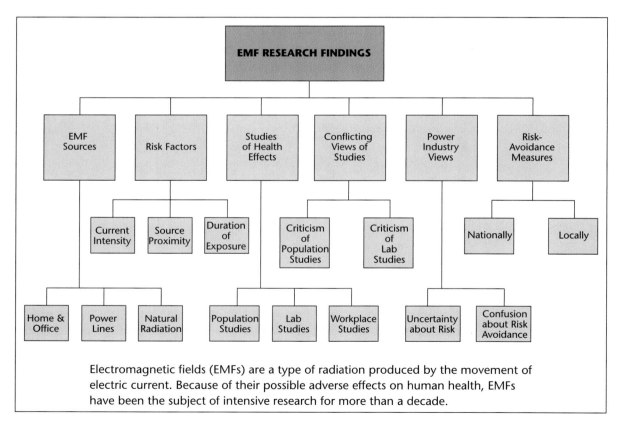

Figure 12.1   Assorted Items Classified by Category

shop sources, bedroom sources, and so on.) Figure 12.2 shows how partition might reveal the parts of a single concept (the electromagnetic spectrum).

In organizing documents, writers use partition and classification routinely, in a process we know as *outlining.*

OUTLINING

With an outline you move from a random listing of items as they occurred to you to a deliberate map that will guide users from point to point. Organize in the sequence in which you expect readers to approach the material.

A Document's Basic Shape

How should you organize to make the document logical from the user's point of view? Begin with the basics. Useful writing of any length—a book, chapter, news article, letter, or memo—typically follows this pattern:

Workplace documents often display this basic shape

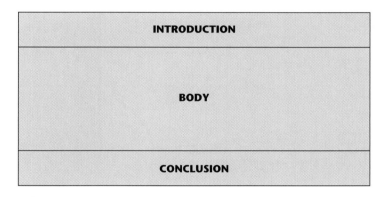

The shape of useful writing

- The *introduction* provides orientation by doing any of these things: explaining the topic's origin and significance and the document's purpose; briefly identifying your intended audience and your information sources; defining specialized terms or general terms that have special meanings in your document; accounting for limitations such as incomplete or questionable data; previewing the major topics to be discussed in the body.

  Some introductions need to be long and involved; others, short and sweet. If you don't know the users well enough to give only what they need, use subheadings so they can choose where to focus.
- The *body* delivers on the promise implied in your introduction ("Show me!"). Here you present your data, discuss your evidence, lay out your case, or tell users what to do and how to do it. Body sections come in all different sizes, depending on how much users need and expect.

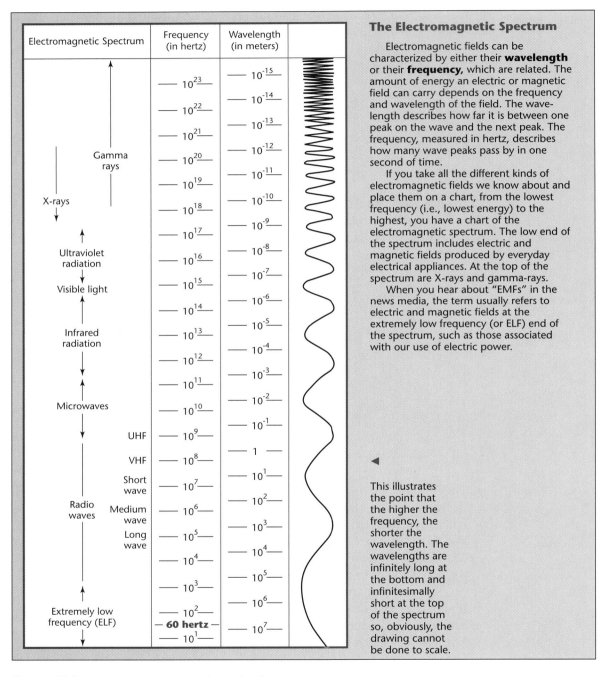

Figure 12.2   One Item Partitioned into Its Components   *Source: U.S. Environmental Protection Agency,* EMF in Your Environment. *Washington, DC: GPO, 1992: 4–5.*

Body sections are titled to reflect their specific purpose: "Description and Function of Parts," for a mechanism description; "Required Steps," for a set of instructions; "Collected Data," for a feasibility analysis.

♦ The *conclusion* has assorted purposes: it might evaluate the significance of the report, reemphasize key points, take a position, predict an outcome, offer a solution, or suggest further study. If the issue is straightforward, the conclusion might be brief and definite. If the issue is complex or controversial, the conclusion might be lengthy and open-ended. A useful conclusion gives readers a clear perspective on the whole document.

Conclusions vary with the document. You might conclude a mechanism description by reviewing the mechanism's major parts and then briefly describing one operating cycle. You might conclude a comparison or feasibility report by offering judgments about the facts you've presented and then recommending a course of action.

NOTE   *A suitable beginning, middle, and ending are essential, but alter your own outline as you see fit. No single form of outline should be followed slavishly. The organization of any document is ultimately determined by the user's needs and expectations. In many cases, specific requirements about a document's organization and style are spelled out in a company's style guide.*

The computer is especially useful for rearranging outlines until they reflect the sequence in which you expect users to approach your message.

The Formal Outline

A simple list usually suffices for organizing short documents or as a tentative outline for longer documents. An author or team rarely begins by developing a formal outline when planning a manuscript. But at some stage (often a *latter* stage) in the writing process, a long, complex document usually calls for a more systematic, formal outline. Here is a formal outline for the report examining the health effects of electromagnetic fields (pages 530–39):

A formal outline using alphanumeric notation

CHILDREN EXPOSED TO EMFs: A RISK ASSESSMENT
I. INTRODUCTION
   A. Definition of electromagnetic fields
   B. Background on the health issues
   C. Description of the local power line configuration
   D. Purpose of this report
   E. Brief description of data sources
   F. Scope of this inquiry
II. DATA SECTION [Body]
   A. Sources of EMF exposure
      1. power lines
      2. home and office

        a. kitchen
        b. workshop [and so on]
      3. natural radiation
      4. risk factors
        a. current intensity
        b. source proximity
        c. duration of exposure
    B. Studies of health effects
      1. population surveys
      2. laboratory measurements
      3. workplace links
    C. Conflicting views of studies
      1. criticism of methodology in population studies
      2. criticism of overgeneralized lab findings
    D. Power industry views
      1. uncertainty about risk
      2. confusion about risk avoidance
    E. Risk-avoidance measures
      1. nationally
      2. locally
III. CONCLUSION
    A. Summary and overall interpretation of findings
    B. Recommendations

NOTE   *Long reports often begin directly with a statement of purpose. For the intended audience (i.e., generalists) of this report, however, the technical topic must first be defined so that users clearly understand the context. Also, each level of division yields at least two items. If you cannot divide a major item into at least two subordinate items, retain only your major heading.*

A formal outline easily converts to a table of contents for the finished report, as shown in Chapter 16.

NOTE   *Because they serve mainly to guide the writer, minor outline headings (such as a and b under II.A.1 above) may be omitted from the table of contents or the report itself. Excessive headings make a document seem fragmented.*

In technical documents, the alphanumeric notation shown above often is replaced by decimal notation:

Decimal notation in
a technical
document

2.0 DATA SECTION
    2.1 Sources of EMF Exposure
      2.1.1 home and office
        2.1.1.1 kitchen
        2.1.1.2 workshop [and so on]

```
        2.1.2 power lines
        2.1.3 natural radiation
        2.1.4 risk factors
              2.1.4.1 current intensity
              2.1.4.2 source proximity
              2.1.4.3 [and so on]
```

The decimal outline makes it easier to refer users to specifically numbered sections of the document ("See 2.1.2"). While both systems achieve the same organizing objective, decimal notation is usually preferred in business, government, and industry.

In some cases, you may wish to expand your *topic outline* into a *sentence outline,* in which each sentence serves as a topic sentence for a paragraph in the report:

A sentence outline

```
2.0  DATA SECTION
     2.1  Although the 2 million miles of power lines crisscrossing the United States have
          been the focus of the EMF controversy, potentially harmful waves also are emit-
          ted by household wiring, appliances, electric blankets, and computer terminals.
          2.1.1 [and so on]
```

Sentence outlines are used mainly in collaborative projects in which various team members prepare different sections of a long document.

NOTE *A survey of the influence of computers on workplace writing found that traditional, formal outlining was giving way to outlining in the form of "notes on audience, purpose, direction, key content points, tone." These outlines were "flexible, sketchy, punctuated by arrows, numbers, or exclamation points; they looked more like lists" (Halpern 179). Outlines needn't be pretty, as long as they help you control your material.*

Not until the final draft of a long document do you compose the finished outline, which serves as a model for your table of contents, as a check on your reasoning, and as a way of revealing to users a clear line of thinking.

Outlining and Reorganizing on a Computer

A word-processing program such as *Microsoft Word* enables you to work on your document and your outline simultaneously. With an "outline view" of the document you can see relationships among ideas at various levels, create new headings, add text beneath headings, and move headings and their subtext (Figure 12.3). You can also *collapse* the outline view to display the headings only.

Switch between "normal view" (to work on your text) and "outline view" (to examine the arrangement of material). You can add or delete headings or text and reorganize whole sections of your document. (*Microsoft Word* 504–05)

As a visual alternative to traditional outlining, many computer graphics programs display prose outlines as tree diagrams.

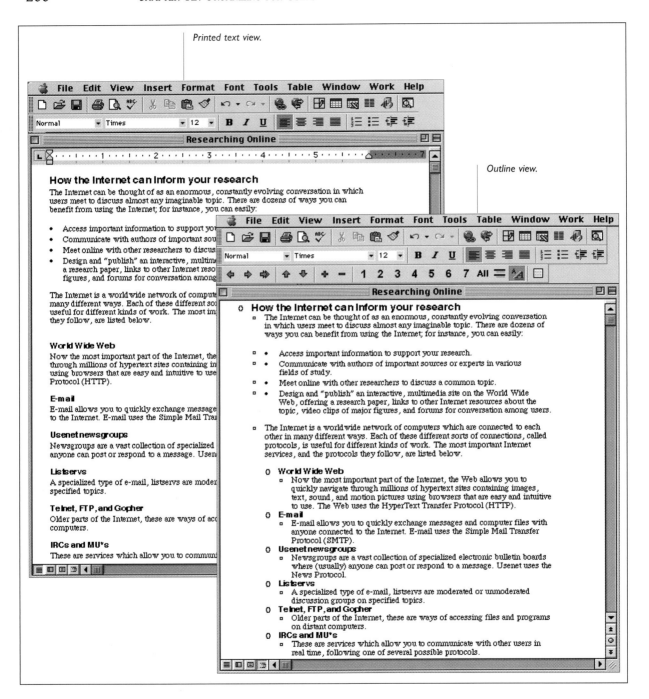

Figure 12.3   Using a Computer's Outline Features   *Source: Munger, David, et al. Researching Online. 2nd ed. New York: Longman, 1998:2.*

Organizing for Cross-Cultural Audiences

Different cultures have different expectations about how information should be organized. For instance, a paragraph in English typically begins with a main idea directly expressed as a topic or orienting sentence and followed by specific support; any digression from this main idea is considered harmful to the paragraph's *unity*. Some cultures, however, consider digression a sign of intelligence or politeness. To native readers of English, the long introductions and digressions in certain Spanish or Russian documents might seem tedious and confusing, but a Spanish or Russian reader might view the more direct organization of English as overly abrupt and simplistic (Leki 151).

Expectations differ even among same-language cultures. British correspondence, for instance, typically expresses the bad news directly up front, instead of taking the indirect approach preferred in the United States. A bad news letter or memo appropriate for a U.S. audience could be considered evasive in Great Britain (Scott and Green 19).

The Report Design Worksheet

As an alternative to the audience and use profile sheet (page 29), the worksheet in Figure 12.4 can supplement your outline and help you focus on your audience and purpose.[1]

STORYBOARDING

When you prepare a long document, a useful organizing tool is the *storyboard,* a sketch of the finished document. Much more specific and visual than an outline, a storyboard maps out each section (or module) of your outline, topic by topic, to help you see the shape and appearance of the entire document in its final form. Working from a storyboard, you can rearrange, delete, and insert material as needed—without having to wrestle with a draft of the entire document.

Storyboarding is especially helpful in collaborative writing, in which various team members prepare various parts of a document and then get together to edit and assemble their material. In such cases, storyboard modules may be displayed on whiteboards, posterboards, or flip charts.

NOTE   *Try creating a storyboard* after *writing a full draft, for a bird's-eye view of the document's organization.*

Figure 12.5 displays one storyboard module based on Section II.A of the outline on page 197. Notice how the module begins with the section title, describes each text block and visual block, and includes a list of suggestions about special considerations.

---

[1] This version is based on a worksheet developed by Professor John S. Harris of Brigham Young University.

REPORT DESIGN WORKSHEET

Preliminary Information

What is to be done? *A report on the health effects of electromagnetic radiation*

Whom is it to be presented to, and when? *Town Meeting, April 1*

Audience Analysis	Primary Audience	Secondary Audience
Position and title:	*town manager*	*selectpersons, various town officials, school board, parents, colleagues, friends*
Relationship to author or organization:	*employer*	
Technical expertise:	*nontechnical (for this topic)*	*nontechnical*
Personal characteristics:	*highly efficient; expects results*	*most have strong views on this issue*
Attitude toward author or organization:	*is preparing my annual performance review*	*friendly and respectful; select board will vote on my contract renewal and pay raise*
Attitude toward subject:	*extremely concerned*	*same*
Effect of report on audience or organization:	*will be read closely and acted upon*	*will be discussed at town meeting*

User's Purpose

Why has audience requested it?	*wants to address any potential hazards without delay*	
What does audience plan to do with it?	*use the data to make an informed decision about action*	*confer with the town manager about the decision*
What should audience know beforehand to understand it as written?	*nothing special; history of the issue is reviewed in report*	*same*
What does audience already know?	*has read and heard very general information about the issue*	*same*
What amount and kinds of detail will audience find significant?	*clear description of the issue and careful review of the evidence*	*same*
What should audience know and/or be able to do after reading it?	*make a decision based on the best evidence available*	*advise the town manager about her decision*

Figure 12.4   Report Design Worksheet

<u>Writer's Purpose</u>

Why am I writing?   *to communicate my research findings*

What effect(s) do I wish to achieve?   *to have my audience conclude that, while we await further research, we should take immediate and inexpensive steps toward risk avoidance and continue to assess the extent of EMF hazards throughout the school*

<u>Design Specifications</u>

Sources of data: *recently published research, including online and internet sources; interviews with local authorities*

Tone: *semiformal*

Point of view: *mostly third person (except for recommendations)*

Needed visuals and supplements: *title page, letter of transmittal, table of contents, informative abstract charts, graphs, and tables*

Appropriate format (letter, memo, etc.):   *formal report format with full heading system*

Basic organization (problem-causes-solution, intro-instructions-summary, etc.):   *causes-possible effects-conclusions and recommendations*

Main items in introduction: *Definition of electromagnetic fields*
Background on the health issue
Description of the local power line configuration
Purpose of report, and intended audience
Data sources
Scope of this inquiry

Main items in body: *Sources of EMF exposure*
Risk factors
Studies of health effects
Conflicting views of studies
Local power company views
Risk-avoidance measures

Main items in conclusion: *Summary of findings*
Overall interpretation of findings
Recommendations

<u>Other Considerations:</u>   *no frills or complex technical data; this audience is interested in the "bottom line" as far as what action they should take*

Figure 12.4   Report Design Worksheet   *Continued*

Figure 12.5
One Module
from a
Storyboard

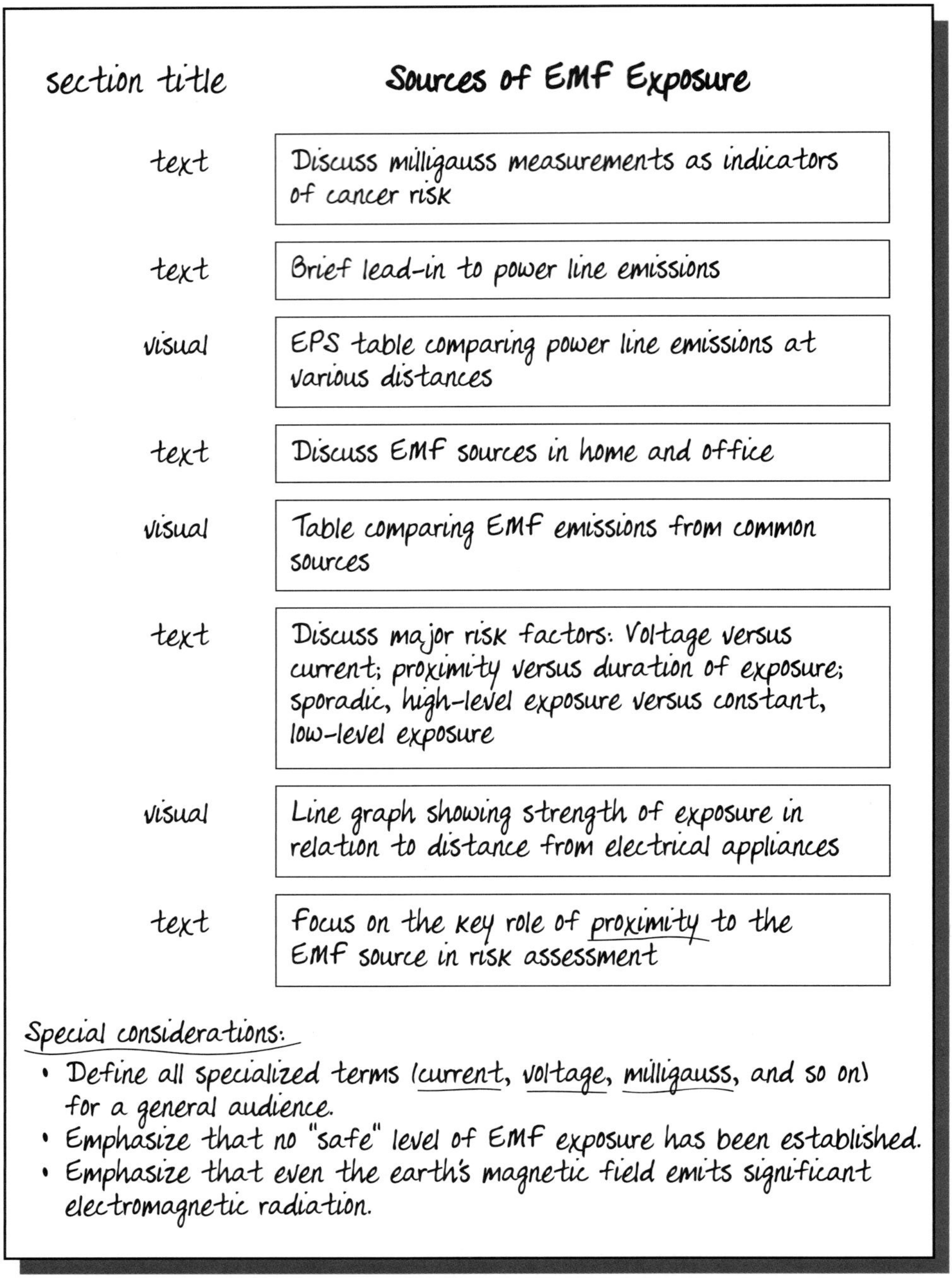

section title   **Sources of EMF Exposure**

text   Discuss milligauss measurements as indicators of cancer risk

text   Brief lead-in to power line emissions

visual   EPS table comparing power line emissions at various distances

text   Discuss EMF sources in home and office

visual   Table comparing EMF emissions from common sources

text   Discuss major risk factors: Voltage versus current; proximity versus duration of exposure; sporadic, high-level exposure versus constant, low-level exposure

visual   Line graph showing strength of exposure in relation to distance from electrical appliances

text   Focus on the key role of proximity to the EMF source in risk assessment

Special considerations:
• Define all specialized terms (current, voltage, milligauss, and so on) for a general audience.
• Emphasize that no "safe" level of EMF exposure has been established.
• Emphasize that even the earth's magnetic field emits significant electromagnetic radiation.

PARAGRAPHING

¶ Users look for orientation, for shapes they can recognize. But a document's shape (introduction, body, conclusion) depends on the smaller shapes of each paragraph.

Although paragraphs can have various shapes and purposes (paragraphs of introduction, conclusion, or transition), our focus here is on *support paragraphs.* Although part of the document's larger design, each of these middle blocks of thought usually can stand alone in meaning and emphasis.

The Support Paragraph

All the sentences in a support paragraph relate to the main point, which is expressed as the *topic sentence:*

Topic sentences

> Computer literacy has become a requirement for all "educated" people.
> A video display terminal can endanger the operator's health.
> Chemical pesticides and herbicides are both ineffective and hazardous.

Each topic sentence introduces an idea, judgment, or opinion. But in order to grasp the writer's exact meaning, users need explanation. Consider the third statement:

> Chemical pesticides and herbicides are both ineffective and hazardous.

Imagine that you are a researcher for the Epson Electric Light Company, assigned this question: Should the company (1) begin spraying pesticides and herbicides under its power lines, or (2) continue with its manual (and nonpolluting) ways of minimizing foliage and insect damage to lines and poles? If you simply responded with the preceding assertion, your employer would have further questions:

> - *Why, exactly, are these methods ineffective and hazardous?*
> - *What are the problems?*
> - *Can you explain?*

To answer these questions and to support your assertion, you need a fully developed paragraph:

Intro. (topic sent.)
Body (2–6)

Conclusion (7–8)

> [1]**Chemical pesticides and herbicides are both ineffective and hazardous.** [2]Because none of these chemicals has permanent effects, pest populations invariably recover and need to be resprayed. [3]Repeated applications cause pests to develop immunity to the chemicals. [4]Furthermore, most of these products attack species other than the intended pest, killing off its natural predators, thus actually increasing the pest population. [5]Above all, chemical residues survive in the environment (and living tissue) for years, often carried hundreds of miles by wind and water. [6]This toxic legacy includes such biological effects as birth deformities, reproductive failures, brain damage, and cancer. [7]Although intended to control pest populations, these chemicals ironically threaten to make the human population their ultimate victims, [8]I therefore recommend we continue our present control methods.

Most standard paragraphs in technical writing have an introduction-body-conclusion structure. They begin with a clear topic (or orienting) sentence stating a generalization. Details in the body support the generalization.

The Topic Sentence

ts

Users look to a paragraph's opening sentences for a framework. When the paragraph's main point is missing, users struggle to grasp your meaning. Read this next paragraph once only.

A paragraph without a topic sentence

> Besides containing several toxic metals, it percolates through the soil, leaching out naturally present metals. Pollutants such as mercury invade surface water, accumulating in fish tissues. Any organism eating the fish—or drinking the water—in turn faces the risk of heavy metal poisoning. Moreover, acidified water can release heavy concentrations of lead, copper, and aluminum from metal plumbing, making ordinary tap water hazardous.

Can you identify the paragraph's main idea? Probably not. Without the topic sentence, you have no framework for understanding. You don't know where to place the emphasis: on polluted fish, on metal poisoning, on tap water?

Now, insert the following sentence at the beginning and reread the paragraph:

The missing topic sentence

> Acid rain indirectly threatens human health.

This orientation makes the message's exact meaning obvious.

The topic sentence should appear *first* in the paragraph, unless you have good reason to place it elsewhere. Think of your topic sentence as the one sentence you would keep if you could keep only one (U.S. Air Force Academy 11). In some instances, a paragraph's main idea may require a "topic statement" consisting of two or more sentences, as in this example:

A topic statement can have two or more sentences

> The most common strip-mining methods are open-pit mining, contour mining, and auger mining. The specific method employed will depend on the type of terrain that covers the coal.

The topic sentence or topic statement should focus and forecast. Don't write *Some pesticides are less hazardous and often more effective than others* when you mean *Organic pesticides are less hazardous and often more effective than their chemical counterparts.* The first topic sentence leads everywhere and nowhere; the second helps us focus, tells us what to expect from the paragraph. Don't write *acid rain poses a danger,* leaving readers to decipher what you mean by *danger.* If you mean that *Acid rain is killing our lakes and polluting our water supplies,* say so.

Paragraph Unity

¶un

A paragraph is unified when every word, phrase, and sentence directly supports the topic sentence.

A unified paragraph

> **Solar power offers an efficient, economical, and safe solution to the Northeast's energy problems.** To begin with, solar power is highly efficient. Solar collectors installed on fewer than thirty percent of roofs in the Northeast would provide more than seventy percent of the area's heating and air-conditioning needs. Moreover, solar heat collectors are economical, operating for up to twenty years with little or no maintenance. These savings recoup the initial cost of installation within only ten years. Most important, solar power is safe. It can be transformed into electricity through photovoltaic cells (a type of storage battery) in a noiseless process that produces no air pollution—unlike coal, oil, and wood combustion. In sharp contrast to its nuclear counterpart, solar power produces no toxic waste and poses no catastrophic danger of meltdown. Thus, massive conversion to solar power would ensure abundant energy and a safe, clean environment for future generations.

One way to destroy unity in the paragraph above would be to veer from the focus on *efficient, economical,* and *safe* with material about the differences between active and passive solar heating, or manufacturers of solar technology, or the advantages of solar power over wind power.

Every topic sentence has a key word or phrase that carries the meaning. In the pesticide-herbicide paragraph (page 205), the key words are *ineffective* and *hazardous.* Anything that fails to advance their meaning throws the paragraph—and the users—off track.

Paragraph Coherence

¶coh

In a coherent paragraph, everything not only belongs, but also sticks together: Topic sentence and support form a *connected line of thought,* like links in a chain.

Paragraph coherence can be damaged by (1) short, choppy sentences, (2) sentences in the wrong order, or (3) insufficient transitions and connectors (Appendix C) for linking related ideas. Here is how the solar energy paragraph might become incoherent:

An incoherent paragraph

> Solar power offers an efficient, economical, and safe solution to the Northeast's energy problems. Unlike nuclear power, solar power produces no toxic waste and poses no danger of meltdown. Solar power is efficient. Solar collectors could be installed on fewer than thirty percent of roofs in the Northeast. These collectors would provide more than seventy percent of the area's heating and air-conditioning needs. Solar power is safe. It can be transformed into electricity. This transformation is made possible by photovoltaic cells (a type of storage battery). Solar heat collectors are economical. The photovoltaic process produces no air pollution.

Here, in contrast, is the original, coherent paragraph with sentences numbered for later discussion and with transitions and connectors shown in boldface. Notice how this version reveals a clear line of thought:

A coherent paragraph

> [1]Solar power offers an efficient, economical, and safe solution to the Northeast's energy problems. [2]**To begin with,** solar power is highly efficient. [3]Solar collectors in-

stalled on fewer than thiry percent of roofs in the Northeast would provide more than seventy percent of the area's heating and air-conditioning needs. [4]**Moreover,** solar heat collectors are economical, operating for up to twenty years with little or no maintenance. [5]**These savings** recoup the initial cost of installation within only ten years. [6]**Most important,** solar power is safe. [7]**It** can be transformed into electricity through photovoltaic cells (a type of storage battery) in a noiseless process that produces no air pollution—unlike coal, oil, and wood combustion. [8]**In sharp contrast** to its nuclear counterpart, solar power produces no toxic waste and poses no danger of catastrophic meltdown. [9]**Thus,** massive conversion to solar power would ensure abundant energy and a safe, clean environment for future generations.

The line of thinking in this paragraph seems easy enough to follow:

1. The topic sentence establishes a clear direction.
2–3. The first reason is given and explained.
4–5. The second reason is given and explained.
6–8. The third and major reason is given and explained.
9. The conclusion sums up and reemphasizes the main point.

To reinforce the logical sequence, related ideas are combined in individual sentences, and transitions and connectors signal clear relationships. The whole paragraph sticks together.

Paragraph Length

¶ lgth

Paragraph length depends on the writer's purpose and the user's capacity for understanding. Writing that contains highly technical information or complex instructions may use short paragraphs, or perhaps a list. In writing that explains concepts, attitudes, or viewpoints, support paragraphs generally run from one hundred to three hundred words.

But word count really means very little. What matters is *how thoroughly the paragraph makes your point.* A flabby paragraph buries users in needless words and details; but a skin-and-bones paragraph may leave them looking for the meat.

Each paragraph requires new decisions. Try to avoid too much of anything. A clump of short paragraphs can make some writing seem choppy and poorly organized, but a stretch of long paragraphs is tiring. Paragraphs of only one or two sentences can focus the user's attention and highlight important ideas.

NOTE

In writing displayed on computer screens, short paragraphs and lists are especially useful because they allow for easy scanning and navigation.

SEQUENCING

Items in logical sequence follow some pattern that reveals a relationship: cause-and-effect, comparison-contrast, and so on. For instance, a progress report usually follows a *chronological* sequence (events in order of occurrence). An argument for a companywide exercise program would likely follow an *emphatic* sequence (benefits in order of importance—least to most, or vice versa).

A single paragraph usually follows one particular sequence. A longer document may use one particular sequence or a combination of sequences. Some common sequences are described below.

Spatial Sequence

A spatial sequence begins at one location and ends at another. It is most useful in describing a physical item or a mechanism. Describe the parts in the sequence users would follow if they were actually looking at the item in the order in which each part functions: left to right, inside to outside. This description of a hypodermic needle proceeds from the needle's base (hub) to its point:

"What does it look like?"

> A hypodermic needle is a slender, hollow steel instrument used to introduce medication into the body (usually through a vein or muscle). It is a single piece composed of three parts, all considered sterile: the hub, the cannula, and the point. The hub is the lower, larger part of the needle that attaches to the necklike opening on the syringe barrel. Next is the cannula (stem), the smooth and slender central portion. Last is the point, which consists of a beveled (slanted) opening, ending in a sharp tip. The diameter of a needle's cannula is indicated by a gauge number; commonly, a 24–25 gauge needle is used for subcutaneous injections. Needle lengths are varied to suit individual needs. Common lengths used for subcutaneous injections are $\frac{3}{8}$, $\frac{1}{2}$, $\frac{5}{8}$, and $\frac{3}{4}$ inch. Regardless of length and diameter, all needles have the same functional design.

Product and mechanism descriptions almost always have some type of visual to amplify the verbal description.

Chronological Sequence

A chronological sequence follows the actual sequence of events. Explanations of how to do something or how something happened generally are arranged according to a strict time sequence: first step, second step, and so on.

"How is it done?"

> Instead of breaking into a jog too quickly and risking injury, take a relaxed and deliberate approach. Before taking a step, spend at least ten minutes stretching and warming up, using any exercises you find comfortable. (After your first week, consult a jogging book for specialized exercises.) When you've completed your warmup, set a brisk pace walking. Exaggerate the distance between steps, taking long strides and swinging your arms briskly and loosely. After roughly one hundred yards at this brisk pace, you should feel ready to jog. Immediately break into a very slow trot: lean your torso forward and let one foot fall in front of the other (one foot barely leaving the ground while the other is on the pavement). Maintain the slowest pace possible, just above a walk. *Do not bolt out like a sprinter!* The biggest mistake is to start fast and injure yourself. While jogging, relax your body. Keep your shoulders straight and your head up, and enjoy the scenery—after all, it is one of the joys of jogging. Keep your arms low and slightly bent at your sides. Move your legs freely from the hips in an action that is easy, not forced. Make your feet perform a heel-to-toe action: land on the heel; rock forward; take off from the toe.

The paragraph explaining how acid rain endangers human health (page 206) offers another example of chronological sequence.

Effect-to-Cause Sequence

Problem-solving analyses typically use a sequence that first identifies a problem and then traces its causes.

"How did this happen?"

> Modern whaling techniques have brought the whale population to the threshold of extinction. In the nineteenth century, invention of the steamboat increased hunters' speed and mobility. Shortly afterward, the grenade harpoon was invented so that whales could be killed quickly and easily from the ship's deck. In 1904, a whaling station opened on Georgia Island in South America. This station became the gateway to Antarctic whaling for the nations of the world. In 1924, factory ships were designed that enabled round-the-clock whale tracking and processing. These ships could reduce a ninety-foot whale to its by-products in roughly thirty minutes. After World War II, more powerful boats with remote sensing devices gave a final boost to the whaling industry. The number of kills had now increased far beyond the whales' capacity to reproduce.

Cause-to-Effect Sequence

A cause-to-effect sequence follows an action to its results. The topic sentence identifies the causes, and the remainder of the paragraph discusses its effects.

"What will happen if I do this?"

> Some of the most serious accidents involving gas water heaters occur when a flammable liquid is used in the vicinity. The heavier-than-air vapors of a flammable liquid such as gasoline can flow along the floor—even the length of a basement—and be explosively ignited by the flame of the water heater's pilot light or burner. Because the victim's clothing frequently ignites, the resulting burn injuries are commonly serious and extremely painful. They may require long hospitalization, and can result in disfigurement or death. *Never, under any circumstances, use a flammable liquid near a gas heater or any other open flame.* (Consumer Product Safety Commission)

Emphatic Sequence

Emphasis makes important things stand out. Reasons offered in support of a specific viewpoint or recommendation often appear in workplace writing, as in the pesticide-herbicide paragraph on page 205 or the solar energy paragraph on page 207. For emphasis, the reasons or examples are usually arranged in decreasing or increasing order of importance.

"What should I remember about this?"

> Although strip mining is safer and cheaper than conventional mining, it is highly damaging to the surrounding landscape. Among its effects are scarred mountains, ruined land, and polluted waterways. Strip operations are altering our country's land at the rate of 5,000 acres per week. An estimated 10,500 miles of streams have been poisoned by silt drainage in Appalachia alone. If strip mining continues at its present rate, 16,000 square miles of U.S. land will eventually be stripped barren.

In this paragraph, the most dramatic example appears last, for greatest emphasis.

Problem-Causes-Solution Sequence

The problem-solving sequence proceeds from description of the problem to diagnosis to solution. After outlining the cause of the problem, this next paragraph explains how the problem has been solved:

"How was the problem solved?"

On all waterfront buildings, the unpainted wood exteriors had been severely damaged by the high winds and sandstorms of the previous winter. After repairing the damage, we took protective steps against further storms. First, all joints, edges, and sashes were treated with water-repellent preservative to protect against water damage. Next, three coats of nonporous primer were applied to all exterior surfaces to prevent paint from blistering and peeling. Finally, two coats of wood-quality latex paint were applied over the nonporous primer. To keep coats of paint from future separation, the first coat was applied within two weeks of the priming coats, and the second within two weeks of the first. Two weeks after completion, no blistering, peeling, or separation has occurred.

Comparison-Contrast Sequence

Workplace writing often requires evaluation of two or more items on the basis of their similarities or differences.

"How do these items compare?"

The ski industry's quest for a binding that ensures good performance as well as safety has led to development of two basic types. Although both bindings improve performance and increase the safety margin, they have different release and retention mechanisms. The first type consists of two units (one at the toe, another at the heel) that are spring-loaded. These units apply their retention forces directly to the boot sole. Thus the friction of boot against ski allows for the kind of ankle movement needed at high speeds over rough terrain, without causing the boot to release. In contrast, the second type has one spring-loaded unit at either the toe or the heel. From this unit a boot plate travels the length of the boot to a fixed receptacle on its opposite end. With this plate binding, the boot has no part in release or retention. Instead, retention force is applied directly to the boot plate, providing more stability for the recreational skier, but allowing for less ankle and boot movement before releasing. Overall, the double-unit binding performs better in racing, but the plate binding is safer.

For comparing and contrasting more specific data on these bindings, two lists would be most effective.

The Salomon 555 offers the following features:

1. upward release at the heel and lateral release at the toe (thus eliminating 80 percent of leg injuries)
2. lateral antishock capacity of 15 millimeters, with the highest available return-to-center force
3. two methods of reentry to the binding: for hard and deep-powder conditions
4. five adjustments
5. (and so on)

The Americana offers these features:

1. upward release at the toe as well as upward and lateral release at the heel
2. lateral antishock capacity of 30 millimeters, with moderate return-to-center force
3. two methods of reentry to the binding
4. two adjustments, one for boot length and another for comprehensive adjustment for all angles of release and elasticity
5. (and so on)

Instead of this block structure (in which one binding is discussed and then the other), the writer might have chosen a point-by-point structure (in which points common to both items, such as "Reentry Methods" are listed together). The point-by-point comparison works best in feasibility and recommendation reports because it offers readers a meaningful comparison between common points.

EXERCISES

1. Locate, copy, and bring to class a paragraph that has the following features:

   • an orienting topic sentence
   • adequate development
   • unity
   • coherence
   • a recognizable sequence
   • appropriate length for its purpose and audience

   Be prepared to identify and explain each of these features in a class discussion.

2. For each of the following documents, indicate the most logical sequence. (For example, a description of a proposed computer lab would follow a spatial sequence.)

   • a set of instructions for operating a power tool
   • a campaign report describing your progress in political fund-raising
   • a report analyzing the weakest parts in a piece of industrial machinery
   • a report analyzing the desirability of a proposed oil refinery in your area
   • a detailed breakdown of your monthly budget to trim excess spending

   • a report investigating the reasons for student apathy on your campus
   • a report evaluating the effects of the ban on DDT in insect control
   • a report on any highly technical subject, written for a general audience
   • a report investigating the success of a no-grade policy at other colleges
   • a proposal for a no-grade policy at your college

COLLABORATIVE PROJECTS

1. Organize into small groups. Choose *one* of these topics, or one your group settles on, and then brainstorm to develop a formal outline for the body of a report. One representative from your group can write the final draft on the board, for class revision.

   • job opportunities in your career field
   • a physical description of the ideal classroom
   • how to organize an effective job search
   • how the quality of your higher educational experience can be improved
   • arguments for and against a formal grading system

- an argument for an improvement you think this college needs most

2. Assume your group is preparing a report titled "The Negative Effects of Strip Mining on the Cumberland Plateau Region of Kentucky." After brainstorming and researching your subject, you all settle on these four major topics:

   - economic and social effects of strip mining
   - description of the strip-mining process
   - environmental effects of strip mining
   - description of the Cumberland Plateau

   Arrange these topics in the most sensible sequence.

      When your topics are arranged, assume that subsequent research and further brainstorming produce this list of subtopics:

   - method of strip mining used in the Cumberland Plateau region
   - location of the region
   - permanent land damage

- water pollution
- lack of educational progress
- geological formation of the region
- open-pit mining
- unemployment
- increased erosion
- auger mining
- natural resources of the region
- types of strip mining
- increased flood hazards
- depopulation
- contour mining

Arrange these subtopics (and perhaps some sub-subtopics) under appropriate topic headings. Use decimal notation to create the body of a formal outline. Appoint one group member to present the outline in class.

   Hint: Assume that your thesis is: "Decades of strip mining (without reclamation) in the Cumberland Plateau have devastated this region's environment, economy, and social structure."

CHAPTER 13

Revising for Readable Style

You might write for a diverse or specific audience, or for experts of nonexperts. But no matter how technically appropriate your document, audience needs are not served unless your style is easy to understand—in a word, *readable*.

A definition of style

What is *writing style,* and how does it influence user reaction to a document? Your writing style is the product of

Elements of style

- the words you choose
- the way in which you put a sentence together
- the length of your sentences
- the way in which you connect sentences
- the tone you convey

One requirement for readable sentences, of course, is correct grammar, punctuation, and spelling. But correctness alone is no guarantee of readability. For example, this response to a job applicant is mechanically correct but inefficient:

Inefficient style

> We are in receipt of your recent correspondence indicating your interest in securing the advertised position. Your correspondence has been duly forwarded for consideration by the personnel office, which has employment candidate selection responsibility. You may expect to hear from us relative to your application as the selection process progresses. Your interest in the position is appreciated.

Notice how hard you worked to extract information that could be expressed this simply:

More efficient

> Your application for the advertised position has been forwarded to our personnel office. As the selection process moves forward, we will be in touch. Thank you for your interest.

Inefficient style makes readers work harder than they should.

Style can be inefficient for many reasons, but it is especially inept when it

Ways in which style goes wrong

- makes the writing impossible to interpret
- takes too long to make the point
- reads like a Dick-and-Jane story from primary school
- uses imprecise or needlessly big words
- sounds stuffy and impersonal

Regardless of the cause, inefficient style results in writing that is less informative, less persuasive. Moreover, inefficient style can be unethical—by confusing or misleading the audience.

To help your audience spend less time reading, you must spend more time revising for a style that is *clear, concise, fluent, exact,* and *likable.*

REVISING FOR CLARITY

A clear sentence requires no more than a single reading. The following suggestions will help you write clear sentences.

amb   **AVOID AMBIGUOUS PHRASING.**   Workplace writing ideally has *one* meaning only, allows for *one* interpretation. Does a person's "suspicious attitude" mean that he is "suspicious" or "suspect"?

Ambiguous phrasing	All managers are not required to submit reports. *(Are some or none required?)*
Revised	Managers are not all required to submit reports.
	or
	Managers are not required to submit reports.
Ambiguous phrasing	Most city workers strike on Friday.
Revised	Most city workers **are planning to strike** Friday.
	or
	Most city workers **typically strike** on Friday.

ref   **AVOID AMBIGUOUS PRONOUN REFERENCES.**   Pronouns (*he, she, it, their,* and so on) must clearly refer to the noun they replace.

Ambiguous referent	Our patients enjoy the warm days while they last. *(Are the patients or the warm days on the way out?)*

Depending on whether the referent (or antecedent) for *they* is *patients* or *warm days,* the sentence can be clarified.

Clear referent	While these warm days last, our patients enjoy them.
	or
	Our terminal patients enjoy the warm days.
Ambiguous referent	Jack resents his assistant because he is competitive. *(Who's the competitive one—Jack or his assistant?)*
Clear referent	Because his assistant is competitive, Jack resents him.
	or
	Because Jack is competitive, he resents his assistant.

(See Appendix C for more on pronoun references, and page 250 for avoiding sexist bias in pronoun use.)

AVOID AMBIGUOUS PUNCTUATION.   A missing hyphen, comma, or other punctuation mark can obscure meaning.

Missing hyphen	Replace the trailer's inner wheel bearings. (*The inner-wheel bearings or the inner wheel-bearings?*)
Missing comma	Does your company produce liquid hydrogen? If so, how[,] and where do you store it? (*Notice how the meaning changes with a comma after "how."*)
	Police surrounded the crowd[,] attacking the strikers. (*Without the comma, the crowd appears to be attacking the strikers.*)

A missing colon after *kill* yields the headline "Moose Kill 200." A missing apostrophe after *Myers* creates this gem: "Myers Remains Buried in Portland."

Exercise 1
Revise each sentence below to eliminate ambiguities in phrasing, pronoun reference, or punctuation.

a. Call me any evening except Tuesday after 7 o'clock.
b. The benefits of this plan are hard to imagine.
c. I cannot recommend this candidate too highly.
d. Visiting colleagues can be tiring.
e. Janice dislikes working with Claire because she's impatient.
f. Our division needs more effective writers.
g. Tell the reactor operator to evacuate and sound a general alarm.
h. If you don't pass any section of the test, your flying days are over.
i. Dial "10" to deactivate the system and sound the alarm.

---

tel   **AVOID TELEGRAPHIC WRITING.**   *Function words* show relationships between the *content words* (nouns, adjectives, verbs, and adverbs) in a sentence. Some examples of function words:

Function words

- ◆ articles (*a, an, the*)
- ◆ prepositions (*in, of, to*)
- ◆ linking verbs (*is, seems, looks*)
- ◆ relative pronouns (*who, which, that*)

Some writers mistakenly try to compress their writing by eliminating these function words.

Ambiguous	Proposal to employ retirees almost dead.
Revised	The proposal to employ retirees **is** almost dead.

Ambiguous	Uninsulated end pipe ruptured. (*What ruptured? The pipe or the end of the pipe?*)
Revised	**The** uninsulated end **of the** pipe is ruptured.
	or
	The uninsulated pipe **on the** end ruptured.

Ambiguous	The reactor operator told management several times she expected an accident. (*Did she tell them once or several times?*)
Revised	The reactor operator told management several times **that** she expected an accident.
	or
	The reactor operator told management **that** several times she expected an accident.

<table>
<tr><td>mod</td></tr>
</table>

AVOID AMBIGUOUS MODIFIERS.   Modifiers explain, define, or add detail to other words or ideas. If a modifier is too far from the words it modifies, the message can be ambiguous.

Misplaced modifier	**Only** press the red button in an emergency. (*Does **only** modify **press** or **emergency?***)
Revised	Press **only** the red button in an emergency.
	or
	Press the red button in an emergency **only**.

Position modifiers to reflect your meaning.

Another problem with ambiguity occurs when a modifying phrase has no word to modify.

| **Dangling modifier** | **Being so well known in the computer industry,** I would appreciate your advice. |

The writer meant to say that the *reader* is well known, but with no word to join itself to, the modifying phrase dangles. Eliminate the confusion by adding a subject:

| **Revised** | Because **you** are so well known in the computer industry, I would appreciate your advice. |

See Appendix C for more on modifiers.

Exercise 2

Revise each sentence below to repair telegraphic writing or to clarify ambiguous modifiers.

    a. Replace main booster rocket seal.
    b. The president refused to believe any internal report was inaccurate.
    c. Only use this phone in a red alert.
    d. After offending our best client, I am deeply annoyed with the new manager.
    e. Send memo to programmer requesting explanation.
    f. Do not enter test area while contaminated.

---

st mod

UNSTACK MODIFYING NOUNS. One noun can modify another noun (as in "software development"). But when two or more nouns modify a noun, the string of densely packed words becomes hard to read and ambiguous.

> **Stacked**   Be sure to leave enough time for a **training session participant** evaluation. (*Evaluation of the session or of the participants?*)

With no function words (articles, prepositions, verbs, relative pronouns) to break up the string of nouns, we cannot make out the relationships among the nouns. What modifies what?

Stacked nouns also deaden your style. Bring your style *and* your audience to life by using action verbs (*complete, prepare, reduce*) and prepositional phrases.

> **Revised**   Be sure to leave enough time **for** participants **to evaluate** the training session.
>
>   *or*
>
> Be sure to leave enough time **to evaluate** participants **in the** training session.

No such problem with ambiguity occurs when *adjectives* are stacked in front of a noun.

> **Clear**   He was a **nervous, angry, confused,** but **dedicated** employee.

The adjectives clearly modify *employee*.

wo

ARRANGE WORD ORDER FOR COHERENCE AND EMPHASIS. In coherent writing, everything sticks together; each sentence builds on the preceding sentence and looks ahead to the following sentence. Sentences generally work best when the beginning looks back at familiar information and the end provides the new (or unfamiliar) information:

Familiar		**Unfamiliar**
My dog	has	fleas.
Our boss	just won	the lottery.
This company	is planning	a merger.

Effective word order

Besides helping a message stick together, the familiar-to-unfamiliar structure emphasizes the new information. Just as every paragraph has a key sentence, every sentence has a key word or phrase that sums up the new information. That key word or phrase usually is emphasized best when it appears at the end of the sentence.

Faulty emphasis	We expect a **refund** because of your error in our shipment.
Correct	Because of your error in our shipment, we expect a **refund**.
Faulty emphasis	In a business relationship, **trust** is a vital element.
Correct	In a business relationship, a vital element is **trust**.

One exception to placing key words last occurs with an imperative statement (a command, an order, an instruction), with the subject [*you*] understood. For instance, each step in a list of instructions should contain an action verb (*insert, open, close, turn, remove, press*). To give readers a forecast, place the verb in that instruction at the beginning.

Correct	**Insert** the diskette before activating the system.
	Remove the protective seal.

With the key word at the beginning of the instruction, users know immediately what action they need to take.

Exercise 3

Revise each sentence below to unstack modifying nouns or to rearrange the word order for clarity and emphasis.

a. Develop online editing system documentation.
b. We need to develop a unified construction automation design.
c. Install a hazardous materials dispersion monitor system.
d. I recommend these management performance improvement incentives.
e. Our profits have doubled since we automated our assembly line.
f. Education enables us to recognize excellence and to achieve it.
g. In all writing, revision is required.
h. We have a critical need for technical support
i. Sarah's job involves fault analysis systems troubleshooting handbook preparation.

---

av   **USE ACTIVE VOICE OFTEN.**   A verb's *voice* signals whether a sentence's subject acts or is acted on. The active voice ("I did it") is more direct, concise, and persuasive than the passive voice ("It was done by me"). In the active voice, the agent performing the action serves as subject:

Active	Agent	Action	Recipient
	Joe	lost	your report.
	Subject	Verb	Object

The passive voice reverses the pattern, making the recipient of an action serve as subject.

Passive	Recipient	Action	Agent
	Your report	was lost	by Joe.
	Subject	Verb	Prepositional phrase

Sometimes the passive eliminates the agent altogether:

Passive	Your report was lost. (*Who lost it?*)

Some writers mistakenly rely on the passive voice because they think it sounds more objective and important. But the passive voice often makes writing wordy, indecisive, evasive, and unethical.

Concise and direct (active)	I underestimated labor costs for this project. (*7 words*)
Wordy and indirect (passive)	Labor costs for this project were underestimated by me. (*9 words*)
Evasive (passive)	Labor costs for this project were underestimated.

Do not evade responsibility by hiding behind the passive voice:

Evasive (passive)	A **mistake** was made in your shipment. (*By whom?*)
	It was decided not to hire you. (*Who decided?*)
	A **layoff** is recommended.

In reporting errors or bad news, use the active voice, for clarity and sincerity. The passive voice creates a weak and impersonal tone.

Weak and impersonal	An offer will be made by us next week.
Strong and personal	We will make an offer next week.

Use the active voice when you want action. Otherwise, your statement will have no power.

Weak passive	If my claim is not settled by May 15, the Better Business Bureau will be contacted, and their advice on legal action will be taken.
Strong active	If you do not settle my claim by May 15, I will contact the Better Business Bureau for advice on legal action.

Notice how this active version emphasizes the new and significant information by placing it at the end.

Use the active voice for giving instructions.

Passive	The bid should be sealed. Care should be taken with the dynamite.
Active	**Seal** the bid. **Be careful** with the dynamite.

Avoid shifts from active to passive voice in the same sentence.

Faulty shift	During the meeting, project members spoke and presentations were given.
Correct	During the meeting, project members spoke and gave presentations.

Unless you have a deliberate reason for choosing the passive voice, prefer the *active* voice for making forceful connections like the one described here:

Why the active voice is preferable

By using the active voice, you direct the reader's attention to the subject of your sentence. For instance, if you write a job application letter that is littered with passive verbs, you fail to achieve an important goal of that letter: to show the readers the important things you have done, and how prepared you are to do important things for them. That strategy requires active verbs, with clear emphasis on *you* and what you have done/are doing. (Pugliano 6)

Exercise 4

The sentences below are wordy, weak, or evasive because of passive voice. Revise each sentence as a concise, forceful, and direct expression in the active voice, to identify the person or agent performing the action.

a. The evaluation was performed by us.
b. Unless you pay me within three days, my lawyer will be contacted.
c. Hard hats should be worn at all times.
d. It was decided to reject your offer.

    e. It is believed by us that this contract is faulty.

    f. Our test results will be sent to you as soon as verification is completed.

---

pv   **USE PASSIVE VOICE SELECTIVELY.**   Passive voice is appropriate in lab reports and other documents in which the agent's identity is immaterial to the message.
Use the passive when your audience does not need to know the agent.

Correct passive	Mr. Jones was brought to the emergency room.
	The bank failure was publicized statewide.

Use the passive voice to focus on events or results when the agent is unknown, unapparent, or unimportant.

Correct passive	All memos in the firm are filed in a database.
	Fred's article was published last week.

Prefer the passive when you want to be indirect or inoffensive (as in requesting the customer's payment or the employee's cooperation, or to avoid blaming someone—such as your boss) (Ornatowski 94).

Active but offensive	**You** have not paid your bill.
	You need to overhaul our filing system.
Inoffensive passive	**This bill** has not been paid.
	Our filing system needs to be overhauled.

Use the passive voice if you need to protect the person behind the action.

Correct passive	The criminal was identified.
	The embezzlement scheme was exposed.

Exercise 5

The sentences below lack proper emphasis because of an improper use of the active voice. Revise each ineffective active as an appropriate passive, to emphasize the recipient rather than the actor.

    a. Joe's company fired him.

    b. Someone on the maintenance crew has just discovered a crack in the nuclear-core containment unit.

    c. A power surge destroyed more than two thousand lines of our new applications program.

    d. You are paying inadequate attention to worker safety.

    e. You are checking temperatures too infrequently.

    f. You did a poor job editing this report.

OS

AVOID OVERSTUFFED SENTENCES.   A sentence crammed with ideas makes details hard to remember and relationships hard to identify.

> **Overstuffed**   Publicizing the records of a private meeting that took place three weeks ago to reveal the identity of a manager who criticized our company's promotion policy would be unethical.

Clear things up by sorting out the relationships.

> **Revised**   In a private meeting three weeks ago, a manager criticized our company's policy on promotion. It would be unethical to reveal the manager's identity by publicizing the records of that meeting. (*Other versions are possible, depending on the writer's intended meaning.*)

Give your readers no more information in one sentence than they can retain and process.

Exercise 6
Unscramble this overstuffed sentence by making shorter, clearer sentences:

A smoke-filled room causes not only teary eyes and runny noses but also can alter people's hearing and vision, as well as creating dangerous levels of carbon monoxide, especially for people with heart and lung ailments, whose health is particularly threatened by secondhand smoke.

REVISING FOR CONCISENESS

Two kinds of wordiness

Writing can display two kinds of wordiness: one kind occurs when readers receive information they don't need (think of an overly detailed weather report during local television news). The other kind occurs when too many words are used to convey information readers *do* need (as in saying "a great deal of potential for the future" instead of "great potential").

Every word in the document should advance your meaning:

> Writing improves in direct ratio to the number of things we can keep out of it that shouldn't be there. (Zinsser 14)

Concise writing conveys the most information in the fewest words. But it does not omit details necessary for clarity.

Use fewer words whenever fewer will do. But remember the difference between *clear writing* and *compressed writing* that is impossible to decipher.

Compressed	Give new vehicle air conditioner compression cut-off system specifications to engineering manager advising immediate action.
Clear	The cut-off system for the air conditioner compressor on our new vehicles is faulty. Give the system specifications to our engineering manager so they will be modified.

First drafts are rarely concise. The following strategies help trim the fat.

w

AVOID WORDY PHRASES.   Each phrase here can be reduced to one word.

Avoiding wordy phrases

at a rapid rate	= rapidly
due to the fact that	= because
the majority of	= most
on a personal basis	= personally
readily apparent	= obvious
a large number	= many
prior to	= before
aware of the fact that	= know
in close proximity	= near

red

ELIMINATE REDUNDANCY.   A redundant expression says the same thing twice, in different words, as in *fellow colleagues.*

a **dead** corpse	**end** result
completely eliminate	cancel **out**
basic essentials	consensus **of opinion**
enter **into**	**utter** devastation
mental awareness	**the month of** August
mutual cooperation	**utmost** perfection

rep

AVOID NEEDLESS REPETITION.   Unnecessary repetition clutters writing and dilutes meaning.

Repetitious	In trauma victims, breathing is restored by **artificial respiration**. Techniques of **artificial respiration** include mouth-to-mouth **respiration** and mouth-to-nose **respiration**.

Repetition in the above passage disappears when sentences are combined.

Concise	In trauma victims, breathing is restored by artificial respiration, either mouth-to-mouth or mouth-to-nose.

NOTE   *Repetition can be useful. Don't hesitate to repeat, or at least rephrase, material (even whole paragraphs in a longer document) if you feel that users need reminders. Effective repetition helps avoid cross-references like this: "See page 23" or "Review page 10."*

Exercise 7
Revise each wordy sentence below to eliminate needless phrases, redundancy, and needless repetition.

    a. I have admiration for Professor Jones.
    b. Due to the fact that we made the lowest bid, we won the contract.
    c. On previous occasions we have worked together.
    d. She is a person who works hard.
    e. We have completely eliminated the bugs from this program.
    f. This report is the most informative report on the project.
    g. Through mutual cooperation, we can achieve our goals.
    h. I am aware of the fact that Sam is trustworthy.
    i. This offer is the most attractive offer I've received.

th   **AVOID *THERE* SENTENCE OPENERS.**   Avoid sentences beginning with *There is* or *There are.*

Weak	**There is a** coaxial cable connecting the antenna to the receiver.
Revised	A coaxial cable connects the antenna to the receiver.

Weak	**There is** a danger of explosion in Number 2 mineshaft.
Revised	Number 2 mineshaft is in danger of exploding.

Dropping these openers places the key words at sentence end, where they are best emphasized.

It   **AVOID SOME *IT* SENTENCE OPENERS.**   Avoid beginning a sentence with *It*—unless the *It* clearly points to a specific referent in the preceding sentence: "This document is excellent. It deserves special recognition."

Weak	**It** was his bad attitude that got him fired.
Revised	His bad attitude got him fired.

Weak	**It** is necessary to complete both sides of the form.
Revised	Please complete both sides of the form.

NOTE   *Of course, in some contexts, proper emphasis would call for a There opener.*

Correct	*People have often wondered about the rationale behind Boris's sudden decision. Actually, there are several good reasons for his dropping out of the program.*

| **pref** | **DELETE NEEDLESS PREFACES.**   Instead of delaying the new information in your sentence, get right to the point. |

| Wordy | **I am writing this letter because** I wish to apply for the position of copy editor. |
| Concise | Please consider me for the position of copy editor. |

| Wordy | **As far as artificial intelligence is concerned,** the technology is only in its infancy. |
| Concise | Artificial intelligence technology is only in its infancy. |

Exercise 8

Revise each sentence below to eliminate *There* and *It* openers and needless prefaces.

   a. There was severe fire damage to the reactor.
   b. There are several reasons why Jane left the company.
   c. It is essential that we act immediately.
   d. It has been reported by Bill that several safety violations have occurred.
   e. This letter is to inform you that I am pleased to accept your job offer.
   f. The purpose of this report is to update our research findings.

| **wv** | **AVOID WEAK VERBS.**   Prefer verbs that express a definite action: *open, close, move, continue, begin.* Avoid weak verbs that express no specific action: *is, was, are, has, give, make, come, take.* |

| **NOTE** | *In some cases, such verbs are essential to your meaning: "Dr. Phillips is operating at 7 A.M." "Take me to the laboratory."* |

All forms of *to be* (*am, are, is, was, were, will, have been, might have been*) are weak. Substitute a strong verb for conciseness:

| Weak and wordy | Please **take into consideration** my offer. |
| Concise | Please **consider** my offer. |

Don't disappear behind weak verbs and their baggage of needless nouns and prepositions.

| Weak | My recommendation **is** for a larger budget. |
| Strong | I **recommend** a larger budget. |

Strong verbs, or action verbs, suggest an assertive, positive, and confident writer. Here are some weak verbs converted to strong:

Avoiding weak verbs

has the ability to	=	can
give a summary of	=	summarize
make an assumption	=	assume
come to the conclusion	=	conclude
take action	=	act
make a decision	=	decide

Exercise 9
Revise each wordy and vague sentence below to eliminate weak verbs.

     a. Our disposal procedure is in conformity with federal standards.
     b. Please make a decision today.
     c. We need to have a discussion about the problem.
     d. I have just come to the realization that I was mistaken.
     e. Your conclusion is in agreement with mine.
     f. This manual gives instructions to end users.

---

to be

DELETE NEEDLESS *TO BE* CONSTRUCTIONS.   Sometimes the *to be* form itself mistakenly appears behind such verbs as *appears, seems,* and *finds.*

> **Wordy**   Your product seems **to be** superior.
>   I consider this employee **to be** highly competent.

prep

AVOID EXCESSIVE PREPOSITIONS.   Needless prepositions make for wordy sentences.

> **Wordy**   The recommendation first appeared **in** the report written **by** the supervisor **in** January **about** that month's productivity.
>
> **Concise**   The recommendation first appeared in the supervisor's productivity report for January.

Each prepositional phrase here can be reduced.

Avoiding needless prepositions

with the exception of	=	except for
in reference to	=	about (*or* regarding)
in order that	=	so
in the near future	=	soon
in the event of	=	if
at the present time	=	now
in the course of	=	during
in the process of	=	during (*or* in)

nom

FIGHT NOUN ADDICTION.   Nouns manufactured from verbs (nominalizations) often accompany weak verbs and needless prepositions.

Weak and wordy	We ask for the **cooperation** of all employees.
Strong and concise	We ask that all employees **cooperate**.

Weak and wordy	Give **consideration** to the possibility of a career change.
Strong and concise	**Consider** a career change.

Besides causing wordiness, a nominalization can be vague—by hiding the agent of an action.

Wordy and vague	A **valid requirement** for immediate action exists. (*Who should take the action? We can't tell.*)
Precise	We **must act** immediately.

Here are nominalizations restored to their verb forms:

Nouns traded for verbs

conduct an investigation of	=	investigate
provide a description of	=	describe
conduct a test of	=	test
make a discovery of	=	discover

Nominalizations drain the life from your style. In cheering for your favorite team, you wouldn't say "Blocking of that kick is a necessity!" instead of "Block that kick!"

NOTE   *Write as you would speak, but avoid slang or overuse of colloquialisms. Also avoid excessive economy. For example, "Employees must cooperate" would not be a desirable alternative to the first example in this section. But, for the final example, "Block that kick" would be.*

Exercise 10
Revise each sentence below to eliminate needless prepositions and *to be* constructions, and to cure noun addiction.

a. Igor seems to be ready for a vacation.
b. Our survey found forty-six percent of users to be disappointed.
c. In the event of system failure, your sounding of the alarm is essential.
d. These are the recommendations of the chairperson of the committee.
e. Our acceptance of the offer is a necessity.
f. Please perform an analysis and make an evaluation of our new system.
g. A need for your caution exists.
h. Power surges are associated, in a causative way, with malfunctions of computers.

---

neg

MAKE NEGATIVES POSITIVE.   A positive expression is easier to understand than a negative one.

Indirect and wordy	Please do not be late in submitting your report.
Direct and concise	Please submit your report on time.

Readers work even harder to translate sentences with multiple negative expressions:

Confusing and wordy	Do **not** distribute this memo to employees who have **not** received a security clearance.
Clear and concise	Distribute this memo only to employees who have received a security clearance.

Some words are directly negative (*no, not, never*); others are indirectly negative. (*except, forget, mistake, lose, uncooperative*). Combined, these negatives make translation even harder.

Confusing and wordy	**Do not neglect** to activate the alarm system. My diagnosis was **not inaccurate.**
Clear and concise	**Be sure** to activate the alarm system. My diagnosis was **accurate.**

The positive versions are more straightforward *and* persuasive.

Some negative expressions, of course, are perfectly correct, as in expressing disagreement.

Correct negatives	This is **not** the best plan. Your offer is **unacceptable.** This project will **never** succeed.

Prefer positives to negatives, though, whenever your meaning allows:

Negatives traded for positives

did not succeed	=	failed
does not have	=	lacks
did not prevent	=	allowed
not unless	=	only if
not until	=	only when
not absent	=	present

cl

CLEAN OUT CLUTTER WORDS.   Clutter words stretch a message without adding meaning. Here are some of the most common: *very, definitely, quite, extremely, rather, somewhat, really, actually, currently, situation, aspect, factor.*

Cluttered	**Actually,** one **aspect** of a business **situation** that could definitely make me **quite** happy would be to have a **somewhat** adventurous partner who **really** shared my **extreme** attraction to risks.
Concise	I seek an adventurous business partner who enjoys risks.

`qual` **DELETE NEEDLESS QUALIFIERS.**   Qualifiers such as *I feel, it seems, I believe, in my opinion,* and *I think* soften the tone and force of a statement. Use qualifiers to express uncertainty or to avoid seeming arrogant or overconfident.

Appropriate qualifier	Despite Frank's poor grades last year he will, **I think,** do well in college. Your product **seems** to be what we need.

But when you are certain, eliminate the qualifier so as not to seem tentative or evasive.

Needless qualifiers	**It seems** that I've made an error. We **appear to** have exceeded our budget. **In my opinion,** this candidate is outstanding.

`NOTE` *In communicating across cultures, keep in mind that a direct, forceful style might be considered offensive (page 252).*

Exercise 11
Revise each sentence below to eliminate inappropriate negatives, clutter words, and needless qualifiers.

   a. Our design must avoid nonconformity with building codes.
   b. Never fail to wear protective clothing.
   c. Do not accept any bids unless they arrive before May 1.
   d. I am not unappreciative of your help.
   e. We are currently in the situation of completing our investigation of all aspects of the accident.
   f. I appear to have misplaced the contract.
   g. Do not accept bids that are not signed.
   h. It seems as if I have just wrecked a company car.

REVISING FOR FLUENCY

Fluent sentences are easy to read because of clear connections, variety, and emphasis. Their varied length and word order eliminate choppiness and monotony. Fluent sentences enhance *clarity,* emphasizing the most important ideas. Fluent sentences enhance *conciseness,* often replacing several short, repetitious sentences with one longer, more economical sentence. To write fluently, use the following strategies.

comb   **COMBINE RELATED IDEAS.**   A series of short, disconnected sentences is not only choppy and wordy, but also unclear.

Disconnected	Jogging can be healthful. You need the right equipment. Most necessary are well-fitting shoes. Without this equipment you take the chance of injuring your legs. Your knees are especially prone to injury. (*5 sentences*)
Clear, concise, and fluent	Jogging can be healthful if you have the right equipment. Shoes that fit well are most necessary because they prevent injury to your legs, especially your knees. (*2 sentences*)

Most sets of information can be combined to form different relationships, depending on what you want to emphasize. Imagine that this set of facts describes an applicant for a junior management position with your company.

- ◆ Roy James graduated from an excellent management school.
- ◆ He has no experience.
- ◆ He is highly recommended.

Assume that you are a personnel director, conveying your impression of this candidate to upper management. To convey a negative impression, you might combine the facts in this way:

Strongly negative emphasis	Although Roy James graduated from an excellent management school and is highly recommended, **he has no experience.**

The *independent* idea (in boldface) receives the emphasis. (See also page 629 on subordination.) But if you are undecided, yet leaning in a negative direction, you might write:

Strongly negative emphasis	Roy James graduated from an excellent management school and is highly recommended, **but** he has no experience.

In this sentence, the ideas before and after *but* are both independent. Joining them with the coordinating word *but* suggests that both sides of the issue are equally important (or "coordinate"). Placing the negative idea last, however, gives it slight emphasis. (See also page 628 on coordination.)

Finally, to emphasize strong support for the candidate, you could say:

Although Roy James has no experience, **he graduated from an excellent management school and is highly recommended.**

Here, the earlier idea is subordinated by *although,* leaving the two final ideas independent.

Caution: Combine sentences only to advance your meaning and simplify the reader's task. Overstuffed sentences with too much information and too many connections can be hard for readers to sort out.

Overstuffed	Our night supervisor's verbal order from upper management to repair the overheated circuit was misunderstood by Leslie Kidd, who gave the wrong instructions to the emergency crew, thereby causing the fire within 30 minutes.
Clearer	Upper management issued a verbal order to repair the overheated circuit. When our night supervisor transmitted the order to Leslie Kidd, it was misunderstood. Kidd gave the wrong instruction to the emergency crew, and the fire began within 30 minutes.

Research indicates that most people can retain between fifteen and twenty-five words at one time (Boyd 18). But even short sentences can be hard to interpret if they include too many details:

Overstuffed	Send three copies of Form 17-e to all six departments, unless Departments A or B or both request Form 16-w instead.

`var` **VARY SENTENCE CONSTRUCTION AND LENGTH.**   Related ideas often need to be linked in one sentence, to emphasize the connections:

Disconnected	The nuclear core reached critical temperature. The loss-of-coolant alarm was triggered. The operator shut down the reactor.
Connected	As the nuclear core reached critical temperature, triggering the loss-of-coolant alarm, the operator shut down the reactor.

But an idea that should stand alone for emphasis needs a whole sentence of its own:

Correct	Core meltdown seemed inevitable.

However, an unbroken string of long or short sentences can bore and confuse readers, as can a series with identical openings:

Boring and repetitive	There are some drawbacks to diesel engines. **They** are difficult to start in cold weather. **They** cause vibration. **They** also give off an unpleasant odor. **They** cause sulfur dioxide pollution.
Varied	Diesel engines have some drawbacks. Most obvious are their noisiness, cold-weather starting difficulties, vibration, odor, and sulfur dioxide emission.

Similarly, when you write in the first person, overusing *I* makes you appear self-centered. (Some organizations require use of the third person, avoiding the first person completely, for all manuals, lab reports, specifications, product descriptions, and so on.)

Do not avoid personal pronouns if they make the writing more readable (say, by eliminating passive constructions).

short **USE SHORT SENTENCES FOR SPECIAL EMPHASIS.** All this talk about combining ideas might suggest that short sentences have no place in good writing. Wrong. Short sentences (even one-word sentences) provide vivid emphasis. They stick in a reader's mind.

Exercise 12
Combine each set of sentences below into one fluent sentence that provides the requested emphasis.

Sentence set	John is a loyal employee. John is a motivated employee. John is short-tempered with his colleagues.
Combined for positive emphasis	Even though John is short-tempered with his colleagues, he is a loyal and motivated employee.
Sentence set	This word processor has many features. It includes a spelling checker. It includes a thesaurus. It includes a grammar checker.
Combined to emphasize thesaurus	Among its many features, such as spelling and grammar checkers, this word processor includes a thesaurus.

a. The job offers an attractive salary.
   It demands long work hours.
   Promotions are rapid.
   (*Combine for negative emphasis.*)
b. The job offers an attractive salary.
   It demands long work hours.
   Promotions are rapid.
   (*Combine for positive emphasis.*)
c. Our office software is integrated.
   It has an excellent database management program.
   Most impressive is its word-processing capability.
   It has an excellent spreadsheet program.
   (*Combine to emphasize the word processor.*)

d. Company X gave us the lowest bid.
Company Y has an excellent reputation.
(*Combine to emphasize Company Y.*)
e. Superinsulated homes are energy efficient.
Superinsulated homes create a danger of indoor air pollution.
The toxic substances include radon gas and urea formaldehyde.
(*Combine for a negative emphasis.*)
f. Computers cannot *think* for the writer.
Computers eliminate many mechanical writing tasks.
They speed the flow of information.
(*Combine to emphasize the first assertion.*)

FINDING THE EXACT WORDS

Too often, language can *camouflage* rather than communicate. People see many reasons to hide behind language, as when they

Situations in which people often hide behind language

- speak for their company but not for themselves
- fear the consequences of giving bad news
- are afraid to disagree with company policy
- make a recommendation some readers will resent
- worry about making a bad impression
- worry about being wrong
- pretend to know more than they do
- avoid admitting a mistake, or ignorance

Inflated and unfamiliar words, borrowed expressions, and needlessly technical terms camouflage meaning. Whether intentional or accidental, poor word choices result in inefficient and often unethical writing that resists interpretation and frustrates the audience.

Following are strategies for finding words that are *convincing, precise,* and *informative.*

simple   **USE SIMPLE AND FAMILIAR WORDS.**   Don't replace technically precise words with nontechnical words that are vague or imprecise. Don't write *a part that makes the computer run* when you mean *central processing unit.* Use the precise term, and define it in a glossary for nontechnical readers:

Glossary entry   **Central processing unit:** the part of the computer that controls information transfer and carries out arithmetic and logical instructions.

Certain technical words may be indispensable in certain contexts, but the nontechnical words can usually be simplified. Instead of *answering in the affirmative,* use *say yes;* instead of *endeavoring to promulgate* a new policy, use *try to announce* it.

Unfamiliar words	Acoustically attenuate the food consumption area.
Revised	Soundproof the cafeteria.

Don't use three syllables when one will do. Generally, trade for less:

More syllables traded for fewer

aggregate	=	total
demonstrate	=	show
endeavor	=	effort, try
frequently	=	often
initiate	=	begin
is contingent upon	=	depends on
multiplicity of	=	many
subsequent to	=	after
utilize	=	use

Trim wherever you can. Most important, choose words you hear and use in everyday speaking—words that are universally familiar.

Don't write *I deem* when you mean *I think, Keep me apprised* instead of *Keep me informed, I concur* instead of *I agree, securing employment* instead of *finding a job,* or *it is cost prohibitive* instead of *we can't afford it.*

Don't write like the author of a report from the Federal Aviation Administration, who recommended that manufacturers of the DC–10 reevaluate *the design of the entire pylon assembly to minimize design factors which are resulting in sensitive and/or critical maintenance and inspection procedures* (25 words, 50 syllables). A plain English translation: *Redesign the pylons so they are easier to maintain and inspect* (11 words, 18 syllables).

Besides being annoying, needlessly big or unfamiliar words can be *ambiguous.*

Ambiguous	Make an improvement in the clerical situation.

Should we hire more clerical personnel or better personnel or should we train the personnel we have? Words chosen to impress readers too often confuse them instead. A plain style is more persuasive because "it leaves no one out" (Cross 6).

Plain English needed	Avoid prolix nebulosity.
Revised	Don't be wordy and vague.

Of course, now and then the complex or more elaborate word best expresses your meaning. For instance, we would not substitute *end* for *terminate* in referring to something with an established time limit.

Correct	Our trade agreement terminates this month.

If a complex word can replace a handful of simpler words—and can sharpen your meaning—use it.

Weak	Six rectangular grooves **around the outside edge** of the steel plate **are needed for** the pressure clamps **to fit into**.
Informative and precise	Six rectangular grooves on the steel plate **perimeter accommodate** the pressure clamps.
Weak	We need a **one-to-one exchange of ideas and opinions**.
Informative and precise	We need a **dialogue**.
Weak	Sexist language **contributes to the ongoing prevalence** of gender stereotypes.
Informative and precise	Sexist language **perpetuates** gender stereotypes.

Exercise 13

Revise each sentence below for straightforward and familiar language.

a. May you find luck and success in all endeavors.
b. I suggest you reduce the number of cigarettes you consume.
c. Within the copier, a magnetic reed switch is utilized as a mode of replacement for the conventional microswitches that were in use on previous models.
d. A good writer is cognizant of how to utilize grammar in a correct fashion.
e. I will endeavor to ascertain the best candidate.
f. In view of the fact that the microscope is defective, we expect a refund of our full purchase expenditure.
g. I wish to upgrade my present employment situation.

---

jarg

When jargon is appropriate

AVOID USELESS JARGON. Every profession has its own "shorthand." Among specialists, technical terms are a precise and economical way to communicate. For example, *stat* (from the Latin "statim" or "immediately") is medical jargon for *Drop everything and deal with this emergency.* For computer buffs, a *glitch* is a momentary power surge that can erase the contents of internal memory; a *bug* is an error that causes a program to run incorrectly. Useful jargon conveys clear meaning to a knowledgeable audience.

When jargon is inappropriate

But technical language can also be used inappropriately. Useless jargon is meaningless to insiders as well as outsiders. In the world of useless jargon people don't *cooperate* on a project; instead, they *interface* or *contiguously optimize their efforts.* Rather than *designing a model,* they *formulate a paradigm.* Instead of *observing limits* or *boundaries,* they *function within specific parameters.*

A popular form of useless jargon is adding *-wise* to nouns, as shorthand for *in reference to* or *in terms of.*

Useless jargon	**Expensewise** and **schedulewise,** this plan is unacceptable.
Revised	In terms of expense and scheduling, this plan is unacceptable.

Writers create another form of useless jargon when they invent verbs from nouns or adjectives by adding an *-ize* ending: Don't invent *prioritize* from *priority;* instead use *to rank priorities.*

Useless jargon's worst fault is that it makes the person using it seem stuffy and pretentious:

Pretentious	Unless all parties interface synchronously within given parameters, the project will be rendered inoperative.
Possible translation	Unless we coordinate our efforts, the project will fail.

Beyond reacting with frustration, readers often conclude that useless jargon is camouflage for a writer with something to hide.

Before using any jargon, think about your specific audience and ask yourself: "Can I find an easier way to say exactly what I mean?" Use jargon only if it *improves* your communication.

> **NOTE**
>
> *If your employer insists on needless jargon or elaborate phrasing, then you have little choice. What is best in matters of style is not always what some people consider appropriate. Use the style your employer or organization expects, but remember that most documents that achieve superior results use plain English.*

acr   **USE ACRONYMS SELECTIVELY.**   Acronyms are another form of specialized shorthand, or jargon. They are formed from the first letters of words in a phrase (as in *LOCA* from *loss of coolant accident*) or from a combination of first letters and parts of words (as in *bit* from *binary digit* or *pixel* from *picture element*).

Computer technology has spawned countless acronyms, including:

```
ISDN   = Integrated Services Digital Network
Telnet = Telephone Network
URL    = Uniform Resource Locator
```

Acronyms *can* communicate concisely—but only when the audience knows their meaning, and only when you use the term often in your document. Whenever you first use an acronym, spell out the words from which it is derived.

An acronym defined | **Modem** ("modulator + demodulator"): a device that converts, or "modulates," computer data in electronic form into a sound signal that can be transmitted via phone line and then reconverted, or "demodulated," into electronic form for the receiving computer.

trite

AVOID TRITENESS.   Writers who rely on worn out phrases (clichés) like the following come across as either too lazy or too careless to find exact ways to say what they mean.

Worn out phrases

make the grade	the chips are down
in the final analysis	not by a long shot
close the deal	last but not least
hard as a rock	welcome aboard
water under the bridge	over the hill
holding the bag	bite the bullet
up the creek	work like a dog

Exercise 14
Revise each sentence below to eliminate useless jargon and triteness.

  a. For the obtaining of the X–33 word processor, our firm will have to accomplish the disbursement of funds to the amount of $6,000.
  b. To optimize your financial return, prioritize your investment goals.
  c. The use of this product engenders a fifty-percent repeat consumer encounter.
  d. We'll have to swallow our pride and admit our mistake.
  e. We wish to welcome all new managers aboard.
  f. Managers who make the grade are those who can take daily pressures in stride.

---

euph

AVOID MISLEADING EUPHEMISMS.   A form of understatement, euphemisms are expressions aimed at politeness or at making unpleasant subjects seem less offensive. Thus, we *powder our nose* or *use the boys' room* instead of *using the bathroom;* we *pass away* or *meet our Maker* instead of *dying.*

When euphemism is appropriate

When euphemisms avoid offending or embarrassing our audience, they are legitimate. Instead of telling a job applicant he or she is *unqualified,* we might say, *Your background doesn't meet our needs.* In addition, there are times when friendliness and interoffice harmony are more likely to be preserved with writing that is not too abrupt, bold, blunt, or emphatic (MacKenzie 2).

Euphemisms are unethical if they understate the truth when only the truth will serve. In the sugarcoated world of misleading euphemisms, bad news disappears:

When euphemism is deceptive

◆ Instead of being *laid off* or *fired,* workers are *surplused* or *deselected,* or the company is *downsized.*
◆ Instead of *lying* to the public, the government *engages in a policy of disinformation.*
◆ Instead of *wars* and *civilian casualties,* we have *conflicts* and *collateral damage.*

Language loses all meaning when *criminals* become *offenders,* when *mistakes* become *teachable moments,* and when people who are just plain *lazy* become *under-*

achievers. Plain talk is always better than deception. If someone offers you a job *with limited opportunity for promotion,* expect a *dead-end job.*

over | **AVOID OVERSTATEMENT.**   Exaggerating destroys credibility. Be cautious when using words such as *best, biggest, brightest, most,* and *worst.*

> Overstated   **Most** businesses have **no** loyalty toward their employees.
> Revised   **Some** businesses have **little** loyalty toward their employees.

> Overstated   You will find our product to be the **best.**
> Revised   You will **appreciate the high quality** of our product.

Be aware of the vast differences in meaning among these words:

Qualify your generalizations

few	rarely
some	sometimes
many	often
most	usually
all	always

Unless you specify *few, some, many,* or *most,* people can interpret your statement to mean *all.*

> Misleading   Assembly-line employees are doing shabby work.

Unless you mean *all,* qualify your generalization with *some,* or *most*—or even better, specify *twenty percent.*

Exercise 15
Revise each sentence below to eliminate euphemism, overstatement, or unsupported generalizations.

    a. I finally must admit that I am an abuser of intoxicating beverages.
    b. I was less than candid.
    c. This employee is poorly motivated.
    d. Most entry-level jobs are boring and dehumanizing.
    e. Clerical jobs offer no opportunity for advancement.
    f. Because of your absence of candor, we can no longer offer you employment.

---

ww | **AVOID IMPRECISE WORDS.**   Even words listed as synonyms have different shades of meaning. Do you mean to say *I'm slender, You're slim, She's lean,* or *He's scrawny?* The wrong choice could be disastrous.

A single wrong word can be offensive, as in this statement by a job applicant:

> **Offensive**   Another attractive feature of your company is its **adequate** training program.

While "adequate" might honestly convey the writer's intended meaning, the word seems inappropriate in this context (an applicant expressing a judgment about a program). Although the program may not have been highly ranked, the writer could have used any of several alternatives (*solid, respectable, growing*—or no modifier at all).

Be especially aware of similar words with dissimilar meanings, as in these examples:

Words often confused

affect/effect	farther/further
all ready/already	fewer/less
almost dead/dying	healthy/healthful
among/between	imply/infer
continual/continuous	invariably/inevitably
eager/anxious	uninterested/disinterested
fearful/fearsome	worse/worst

Don't write *Skiing is healthy* when you mean that skiing promotes good health. *Healthy* means to be in a state of health. Healthful behaviors keep us healthy.

Be on the lookout for imprecisely phrased (and therefore illogical) comparisons.

> **Imprecise**   Your bank's interest rate is higher than BusyBank. (Can a rate be higher than a bank?)
>
> **Precise**   Your bank's interest rate is higher than BusyBank's.

Imprecision can create ambiguity. For instance, is *send us more personal information* a request for more information that is personal or for information that is more personal? Does your client expect *fewer* or *less* technical details in your report?

Precision ultimately enhances conciseness, when one exact word replaces multiple inexact words.

> **Wordy and less exact**   I have **put together** all the financial information.
> **Keep doing** this exercise for ten seconds.
>
> **Concise and more exact**   I have **assembled** all the. . . .
> **Continue** this exercise. . . .

`spec`   **BE SPECIFIC AND CONCRETE.**   General words name broad classes of things, such as *job, computer,* or *person.* Such terms usually need to be clarified by more specific ones.

General terms traded for specific terms

job	=	senior accountant for Softbyte Press
computer	=	Macintosh PowerBook G3
person	=	Sarah Jones, production manager

The more specific your words, the sharper your meaning.

How the level of generality affects clarity

General ↑ structure
dwelling
vacation home
log cabin

Specific ↓ log cabin in Vermont
a three-room log cabin on the banks of the Battenkill River

Notice how the picture becomes more vivid as we move to lower levels of generality.

Abstract words name qualities, concepts, or feelings (*beauty, luxury, depression*) whose exact meaning must be clarified by *concrete* words—words that name things we can know through our five senses.

Abstract terms traded for concrete terms

a **beautiful** view	=	snowcapped mountains, a wilderness lake, pink granite ledge, ninety-foot birch trees
a **luxurious** condominium	=	imported tiles, glass walls, oriental rugs
a **depressed** worker	=	suicidal urge, insomnia, feelings of worthlessness, no hope for improvement

Informative writing *tells* and *shows.*

General   One of our **workers** was **injured** by a **piece of equipment recently.**

The boldface words only *tell* without showing.

Specific   Alan Hill suffered a **broken thumb** while working on a **lathe yesterday.**

Choose informative words that express exactly what you mean. Don't write *thing* when you mean *lever, switch, micrometer,* or *disk.*

In some instances, of course, you may wish to generalize for the sake of diplomacy. Instead of writing *Bill, Mary, and Sam have been tying up the office phones with personal calls,* you might prefer to generalize: *Some employees have been. . . .* The second version gets your message across without pointing the finger.

When you can, provide solid numbers and statistics that get your point across:

General   In 1972, thousands of people were killed or injured on America's highways. Many families had at least one relative who was

> a casualty. After the speed limit was lowered to 55 miles per hour in late 1972, the death toll began to drop.

Specific   In 1972, 56,000 people died on America's highways; 200,000 were injured; 15,000 children were orphaned. In that year, if you were a member of a family of five, chances are that someone related to you by blood or law was killed or injured in an auto accident. After the speed limit was lowered to 55 miles per hour in late 1972, the death toll dropped steadily to 41,000 in 1975.

Most good writing offers both general and specific information. The most general material appears in the topic statement and sometimes in the conclusion because these parts, respectively, set the paragraph's direction and summarize its content.

Exercise 16
Revise each sentence below to make it more precise or informative.

a. Our outlet does more business than Chicago.
b. Anaerobic fermentation is used in this report.
c. Loan payments are due bimonthly.
d. Your crew damaged a piece of office equipment.
e. His performance was admirable.
f. This thing bothers me.

Analogy versus comparison

USE ANALOGIES TO SHARPEN THE IMAGE.   Ordinary comparison shows similarities between two things *of the same class* (two computer keyboards, two methods of cleaning dioxin-contaminated sites). Analogy shows some essential similarity between two things of *different classes* (writing and computer programming, computer memory and post office boxes).

Analogies are good for emphasizing a point (*Some rain is now as acidic as vinegar*). They are especially useful in translating something abstract, complex, or unfamiliar, as long as the easier subject is broadly familiar to readers. Analogy therefore calls for particularly careful analyses of audience.

Analogies can save words and convey vivid images. *Collier's Encyclopedia* describes the tail of an eagle in flight as "spread like a fan." The following sentence from a description of a trout feeder mechanism uses an analogy to clarify the positional relationship between two working parts:

Analogy   The metal rod is inserted (and centered, **crosslike**) between the inner and outer sections of the clip.

Without the analogy *crosslike,* we would need something like this to visualize the relationship:

Missing analogy	The metal rod is inserted, **perpendicular to the long plane and parallel to the flat plane,** between the inner and outer sections of the clip.

Besides naming things, analogies help *explain* things. The following analogy from the *Congressional Research Report* helps us understand something unfamiliar (dangerous levels of a toxic chemical) by comparing it to something more familiar (human hair).

Analogy	A dioxin concentration of 500 parts per trillion is lethal to guinea pigs. One part per trillion is roughly equal to the thickness of a human hair compared to the distance across the United States.

ADJUSTING YOUR TONE

tone

How tone is created

Your tone is your personal stamp—the personality that takes shape between the lines. The tone you create depends on (1) the distance you impose between yourself and the reader, and (2) the attitude you show toward the subject.

Assume, for example, that a friend is going to take over a job you've held. You've decided to write your friend instructions for parts of the job. Here is your first sentence:

Informal

> Now that you've arrived in the glamorous world of office work, put on your track shoes; this is no ordinary manager-trainee job.

This sentence imposes little distance between you and reader (it uses the direct address, "you," and the humorous suggestion to "put on your track shoes"). The ironic use of "glamorous" suggests that you mean just the opposite: that the job holds little glamor.

For a different reader (say, the recipient of a company training manual), you would have chosen some other opening:

Semiformal

> As a manager trainee at GlobalTech, you will work for many managers. In short, you will spend little of your day seated at your desk.

The tone now is serious, no longer intimate, and you express no distinct attitude toward the job. For yet another audience (clients or investors who will read an annual report), you again might alter the tone:

Formal

> Manager trainees at GlobalTech are responsible for duties that extend far beyond desk work.

Here the businesslike shift from second- to third-person address makes the tone too impersonal for any document addressed to the trainees themselves.

We already know how tone works in speaking. When you meet someone new, for example, you respond in a tone that defines your relationship:

Tone announces interpersonal distance

Honored to make your acquaintance. [*formal tone—greatest distance*]
How do you do? [*formal*]
Nice to meet you. [*semiformal—medium distance*]
Hello. [*semiformal*]
Hi. [*informal—least distance*]
What's happening? [*informal—slang*]

Each of these greetings is appropriate in some situations, inappropriate in others. Whichever tone you decide on, be consistent throughout your document.

Inconsistent tone   My office isn't fit for a pig [*too informal*]; it is ungraciously unattractive [*too formal*].

Revised   The shabbiness of my office makes it an unfit place to work.

In general, lean toward an informal tone without using slang.

In addition to setting the distance between writer and reader, your tone implies your *attitude* toward the subject and the reader.

Tone announces attitude

We dine at seven.
Dinner is at seven.
Let's eat at seven.
Let's chow down at seven.
Let's strap on the feedbag at seven.
Let's pig out at seven.

The words you choose tell readers a great deal about where you stand.

One problem with tone occurs when your attitude is unclear. Say *I enjoyed the fiber optics seminar* instead of *My attitude toward the fiber optics seminar was one of high approval.* Say *Let's liven up our dull relationship* instead of *We should inject some rejuvenation into our lifeless liaison.*

GUIDELINES

for Finding the Right Tone

1. Use a formal or semiformal tone in writing for superiors, professionals, or academics (depending on what you think the reader expects).
2. Use a semiformal or informal tone in writing for colleagues and subordinates (depending on how close you feel to your reader).
3. Use an informal tone when you want your writing to be conversational, or when you want it to sound like a person talking.
4. Above all, find out what the preferences are in your organization.

In writing a memo about an upcoming meeting to review the reader's job evaluation, would you invite this person to *discuss* the evaluation, *talk it over, have a chat,* or *chew the fat?* Decide how casual or serious your attitude should be. Use the following strategies for making your tone conversational and appropriate.

USE AN OCCASIONAL CONTRACTION.   Unless you have reason to be formal, use (but do not overuse) contractions. Balance an *I am* with an *I'm,* a *you are* with a *you're,* and *it is* with an *it's.*

Missing contraction	Do not be wordy and vague.
Revised	Don't be wordy and vague.

Generally, use contractions only with pronouns, not with nouns or proper nouns (names).

Awkward contractions	Barbara'll be here soon.
	Health's important.
Ambiguous contractions	The dog's barking.
	Bill's skiing.

These ambiguous contractions could be confused with possessive constructions.

ADDRESS READERS DIRECTLY.   Use the personal pronouns *you* and *your* to connect with readers.

Impersonal tone	Students at our college will find the faculty always willing to help.
Personal tone	As a student at our college, **you** will find the faculty always willing to help.

Readers relate better to something addressed directly to them.

Caution: Use *you* and *your* only to correspond *directly* with the reader, as in a letter, memo, instructions, or some form of advice, encouragement, or persuasion. By using *you* and *your* when your subject and purpose call for first or third person, you might write something like this:

Wordy and awkward	**When you** are in northern Ontario, **you** can see wilderness lakes everywhere around **you.**
Appropriate	Wilderness lakes are everywhere in northern Ontario.

Exercise 17

The sentences below suffer from pretentious language, unclear expression of attitude, missing contractions, or indirect address. Adjust the tone.

a. Further interviews are a necessity to our ascertaining the most viable candidate.
b. Do not submit the proposal if it is not complete.
c. Employees must submit travel vouchers by May 1.
d. Persons taking this test should use the HELP option whenever they need it.
e. I am not unappreciative of your help.
f. My disapproval is far more than negligible.

USE *I* AND *WE* WHEN APPROPRIATE.   Instead of disappearing behind your writing, use *I* or *We* when referring to yourself or your organization.

Distant	The writer would like a refund.
Revised	I would like a refund.

A message becomes doubly impersonal when both writer and reader disappear.

Impersonal	The requested report will be sent next week.
Personal	**We** will send the report **you** requested next week.

PREFER THE ACTIVE VOICE.   Because the active voice is more direct and economical than the passive voice, it generally creates a less formal tone. (Review pages 220–23.)

Passive and impersonal	Travel expenses cannot be reimbursed unless receipts are submitted.
Active and personal	We cannot reimburse your travel expenses unless you submit receipts.

Exercise 18

These sentences have too few *I* or *We* constructions or too many passive constructions. Adjust the tone.

a. Payment will be made as soon as an itemized bill is received.
b. You will be notified.
c. Your help is appreciated.
d. Our reply to your bid will be sent next week.
e. Your request will be given our consideration.
f. My opinion of this proposal is affirmative.
g. This writer would like to be considered for your opening.

EMPHASIZE THE POSITIVE.   Whenever you offer advice, suggestions, or recommendations, try to emphasize benefits rather than flaws.

Critical tone	Because of your division's lagging productivity, a management review may be needed.

Encouraging tone   A management review might help boost productivity in your division.

How tone can be too informal

AVOID AN OVERLY INFORMAL TONE.   We generally do not write in the same way we would speak to friends at the local burger joint or street corner. Achieving a conversational tone does not mean lapsing into substandard usage, slang, profanity, or excessive colloquialisms. *Substandard usage* ("He ain't got none," "I seen it today," "She brang the book") ignores standards of educated expression. *Slang* ("hurling," "belted," "bogus," "bummed") usually has specific meaning only for members of a particular in-group. *Profanity* ("This idea sucks," "pissed off," "What the hell") not only displays contempt for the audience but also triggers contempt for the person using it. *Colloquialisms* ("O.K.," "a lot," "snooze," "in the bag") are understood more widely than slang, but tend to appear more in speaking than in writing.

How tone can offend

Tone is considered offensive when it violates the reader's expectations: when it seems disrespectful, tasteless, distant and aloof, too chummy, casual, or otherwise inappropriate for the topic, the reader, and the situation.

When to use an academic tone

A formal or academic tone is perfectly appropriate in countless writing situations: a research paper, a job application, a report for the company president. In a history essay, for example, you would not refer to George Washington and Abraham Lincoln as "those dudes, George and Abe." Whenever you begin with rough drafting or brainstorming, your tone might be overly informal and is likely to require some adjustment during subsequent drafts.

Slang and profanity are almost always inappropriate in school or workplace writing. The occasional colloquial expression, however, helps soften the tone of any writing—as long as the situation calls for a measure of informality.

bias

AVOID BIAS.   If people expect an impartial report, try to keep you own biases out of it. Imagine, for instance, that you have been sent to investigate the causes of an employee-management confrontation at your company's Omaha branch. Your initial report, written for the New York central office, is intended simply to describe what happened. Here is how an unbiased description might read:

A factual account

> At 9:00 A.M. on Tuesday, January 21, eighty women employees set up picket lines around the executive offices of our Omaha branch, bringing business to a halt. The group issued a formal protest, claiming that their working conditions were repressive, their salary scale unfair, and their promotional opportunities limited. The women demanded affirmative action, insisting that the company's hiring and promotional policies and wage scales be revised. The demonstration ended when Garvin Tate, vice president in charge of personnel, promised to appoint a committee to investigate the group's claims and to correct any inequities.

Notice the absence of implied judgments; the facts are presented objectively. A less impartial version of the event, from a protestor's point of view, might read like this:

A biased version

Last Tuesday, sisters struck another blow against male supremacy when eighty women employees paralyzed the company's repressive and sexist administration for more than six hours. The timely and articulate protest was aimed against degrading working conditions, unfair salary scales, and lack of promotional opportunities for women. Stunned executives watched helplessly as the group organized their picket lines, determined to continue their protest until their demands for equal rights were addressed. An embarrassed vice president quickly agreed to study the group's demands and to revise the company's discriminatory policies. The success of this long-overdue confrontation serves as an inspiration to oppressed women employees everywhere.

Judgmental words (*male supremacy, degrading, paralyzed, articulate, stunned, discriminatory*) inject the writer's attitude, even though it isn't called for. In contrast to this bias, the following version patronizingly defends the status quo:

A biased version

Our Omaha branch was the scene of an amusing battle of the sexes last Tuesday, when a group of irate feminists, eighty strong, set up picket lines for six hours at the company's executive offices. The protest was lodged against supposed inequities in hiring, wages, working conditions, and promotion for women in our company. The radicals threatened to surround the building until their demands for "equal rights" were met. A bemused vice president responded to this carnival demonstration with patience and dignity, assuring the militants that their claims and demands—however inaccurate and immoderate—would receive just consideration.

Again, qualifying adjectives and superlatives slant the tone.

Being unbiased, of course, doesn't mean remaining "neutral" about something you know to be wrong or dangerous (Kremers 59). If, for instance, you conclude that the Omaha protest was clearly justified, say so. In situations in which your opinion is expected, announce where you stand.

sexist

AVOID SEXIST USAGE.   Sexist usage refers to doctors, lawyers, and other professionals as *he* or *him,* while referring to nurses, secretaries, and homemakers as *she* or *her.* In this traditional stereotype, males do the jobs that really matter and that pay higher wages, whereas females serve only as support and decoration. When females do invade traditional "male" roles, we might express our surprise at their boldness by calling them *female executives, female sportscasters, female surgeons,* or *female hockey players.* Likewise, to demean males who have settled for "female" roles, we sometimes refer to *male secretaries, male nurses, male flight attendants,* or *male models.*

offen

AVOID OFFENSIVE USAGES OF ALL TYPES.   Enlightened communication respects all people in reference to their specific cultural, racial, ethnic, and national background; sexual and religious orientation; age or physical condition. References to individuals and groups should be as neutral as possible; no matter how inadvertent, any expression that seems condescending or judgmental or that violates

GUIDELINES

for Nonsexist Usage

1. Use neutral expressions:

chair, or chairperson	rather than	chairman
businessperson	rather than	businessman
supervisor	rather than	foreman
letter carrier	rather than	postman
homemaker	rather than	housewife
humanity, or humankind	rather than	mankind
actor	rather than	actor vs. actress

2. Rephrase to eliminate the pronoun, if you can do so without altering your original meaning.

>**Sexist**   A writer will succeed if **he** revises.
>**Revised**   A writer who revises succeeds.

3. Use plural forms.

>**Sexist**   A writer will succeed if **he** revises.
>**Revised**   Writers will succeed if **they** revise (but *not* A writer will succeed if **they** revise.)

When using a plural form, don't create an error in pronoun-referent agreement by having the *plural* pronoun *they* or *their* refer to a *singular* referent (as in **Each writer** *should do* **their** *best*).

4. When possible (as in direct address) use *you*: **You** *will succeed if* **you** *revise.* But use this form *only* when addressing someone directly. (See page 246 for discussion.)

5. Use occasional pairings (*him* or *her, she* or *he, his* or *hers*): *A writer will succeed if* **she or he** *revises.* But note that overuse of such pairings can be awkward: *A writer should do* **his or her** *best to make sure that* **he or she** *connects with* **his or her** *readers.*

6. Use feminine and masculine pronouns alternately.

>**Alternating**   An effective writer always focuses on *her* audience. The
>**pronouns**   writer strives to connect with all *his* readers.

7. Drop diminutive endings such as *-ess* and *-ette* used to denote females (*poetess, drum majorette, actress*).

8. Use *Ms.* instead of *Mrs.* or *Miss,* unless you know that person prefers one of the traditional titles. Or omit titles completely: *Roger Smith and Jane Kelly; Smith and Kelly.*

9. In quoting sources that have ignored present standards for nonsexist usage, consider these options:
   - Insert [*sic*] ("thus" or "so") following the first instance of sexist terminology in a particular passage.
   - Use ellipses to omit sexist phrasing.
   - Paraphrase instead of quoting directly.

the reader's sense of appropriateness is offensive. Detailed guidelines for reducing biased usage appear in these two works:

Schwartz, Marilyn, et al. *Guidelines for Bias-Free Writing.* Bloomington: Indiana UP, 1995.

Publication Manual of the American Psychological Association, 4th ed. Washington, DC: American Psychological Association, 1994.

GUIDELINES

for Inoffensive Usage

Below is a sampling of suggestions adapted from the works listed above:

1. When referring to members of a particular culture, be as specific as possible about that culture's identity: Instead of *Latin American* or *Asian* or *Hispanic,* for instance, prefer *Cuban American* or *Korean* or *Nicaraguan.* Instead of *American workforce,* specify *U.S. workforce* when referring to the United States.

   Avoid judgmental expressions: Instead of *third-world* or *undeveloped nations* or the *Far East,* prefer *developing* or *newly industrialized nations* or *East Asia.* Instead of *nonwhites,* refer to *people of color.*

2. When referring to someone who has a disability, avoid terms that could be considered pitying or overly euphemistic, such as *victims, unfortunates, challenged,* or *differently abled.* Focus on the individual instead of the disability: Instead of *blind person* or *amputee,* refer to a *person who is blind* or a *person who has lost an arm.*

   In general usage, avoid expressions that demean those who have medical conditions: *retard, mental midget, insane idea, lame excuse, the blind leading the blind, able-bodied workers,* and so on.

3. When referring to members of a particular age group, prefer *girl* or *boy* for people of age fourteen or under; *young person, young adult, young man,* or *young woman* for those of high-school age; and *woman* or *man* for those of college age. (*Teenager* or *juvenile* carries certain negative connotations.) Instead of *the elderly,* prefer *older persons.*

Exercise 19

The sentences below suffer from negative emphasis, excessive informality, biased expressions, or offensive usage. Adjust the tone.

   a. If you want your workers to like you, show sensitivity to their needs.
   b. By not hesitating to act, you prevented my death.
   c. The union has won its struggle for a decent wage.

    d. The group's spokesman demanded salary increases.
    e. Each employee should submit his vacation preferences this week.
    f. While the girls played football, the men waved pom-poms.
    g. Aggressive management of this risky project will help you avoid failure.
    h. The explosion left me blind as a bat for nearly an hour.
    i. This dude would be an excellent employee if only he could learn to chill out.

Exercise 20

Find examples of overly euphemistic language (such as "chronologically challenged") or of insensitive language. Discuss your examples in class.

CONSIDERING THE CULTURAL CONTEXT

The style guidelines in this chapter apply specifically to standard English in North America. But practices and preferences differ widely in different cultural contexts.

Certain cultures might prefer long sentences and technical language, to convey an idea's full complexity. Other cultures value expressions of respect, politeness, praise, and gratitude more than clarity or directness (Hein 125–26; Mackin 349–50).

Writing in non-English languages tends to be more formal than in English, and may rely heavily on passive voice (Weymouth 144). French readers, for example, may prefer an elaborate style that reflects sophisticated and complex modes of thinking. In contrast, our "plain English," conversational style might connote simple-mindedness, disrespect, or incompetence (Thrush 277).

In translation or in a different cultural context, certain words carry offensive or unfavorable connotations. For example, certain cultures use "male" and "female" in referring only to animals (Coe, "Writing for Other Cultures" 17). Other notable disasters (Gesteland 20; Victor 44):

- The Chevrolet *Nova*—meaning "doesn't go" in Spanish
- The Finnish beer *Koff*—for an English-speaking market
- Colgate's *Cue* toothpaste—an obscenity in French
- A brand of bicycle named *Flying Pigeon*—imported for a U.S. market

Idioms ("strike out," "ground rules") hold no logical meaning for other cultures. Slang ("bogus," "fat city") and colloquialisms ("You bet," "Gotcha") can strike readers as being too informal and crude.

Offensive writing (including inappropriate humor) can alienate audiences—toward you *and* your culture (Sturges 32).

LEGAL AND ETHICAL IMPLICATIONS OF WORD CHOICE

Chapter 5 (page 73) discusses how workplace writing is regulated by laws against libel, deceptive advertising, and defective information. One common denominator among these violations resides in imprecise or inappropriate word choice. We are all accountable for the words we use—intentionally or not—in framing the audience's perception and understanding.

Situations in which word choice has ethical or legal consequences

- *Assessing risk.* Is the investment you are advocating "a sure thing" or merely "a good bet," or even "risky"? Are you announcing a "caution," a "warning," or a "danger"? Should methane levels in mineshaft #3 "be evaluated" or do "they pose a definite explosion risk"? Never underestimate the risks.
- *Offering a service or product.* Are you proposing to "study the problem" or "explore solutions to the problem" or "eliminate the problem"? Do you "stand behind" your product or do you "guarantee" it? Never promise more than you can deliver.
- *Giving instructions.* Before inserting the widget between the grinder blades, should I "switch off the grinder" or "disconnect the grinder from its power source" or "trip the circuit breaker," or do all three? Triple-check the clarity of your instructions.
- *Comparing your product with competing products.* Instead of referring to a competitor's product as "inferior" or "second-rate" or "substandard," talk about your own "first-rate product" that "exceeds (or meets) standards." Never run down the competition.
- *Evaluating an employee* (Clark T. 75–76). In a personnel evaluation, don't refer to the employee as a "troublemaker" or as "unprofessional," or "too abrasive" or "too uncooperative" or "incompetent" or "too old" for the job. Focus on the specific requirements of this job, and offer *factual* instances in which these requirements have been violated: "Our monitoring software recorded five visits by this employee to X-rated Web sites during working hours." or "This employee arrives late for work on average twice weekly, has failed to complete assigned projects on three occasions, and has difficulty working with others." Instead of expressing personal judgments, offer the facts. Otherwise, you risk violating federal laws against discrimination and you invite costly libel or anti-discrimination suits.

AVOIDING RELIANCE ON AUTOMATED TOOLS

Many of the strategies in this chapter could be executed rapidly with word-processing software. By using the *global search and replace function* in some programs, you can command the computer to search for ambiguous pronoun references, overuse of passive voice, *to be* verbs, *There* and *It* sentence openers, negative constructions, clutter words, needless prefaces and qualifiers, overly technical language, jargon, sexist language, and so on. With an online dictionary or thesaurus, you can check definitions or see a list of synonyms for a word you have used in your writing.

The limits of automation

But these editing aids can be extremely imprecise. They can't eliminate the writer's burden of *choice*. None of the rules offered in this chapter applies universally. Ultimately, the informed writer's sensitivity to meaning, emphasis, and tone—the human contact—determines the effectiveness of any document.

Visual, Design, and Usability Elements

Designing
Visuals

A visual is any pictorial representation used to clarify a concept, emphasize a particular meaning, illustrate a point, or analyze ideas or data. Besides saving space and words, visuals help people process, understand, and remember information. Because they offer powerful new ways of looking at data, visuals reveal trends, problems, and possibilities that otherwise might remain buried in lists of facts and figures.

In printed or online documents, in oral presentations or multimedia programs, visuals are a staple of communication today. This chapter covers four main types of visuals: tables, graphs, charts, and illustrations.

WHY VISUALS ARE ESSENTIAL

Audiences expect more than just raw information; they want the information processed for their understanding. They want to feel intelligent, to understand the message at a glance. Visuals help us answer many of the questions people ask as they process information:

TYPICAL AUDIENCE QUESTIONS IN PROCESSING INFORMATION

- *Which information is most important?*
- *Where, exactly, should I focus?*
- *What do these numbers mean?*
- *What should I be thinking or doing?*
- *What should I remember about this?*
- *What does it look like?*
- *How is it organized?*
- *How is it done?*
- *How does it work?*

More receptive to images than to words, most people resist unbroken pages of printed text. Visuals help diminish this resistance in several ways:

How visuals help neutralize reader resistance

- *Visuals enhance comprehension by displaying abstract concepts in concrete, geometric shapes.* "How does the metric system work?" (see Figure 14.1). "How do lasers work?" (Figure 14.34).
- *Visuals make meaningful comparisons possible.* "How do countries compare in terms of nuclear power generation?" (Figure 14.2). "How does one pound compare with one kilogram?" (Figure 14.1).
- *Visuals depict relationships.* "How does seasonal change affect the rate of construction in our county?" (Figure 14.11). "What is the relationship between Fahrenheit and Celsius temperature?" (Figure 14.1).
- *Visuals serve as a universal language.* In the global workplace, carefully designed visuals can transcend cultural and language differences, and thus facilitate international communication (Figure 14.1).
- *Visuals provide emphasis.* To emphasize the change in death rates for heart disease and cancer since 1980, a table (Table 14.1) or bar graph (Figure 14.4) would be more vivid than a prose statement.

All You Need to Know About Metric
(For Your Everyday Life)

10

Metric is based on the Decimal system

The metric system is simple to learn. For use in your everyday life you need to know only ten units. You also need to get used to a few new temperatures. Of course, there are other units which most persons will not need to learn. There are even some metric units with which you are already familiar; those for time and electricity are the same as you use now.

BASIC UNITS

METER: a little longer than a yard (about 1.1 yards)
LITER: a little larger than a quart (about 1.06 quarts)
GRAM: a little more than the weight of a paper clip

(comparative sizes are shown)

1 METER

1 YARD

25 DEGREES FAHRENHEIT

COMMON PREFIXES
(to be used with basic units)

milli: one-thousandth (0.001)
centi: one-hundredth (0.01)
kilo: one-thousand times (1000)
For example
1000 millimeters = 1 meter
100 centimeters = 1 meter
1000 meters = 1 kilometer

1 LITER

1 QUART

MILK MILK

OTHER COMMONLY USED UNITS

millimeter: 0.001 meter diameter of a paper clip wire
centimeter: 0.01 meter a little more than the width of a paper clip (about 0.4 inch)
kilometer: 1000 meters somewhat farther than 1/2 mile (about 0.6 mile)
kilogram: 1000 grams a little more than 2 pounds (about 2.2 pounds)
milliliter: 0.001 liter five of them make a teaspoon

OTHER USEFUL UNITS
hectare: about 2 1/2 acres
metric ton: about one ton

25 DEGREES CELSIUS

BUTTER

1 POUND

WEATHER UNITS: **FOR TEMPERATURE**   **FOR PRESSURE**
degrees celsius   kilopascals are used
100 kilopascals = 29.5 inches of Hg (14.5 psi)

°C	−40	−20	0	20	37	60	80	100
°F	−40	0	32	80	98.6		160	212

water freezes body temperature water boils

BUTTER

1 KILOGRAM

Figure 14.1   Visuals that Clarify and Simplify   *Source: National Institute of Standards and Technology, 1992.*

Figure 14.2
A Graph
Displaying the
"Big Picture"
*Source: U.S. Bureau
of the Census,
Statistical Abstract
of the United
States: 1997 (117th
edition): 580.*

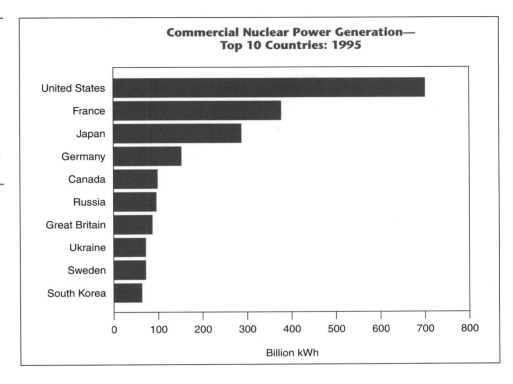

**Commercial Nuclear Power Generation—
Top 10 Countries: 1995**

- United States
- France
- Japan
- Germany
- Canada
- Russia
- Great Britain
- Ukraine
- Sweden
- South Korea

0   100   200   300   400   500   600   700   800

Billion kWh

- ◆ *Visuals focus and organize information, making it easier to remember and interpret.* A simple table, for instance, can summarize a long and difficult printed passage, as in the example that follows.

Assume that you are researching recent death rates for heart disease and cancer. From various sources, you collect these data:

Technical data in
printed form can be
hard to interpret

1. In 1970, 419 males and 309 females per 100,000 people died of heart disease; 172 males and 135 females died of cancer.
2. In 1980, 369 males and 305 females per 100,000 people died of heart disease; 205 males and 164 females died of cancer.
3. In 1990 . . . .

In the written form above, numerical information is repetitious, tedious, and hard to interpret. As the amount of numerical data increases, so does our difficulty in processing this material. Arranged in Table 14.1, these statistics become easier to compare and comprehend.

Along with your visual, analyze or interpret the important trends or key points you want your readers to recognize:

Table 14.1
Data Displayed in
a Table

Death Rates for Heart Disease and Cancer 1970–1994				
	Number of Deaths (per 100,000)			
	Heart Disease		Cancer	
Year	Male	Female	Male	Female
1970	419	309	172	135
1980	369	305	205	164
1990	298	280	221	186
1994	284	279	221	191
% change, 1980–94	−32.2	−9.7	+28.5	+41.5

Source: Adapted from Statistical Abstract of the United States: 1997 *(117th ed.). Washington: GPO: 100, 101.*

A caption explaining
the numerical
relationships

As Table 14.1 indicates, both male and female death rates from heart disease decreased from 1970 to 1994, but males showed a sizably larger decrease. Cancer deaths during this period increased for both groups, with females showing the largest increase.

Besides their value as presentation devices, visuals help us analyze information. Table 14.1 is one example of how visuals enhance critical thinking by helping us to identify and interpret crucial information and to discover meaningful connections.

WHEN TO USE A VISUAL

Translate your writing into visuals whenever they make your point more clearly than prose. Use visuals to *clarify* and to enhance your discussion, not to *decorate* it. Use a visual display to direct the audience's focus or to help them remember something, as in the following situations (Dragga and Gong 46–48):

Use visuals in
situations like these

- when you want to instruct or persuade
- when you want to draw attention to something immediately important
- when you expect the document to be consulted randomly or selectively (e.g., a manual or other reference work) instead of being read in its original sequence (e.g., a memo or letter)
- when you expect the audience to be relatively less educated, less motivated, or less familiar with the topic
- when you expect the audience to be distracted

Never waste space with needless visuals.

NOTE

One alternative approach to the writing process is to begin with one or more key visuals and then compose the text to introduce and interpret the visual.

WHAT TYPES OF VISUALS TO CONSIDER

The following overview sorts visual displays into four categories: tables, graphs, charts, and graphic illustrations. Each type of visual offers a new way of seeing, a different perspective.

TABLES DISPLAY ORGANIZED LISTS OF DATA.   Tables display data (as numbers or words) in rows and columns for comparison. Use tables to present exact numerical values and to organize data so that people can sort out relationships for themselves. Complex tables are usually reserved for more specialized audiences.

Numerical tables present data for analysis, interpretation, and exact comparison.

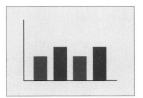

TABLE 1 Charting the Lesson

Lesson	Page	Page	Page
A	1	3	6
B	2	2	5
C	3	1	4

Prose tables organize verbal descriptions, explanations, or instructions.

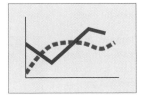

TROUBLESHOOTING

Problem	Cause	Solution
• power	• cord	• plug-in
• light	• bulb	• replace
• flicker	• tube	• replace

GRAPHS DISPLAY NUMERICAL RELATIONSHIPS.   Graphs translate numbers into shapes, shades, and patterns by plotting two or more data sets on a coordinate system. Use graphs to sort out or emphasize specific numerical relationships or to focus on one aspect of the data at a time. Graphs display, at a glance, the approximate values, the point being made about those values, or the relationship being emphasized.

Bar graphs often show comparisons.

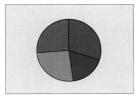

Line graphs often show changes over time.

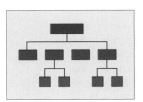

CHARTS DISPLAY THE PARTS OF A WHOLE.   Charts depict relationships without the use of a coordinate system by using circles, rectangles, arrows, connecting lines, and other designs.

Pie charts show the parts or percentages of a whole.

Organization charts show the links among departments, management structures, or other elements of a company.

Flowcharts trace the steps (or decisions) in a procedure or stages in a process.

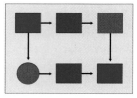

Tree charts show how the parts of an idea or a concept interrelate.

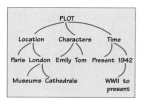

Gantt charts show when each phase of a project is to begin and end.

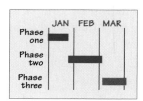

Pictograms use icons (or isotypes) to symbolize the items being displayed or measured.

GRAPHIC ILLUSTRATIONS DEPICT ACTUAL OR VIRTUAL VIEWS.   Graphic illustrations are pictorial devices for helping readers visualize what something looks like, how it works, how it's done, how it happens, or where it's located. Certain diagrams present views that could not be captured by photographing or observing the object.

Representational diagrams present a realistic but simplified view, usually with essential parts labeled.

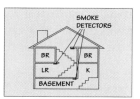

Exploded diagrams show the item pulled apart, to reveal its assembly.

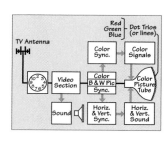

Cutaway diagrams eliminate outer layers to reveal inner parts.

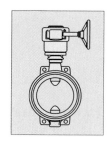

Block or schematic diagrams present the conceptual elements of a principle, process, or system to depict *function* instead of appearance.

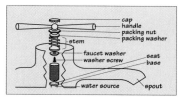

Maps help readers to visualize a specific location or to comprehend data about a specific geographic region.

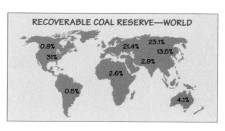

Photographs present an actual picture of the item, process, or procedure.

HOW TO SELECT VISUALS FOR YOUR PURPOSE AND AUDIENCE

You will usually have more than one way to display information in a visual format. To select the most effective display, carefully consider your specific purpose and the abilities and preferences of your audience.

QUESTIONS ABOUT A VISUAL'S PURPOSE AND INTENDED AUDIENCE

What is my purpose?

- *What do I want the audience to do or think (know facts and figures, follow directions, make a judgment, understand how something works, perceive a relationship, identify something, see what something looks like, pay attention)?*
- *Do I want users to focus on one or more exact values, compare two or more values, or synthesize a range of approximate values?*

Who is my audience?

- *What is their technical background on this topic?*
- *What is their level of interest in this topic?*

- *Would they prefer the raw data or interpretations of the data?*
- *Are they accustomed to interpreting visuals?*

Which type of visual might work best in this situation?

- *What forms of information should this visual depict (numbers, shapes, words, pictures, symbols)?*
- *Which visual display would be most compatible with the type of judgment, action, or understanding I seek from this audience?*
- *Which visual display would this audience find most accessible?*

Here are a few examples of the choices you must consider in selecting visuals:

Choices to consider in selecting visuals

- If you merely want the audience to know facts and figures, a table might suffice, but if you want them to make a particular judgment about these data, a bar graph, line graph, or pie chart might be preferable.
- To depict the operating parts of a mechanism, an exploded or cutaway diagram might be preferable to a photograph.
- Expert audiences tend to prefer numerical tables, flowcharts, schematics, and complex graphs or diagrams they can interpret for themselves.
- General audiences tend to prefer basic tables, graphs, diagrams, and other visuals that direct their focus and interpret key points extracted from the data.

Although several alternatives might be possible, one particular type of visual (or a combination) is usually superior for a given purpose and audience. The best option, however, may not always be available to you. Your particular audience or organization may express its own preferences. Or your choices may be limited by lack of equipment (software, scanners, digitizers), insufficient personnel (graphic

PREFERRED DISPLAYS FOR SPECIFIC VISUAL PURPOSES

Purpose . . . . .Preferred Visual

◆ Organize numerical data . . . . .Table
◆ Show comparative data . . . . .Table, bar graph, line graph
◆ Show a trend .Line graph
◆ Interpret or emphasize data . . . . .Bar graph, line graph, pie chart, map
◆ Introduce an unfamiliar object . . . . .Photo, representational diagram
◆ Display a project schedule . . . . .Gantt chart
◆ Show how parts are assembled . . . . . .Photo, exploded diagram
◆ Show how something is organized . . . .Organization chart, map
◆ Give instructions . . . . .Prose table, photo, diagrams, flowchart
◆ Explain a process . . . . .Flowchart, block diagram
◆ Clarify a concept or principle . . . . .Block or schematic diagram, tree chart
◆ Describe a mechanism . . . . .Photo, representational diagram, or cutaway diagram

designers, technical illustrators), or insufficient budget. In any case, your basic task is to enable the intended audience to interpret the visual correctly.

TABLES

Tables can display exact quantities, compare sets of data, and present information systematically and economically. Numerical tables such as Table 14.1 present *quantitative information* (data that can be measured). In contrast, prose tables present *qualitative information* (brief descriptions, explanations, or instructions). Table 14.2, for example, combines numerical data, probability estimates, comparisons, and instructions—all organized for the smoker's understanding of radon gas risk in the home.

No table should be overly complex for its audience. An otherwise impressive-looking table, such as Table 14.3, is hard for nonspecialists to interpret because it presents too much information at once. We can see how an unethical writer might use a complex table to bury numbers that are questionable or embarrassing (R. Williams 12). Can you discover any hidden facts in Table 14.3? (For instance, which industry has been slowest in cleaning up its act?) Users need to understand how the table is organized, where to find what they need, and how to interpret the information they find (Hartley 90).

Tables are constructed with various tools: (a) tab markers and tab keys on a typewriter or word processor, (b) row-and-column displays in a spreadsheet program, (c) the "Table" command in better word-processing programs. The "Table" command option offers a full range of table editing features: cut and paste, adjust spacing, insert text between rows, add rows or columns, adjust column width, and so on. Whichever options you employ, follow the suggestions on page 267.

**Table 14.2
A Prose Table**
Source: Home
Buyer's and Seller's
Guide to Radon.
*Washington: GPO,
1993.*

	Radon Risk if You Smoke		
Radon level	If 1,000 people who smoked were exposed to this level over a lifetime . . .	The risk of cancer from radon exposure compares to . . .	WHAT TO DO: Stop Smoking and . . .
20 pCi/L[a]	About 135 people could get lung cancer	←100 times the risk of drowning	Fix your home
10 pCi/L	About 71 people could get lung cancer	←100 times the risk of dying in a home fire	Fix your home
8 pCi/L	About 57 people could get lung cancer		Fix your home
4 pCi/L	About 29 people could get lung cancer	←100 times the risk of dying in an airplane crash	Fix your home
2 pCi/L	About 15 people could get lung cancer	←2 times the risk of dying in a car crash	Consider fixing between 2 and 4 pCi/L
1.3 pCi/L	About 9 people could get lung cancer	(Average indoor radon level)	(Reducing radon levels below 2 pCi/L is difficult)
0.4 pCi/L	About 3 people could get lung cancer	(Average outdoor radon level)	

Note: If you are a former smoker, your risk may be lower.
[a]picocuries per liter

**Table 14.3
A Complex Table**
*Source: U.S.
Environmental
Protection Agency,
1994 Toxics
Release Inventory,
June 1996.*

Toxic Release Inventory, by Industry and Source: 1990 to 1994									
In millions of pounds. Based on reports from almost 23,000 manufacturing facilities which have 10 or more full-time employees and meet established thresholds for using the list of more than 300 chemicals covered.									
INDUSTRY	1987 SIC[1] code	1990	1991	1992	1993	1994 Total[2]	1994 Air[3] nonpoint	1994 Air[4] point	1994 Water
Total	(X)	2,603.7	2,684.3	2,449.6	2,157.4	1,976.9	350.0	991.0	47.0
Food and kindred prod.	20	9.9	12.0	11.9	12.0	10.3	2.5	6.3	0.1
Tobacco products	21	1.0	0.6	0.6	0.6	1.0	0.1	0.9	–
Textile mill products	22	23.4	22.2	19.1	17.6	15.9	3.2	12.5	0.1
Apparel and other textile prod.	23	1.1	1.3	1.3	1.0	1.3	0.3	1.0	–
Lumber and wood products	24	33.6	30.2	30.0	29.8	31.7	3.7	27.9	–
Furniture and fixtures	25	57.8	652.8	53.2	54.0	50.6	6.4	44.1	–
Paper and allied products	26	203.6	207.2	199.1	179.8	218.6	18.4	186.6	8.9
Printing and publishing	27	51.4	46.4	40.4	35.9	34.2	19.9	14.3	–
Chemical and allied products	28	1,007.9	960.5	991.3	874.4	700.7	87.5	220.4	33.5
Petroleum and coal products	29	62.4	56.5	61.7	50.9	43.8	28.4	13.6	0.5
Rubber and misc. plastic prod.	30	154.5	134.2	121.1	111.0	111.6	30.7	80.6	–
Leather and leather products	31	9.5	7.8	7.2	4.4	3.6	1.0	2.6	–
Stone, clay, glass products	32	20.5	16.4	14.3	14.3	12.4	2.1	8.9	–
Primary metal industries	33	438.6	386.8	341.2	304.6	293.8	21.6	68.7	1.7
Fabricated metals products	34	125.6	109.1	100.6	88.6	86.1	29.5	55.8	0.1
Industrial machinery and equip.	35	48.0	38.8	33.0	26.5	23.5	7.1	16.1	0.1
Electronic, electric equip.	36	73.4	60.2	47.1	32.9	29.0	6.1	22.7	0.1
Transportation equipment	37	157.5	137.5	125.3	123.6	119.7	29.5	89.9	0.1
Instuments and related prod.	38	38.9	34.9	29.1	22.5	15.7	4.2	11.3	0.3
Misc. manufacturing industries	39	21.6	18.4	16.9	15.2	13.7	3.9	9.8	–
Multiple codes	20–39	251.7	203.6	191.8	137.2	142.9	39.0	70.3	1.3
No codes	20–39	11.9	26.8	13.6	20.1	16.9	5.2	6.4	0.2

* Represents or rounds to zero. (X) Not applicable. [1]Standard Industrial Classification, see text, section 13. [2]Includes other releases not shown separately. [3]Fugitive. [4]Stack.

♦ Use a table only when you are reasonably sure it will enlighten—rather than frustrate—the audience. For laypersons, use fewer tables and keep them simple.

♦ Try to limit the table to one page. Otherwise, write "continued" at the bottom, and begin the second page with the full title, "continued," and the original column headings.

♦ If the table is too wide for the page, turn it 90 degrees so that the left-hand side is at the bottom of the page. Or divide the data into two tables. (Few readers may bother rotating the page to read the table broadside.)

♦ In your discussion, refer to the table by number, and explain what readers should be looking for. Or include a prose caption with the table. Specifically, introduce the table, show it, and then interpret it.

For information about creating tables, see How to Construct a Table and the accompanying Table 14.4 on page 267.

Tables work well for displaying exact values, but readers find graphs or charts easier to interpret. Geometric shapes (bars, curves, circles) are generally easier to remember than lists of numbers (Cochran et al. 25).

Any visual other than a table is usually categorized as a *figure,* and titled as such (*Figure 1 Aerial View of the Panhandle Site*). Figures covered in this chapter include graphs, charts, and illustrations.

Like all other components in the document, visuals are designed with audience and purpose in mind (Journet 3). An accountant doing an audit might need a table listing exact amounts, whereas the average public stockholder reading an annual report would prefer the "big picture" in an easily grasped bar graph or pie chart (Van Pelt 1). Similarly, an audience of scientists might find a table like 14.3 perfectly appropriate, but a less specialized audience (say, environmental groups) might prefer the clarity and simplicity of a graph like Figure 14.2.

GRAPHS

Graphs translate numbers into pictures. Plotted as a set of points (a *series*) on a coordinate system, a graph shows the relationship between two variables.

Graphs have a horizontal and a vertical axis. The horizontal axis shows categories (the independent variables) to be compared, such as years within a period (1970, 1978, 1986). The vertical axis shows the range of values (the dependent variables) for comparing or measuring the categories, such as the number of people who died from heart failure in a specific year. A dependent variable changes according to activity in the independent variable (e.g., a decrease in quantity over a set time, as in Figure 14.3). In the equation $y = f(x)$, x is the independent variable and y is the dependent variable.

Graphs are especially useful for displaying comparisons, changes over time, patterns, or trends. When you decide to use a graph, choose the best type for your purpose: bar graph or line graph.

How to Construct a Table

TABLE 14.4 ■ Science and Engineering Graduates in 1993 and 1994: 1995 Career Status

| DEGREE AND FIELD | Graduates 1993 and 1994 (1,000) | 1995—PERCENT DISTRIBUTION | | | | Median salary ($1,000) |
| | | In school[a] | Employed | | Not employed | |
			In S&E[b]	In other		
All science fields	**580.2**	**25**	**10**	**59**	**6**	**22.8**
Computer science/math	69.2	13	32	51	4	29.0
Life sciences	121.1	37	10	47	5	21.8
Physical sciences	33.2	39	27	30	4	25.5
Social sciences	356.7	X	5	67	7	21.2
All engineering fields[c]	**118.4**	**15**	**62**	**20**	**4**	**33.5**
Civil	18.1	13	67	17	3	31.0
Electrical/electronics	38.6	11	64	21	4	35.0
Industrial	6.4	9	59	28	3	34.0
Mechanical	28.9	12	66	17	4	31.5

[a]Full-time grad. students. [b]Science & engineering. [c]Other fields not shown. (X) Not available.

Source: National Science Foundation/SRS, National Survey of Recent College Graduates: 1995.

Statistical Abstract of the United States: 1997 (117th edition). Washington: GPO: 611.

1. Number the table in its order of appearance and provide a title that describes exactly what is being compared or measured.

2. Label stub, row, and column heads (*Degree and Field, Median salary, Computer science*) so readers know what they are looking at.

3. Stipulate all units of measurement using familiar symbols and abbreviations ($, hr., no.). Define specialized symbols or abbreviations (*Å* for *angstrom, db* for *decibel*) in a footnote.

4. Compare data vertically (in columns) instead of horizontally (in rows). Columns are easier to compare than rows. Try to include row or column averages or totals, as reference points for comparing individual values.

5. Use horizontal rules to separate headings from data. In a complex table, use vertical rules to separate columns. In a simple table, use as few rules as clarity allows.

6. List the items in a logical order (alphabetical, chronological, decreasing cost). Space listed items so they are not too cramped or too far apart for easy comparison. Keep prose entries as brief as clarity allows.

7. Convert fractions to decimals, and align decimals vertically. Keep decimal places for all numbers equal. Round insignificant decimals to the nearest whole number.

8. Use *x, NA,* or a dash to signify any omitted entry, and explain the omission in a footnote ("Not available," "Not applicable").

9. Use footnotes to explain entries, abbreviations, or omissions. Label footnotes with lowercase letters so readers do not confuse the notation with the numerical data.

10. Cite data sources beneath any footnotes. When adapting or reproducing a copyrighted table for a work to be published, obtain written permission from the copyright holder.

Figure 14.3
A Simple Bar Graph

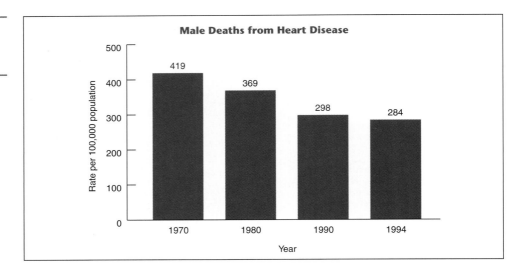

Bar Graphs

Easily understood by most people, bar graphs show discrete comparisons, as on a year-by-year or month-by-month basis. Each bar represents a specific quantity. Use bar graphs to focus on one value or compare values that change over equal time intervals (expenses calculated at the end of each month, sales figures totaled at yearly intervals). Use a bar graph only to compare values that are noticeably different.

SIMPLE BAR GRAPHS. The simple bar graph in Figure 14.3 displays one relationship extracted from the data in Table 14.1, the rate of male deaths from heart disease. To aid interpretation, you can record exact values above each bar—but only if the audience needs exact numbers.

MULTIPLE-BAR GRAPHS. A bar graph can display two or three relationships simultaneously, each relationship plotted as a separate series. Figure 14.4 displays two comparisons from Table 14.1, the rate of male deaths from both heart disease and cancer.

Whenever a graph shows more than one relationship (or series), each series of numbers is represented by a different pattern, color, shade, or symbol, and the patterns are identified by a *legend*. The more relationships a graph displays, the harder they are to interpret. As a rule, plot no more than three series of numbers on one graph.

HORIZONTAL-BAR GRAPHS. To make a horizontal-bar graph, turn a vertical-bar graph (and scales) on its side. Horizontal-bar graphs are good for displaying a large series of bars arranged in order of increasing or decreasing value, as in Figure 14.5. The horizontal format leaves room for labeling the categories horizontally (*Service,* and so on). A vertical-bar graph leaves no room for labeling.

**Figure 14.4
A Multiple-Bar
Graph**

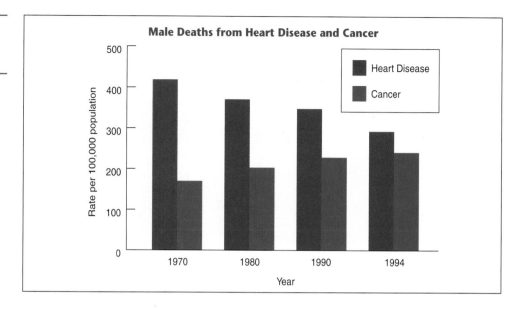

STACKED-BAR GRAPHS. Instead of side-by-side clusters of bars, you can display multiple relationships by stacking bars. Stacked-bar graphs are especially useful for showing how much each item contributes to the whole. Figure 14.6 displays other comparisons from Table 14.1.

Display no more than four or five relationships in a stacked-bar graph. Excessive subdivisions and patterns create visual confusion.

100-PERCENT BAR GRAPH. A type of stacked bar graph, the 100-percent bar graph is useful for showing the relative values of the parts that make up the 100-

**Figure 14.5
A Horizontal-Bar
Graph**
*Source: Bureau of
Labor Statistics.*

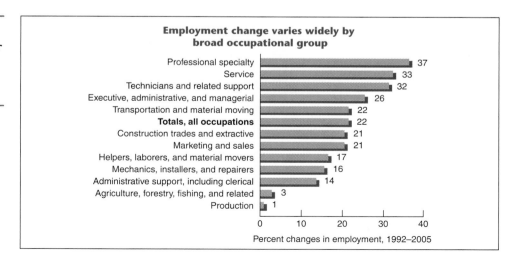

**Figure 14.6
A Stacked-Bar
Graph**

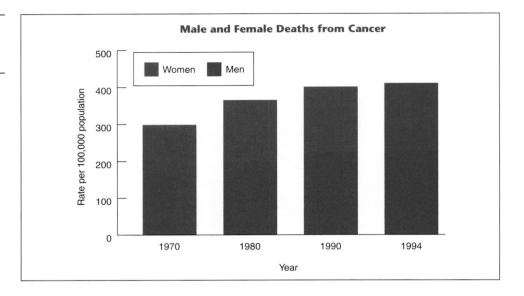

percent value, as in Figure 14.7. Like any bar graph, the 100-percent graph can have either horizontal or vertical bars.

Notice how bar graphs become harder to interpret as bars and patterns increase. For a less sophisticated audience, the two separate data depictions in Figure 14.7 might be easier to interpret in pie charts (pages 276, 277).

DEVIATION BAR GRAPHS.   The deviation bar graph can display both positive and negative values, as in Figure 14.8. Notice how the vertical axis extends to the

**Figure 14.7
A 100-percent
Bar Graph**
*Source: U.S. Bureau
of the Census.*

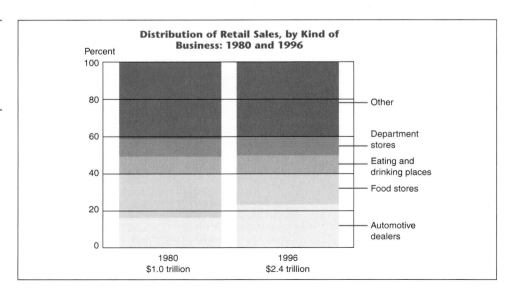

Figure 14.8
A Deviation Bar Graph
Source: Bureau of Labor Statistics.

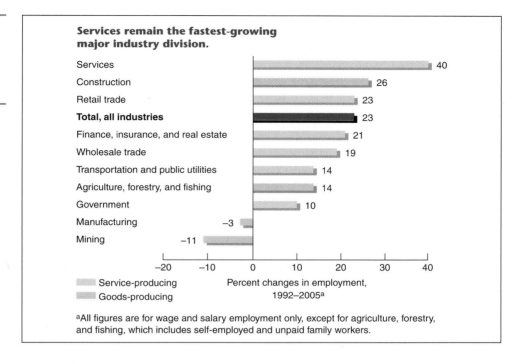

Services remain the fastest-growing major industry division.

Industry	Percent change
Services	40
Construction	26
Retail trade	23
Total, all industries	23
Finance, insurance, and real estate	21
Wholesale trade	19
Transportation and public utilities	14
Agriculture, forestry, and fishing	14
Government	10
Manufacturing	−3
Mining	−11

Service-producing
Goods-producing

Percent changes in employment, 1992–2005[a]

[a]All figures are for wage and salary employment only, except for agriculture, forestry, and fishing, which includes self-employed and unpaid family workers.

negative side of the zero baseline, following the same incremental division as the positive side.

3-D BAR GRAPHS.   Graphics software makes shading and rotating images to produce three-dimensional views easy. The 3-D perspectives in Figure 14.9 engage our attention and add visual emphasis to the data.

> NOTE

Although 3-D graphs can enhance and dramatize a presentation, an overly complex graph can be misleading or hard to interpret. Use 3-D only when a two-dimensional version will not serve as well. Never sacrifice clarity and simplicity for the sake of visual effect.

HOW TO DISPLAY A BAR GRAPH.   After you decide on a type of bar graph, follow the suggestions below for achieving a user-friendly display.

How to display a bar graph

- Keep the graph simple and easy to read. Avoid plotting more than three types of bars in each cluster. Avoid needless visual details.
- Number your scales in units the audience will find familiar and easy to follow. Units of 1 or multiples of 2, 5, or 10 are best (Lambert 45). Space the numbers equally.

The age distribution of the labor force will continue to shift.

Percent distribution by age of the civilian labor force

Age

55 years and over

45 to 54 years

35 to 44 years

25 to 34 years

16 to 24 years

	1979	1992	2005
55 years and over	14	12	14
45 to 54 years	16	18	24
35 to 44 years	19	27	25
25 to 34 years	27	28	21
16 to 24 years	24	16	16

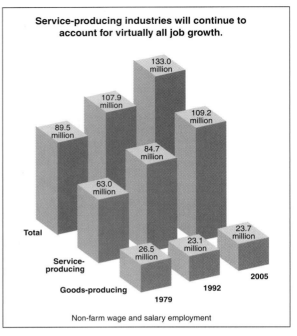

Service-producing industries will continue to account for virtually all job growth.

133.0 million
107.9 million
109.2 million
89.5 million
84.7 million
63.0 million
26.5 million
23.1 million
23.7 million

Total
Service-producing
Goods-producing

1979   1992   2005

Non-farm wage and salary employment

Figure 14.9   3-D Bar Graphs.   *Source: Bureau of Labor Statistics.*

♦ Label both scales to show what is being measured or compared. If space allows, keep all labels horizontal for easier reading.

♦ Label each bar or cluster of bars at its base.

♦ Use *tick marks* to show the points of division on your scale. If the graph has many bars, extend the tick marks into *grid lines* to help readers relate bars to values.

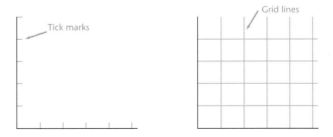

♦ To avoid confusion, make all bars the same width (unless you are overlapping them). If you must produce your graphs by hand, use graph paper to keep bars and increments evenly spaced.

♦ In a multiple-bar graph, use a different pattern, color, or shade for each bar in a cluster. Provide a legend identifying each pattern, color, or shade.

♦ If you are trying for emphasis, be aware that darker bars are seen as larger, closer, and more important than lighter bars of the same size (Lambert 93).

♦ Cite data sources beneath the graph. When adapting or reproducing a copyrighted graph for a work to be published, you must obtain written permission from the copyright holder.

♦ In your discussion, refer to the graph by number ("Figure 1"), and explain what the user should be looking for. Or include a prose caption along with the graph.

Many computer graphics programs automatically employ most of these design features. Anyone producing visuals, however, should know all the conventions.

Line Graphs

A line graph can accommodate many more data points than a bar graph (e.g., a twelve-month trend, measured monthly). Line graphs help readers synthesize large bodies of information in which exact quantities need not be emphasized. Whereas bar graphs display quantitative differences among items (cities, regions, yearly or monthly intervals), line graphs display data whose value changes over time, as in a trend, forecast, or other change during a specified time (profits, losses, growth). Some line graphs depict cause-and-effect relationships (e.g., how seasonal patterns affect sales or profits).

SIMPLE LINE GRAPHS.   A simple line graph, as in Figure 14.10, uses one line to plot time intervals on the horizontal scale and values on the vertical scale. The relationship depicted here would be much harder to express in words alone.

Figure 14.10
A Simple Line
Graph

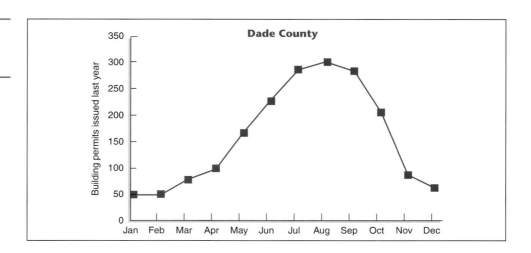

Figure 14.11
A Multiple-Line
Graph

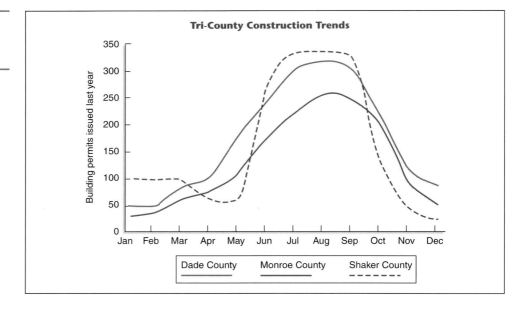

MULTIPLE-LINE GRAPHS. A multiple-line graph displays several relationships simultaneously, as in Figure 14.11.

For legibility, use no more than three or four curves in a single graph. Explain the relationships readers are supposed to see.

A caption explaining
the visual
relationships

> Building permits issued in all three counties increased steadily as the weather warmed, but Shaker's increase was more erratic. Its permits declined for April–May, but then surpassed Dade's and Monroe's for June–September.

DEVIATION LINE GRAPHS. Extend your vertical scale below the zero baseline to display positive and negative values in one graph, as in Figure 14.12. Mark values below the baseline in intervals parallel to those above it.

BAND OR AREA GRAPHS. For emphasis and appeal, fill the area beneath each plotted line with a pattern. Figure 14.13 is another version of the Figure 14.10 line graph.

The multiple bands in Figure 14.14 depict relationships among sums instead of the direct comparisons depicted in the equivalent Figure 14.11 line graph.

Despite their visual appeal, multiple-band graphs are easy to misinterpret: In a multiple-*line* graph, each line depicts its own distance from the zero baseline. But in a multiple-*band* graph, the very top line depicts the *total,* with each band below it being a part of that total (like stacked-bar graph segments). Always clarify these relationships for users.

Figure 14.12
A Deviation Line Graph
Source: Chart prepared by U.S. Bureau of the Census.

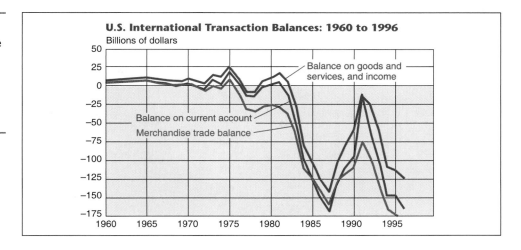

U.S. International Transaction Balances: 1960 to 1996

HOW TO DISPLAY A LINE GRAPH. Follow the suggestions for displaying a bar graph (page 271), with these additions:

How to display a line graph

- ◆ Display no more than three or four lines on one graph.
- ◆ Mark each individual data point used in plotting each line.
- ◆ Make each line visually distinct (using color, symbols, and so on).
- ◆ Label each line so users know what each one represents.
- ◆ Avoid grid lines that users could mistake for plotted lines.

Figure 14.13
A Simple Band Graph

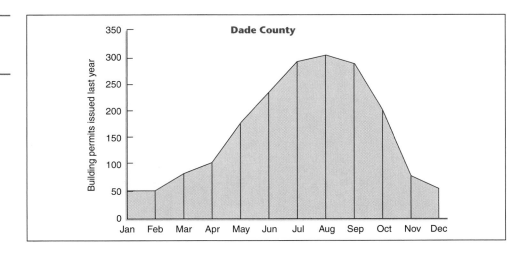

Figure 14.14
A Multiple-Band
Graph

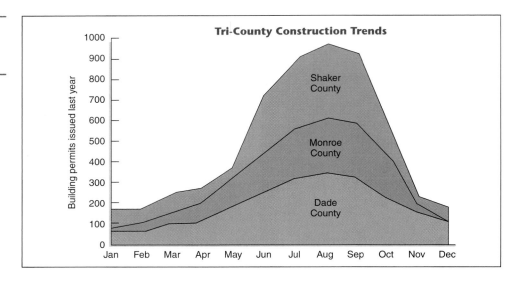

CHARTS

The terms *chart* and *graph* are often used interchangeably. But a chart is more precisely a figure that displays relationships (quantitative or cause-and-effect) that are not plotted on a coordinate system. Commonly used charts include pie charts, organizational charts, flowcharts, tree charts, Gantt charts, and pictorial charts (pictograms).

Figure 14.15
A Simple Pie
Chart

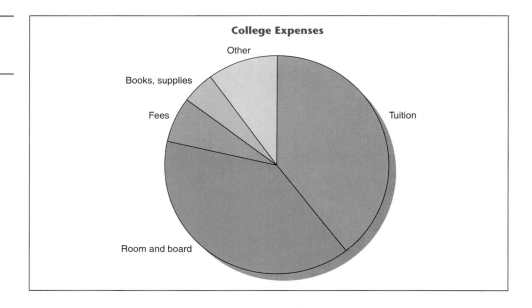

Figure 14.16
Two Other
Versions of
Figure 14.15

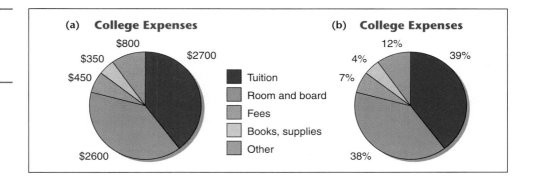

Pie Charts

Usually considered easy for users to understand, a pie chart depicts the percentages or proportions of the parts that make up a whole. In a pie chart, readers can compare the parts to each other as well as to the whole (to show how much was spent on what, how much income comes from which sources, and so on). Figure 14.15 shows a simple pie chart. Figure 14.16 shows two other versions of the pie chart in Figure 14.15. Version (a) displays dollar amounts, and version (b) the percentage relationships among these dollar amounts.

For displaying pie charts, follow these suggestions:

How to display a
pie chart

◆ Be sure the parts add up to 100 percent.
◆ If you must produce your charts by hand, use a compass and protractor for precise segments. Each 3.6-degree segment equals 1 percent. Include no less

Figure 14.17
An Organization
Chart

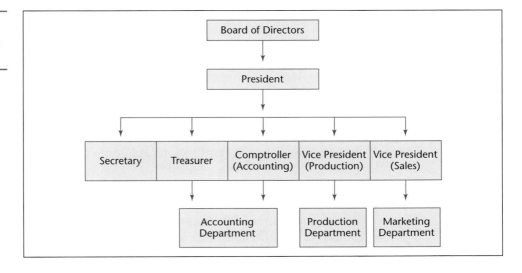

Figure 14.18
A Flowchart for
Producing a
Textbook
*Source: Adapted
from Harper & Row
Author's Guide.*

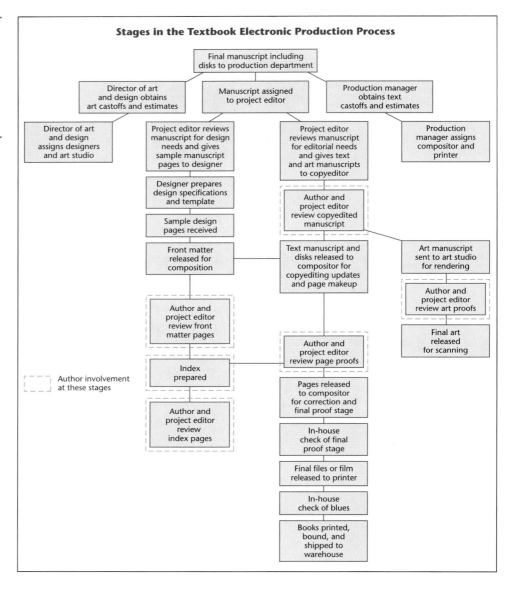

Stages in the Textbook Electronic Production Process

Final manuscript including disks to production department

Director of art and design obtains art castoffs and estimates

Manuscript assigned to project editor

Production manager obtains text castoffs and estimates

Director of art and design assigns designers and art studio

Project editor reviews manuscript for design needs and gives sample manuscript pages to designer

Project editor reviews manuscript for editorial needs and gives text and art manuscripts to copyeditor

Production manager assigns compositor and printer

Designer prepares design specifications and template

Author and project editor review copyedited manuscript

Sample design pages received

Front matter released for composition

Text manuscript and disks released to compositor for copyediting updates and page makeup

Art manuscript sent to art studio for rendering

Author and project editor review art proofs

Author and project editor review front matter pages

Final art released for scanning

Author involvement at these stages

Index prepared

Author and project editor review page proofs

Author and project editor review index pages

Pages released to compositor for correction and final proof stage

In-house check of final proof stage

Final files or film released to printer

In-house check of blues

Books printed, bound, and shipped to warehouse

than two and no more than eight segments. A pie chart containing more than eight segments can be hard to interpret, especially if the segments are small (Hartley 96).

◆ Combine small segments under the heading "Other."
◆ Locate your first radial line at 12 o'clock and then move clockwise from large to small (except for "Other," usually the final segment).
◆ For easy reading, keep all labels horizontal.

Figure 14.19
An Outline
Converted to a
Tree Chart

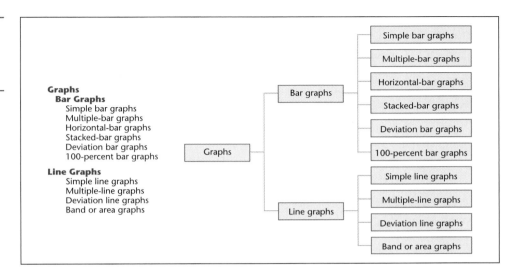

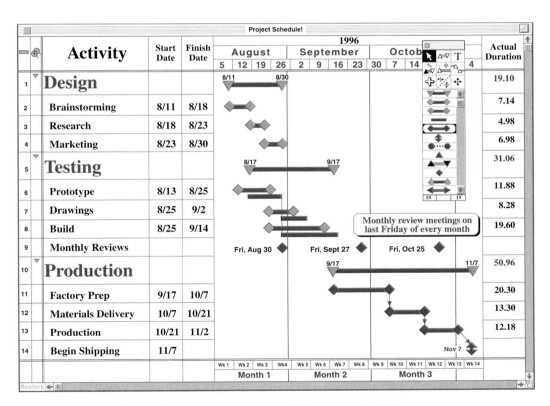

Figure 14.20   A Gantt Chart   *Source: Courtesy of AEC Software, © 1996.*

Organization Charts

An organization chart divides an organization into its administrative or managerial parts. Each part is ranked according to its authority and responsibility in relation to other parts and to the whole, as in Figure 14.17.

Flowcharts

A flowchart traces a procedure or process from beginning to end. In displaying the steps in a manufacturing process, the flowchart would begin with the raw materials and proceed to the finished product. Figure 14.18 traces the procedure for producing a textbook. (Other flowchart examples appear on pages 103, 309, and elsewhere throughout the text.)

Tree Charts

Whereas flowcharts display the steps in a process, tree charts show how the parts of an idea or concept relate to each other. Figure 14.19 displays part of an outline for this chapter so that users can better visualize relationships. The tree chart seems clearer and more interesting than the prose listing.

Gantt Charts

Named for engineer H. L. Gantt (1861–1919), a Gantt chart depicts progress as a function of time. A series of bars or lines (time lines) indicates start-up and completion dates for each phase or task in a project, relative to the other phases or tasks.

Figure 14.21
A Pictogram
Source: U.S. Bureau of the Census.

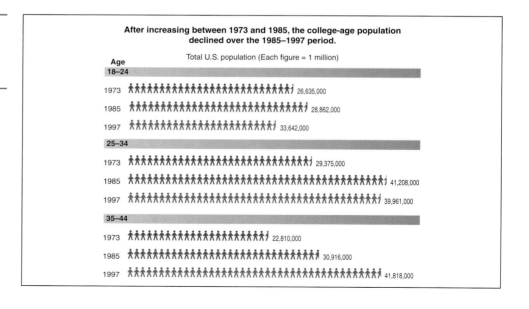

Gantt charts are especially useful for planning a project (as in a proposal) and for tracking it (as in a progress report). Notice that the Gantt chart in Figure 14.20 depicts tasks whose time lines can be simultaneous, overlapping, or consecutive.

Pictograms

Pictograms depict numerical relationships with icons or symbols (cars, houses, smokestacks) of the items being measured, instead of using bars or lines. Each symbol represents a stipulated quantity, as in Figure 14.21. Many graphics programs provide an assortment of predrawn symbols.

Use pictograms when you want to make information more appealing for nontechnical audiences.

GRAPHIC ILLUSTRATIONS

Illustrations consist of diagrams, maps, drawings, and photographs depicting relationships that are physical rather than numerical. Good illustrations help people understand and remember the material (Hartley 82). Consider this information from a government pamphlet, explaining the operating principle of the seat belt:

> The safety-belt apparatus includes a tiny pendulum attached to a lever, or locking mechanism. Upon sudden deceleration, the pendulum swings forward, activating the locking device to keep passengers from pitching into the dashboard.

Visualizing the mechanism without the illustration in Figure 14.22 is difficult. Clear and uncluttered, a good diagram eliminates unnecessary details and focuses only on essential material. The following pages include examples of some commonly used diagrams.

Figure 14.22
A Diagram of a Safety-Belt Locking Mechanism
Source: Safety Belts. *U.S. Department of Transportation.*

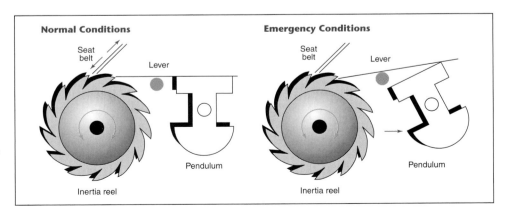

Figure 14.23
An Exploded
Diagram of a
Brace for a
Basketball Hoop
*Source: Courtesy of
Spalding.*

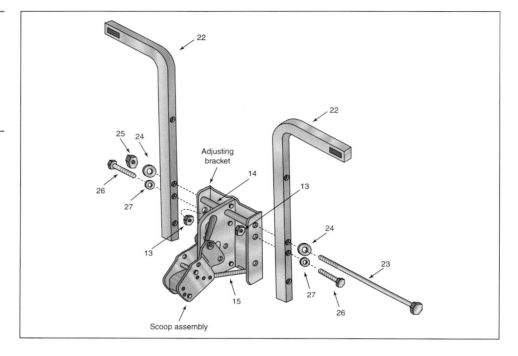

Diagrams

Exploded diagrams, like that of a brace for an adjustable basketball hoop in Figure 14.23, show how the parts of an item are assembled; they often appear in repair or maintenance manuals. Notice how all parts are numbered for the user's easy reference to the written instructions.

Cutaway diagrams show the item with its exterior layers removed in order to reveal interior sections, as in Figure 14.24. Unless the specific viewing perspective is immediately recognizable (as in Figure 14.24), name the angle of vision: "top view," "side view," and so on.

Block diagrams are simplified sketches that represent the relationship between the parts of an item, principle, system, or process. Because block diagrams are designed to illustrate *concepts* (such as current flow in a circuit), the parts are represented as symbols or shapes. The block diagram in Figure 14.25 illustrates how any process can be controlled automatically through a feedback mechanism. Figure 14.26 shows the feedback concept applied as the cruise-control mechanism on a motor vehicle.

It is easy to create impressive-looking visuals by using electronic drawing programs, clip art, and image banks. However, specialized diagrams generally require the services of graphic artists or technical illustrators. The client requesting or commissioning the visual provides the art professional with an *art brief* (often prepared by writers and editors) that spells out the visual's purpose and specifications.

Figure 14.24
Cutaway Diagram
of a Surgical
Procedure
Source: Transsphen-
oidal Approach
for Pituitary Tumor,
© 1986 by The
Ludann Co., Grand
Rapids, MI.

THE OPERATION

Incision

Transsphenoidal surgery is performed with the patient under general anesthesia and positioned on his back. The head is fixed in a special headrest, and the operation is monitored on a special x-ray machine (fluoroscope).

In the approach illustrated here (not used by all surgeons), a small incision is made in one side of the nasal septum **(Fig. 2)**. Part of the septum is then removed to provide access to the sphenoid sinus cavity **(Fig. 3)**.

Figure 2
Incision into nasal septum

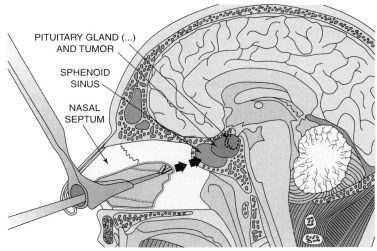

PITUITARY GLAND (...)
AND TUMOR

SPHENOID
SINUS

NASAL
SEPTUM

Figure 3
Removal of nasal septum to reach pituitary chamber

For example, part of the brief addressed to the medical illustrator for Figure 14.24 might read as follows:

An art brief for
Figure 14.24

- ◆ **Purpose:** to illustrate transsphenoidal adenomectomy for laypersons
- ◆ **View:** full cutaway, axial
- ◆ **Range:** descending from cranial apex to a horizontal plane immediately below the upper jaw and second cervical vertebra
- ◆ **Depth:** medial cross-section
- ◆ **Structures omitted:** cranial nerves, vascular and lymphatic systems

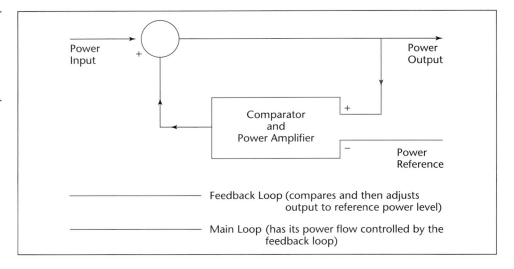

- ◆ **Structures included:** gross anatomy of bone, cartilage, and soft tissue—delineated by color, shading, and texture
- ◆ **Structures highlighted:** nasal septum, sphenoid sinus, and sella turcica, showing the pituitary embedded in a 1.5 cm tumor invading the sphenoid sinus via an area of erosion at the base of the sella

Maps

Besides being visually engaging and easily remembered, maps are especially useful for showing comparisons and for helping users to *visualize* position, location, and relationships among complex data. Consider how Figure 14.27 conveys important statistical information in a format that is both accessible and understandable. Color enhances the percentage comparisons.

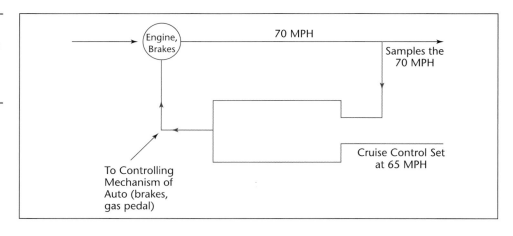

Figure 14.27
A Map Rich in Statistical Significance
Source: Bureau of the Census.

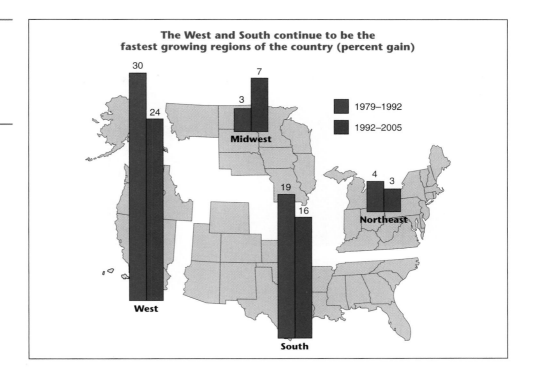

Photographs

Photographs are especially useful for showing what something looks like (Figure 14.28) or how something is done (Figure 14.29). But no matter how visually engaging, a photograph is hard to interpret if it includes needless details or fails to identify or emphasize the important material. One graphic design expert offers this practical advice for using photos in technical documents:

> To use pictures as tools for communication, pick them for their capacity to carry meaning, not just for their prettiness as photographs, . . . [but] for their inherent significance to the [document]. (White, *Great Pages* 110, 122)

Specialized photographs often require the services of a professional who knows how to use angles, lighting, and special film to obtain the desired focus and emphasis.

Whenever you plan to include photographs in a document or presentation, follow these suggestions:

How to display a photograph

◆ Try to simulate the approximate angle of vision readers would have in identifying or viewing the item or, for instructions, in doing the procedure (Figure 14.30).

Figure 14.28   A Photograph that Shows the Appearance of Something

A Fixed-Platform Oil Rig   *Source: SuperStock.*

Figure 14.29   A Photograph that Shows How Something Is Done

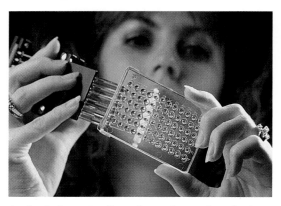

Antibody Screening Procedure   *Source: SuperStock.*

Figure 14.30   A Photograph that Shows a Realistic Angle of Vision

Titration in Measuring Electron-Spin Resonance
Source: SuperStock.

Figure 14.31   A Photograph that Needs to Be Cropped

Replacing the Microfilter Activation Unit
Source: SuperStock.

Figure 14.32 The Cropped Version of Figure 14.31

Source: SuperStock.

Figure 14.33 A Photograph of a Complex Mechanism

Sapphire Tunable Laser   *Source: SuperStock.*

Figure 14.34 A Simplified Diagram of Figure 14.33

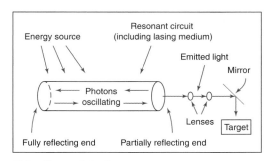

Major Parts of the Laser

Figure 14.35 A Photograph with Essential Features Labeled

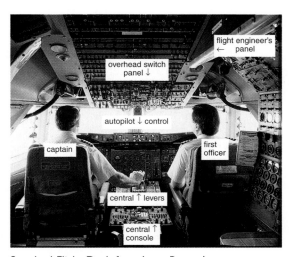

Standard Flight Deck for a Long-Range Jet
Source: SuperStock.

- Trim (or crop) the photograph to eliminate needless details (Figure 14.31 and 14.32).
- For emphasizing selected features of a complex mechanism or procedure, consider using diagrams in place of photographs or as a supplement (Figures 14.33 and 14.34).
- Label all the parts readers need to identify (Figure 14.35).
- For an image unfamiliar to readers, provide a sense of scale by including a person, a ruler, or a familiar object (such as a hand) in the photo.
- If your document will be published, obtain a signed release from any person depicted in the photograph and written permission from the copyright holder. Beneath the photograph, cite the photographer and the copyright holder.
- In your discussion, refer to the photograph by figure number and explain what readers should be looking for. Or include a prose caption.

Digital-imaging technology allows photographs to be scanned and stored electronically. These stored images can then be downloaded, edited, and altered.

NOTE   *Such capacity for altering photographic content creates unlimited potential for distortion and raises ethical questions about digital manipulation as well as legal questions about copyright infringement (page 125).*

COMPUTER GRAPHICS

Computer technology transfers many of the tasks formerly performed by graphic designers and technical illustrators to individuals with little formal training in graphic design. In whatever career you anticipate, you may be expected to produce high-quality graphics for conferences, presentations, and in-house publications.

Selecting Design Options

Computer graphics systems allow you to experiment with scales, formats, colors, perspectives, and patterns. By testing design options on-screen, you can revise and enhance your visual repeatedly until it achieves your exact purpose.

Here is a sampling of design options:

Options for electronic enhancement of visuals

- Update charts and graphs whenever the data change. The software will calculate the new data and plot the relationships.
- Edit your graphics on-screen, adding, deleting, or moving material as needed.
- Create your image at one scale, and later specify a different scale for the same image.
- Overlay images in one visual.
- Adjust bar width and line thickness.
- Fill a shape with a color or pattern.
- Scan, edit, and alter pages, photos, or other images.

Figure 14.36
A Clip-art Image
*Source: Desktop
Art®; Business 1
© Dynamic
Graphics, Inc.*

Figure 14.37
A Customized
Image
*Source: Professor R.
Armand Dumont.*

Most of these options are available via simple keystrokes or pull-down menus.

Using Clip Art

Clip art is a generic term for collections of ready-to-use images (of computer equipment, maps, machinery, medical equipment, and so on), all stored electronically. Various clip art packages allow you to import into your document countless images like the one in Figure 14.36. Using a drawing program, you can enlarge, enhance, or customize the image, as in Figure 14.37.

NOTE

Although handy, clip art often has a generic or crude appearance that makes a document look unprofessional. Consider using clip art for icons only, for in-house documents, or for situations in which time or budget preclude using original artwork (Menz 5).

One form of clip art that is especially useful in technical writing is in the *icon* (an image with all nonessential background removed). Icons convey a specific idea visually as in Figure 14.38. Icons appear routinely in computer documentation and in other types of instructions because the images provide users with an immediate signal of the desired action.

Figure 14.38
Icons
*Source: Desktop
Art®; Business 1 and
Health Care 1, ©
Dynamic Graphics,
Inc.*

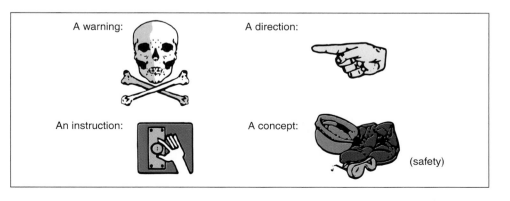

Figure 14.39
Color Used as a
Visualizing Tool
*Source: Courtesy of
National Audubon
Society.*

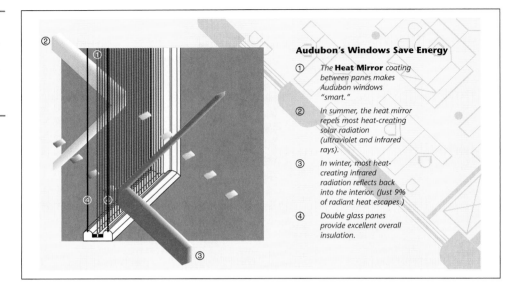

Whenever you use an icon, be sure it is "intuitively recognizable" to users ("Using Icons" 3). Otherwise, users are likely to misinterpret its meaning—in some cases with disastrous results.

Keep in mind that certain icons have offensive connotations in certain cultures. Hand gestures, for example, are especially problematic: some Arab cultures consider the left hand unclean; a pointing index finger—on either hand—as in Figure 14.38, is a sign of rudeness in Venezuela or Sri Lanka (Bosley 5–6).

Using Color

Color often makes a presentation more esthetically pleasing, more interesting to look at. But color serves purposes beyond visual appeal. Used effectively in a visual, color draws and directs users' attention and helps them identify the various elements. In Figure 14.20, for example, color helps users sort out the key schedule elements of a Gantt chart for a major project: activities, time lines, durations, and meetings.

Color can help clarify a concept or dramatize how something works. In Figure 14.39, for example, the use of bright colors against a darker, duller background enables users to *visualize* the "heat mirror" concept.

Color can help depict complex relationships in ways that are understandable. In Figure 14.40, an area map using three distinctive colors allows users to make any number of comparisons at a glance.

Along with shape, type style, and position of elements on a printed page, color can guide users through the material. Used effectively on a printed page, color helps organize the user's understanding, provides orientation, and emphasizes im-

Figure 14.40
Colors Used to
Show
Relationships
*Source: U.S.
Department of
Commerce;
Economics and
Statistics
Administration,
Bureau of the
Census*

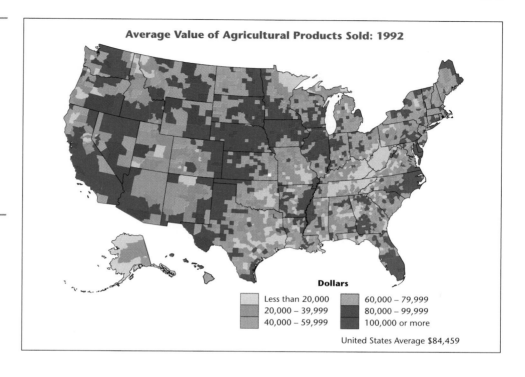

Average Value of Agricultural Products Sold: 1992

Dollars

Less than 20,000	60,000 – 79,999
20,000 – 39,999	80,000 – 99,999
40,000 – 59,999	100,000 or more

United States Average $84,459

portant material. Following are just a few possible uses of color (White, *Color* 39–44; Keyes 647–49).

USE COLOR TO ORGANIZE.   Users look for ways of organizing their understanding of a document (Figure 14.41). Color can reveal structure and break material up into discrete blocks that are easier to locate, process, and digest:

Figure 14.41
Color Used for
Organization

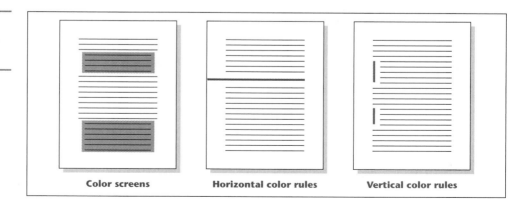

Color screens   **Horizontal color rules**   **Vertical color rules**

Figure 14.42
Color Used for
Orientation

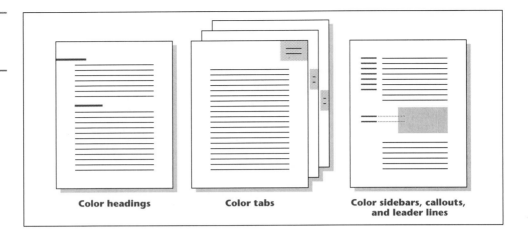

Color headings   Color tabs   Color sidebars, callouts,
                                            and leader lines

How color reveals
organization

- A color background screen can set off like elements such as checklists, instructions, or examples.
- Horizontal rules can separate blocks of text, such as sections of a report or areas of a page.
- Vertical rules can set off examples, quotations, captions, and so on.

USE COLOR TO ORIENT.   Users look for signposts that help them find their place or find what they need (Figure 14.42):

How color provides
orientation

- Color can help headings stand out from the text and differentiate major from minor headings.
- Color tabs and boxes can serve as location markers.
- Color sidebars (for marginal comments), callouts (for labels), and leader lines (for connecting a label to its referent) can guide the eyes.

USE COLOR TO EMPHASIZE.   Users look for places to focus their attention in a document (Figure 14.43).

How color
emphasizes

- Color type can highlight key words or ideas.
- Color can call attention to cross-references or to links on a Web page.
- A color, ruled box can frame a warning, caution, note, or hint.

HOW TO INCORPORATE COLOR.   To use color effectively, follow these suggestions:

How to use color
for greatest effect

- "Color gains impact when it is used selectively. It loses impact when it is overused" (*Aldus Guide* 39). Use color sparingly, and use no more than three or four distinct colors—including black and white (White, *Great Pages* 76).

Figure 14.43
Color Used for
Emphasis

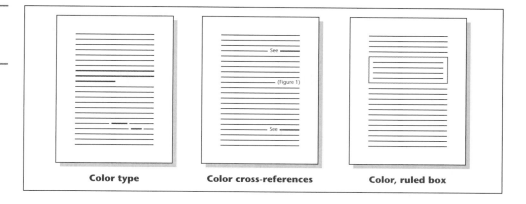

Color type   Color cross-references   Color, ruled box

◆ Apply color consistently to elements throughout your document. Inconsistent use of color can distort users' perception of the relationships (Wickens 117).

◆ Make color redundant. Be sure all elements are first differentiated in black and white: by shape, location, texture, type style, or type size. Different readers perceive colors differently or, in some cases, not at all. A sizable percentage of readers have impaired color vision (White, *Great Pages* 76).

◆ Use a darker color to make a stronger statement. The darker the color the more important the material. Readers perceive the sizes of variously colored objects differently. Darker items can seem larger and closer than lighter objects of identical size.

◆ Make color type larger than text type. Try to avoid color for text type, or use a high-contrast color (dark against a light background). Color is less visible on the page than black ink on a white background. The smaller the image or the thinner the rule, the stronger or brighter the color should be (White, *Editing* 229, 237).

◆ For contrast in a color screen, use a very dark type against a very light background, say a ten-to twenty-percent screen (Gribbons 70). The larger the screen area, the lighter the background color should be (Figure 14.44).

NOTE   *A color's connotations can vary from culture to culture. In the United States for example, red signifies danger and green traditionally signifies safety. But in Ireland, green or
orange carry political connotations in certain contexts. In Muslim cultures, green is a
holy color (Cotton 169).*

Figure 14.44
A Color-Density
Chart

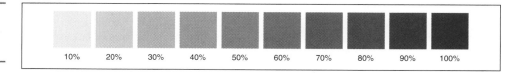

10%   20%   30%   40%   50%   60%   70%   80%   90%   100%

Using Web Sites for Graphics Support

The World Wide Web offers a growing array of visual resources. Following is a sample of useful Web sites (Martin 135–36).

◆ Image banks such as Image Club, Graphics Stock Solution, Index Stock, and the Internet Font Directory offer clip art, photo, and font databases that can be browsed and from which material can be purchased online and downloaded to a personal computer.

◆ Royalty-free images are available from Xoom and the Digital Picture Archive.

◆ Samples of computer-generated and traditional art work can be browsed and purchased via Artists on Line.

◆ Online versions of graphics magazines offer helpful design suggestions.

◆ Graphic design firms offer samples of work that show innovative design ideas.

Whether you seek design ideas, tools for creating your own visuals, or electronic catalogs of completed artwork, the Web is a good bet.

NOTE *Be extremely cautious about downloading visuals (or any material, for that matter) from the Web and then using them. Review the copyright law (pages 125–26). Originators of any work on the Web own the work and the copyright.*

HOW TO AVOID VISUAL DISTORTION

Although you are perfectly justified in presenting data in its best light, you are ethically responsible for avoiding misrepresentation. Any one set of data can support contradictory conclusions. Even though your numbers may be accurate, your visual display could be misleading.

Present the Real Picture

Visual relationships in a graph should accurately portray the numerical relationships they represent. Begin the vertical scale at zero. Never compress the scales to reinforce your point.

Notice how visual relationships in Figure 14.45 become distorted when the value scale is compressed or fails to begin at zero. In version A, the bars accurately depict the numerical relationships measured from the value scale. In version B, item Z (400) is depicted as three times X (200). In version C, the scale is overly compressed, causing the shortened bars to understate the quantitative differences.

Deliberate distortions are unethical because they imply conclusions contradicted by the actual data.

Figure 14.45
An Accurate Bar
Graph and Two
Distorted
Versions

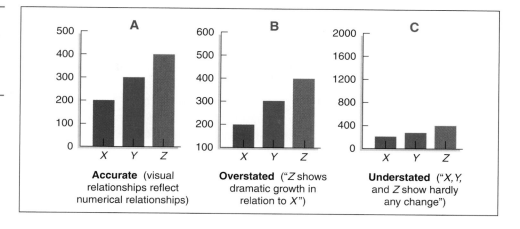

Present the Complete Picture

Without getting bogged down in needless detail, an accurate visual includes all essential data. Figure 14.46 shows how distortion occurs when data that would provide a complete picture are selectively omitted. Version A accurately depicts the numerical relationships measured from the value scale. In version B, too few points are plotted.

Decide carefully what to include and what to leave out of your visual display.

Figure 14.46
An Accurate Line
Graph and a
Distorted Version

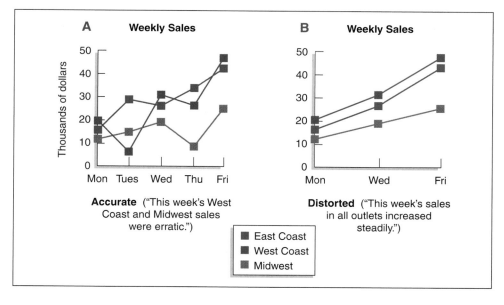

GUIDELINES

for Fitting
Visuals with
Printed Text

To ensure that visual and verbal elements in your document complement each other, follow these suggestions:

1. *Place the visual where it will best serve your readers.* If it is central to your discussion, place the visual as close as possible to the material it clarifies. (Achieving proximity often requires that you ignore the traditional "top or bottom" design rule for placing visuals on a page.) If the visual is peripheral to your discussion or of interest to only a few readers, place it in an appendix so that interested readers can refer to it. Tell readers when to consult the visual and where to find it.

2. *Never refer to a visual that readers cannot easily locate.* In a long document, don't be afraid to repeat a visual if you discuss it again later.

3. *Never crowd a visual into a cramped space.* Set the visual off by framing it with plenty of white space, and position it on the page for balance. To save space and to achieve proportion with the surrounding text, carefully consider the size of each visual and the amount of space it will occupy.

4. *Number the visual and give it a clear title and labels.* Your title should tell readers what they are seeing. Label all the important material and cite the source of data or of graphics.

5. *Match the visual to your audience.* Don't make it too elementary for specialists or too complex for nonspecialists. Your intended audience should be able to interpret the visual correctly.

6. *Introduce and interpret the visual.* In your introduction, tell readers what to expect:

   Informative   As Table 2 shows, operating costs have increased 7 percent annually since 1980.

   Uninformative   See Table 2.

   Visuals alone make ambiguous statements (Girill, "Technical Communication and Art" 35); pictures need to be interpreted. Instead of leaving readers to struggle with a page of raw data, explain the relationships displayed. Follow the visual with a discussion of its important features:

   Informative   This cost increase means that....

   Always tell readers what to look for and what it means.

7. *Use prose captions to explain important points made by the visual.* Captions help readers interpret a visual. Use a smaller type size so that captions don't compete with text type (*Aldus Guide* 35).

8. *Never include excessive information in one visual.* Any visual that contains too many lines, bars, numbers, colors, or patterns will overwhelm readers. In place of one complicated visual, use two or more straightforward ones.

9. *Be sure the visual can stand alone.* Even though it repeats or augments information already in the text, the visual should contain everything users will need to interpret it correctly.

Don't Mistake Distortion for Emphasis

When you want to emphasize a point (a sales increase, a safety record, etc.), be sure your data support the conclusion implied by your visual. For instance, don't use inordinately large visuals to emphasize good news or small ones to downplay bad news (R. Williams 11). When using clip art, pictograms, or drawn images to dramatize a comparison, be sure the relative size of the images or icons reflects the quantities being compared.

A visual accurately depicting a one hundred-percent increase in phone sales at your company might look like version A in Figure 14.47. Version B overstates the good news by depicting the larger image four times the size, instead of twice the size, of the smaller. Although the larger image is twice the height, it is also twice the *width,* so the total area conveys the visual impression that sales have *quadrupled.*

Visuals have their own rhetorical and persuasive force, which we can use to advantage—for positive or negative purposes, for the reader's benefit or detriment (Van Pelt 2). Avoiding visual distortion is ultimately a matter of ethics.

For additional guidance, use the Planning Sheet in Figure 14.48, and the Checklist on page 299.

Figure 14.47
An Accurate
Pictogram and a
Distorted Version

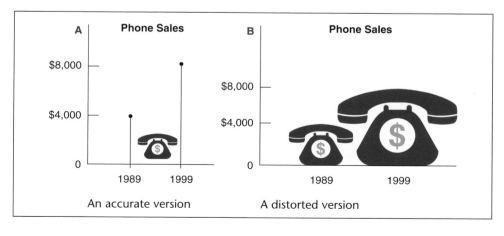

Focusing on Your Purpose

- What is this visual's purpose (to instruct, persuade, create interest)? _____

- What forms of information (numbers, shapes, words, pictures, symbols) will this visual depict? _____

- What kind of relationship(s) will the visual depict (comparison, cause-effect, connected parts, sequence of steps)? _____

- What judgment, conclusion, or interpretation is being emphasized (that profits have increased, that toxic levels are rising, that X is better than Y, that time is being wasted)? _____

- Is a visual needed at all? _____

Focusing on Your Audience

- Is this audience accustomed to interpreting visuals? _____

- Is the audience interested in specific numbers or an overall view? _____

- Should the audience focus on one exact value, compare two or more values, or synthesize a range of approximate values (Wickens 121)? _____

- Which type of visual will be most accurate, representative, accessible, and compatible with the type of judgment, action, or understanding expected from the audience? _____

- In place of one complicated visual, would two or more straightforward ones be preferable? _____

Focusing on Your Presentation

- What enhancements, if any, will increase audience interest (colors, patterns, legends, labels, varied typefaces, shadowing, enlargement or reduction of some features)? _____

- Which medium—or combination of media—will be most effective for presenting this visual (slides, transparencies, handouts, large-screen monitor, flip chart, report text)? _____

- To achieve the greatest utility and effect, where in the presentation does this visual belong? _____

Figure 14.48   A Planning Sheet for Preparing Visuals

Usability Checklist FOR VISUALS

(Numbers in parentheses refer to the first page of discussion.)

Content

☐ Does the visual serve a legitimate purpose (clarification, not mere ornamentation) in the document? (260)

☐ Is the visual titled and numbered? (296)

☐ Is the level of complexity appropriate for the audience? (296)

☐ Are all patterns in the visual identified by label or legend? (272)

☐ Are all values or units of measurement specified (grams per ounce, millions of dollars)? (267)

☐ Are the numbers accurate and exact? (267)

☐ Do the visual relationships represent the numerical relationships accurately? (294)

☐ Are explanatory notes added as needed? (267)

☐ Are all data sources cited? (267)

☐ Has written permission been obtained for reproducing or adapting a visual from a copyrighted source in any type of work to be published? (267)

☐ Is the visual introduced, discussed, interpreted, integrated with the text and referred to by number? (296)

☐ Can the visual itself stand alone in terms of meaning? (297)

Arrangement

☐ Is the visual easy to locate? (296)

☐ Do all design elements (title, line thickness, legends, notes, borders, white space) work to achieve balance? (297)

☐ Is the visual positioned on the page to achieve balance? (296)

☐ Is the visual set off by adequate white space or borders? (296)

☐ Is a broadside visual turned 90 degrees so that the left-hand side is at the bottom of the page? (266)

☐ Is the visual in the best report location? (296)

Style

☐ Is this the best type of visual for your purpose and audience? (263)

☐ Are all decimal points in each column vertically aligned? (267)

☐ Is the visual uncrowded and uncluttered? (296)

☐ Is the visual engaging (patterns, colors, shapes), without being too busy? (297)

☐ Is the visual in good taste? (292)

☐ Is the visual ethically acceptable? (294)

EXERCISES

1. The following statistics are based on data from three colleges in a large western city. They give the number of applicants to each college over six years.

   • In 1994, X college received 2,341 applications for admission, Y College received 3,116, and Z College 1,807.

   • In 1995, X College received 2,410 applications for admission, Y College received 3,224, and Z College 1,784.

   • In 1996, X College received 2,689 applications for admission, Y College received 2,976, and Z College 1,929.

   • In 1997, X College received 2,714 applications for admission, Y College received 2,840, and Z College 1,992.

   • In 1998, X College received 2,872 applications for admission, Y College received 2,615, and Z College 2,112.

   • In 1999, X College received 2,868 applications for admission, Y College received 2,421, and Z College 2,267.

Illustrate these data in a line graph, a bar graph, and a table. Which version seems most effective for a reader who (a) wants to exact figures, (b) wonders how overall enrollments are changing, or (c) wants to compare enrollments at each college in a certain year? Include a caption interpreting each of these versions.

2. Devise a flowchart for a process in your field or area of interest. Include a title and a brief discussion.

3. Devise an organization chart showing the lines of responsibility and authority in an organization where you work.

4. Devise a pie chart to depict your yearly expenses. Title and discuss the chart.

5. Obtain enrollment figures at your college for the past five years by sex, age, race, or any other pertinent category. Construct a stacked-bar graph to illustrate one of these relationships over the five years.

6. Keep track of your pulse and respiration at thirty-minute intervals over a four-hour period of changing activities. Record your findings in a line graph, noting the times and specific activities below your horizontal coordinate. Write a prose interpretation of your graph and give the graph a title.

7. In textbooks or professional journal articles, locate each of these visuals: a table, a multiple-bar graph, a multiple-line graph, a diagram, and a photograph. Evaluate each according to the revision checklist, and discuss the most effective visual in class.

8. We have discussed the importance of choosing an appropriate scale for your graph, and the most effective form for presenting your data. Study this presentation carefully:

> Strong evidence now indicates that not only nicotine and tar in cigarette smoke can be lethal, but also carbon monoxide. Much of cigarette smoke is carbon monoxide. The bar graph in Figure 14.49 lists the ten leading U.S. cigarette brands according to the carbon monoxide given off per pack of inhaled cigarettes.

Is the scale effective? If not, why not? Can these data be presented effectively in a bar graph? Present the same data in some other form that seems most effective.

9. Choose the most appropriate visual for illustrating each of these relationships. Justify each choice in a short paragraph.

a. A comparison of three top brands of fiberglass skis, according to cost, weight, durability, and edge control.

b. A breakdown of your monthly budget.

c. The changing cost of an average cup of coffee, as opposed to that of an average cup of tea, over the past three years.

d. The percentage of college graduates finding desirable jobs within three months after graduation, over the last ten years.

e. The percentage of college graduates finding desirable jobs within three months after graduation, over the last ten years—by gender.

f. An illustration of automobile damage for an insurance claim.

g. A breakdown of the process of radio wave transmission.

h. A comparison of five cereals on the basis of cost and nutritive value.

i. A comparison of the average age of students enrolled at your college in summer, day, and evening programs, over the last five years.

j. Comparative sales figures for three items made by your company.

10. *Computer graphics:* Compose and enhance one or more visuals electronically. You might begin by looking through a recent edition of the *Statistical Abstract of the United States* (in the government documents, reserve, or reference section of your library). From the *Abstract,* or from a source you prefer, select a body of numerical data that will interest your classmates. After completing the Visual Plan Sheet (page 298), compose one or more visuals to convey a message about your data to make a point, as in these examples:

Figure 14.49
Ten Leading U.S.
Cigarette Brands
in Order of CO
Content

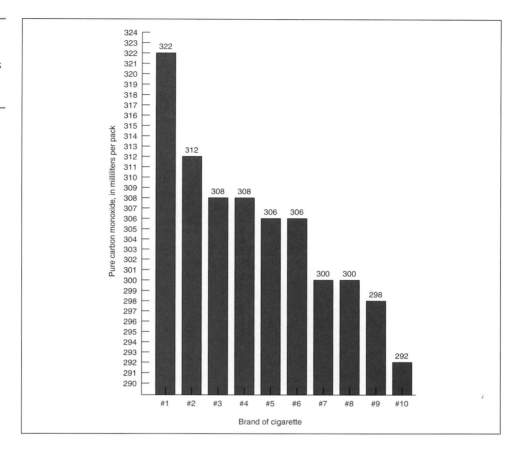

- Consumer buying power has increased or decreased since 1990.
- Defense spending, as a percentage of the federal budget, has increased or decreased since 1990.
- Average yearly temperatures across the United States are rising or falling.

Experiment with formats and design options, and enhance your visual(s) as appropriate. Add any necessary prose explanations.

Be prepared to present your visual message in class, using either an overhead or opaque projector or a large-screen monitor.

11. Revise the layout of Table 14.5 according to the guidelines on page 267, and explain to readers

the significant comparisons in the table. (*Hint:* The unit of measurement is percentage.)

12. Display each of these sets of information in the visual format most appropriate for the stipulated audience. Complete the Visual Plan Sheet (page 298) for each visual. Explain why you selected the type of visual as most effective for that audience. Include with each visual a brief prose passage interpreting and explaining the data.

a. (For general readers.) The 1997 *Statistical Abstract of the United States* breaks down energy consumption (by source) into these percentages: In 1970, coal, 18.5; natural gas, 32.8; hydro and geothermal, 3.1; nuclear,

Table 14.5
An Example of a
Poor Layout

	Educational Attainment of Persons 25 Years Old and Over	
	Highest level completed	
Year	High school (4 years or more)	College (4 years or more)
1970	52.3164	10.7431
1980	66.5432	16.2982
1982	71.0178	17.7341
1984	73.3124	19.1628
1993	80.2431	21.9316
1996	81.7498	23.6874

Source: Adapted from Statistical Abstract of the United States: 1997 (117th ed.). *Washington, DC: GPO: 159.*

1.2; oil, 44.4. In 1980, coal, 20.3; natural gas, 26.9; hydro and geothermal, 3.8; nuclear, 4.0; oil, 45.0. In 1990, coal, 23.5; natural gas, 23.8; hydro and geothermal, 7.3; nuclear, 4.1; oil, 41.3.

b. (For experienced investors in rental property.) As an aid in estimating annual heating and air-conditioning costs, here are annual maximum and minimum temperature averages from 1961 to 1997 for five Sunbelt cities (in Fahrenheit degrees): In Jacksonville, the average maximum was 78.7; the minimum was 57.2. In Miami, the maximum was 82.6; the minimum was 67.8. In Atlanta, the maximum was 71.3; the minimum was 51.1. In Dallas, the maximum was 76.9; the minimum was 55. In Houston, the maximum was 77.5; the minimum was 57.4. (From U.S. National Oceanic and Atmospheric Administration.)

c. (For the student senate.) Among the students who entered our school four years ago, here are the percentages of those who graduated, withdrew, or are still enrolled: In Nursing, 71 percent graduated; 27.9 percent withdrew; 1.1 percent are still enrolled. In Engineering, 62 percent graduated; 29.2 percent withdrew; 8.8 percent are still enrolled. In Business, 53.6 percent graduated; 43 percent withdrew; 3.4 percent are still enrolled. In Arts and Sciences, 27.5 percent graduated; 68 percent withdrew; 4.5 percent are still enrolled.

d. (For the student senate.) Here are the enrollment trends from 1987 to 1999 for two colleges in our university. In Engineering: 1987, 455 students enrolled; 1988, 610; 1989, 654; 1990, 758; 1991, 803; 1992, 827; 1993, 1,046; 1994, 1,200; 1995, 1,115; 1996, 1,075; 1997, 1,116; 1998, 1,145; 1999, 1,177. In Business: 1987, 922; 1988, 1,006; 1989, 1,041; 1990, 1,198; 1991, 1,188; 1992, 1,227; 1993, 1,115; 1994, 1,220; 1995, 1,241; 1996, 1,366; 1997, 1,381; 1998, 1,402; 1999, 1,426.

13. Anywhere on campus or at work, locate at least one visual that needs revision for accuracy, clarity, appearance, or appropriateness. Look in computer manuals, lab manuals, newsletters, financial aid or admissions or placement brochures, student or faculty handbooks, newspapers, or textbooks. Use the Visual Plan Sheet (page 298) and the Checklist (page 299) as guides to revise and enhance the visual. Submit

to your instructor a copy of the original, along with a memo explaining your improvements. Be prepared to discuss your revision in class.

COLLABORATIVE PROJECTS

1. Assume that your technical writing instructor is planning to purchase five copies of a graphics software package for students to use in designing their documents. The instructor has not yet decided which general-purpose package would be most useful. Your group's task is to test one package and to make a recommendation.

   In small groups, visit your school's microcomputer lab and ask for a listing of the graphics packages that are available to students and faculty. Select one package and learn how to use it. Design at least four representative visuals. In a memo or presentation to your instructor and classmates, describe the package briefly and tell what it can do. Would you recommend purchasing five copies of this package for general-purpose use by writing students? Explain. Submit your report, along with the sample graphics you have composed. Appoint one member to present your group's recommendation in class.

   Do the same assignment, comparing various clip art packages. Which package offers the best image selection for writers in your specialty?

2. Compile a list of twelve World Wide Web sites that offer graphics support by way of advice, image banks, design ideas, artwork catalogs, and the like. Provide the address for each site, along with a description of the resources offered and their appropriate cost. Report your findings in the format stipulated by your instructor.

3. *Creating icons:* The U.S. Department of Energy has created a facility in New Mexico for storing nuclear defense by-products and waste materials that will remain radioactive for more than ten thousand years. How can future generations be warned against drilling in this area or disturbing the material buried more than two thousand feet underground—especially when these distant generations might differ radically in use of language and symbols or in technological and scientific awareness? What system of icons would most likely endure through the centuries and convey a clear warning to any culture at any time? Among the solutions proposed:

   - Covering the entire site with giant concrete blocks (30–40 feet square), placed close enough together to prevent any useful human activity in the spaces. Furthermore, black-dyed concrete would absorb the desert sun, creating unbearable heat.
   - Covering the site with huge stone "thorns."
   - Erecting fifty-foot-high stone pillars over the site, with an entrance chamber at the surface that would contain an array of warning symbols, icons, and drawings.

   Your group's assignment is to improve on these proposals by devising a system of symbolic obstacles and visual warnings. Prepare a detailed proposal that includes a full set of visuals depicting your plan.

   Note: You may wish to consult the article from which this assignment was adapted: Kliewer, Gary. "The 10,000-Year Warning." *The Futurist* Sept. 1992: 17–19.

CHAPTER 15

Designing Pages and Documents

PAGE DESIGN IN WORKPLACE WRITING

DESKTOP PUBLISHING

CREATING A USABLE DESIGN

AUDIENCE CONSIDERATIONS IN PAGE DESIGN

DESIGNING ON-SCREEN PAGES

Usability Checklist for Page Design

Page design determines the look of a page, the layout of words and graphics. An audience's first impression tends to be a purely visual, esthetic judgment. Well-designed pages invite users in, guide them through the material, and help them understand and remember it.

PAGE DESIGN IN WORKPLACE WRITING

Page design becomes especially significant when we consider these realities about writing and reading in the workplace:

Why usable page design is essential

- *Technical documents are designed differently from most other forms of writing.* Instead of an unbroken sequence of paragraphs, users look for charts, diagrams, lists, various typesizes and fonts, different headings, and other aids to navigation.
- *Technical documents rarely get users' undivided attention.* Busy users often only skim a document, or they refer to certain sections during a meeting or presentation. Amid frequent distractions, people must be able to leave the document and return to it easily.
- *People read work-related documents only because they have to.* If they have easier ways of getting the information, people will use them.
- *As computers generate more and more paper, any document competes for audience attention.* People resist any document that appears overwhelming. They expect a usable design that announces how the document is organized, where they are in the document, which items matter most, and how the items relate.

Having decided at a glance whether your document is inviting and accessible, users will draw conclusions about the value of your information, the quality of your work, and your credibility.

Notice how the information in Figure 15.1 resists interpretation. Without design cues, we have no way of grouping this information into organized units of meaning. Figure 15.2 shows the same information after a design overhaul.

DESKTOP PUBLISHING

Planning, drafting, and writing at a workstation, each employee is increasingly responsible for all stages in document preparation. Because they combine word processing, typesetting, and graphics, desktop publishing (DTP) systems mean less reliance on typists, print shops, and graphic artists.

Using page design software, optical scanners, and laser printers, one person, or a few working collaboratively, can control the entire production cycle: designing, illustrating, laying out, and printing the final document (Cotton 36–47):

Figure 15.1
Ineffective Page
Design
*Source: U.S.
Department of
Energy.*

Sunspaces

Either as an addition to a home or as an integral part of a new home,

sunspaces have gained considerable popularity.

A sunspace should face within 30 degrees of true south. In the

winter, sunlight passes through the windows and warms the darkened

surface of a concrete floor, brick wall, water-filled drums, or other storage

mass. The concrete, brick, or water absorbs and stores some of the heat

until after sunset, when the indoor temperature begins to cool. The heat

not absorbed by the storage elements can raise the daytime air

temperature inside the sunspace to as high as 100 degrees Fahrenheit. As

long as the sun shines, this heat can be circulated into the house by

natural air currents or drawn in by a low-horsepower fan.

To be considered a passive solar heating system, any sunspace must

consist of these parts: a collector, such as a double layer of glass or plastic;

an absorber, usually the darkened surface of the wall, floor, or water-filled

containers inside the sunspace; a storage mass, normally concrete, brick,

or water, which retains heat after it has been absorbed; a distribution

system, the means of getting the heat into and around the house by fans

or natural air currents; and a control system, or heat-regulating device,

such as movable insulation, to prevent heat loss from the sunspace at

night. Other controls include roof overhangs that block the summer sun,

and thermostats that activate fans.

Some specific
features of DTP
systems

- ◆ Text can be typed or scanned into the program and then edited, checked for spelling and grammar, displayed in columns or other spatial arrangements, set in a variety of sizes and fonts—or sent to a print shop electronically.
- ◆ Page highlights and orienting devices can be added: headings, ruled boxes, vertical or horizontal rules, colored background screens, marginal sidebars or labels, page locator tabs, shadowing, shading, and so on.

Sunspaces

Either as an addition to a home or as an integral part of a new home, sunspaces have gained considerable popularity.

How Sunspaces Work

A sunspace should face within 30 degrees of true south. In the winter, sunlight passes through the windows and warms the darkened surface of a concrete floor, brick wall, water-filled drums, or other storage mass. The concrete, brick, or water absorbs and stores some of the heat until after sunset, when the indoor temperature begins to cool.

The heat *not* absorbed by the storage elements can raise the daytime air temperature inside the sunspace to as high as 100 degrees Fahrenheit. As long as the sun shines, this heat can be circulated into the house by natural air currents or drawn in by a low-horsepower fan.

The Parts of a Sunspace

To be considered a passive solar heating system, any sunspace must consist of these parts:

1. A *collector,* such as a double layer of glass or plastic.

2. An *absorber,* usually the darkened surface of the wall, floor, or water-filled containers inside the sunspace.

3. A *storage mass,* normally concrete, brick, or water, which retains heat after it has been absorbed.

4. A *distribution system,* the means of getting the heat into and around the house (by fans or natural air currents).

5. A *control system* (or heat-regulating device), such as movable insulation, to prevent heat loss from the sunspace at night. Other controls include roof overhangs that block the summer sun, and thermostats that activate fans.

Figure 15.2 Effective Page Design

Some specific
features of DTP
system

- Images can be drawn directly or imported into the program via scanners; charts, graphs, and diagrams can be created via graphics programs. These visuals can then be enlarged, reduced, cropped, and pasted electronically into the text pages.
- At any point in the process, entire pages can be viewed and evaluated for visual appeal, accessibility, or emphasis, and then revised.
- All work at all stages can be stored in the computer for later use, adaptation, or upgrading. Documents or parts of documents used repeatedly (*boilerplate*) can be retrieved when needed, modified, or inserted in some other document.

Using *groupware* (group authoring systems), writers from different locations can publish documents collaboratively.

CREATING A USABLE DESIGN

Approach your design decisions from the top down. First, consider the overall look of your pages; next, the shape of each paragraph; and finally, the size and style of individual words and letters (Kirsh 112). Figure 15.3 depicts how design considerations move from large matters to small.

NOTE   *All design considerations are influenced by the budget for a publication. For instance, adding a single color, say, to major heads, can double the printing cost.*

If your organization prescribes no specific guidelines, the following design principles should serve in most situations.

Shaping the Page

In shaping a page, consider its look, feel, and overall layout. The following suggestions will help you shape appealing and usable pages.

USE THE RIGHT PAPER AND INK.   For routine documents (memos, letters, in-house reports) print in black ink, on 8 ½" x 11" low-gloss, white paper. Use rag-bond paper (20 pound or heavier) with a high fiber content (twenty-five percent minimum). Shiny paper produces glare that tires the eyes. Flimsy or waxy paper feels inferior.

For documents that will be published (manuals, marketing literature), the grade and quality of paper are important considerations. Paper varies in weight, grain, and finish—from low-cost newsprint, with noticeable wood fiber, to wood-free, specially coated paper with custom finishes. Choice of paper depends on the artwork to be included, the type of printing, and the intended aesthetic effect: for

Figure 15.3
A Flowchart for
Decisions in Page
Design

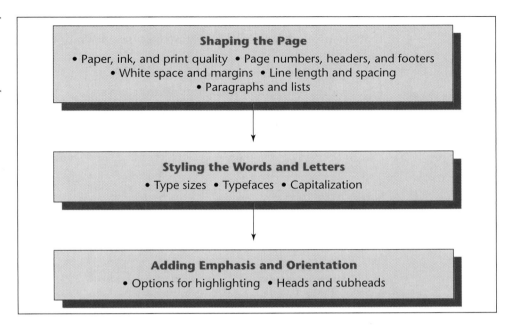

Shaping the Page
- Paper, ink, and print quality • Page numbers, headers, and footers
- White space and margins • Line length and spacing
- Paragraphs and lists

Styling the Words and Letters
- Type sizes • Typefaces • Capitalization

Adding Emphasis and Orientation
- Options for highlighting • Heads and subheads

example, you might choose specially coated, heavyweight, glossy paper for an elegant effect in an annual report (Cotton 73).

USE HIGH-QUALITY TYPE OR PRINT. Print hard copy on an inkjet or laser printer. If your inkjet's output is blurry, consider purchasing special inkjet paper for the best output quality.

USE CONSISTENT PAGE NUMBERS, HEADERS, AND FOOTERS. For a long document, count your title page as page i, without numbering it, and number all front-matter pages, including the table of contents and abstract, with lowercase roman numerals (ii, iii, iv). Number the first text page and subsequent pages with arabic numerals (1, 2, 3). Along with page numbers, *headers* or *footers* appear in the top or bottom page margins, respectively. These provide chapter or article titles, authors' names, dates, or other publication information. (See, for example, the headers on the pages in this book.)

USE ADEQUATE WHITE SPACE. White space is all the space not filled by text or images. White space divides printed areas into small, digestible chunks. For instance, it separates sections in a document, headings and visuals from text, paragraphs on a page. White space can be designed to improve a document's appearance, clarity, and emphasis.

Well-designed white space imparts a shape to the whole document, a shape that orients users and lends a distinctive visual form to the printed matter by:

1. keeping related elements together
2. isolating and emphasizing important elements
3. providing breathing room between blocks of information

Use white space to orient the readers

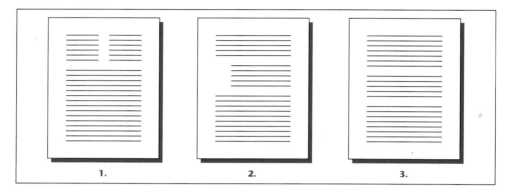

1.   2.   3.

Pages that look uncluttered, inviting, and easy to follow convey an immediate sense of user-friendliness.

PROVIDE AMPLE AND APPROPRIATE MARGINS.   Small margins crowd the page and make the material look difficult. On your 8½-by-11-inch page, leave margins of at least 1 or 1½ inches. If the manuscript is to be bound in some kind of cover, widen the inside margin to two inches.

Choose between *unjustified* text (uneven or "ragged" right margins) and *justified* text (even right margins). Each arrangement creates its own "feel."

Justified lines

To make the right margin even in justified text, the spaces vary between words and letters on a line, sometimes creating channels or rivers of white space. The eyes are then forced to adjust continually to these space variations within a line or paragraph. Because each line ends at an identical vertical space, the eyes must work harder to differentiate one line from another (Felker 85). Moreover, to preserve the even margin, words at line's end are often hyphenated, and frequently hyphenated line endings can be distracting.

Unjustified lines

Unjustified text, on the other hand, uses equal spacing between letters and words on a line, and an uneven right margin (as traditionally produced by a typewriter). For some readers, a ragged right margin makes reading easier. These differing line lengths can prompt the eye to move from one line to another (Pinelli 77). In contrast to justified text, an unjustified page looks less formal, less distant, and less official.

Justified text seems preferable for books, annual reports, and other formal materials. Unjustified text seems preferable for more personal forms of communication such as letters, memos, and in-house reports.

KEEP LINE LENGTH REASONABLE.   Long lines tire the eyes. The longer the line, the harder it is for the reader to return to the left margin and locate the beginning of the next line (White, *Visual Design* 25).

Notice how your eye labors to follow this apparently endless message that seems to stretch in lines that continue long after your eye was prepared to move down to the next line. After reading more than a few of these lines, you begin to feel tired and bored and annoyed, without hope of ever reaching the end.

Short lines force the eyes back and forth (Felker 79). "Too-short lines disrupt the normal horizontal rhythm of reading" (White, *Visual Design* 25).

Lines that are too
short cause your eye
to stumble from one
fragment to another
at a pace that too
soon becomes
annoying, if not
nauseating.

A reasonable line length is sixty to seventy characters (or nine to twelve words) per line for an 8$^1/_2$-by-11-inch single-column page. The number of characters will depend on print size. Longer lines call for larger type and wider spacing between lines (White, *Great Pages* 70).

Line length, of course, is affected by the number of columns (vertical blocks of print) on your page. Two-column pages often appear in newsletters and brochures, but research indicates that single-column pages work best for complex, specialized information (Hartley 148).

KEEP LINE SPACING CONSISTENT.   For any document likely to be read completely (letters, memos, instructions), single-space within paragraphs and double-space between. Instead of indenting the first line of single-spaced paragraphs, separate them with a line of space.

For longer documents likely to be read selectively (proposals, formal reports) increase line spacing within paragraphs. Indent these paragraphs or separate them with an extra line of space. This open spacing allows users to scan a long document, and to locate quickly what they need.

TAILOR EACH PARAGRAPH TO ITS PURPOSE.   Users often skim a long document to find what they want. Most paragraphs, therefore, begin with a topic sentence forecasting the content.

Shape each
paragraph

Use a long paragraph (no more than fifteen lines) for clustering material that is closely related (such as history and background, or any body of information best understood in one block).

Use short paragraphs for making complex material more digestible, for giving step-by-step instructions, or for emphasizing vital information.

Instead of indenting a series of short paragraphs, separate them by inserting an extra line of space (as here).

Avoid "orphans," leaving a paragraph's opening line on the bottom of a page, or "widows," leaving a paragraph's closing line on the top of the page.

MAKE LISTS FOR EASY READING.   Users prefer information in list form rather than in continuous prose paragraphs (Hartley 51). Types of items you might list: advice or examples, conclusion and recommendations, criteria for evaluation, errors to avoid, materials and equipment for a procedure, parts of a mechanism, or steps or events in a sequence. Notice how the preceding information becomes easier to grasp and remember when displayed in the list below.

Types of items you might list:

◆ advice or examples
◆ conclusions and recommendations
◆ criteria for evaluation
◆ errors to avoid
◆ materials and equipment for a procedure
◆ parts of a mechanism
◆ steps or events in a sequence

A list of brief items usually needs no punctuation at the end of each line. A list of full sentences or questions requires appropriate punctuation after each item.

Depending on the list's contents, set off each item with some kind of visual or verbal signal. If the items require a strict sequence or chronology (say, parts of a mechanism or a set of steps), use arabic numbers (1, 2, 3) or the words *First, Second, Third,* and so on. If the items require no strict sequence (as in the list above), use dashes, asterisks, or bullets. For a checklist, use open boxes.

Introduce your list with an explanation. Phrase all listed items in parallel grammatical form. If the items suggest no strict sequence, try to impose some logical ranking (most to least important, alphabetical, or some such). Set off the list with extra white space above and below.

NOTE

Keep in mind that a document with too many lists appears busy, disconnected, and splintered (Felker 55). And long lists could be used by unethical writers to camouflage bad or embarrassing news (R. Williams 12).

Use lists to help readers organize their understanding

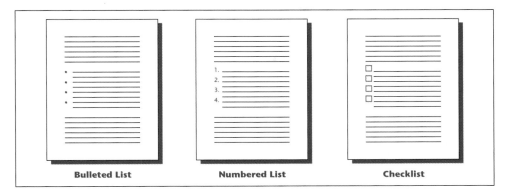

| Bulleted List | Numbered List | Checklist |

Styling the Words and Letters

In styling words and letters, we consider typographic choices that will make the text easy to read.

USE STANDARD TYPE SIZES.   Word-processing programs offer various type sizes:

Select the appropriate type size

9 point

10 point

12 point

14 point

18 point

24 point

Standard type sizes for manuscripts run from 10 to 12 point. Use larger or smaller sizes for headings, titles, captions (brief explanation of a visual), sidebars (marginal comments), or special emphasis. Use a consistent type size for similar elements throughout the document.

SELECT APPROPRIATE FONTS.   A font, or typeface, is the style of individual letters and characters. Each font has its own *personality:* "The typefaces you select for . . . [heads], subheads, body copy, and captions affect the way readers experience your ideas" (*Aldus Guide* 24).

Particular fonts can influence reading speed by as much as thirty percent (Chauncey 26).

Word-processing programs offer a variety of fonts like the examples below, listed by name.

Select a font for its personality

11-point New York

11-point Courier

11-point Palatino

11-point Geneva

11-point Monaco

11-point Chicago

11-point Helvetica

11-point Times

For visual unity, use different sizes and versions (**bold,** *italic,* SMALL CAPS) of the same font throughout your document—except possibly for headings, captions, sidebars, or visuals. In any case, try to use no more than two different typeface families throughout a document.

All fonts divide into two broad categories: *serif* and ***sans serif.*** Serifs are the fine lines that extend horizontally from the main strokes of a letter.

Decide between serif and sans serif type

Serif type makes body copy more readable because the horizontal lines "bind the individual letters" and thereby guide the reader's eyes from letter to letter—as in the type you are now reading (White, *Visual Design* 14).

In contrast, sans serif type is purely vertical (like this). Clean looking and "businesslike," sans serif is considered ideal for technical material (numbers, equations, etc.), marginal comments, headings, examples, tables, and captions to pictures and visuals, and any other material set off from the body copy (White, *Visual Design* 16).

Font preferences are culturally determined

Europeans generally prefer sans serif fonts throughout their documents, and other cultures have their own preferences as well. Learn all you can about the design conventions of the culture you are addressing.

NOTE   *Except for special emphasis, use conservative fonts; the more ornate ones are harder to read and inappropriate for most workplace documents.*

AVOID SENTENCES IN FULL CAPS. Sentences or long passages in full capitals (uppercase letters) are hard to read because all uppercase letters and words have the same visual outline (Felker 87). The longer the passage, the harder readers work to grasp your emphasis.

MY DOG HAS MANY FLEAS.

My dog has many fleas.

FULL CAPS *are good for emphasis but long passages are hard to read*

Hard   ACCORDING TO THE NATIONAL COUNCIL ON RADIA-TION PROTECTION, YOUR MAXIMUM ALLOWABLE DOSE OF LOW-LEVEL RADIATION IS 500 MILLIREMS PER YEAR.

Easier   According to the National Council on Radiation Protection, your MAXIMUM allowable dose of low-level radiation is 500 millirems per year.

Lowercase letters take up less space, and the distinctive shapes make each word easier to recognize and remember (Benson 37).

Use full caps as section headings (INTRODUCTION) or to highlight a word or phrase (WARNING: NEVER TEASE THE ALLIGATOR). As with other highlighting options discussed below, use full caps sparingly.

Highlighting for Emphasis

Effective highlighting helps users distinguish important from less important elements. Highlighting options include fonts, type sizes, white space, and other graphic devices that:

Purposes of
highlighting

◆ emphasize key points
◆ make headings prominent
◆ separate sections of a long document
◆ set off examples, warnings, and notes

On a typewriter, you can highlight with <u>underlining</u>, FULL CAPS, dashes, parentheses, and asterisks.

> You can indent to set off examples, explanations, or any material that should be distinguished from other elements in your document.

Using ruled (or typed) horizontal lines, you can separate sections in a long document:

Using ruled lines, broken lines, or ruled boxes, you can set off crucial information such as a warning or a caution:

Caution: A document with too many highlights can appear confusing, disorienting, and tasteless.

In adding emphasis and orientation, consider design elements that will direct readers' focus and help them navigate the text. See pages 291–93 for more on background screens, ruled lines, and ruled boxes.

Word processors offer highlighting options that include **boldface,** *italics,* SMALL CAPS, varying type sizes and fonts, and color. For specific highlighted items, some options are better than others:

Not all highlighting is equal

Boldface works well for emphasizing a single sentence or brief statement, and is perceived by readers as being "authoritative" *(Aldus Guide* 42).

Italics suggest a more subtle or "refined" emphasis than boldface (Aldus Guide 42). Italics can highlight words, phrases, book titles, or anything else you would have underlined on a typewriter. But multiple lines (like these) of italic type are hard to read.

SMALL CAPS WORK FOR HEADINGS AND SHORT PHRASES. BUT ANY LONG STATEMENT ALL IN CAPS IS HARD TO READ.

Small type sizes (usually sans serif) work well for captions and credit lines and as labels for visuals or to set off other material from the body copy.

Large type sizes and dramatic typefaces are both hard to miss and hard to digest. Be conservative—unless you really need to convey forcefulness.

Color is appropriate only in some documents, and only when used sparingly. (Pages 290–93) discuss how color can influence audience perception and interpretation of a message.

Whichever highlights you select for a document, be consistent. Make sure that all headings at one level are highlighted identically, that all warnings and cautions are set off identically, and so on. And *never* combine too many highlights.

Using Headings for Access and Orientation

Readers of a long document often look back or jump ahead to sections that interest them most. Headings announce how a document is organized, point readers to what they need, and divide the document into accessible blocks or "chunks." An informative heading can help a person decide whether a section is worth reading (Felker 17). Besides cutting down on reading and retrieval time, headings help readers remember information (Hartley 15).

MAKE HEADINGS INFORMATIVE. A heading should be informative but not wordy. Informative headings orient readers, showing them what to expect. Vague or general headings can be more misleading than no headings at all (Redish et al. 144). Whether your heading takes the form of a phrase, a statement, or a question, be sure it advances meaning.

Uninformative heading	Document Formatting

What should we expect here: specific instructions, an illustration, a discussion of formatting policy in general? We can't tell.

Informative versions	How to Format Your Document
	Format Your Document for Usability
	How Do I Format My Document?

When you use questions as headings, phrase the questions the way readers might ask them.

MAKE HEADINGS SPECIFIC AS WELL AS COMPREHENSIVE. Focus the heading on a specific topic. Do not preface a discussion of the effects of acid rain on lake trout with a broad heading such as "Acid Rain." Use instead "The Effects of Acid Rain on Lake Trout."

Also, provide enough headings to contain each discussion section. If chemical, bacterial, and nuclear wastes are three *separate* discussion items, provide a heading for each. Do not simply lump them under the sweeping heading "Hazardous Wastes." If you have prepared an outline for your document, adapt major and minor headings from it.

MAKE HEADINGS GRAMMATICALLY CONSISTENT. All major topics or all minor topics in a document share equal rank; to emphasize this equality, express topics at the same level in identical—or parallel—grammatical form.

Nonparallel headings	How to Avoid Damaging Your Disks:

  1. Clean Disk Drive Heads
  2. Keep Disks Away from Magnets
  3. Writing on Disk Labels with a Felt-Tip Pen
  4. It Is Crucial That Disks Be Kept Away from Heat
  5. Disks Should Be Kept Out of Direct Sunlight
  6. Keep Disks in Their Protective Jackets

In items 3, 4, and 5, the lack of parallelism (no verbs in the imperative mood) obscures the relationship between individual steps and causes confusion. This next version emphasizes the equal rank of these items.

Parallel headings

3. Write on Disk Labels with a Felt-Tip Pen
4. Keep Disks Away from Heat
5. Keep Disks Out of Direct Sunlight

Parallelism helps make a document readable and accessible.

MAKE HEADINGS VISUALLY CONSISTENT.   "Wherever heads are of equal importance, they should be given similar visual expression, because the regularity itself becomes an understandable symbol" (White, *Visual Design* 104). Use identical type size and typeface for all headings at a given level.

LAY OUT HEADINGS BY LEVEL.   Like a road map, your headings should clearly announce the large and small segments in your document. (Use the logical divisions from your outline as a model for heading layout.) Think of each heading at a particular level as an "event in a sequence" (White *Visual Design* 95).

Headings vary in positioning and highlighting, depending on their level (Figure 15.4). Follow these suggestions for using headlines effectively:

How to provide effective headings

- *Ordinarily, use no more than four levels of heading (section, major topic, minor topic, subtopic).* Excessive heads and subheads make a document seem cluttered or fragmented.
- *To divide logically, be sure each higher-level heading yields at least two lower-level headings.*
- *Insert one additional line of space above each heading.* For double-spaced text, triple-space before the heading, and double-space after; for single-spaced text, double-space before the heading, and single-space after.

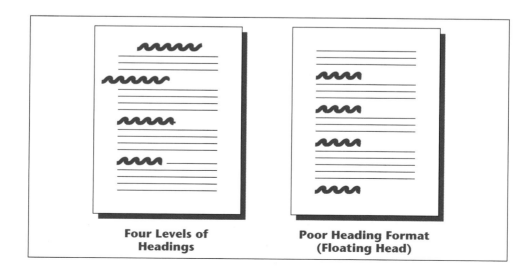

Four Levels of Headings   **Poor Heading Format (Floating Head)**

Section Heading

Section headings in boldface and enlarged type are more appealing and readable than headings in full caps. Use a type size roughly 4 points larger than body copy (say, 16-point section heads for 12-point body copy). Avoid overly large heads, and use no other highlights. Set these and all lower heads one extra line space below any preceding text.

Major Topic Heading

Major topic heads abut the left margin (flush left), and each important word begins with an uppercase letter. Use boldface and a type size roughly 2 points larger than body copy, with no other highlights.

Minor Topic Heading

Minor topic heads are also set flush left. Use boldface, italics (optional), and the same type size as in the body copy, with no other highlights.

Subtopic Heading. Instead of indenting the first line of body copy, place subtopic heads flush left on the same line as following text, and set off by a period. Use boldface and the same type size as in the body copy, with no other highlights.

Figure 15.4   Recommended Format for Word-Processed Headings

- *Never begin the sentence right after the heading with "this," "it," or some other pronoun referring to the heading.* Make the sentence's meaning independent of the heading.
- *Never leave a heading floating as the final line of a page.* Unless two lines of text can fit below the heading, carry it over to the top of the next page.
- *Use different type sizes to reflect levels of heads.* Readers often equate large type size with importance (White, *Visual Design* 95; Keyes 641). Set major heads in a larger type size.

When headings show the relationships among all the parts, readers can grasp at a glance how a document is organized.

AUDIENCE CONSIDERATIONS IN PAGE DESIGN

Like any writing decisions, page design choices are by no means random. An effective writer designs a document for specific use by a specific audience.

In deciding on a format, work from a detailed audience and use profile (Wight 11). Know your audience and their intended use of your information. Design a document to meet their particular needs and expectations, as in these examples:

How users' needs determine page design

- If people will use your document for reference only (as in a repair manual), use plenty of headings.
- If users will follow a sequence of steps, show that sequence in a numbered list.
- If users will need to evaluate something, give them a checklist of criteria (as in this book at the end of most chapters).
- If users need a warning, highlight the warning so that it cannot possibly be overlooked.
- If users have asked for your one-page report or résumé, save space by using the 10-point type size.
- If users will be facing complex information or difficult steps, widen the margins, increase all white space, and shorten the paragraphs.

Regardless of the audience, never make the document look "too intellectually intimidating" (White, *Visual Design* 4).

Consider also your audience's specific cultural expectations. For instance, Arabic and Persian text is written from right to left instead of left to right (Leki 149). In other cultures, readers move up and down the page, instead of across. Be aware that a particular culture might be offended by certain icons or by a typeface that seems too plain or too fancy (Weymouth 144). Ignoring a culture's design conventions can be interpreted as a sign of disrespect.

NOTE *Remember that even the most brilliant page design cannot redeem a document with worthless content, chaotic organization, or unreadable style. The value of any document ultimately depends on elements beneath the visual surface.*

DESIGNING ON-SCREEN PAGES

To be read on a computer screen, pages must accommodate small screen size, reduced resolution, and reader resistance to scrolling—among other restrictions. Some special design requirements of on-screen pages:

Elements of on-screen page design

- Sentences and paragraphs are short and more concise than their hard copy equivalents.

- Sans serif type is preferred for on-screen readability.
- The main point usually appears close to the top of each screen.
- Each "page" often stands alone as a discrete "module," or unit of meaning. Instead of a traditional introduction-body-conclusion sequence of pages, material is displayed in screen-sized chunks, each linked as hypertext.
- Links, navigation bars, hot buttons, and help options are displayed on each page.

See Chapter 20 for specific features of on-screen page design.

Special authoring software such as *Adobe FrameMaker®* or *RoboHelp®* automatically converts hard copy document format to various on-screen formats, chunked and linked for easy navigation.

NOTE   *Provide margins for on-screen pages, so that your text won't bump against (or run off) the edge of the user's screen. Also, avoid using underlines for emphasis because these might be confused with hyperlinks (Munger).*

Usability Checklist FOR PAGE DESIGN

(Numbers in parentheses refer to the first page of discussion.)

- ☐ Is the paper white, low-gloss, rag bond, with black ink? (308)
- ☐ Is all type or print neat and legible? (309)
- ☐ Does the white space adequately orient the readers? (309)
- ☐ Are the margins ample? (310)
- ☐ Is line length reasonable? (311)
- ☐ Is the right margin unjustified? (310)
- ☐ Is line spacing appropriate and consistent? (311)
- ☐ Does each paragraph begin with a topic sentence? (312)
- ☐ Does the length of each paragraph suit its subject and purpose? (312)
- ☐ Are all paragraphs free of "orphan" lines or "widows"? (312)
- ☐ Do parallel items in strict sequence appear in a numbered list? (312)
- ☐ Do parallel items of any kind appear in a list whenever a list in appropriate? (312)
- ☐ Are pages numbered consistently? (309)
- ☐ Is the body type size 10 to 12 points? (313)
- ☐ Do full caps highlight only words or short phrases? (314)
- ☐ Is the highlighting consistent, tasteful, and subdued? (315)
- ☐ Are all format patterns distinct enough so that readers will find what they need? (316)
- ☐ Are there enough headings for readers to know where they are in the document? (316)
- ☐ Are headings informative, comprehensive, specific, parallel, and visually consistent? (317)
- ☐ Are headings clearly differentiated according to level? (318)
- ☐ Is the overall design inviting without being overwhelming? (320)
- ☐ Does this design respect the cultural conventions of the audience? (320)

EXERCISES

1. Find an example of effective page design, in a textbook or elsewhere. Photocopy a selection (two or three pages), and attach a memo explaining to your instructor and classmates why this design is effective. Be specific in your evaluation. Now do the same for an example of bad page design. Bring your examples and explanations to class, and be prepared to discuss why you chose them.

2. These are headings from a set of instructions for listening. Rewrite the headings to make them parallel.

   • You Must Focus on the Message
   • Paying Attention to Nonverbal Communication
   • Your Biases Should Be Suppressed
   • Listen Critically
   • Listen for Main Ideas
   • Distractions Should Be Avoided
   • Provide Verbal and Nonverbal Feedback
   • Making Use of Silent Periods
   • Are You Allowing the Speaker Time to Make His or Her Point?
   • Keeping an Open Mind Is Important

3. Using the checklist on page 321, redesign an earlier assignment or a document you've prepared on the job. Submit to your instructor the revision and the original, along with a memo explaining your improvements. Be prepared to discuss your format design in class.

4. Anywhere on campus or at work, locate a document with a design that needs revision. Candidates include career counseling handbooks, financial aid handbooks, student or faculty handbooks, software or computer manuals, medical information, newsletters, or registration procedures. Redesign the document or a two- to five-page selection from it. Submit to your instructor a copy of the original, along with a memo explaining your improvements. Be prepared to discuss your revision in class.

COLLABORATIVE PROJECT

Working in small groups, redesign a document you select or your instructor provides. Prepare a detailed explanation of your group's revision. Appoint a group member to present your revision to the class, using an opaque or overhead projector, a large-screen monitor, or photocopies.

Adding Document Supplements

PURPOSE OF SUPPLEMENTS

COVER

TITLE PAGE

LETTER OF TRANSMITTAL

TABLE OF CONTENTS

LIST OF TABLES AND FIGURES

INFORMATIVE ABSTRACT

GLOSSARY

APPENDIXES

DOCUMENTATION

Supplements are reference items that make a long document more accessible, to accommodate users with various interests: The title page, letter of transmittal, table of contents, and abstract summarize the content of the document. The glossary, appendixes, and list of works cited can either provide supporting data or help readers follow technical sections. Readers can refer to any of these supplements or skip them altogether, according to their needs.

PURPOSE OF SUPPLEMENTS

Supplements address these workplace realities:

Why supplements
are essential

- *Confronted by a long document, users often try to avoid reading the whole thing.* Instead they look for only what they need to complete their specific task.
- *Different people often use a document for different purposes.* Some look for an overview; others want details or only conclusions and recommendations, or the "bottom line." Technical personnel might focus on the body of a highly specialized report and on the appendixes for supporting data (maps, formulas, calculations). Executives and managers might read only the transmittal letter, the abstract, and the conclusions and recommendations.

Carefully planned supplements give everyone what they need.

Report supplements are classified in two groups:

1. *supplements that precede your report* (front matter): cover, title page, letter of transmittal, table of contents (and of figures), and abstract
2. *supplements that follow your report* (end matter): glossary, appendix(es), footnotes, endnote pages

COVER

Use a sturdy, plain cover with page fasteners. With the cover on, the open pages should lie flat. Use covers only for long documents.

Center the report title and your name four to five inches below the top of your page:

THE FEASIBILITY OF A TECHNICAL MARKETING CAREER:
AN ANALYSIS
by
Richard B. Larkin, Jr.

TITLE PAGE

The title page lists the report title, author's name, name of person(s) or organization to whom the report is addressed, and date of submission.

How to prepare a title page

TITLE.   The title announces the report's purpose and subject. The previous title (given as an example for the cover) is clear, accurate, comprehensive, and specific. But even slight changes can distort this title's signal.

An unclear title   A TECHNICAL MARKETING CAREER

This version is unclear about the report's purpose. Is the report *describing* the career, *proposing* the career, *giving instructions* for career preparation? Insert descriptive words ("analysis," "instructions," "proposal," "feasibility," "description," "progress") that accurately state your purpose.

PLACEMENT OF TITLE PAGE ITEMS.   Do not number your title page but count it as page i of the front matter. Center the title and all other lines (Figure 16.1).

LETTER OF TRANSMITTAL

For college reports, the letter of transmittal usually follows the title page and is bound as part of the report. For workplace reports, the letter usually precedes the title page. Include a letter of transmittal with any formal report or proposal addressed to a specific reader. As a gesture of courtesy, your letter might

What to include in a letter of transmittal

- ◆ acknowledge those who helped with the report
- ◆ refer to sections of special interest: unexpected findings, key visuals, major conclusions, special recommendations, and the like
- ◆ discuss the limitations of your study, or any problems gathering data
- ◆ discuss the need and approaches for follow-up investigations
- ◆ describe any personal (or off-the-record) observations
- ◆ suggest some special uses for the information
- ◆ urge the reader to immediate action

The letter of transmittal can be tailored to a particular audience, as is Richard Larkin's in Figure 16.2. If a report is being sent to a number of people who are variously qualified and bear various relationships to the writer, individual letters of transmittal may vary, within the following basic structure:

INTRODUCTION.   Open with reference to the user's original request. Briefly review the reasons for your report or include a descriptive abstract. Maintain a con-

Feasibility Analysis
of a Career
in Technical Marketing

for

Professor J. M. Lannon
Technical Writing Instructor
University of Massachusetts
North Dartmouth, Massachusetts

by

Richard B. Larkin, Jr.
English 266 Student

May 1, 20XX

Figure 16.1   Title Page for a Formal Report

165 Hammond Way
Hyannis, MA 02457
April 29, 20XX

John Fitton
Placement Director
University of Massachusetts
North Dartmouth, MA 02747

Dear Mr. Fitton:

Here is my analysis of the feasibility of a career in technical marketing and sales. In preparing my report, I've learned a great deal about the requirements and modes of access to this career, and I believe my information will help other students as well.

Although committed to their specialities, some technical and science graduates seem interested in careers in which they can apply their technical knowledge to customer and business problems. Technical marketing may be an attractive choice of career for those who know their field, who can relate to different personalities, and who are good communicators.

Technical marketing is competitive and demanding, but highly rewarding. In fact, it is an excellent route to upper-management and executive positions. Specifically, marketing work enables one to develop a sound technical knowledge of a company's products, to understand how these products fit into the marketplace, and to perfect sales techniques and interpersonal skills. This is precisely the kind of background that paves the way to top-level jobs.

I've enjoyed my work on this project, and would be happy to answer any questions.

Sincerely,

Richard B. Larkin

Richard B. Larkin, Jr.

Figure 16.2   Letter of Transmittal for a Formal Report

fident and positive tone throughout. Indicate pride and satisfaction in your work. Avoid implied apologies, such as "I hope this report meets your expectations."

BODY.   In the letter body, include items from the prior list of possibilities (acknowledgments, special problems). Although your informative abstract will summarize major findings, conclusions, and recommendations, your letter gives a brief and personal overview of the *entire project.*

CONCLUSION.   State your willingness to answer questions or discuss findings. End positively: "I believe that the data in this report are accurate, that they have been analyzed rigorously and impartially, and that the recommendations are sound."

TABLE OF CONTENTS

Compose this supplement by assigning page numbers to headings from your outline. Keep in mind, however, that not all levels of outline headings appear in your table of contents or your report. Excessive headings can fragment the discussion. Follow these guidelines:

How to prepare a table of contents

- ◆ List front matter (transmittal letter, abstract), numbering the pages with lowercase roman numerals. (The title page, though not listed, is counted page i.) List glossary, appendix, and endnotes; number these pages with arabic numerals, continuing the page sequence of your report proper, in which page 1 is the first page of report text.
- ◆ Include no headings in the table of contents not listed as headings or subheadings in the report; the report may, however, contain subheadings not listed in the table of contents.
- ◆ Phrase headings in the table of contents exactly as in the report.
- ◆ List headings at various levels in varying type styles and indention.
- ◆ Use *leader lines* (........) to connect headings to page numbers. Align rows of dots vertically, each above the other.

Figure 16.3 shows the table of contents for Richard Larkin's feasibility analysis.

LIST OF TABLES AND FIGURES

Following the table of contents is a list of tables and figures. When a report has more than four or five visuals, place this table on a separate page. Figure 16.4 shows the list of tables and figures for Larkin's report.

iii

CONTENTS

Figure 16.3   Table of Contents for a Formal Report

Figure 16.4   A List of Tables and Figures for a Formal Report

INFORMATIVE ABSTRACT

For readers who lack the time or willingness to read an entire report, the informative abstract (page 183) is the most important part of the document. The abstract is always written *after* the report proper. If you cannot effectively summarize your report, it probably needs revision.

Follow these suggestions for preparing your abstract:[1]

How to prepare an informative abstract

- ◆ Make sure your abstract stands alone in terms of meaning.
- ◆ Write for a general audience. Readers of the abstract are likely to vary in expertise, perhaps more than those who read the report itself; therefore, translate all technical data.
- ◆ Add no new information. Simply summarize the report.
- ◆ Present your information in the following sequence:

    a. Identify the issue or need that led to the report.
    b. Offer the major findings from the report body.
    c. Include a condensed conclusion and recommendations, if any.

The informative abstract in Figure 16.5 accompanies Richard Larkin's report.

GLOSSARY

A glossary is an alphabetical listing of special terms and definitions, following your report. A glossary makes key definitions available to laypersons without interrupting technical readers. If fewer than five terms need defining, place them in

---

[1]My thanks to Professor Edith K. Weinstein for these suggestions.

ABSTRACT

The feasibility of technical marketing as a career is based on a college graduate's interests, abilities, and expectations, as well as on possible entry options.

Technical marketing is a feasible career for anyone who is motivated, who can communicate well, and who knows how to get along. Although this career offers job diversity and excellent potential for income, it entails almost constant travel, competition, and stress.

College graduates enter technical marketing through one of four options: entry-level positions that offer hands-on experience, formal training programs in large companies, prior experience in one's specialty, or graduate programs. The relative advantages and disadvantages of each option can be measured in resulting immediacy of income, rapidity of advancement, and long-term potential.

Anyone considering a technical marketing career should follow these recommendations:

• Speak with people who work in the field.
• Weigh the implications of each entry option carefully.
• Consider combining two or more options.
• Choose options for personal as well as professional benefits.

Figure 16.5   Informative Abstract

the report introduction as working definitions, or use footnote definitions. If you use a separate glossary, announce its location: "(see the glossary at the end of this report)."

Follow these suggestions for preparing a glossary:

How to prepare a glossary

◆ Define all terms unfamiliar to an intelligent layperson.
◆ Define all terms that have a special meaning in your report (e.g., "In this report, a small business is defined as").
◆ Define all terms by giving their class and distinguishing features, unless some terms need expanded definitions.
◆ List the glossary and its first page number in the table of contents.
◆ List all terms in alphabetical order. Highlight each term and use a colon to separate it from its definition.
◆ Define only terms that need explanation. When in doubt, overdefining is safer than underdefining.

GLOSSARY

Analgesic: a medication given to relieve pain during the first stage of labor.

Cervix: the neck-shaped anatomical structure that forms the mouth of the uterus.

Dilation: cervical expansion occurring during the first stage of labor.

Episiotomy: an incision of the outer vaginal tissue, made by the obstetrician just before the delivery, to enlarge the vaginal opening.

First stage of labor: the stage in which the cervix dilates and the baby remains in the uterus.

Induction: the stimulating of labor by puncturing the membranes around the baby or by giving an oxytoxic drug (uterine contractant), or both.

Figure 16.6   A Glossary (Partial)

- On first use, place an asterisk in the text by each item defined in the glossary.

Figure 16.6 shows part of a glossary for a comparative analysis of two natural childbirth techniques, written by a nurse practitioner for expectant mothers and student nurses.

APPENDIXES

The appendix is a catchall for items that are important but difficult to integrate into the body of a report: Figure 16.7 shows an appendix to a budget proposal.

What an appendix might include

- complex formulas
- details of an experiment
- interview questions and responses
- long quotations (one or more pages)
- maps
- material more essential to secondary readers than to primary readers
- photographs
- related correspondence (letters of inquiry, and so on)
- sample questionnaires and tabulated responses
- sample tests and tabulated results
- some visuals occupying more than one full page
- statistical or other measurements
- texts of laws and regulations

APPENDIX A

Table 1 Allocations and Performance of Five Massachusetts College Newspapers

	Stonehorse College	Alden College	Simms University	Fallow State	UMD
Enrollment	1,600	1,400	3,000	3,000	5,000
Fee paid (per year)	$65.00	$85.00	$35.00	$50.00	$65.00
Total fee budget	$88,000	$119,000	$105,000	$150,000	$334 429.28
Newspaper budget	$10,000	$6,000	$25,300	$37,000	$21,500 $25.337.14[a]
Yearly cost per student	$6.25	$4.29	$8.43	$12.33	$4.06 $5.20[a]
Publication rate	Weekly	Every third week	Weekly	Weekly	Weekly
Average no. of pages	8	12	18	12	20
Average total pages	224	120	504	336	560 672[a]
Yearly cost per page	$44.50	$50.00	$50.50	$110.11	$38.25 $38.69[a]

[a]These figures are next year's costs for the UMD *Torch*.

Source: Figures were quoted by newspaper business managers in April 20XX.

Figure 16.7 An Appendix

Do not stuff appendixes with needless information or use them unethically for burying bad or embarrassing news that belongs in the report proper. Follow these suggestions for preparing an appendix:

How to prepare an appendix

- ◆ Include only relevant material.
- ◆ Use a separate appendix for each major item.
- ◆ Title each appendix clearly: "Appendix A: Projected Costs."
- ◆ Use appendixes sparingly. Four or five appendixes in a ten-page report indicates a poorly organized document.
- ◆ Limit an appendix to a few pages, unless more length is essential.
- ◆ Mention the appendix early in the introduction, and refer readers to it at appropriate points in the report: "(see Appendix A)."

Remember that users should be able to understand your report without having to turn to the appendix. Distill essential facts from your appendix and place them in your report text.

| **Improper reference** | The whale population declined drastically between 1986 and 1987 (see Appendix B for details). |
| **Proper reference** | The whale population declined by 16 percent from 1986 to 1987 (see Appendix B for statistical breakdown). |

DOCUMENTATION

The endnote or works cited pages list each of your outside references in alphabetical order or in the same numerical order as they are cited in the report proper. See Appendix A for discussion.

EXERCISES

1. These titles are intended for investigative, research, or analytical reports. Revise each inadequate title to make it clear and accurate.

   a. The Effectiveness of the Prison Furlough Program in Our State
   b. Drug Testing on the Job
   c. The Effects of Nuclear Power Plants
   d. Woodburning Stoves
   e. Interviewing
   f. An Analysis of Vegetables (for a report assessing the physiological effects of a vegetarian diet)
   g. Wood as a Fuel Source
   h. Oral Contraceptives
   i. Lie Detectors and Employees

2. Prepare a title page, letter of transmittal (for a specific reader who can use your information in a definite way), table of contents, and informative abstract for a report you have written earlier.

3. Find a short but effective appendix in one of your textbooks, in a journal article in your field, or in a report from your workplace. In a memo to your instructor and classmates, explain how the appendix is used, how it relates to the main text, and why it is effective. Attach a copy of the appendix to your memo. Be prepared to discuss your evaluation in class.

Testing the Usability of Your Document

A *usable* document is one that is safe, dependable, and easy to use.

WHY USABILITY TESTING IS ESSENTIAL

Companies routinely measure the usability of their products—including the **documentation** that accompanies the product (warnings, explanations, instructions for assembling or operating and so on). To keep their customers—and to avoid lawsuits—companies go to great lengths to identify flaws or to anticipate all the ways a product might fail or be misused.

The purpose of usability testing is to keep what works in a product or a document and to fix what doesn't. Figure 17.1, for example, is designed to assess user needs and preferences for the *Statistical Abstract of the United States,* a widely consulted publication updated yearly.

HOW USABILITY TESTING IS DONE

Usability testing usually occurs at two levels (Petroski 90): (1) *alpha testing,* by the product's designers or the document's authors, and (2) *beta testing,* by the actual users of the product or document.

For a document to achieve its objectives—to be *usable*—users must be able to do at least three things (Coe, *Human Factors* 193; Spencer 74):

What a usable document enables readers to do

- ◆ easily locate the information they need,
- ◆ understand the information immediately, and
- ◆ use the information successfully.

To measure usability in a set of instructions, for instance, we ask this question: *How effectively do these instructions enable all users to carry out the task safely, efficiently, and accurately?*

Ideally, usability tests for documents occur in a setting that simulates the actual situation, with people who will actually use the document (Redish and Schell 67; Ruhs 8). But even in a classroom setting we can reasonably assess a document's effectiveness on the basis of usability criteria discussed below.

HOW USABILITY CRITERIA ARE DETERMINED

Usability criteria for workplace documents are based on answers to these questions about the task, the user, and the setting:

Questions for assessing usability

- ◆ *What specific task is being documented?*
- ◆ *What do we know about the specific users' abilities and limitations?*
- ◆ *In what setting will the document be read/used?*

Only after these questions are answered can we decide on criteria for evaluating the effectiveness of a specific document (Figure 17.2).

FORM **S-555**
(8-3-94)

STATISTICAL ABSTRACT SURVEY

U.S. DEPARTMENT OF COMMERCE
BUREAU OF THE CENSUS

Please take a few minutes to answer the questions below. Your voluntary cooperation will help us continue to serve your needs as data users. When completed, please refold, **apply tape to the open edges at the top,** and drop in the mail. Thank you.

1a. Which sections do you refer to frequently? *Mark (x) all that apply.*

☐ Population
☐ Vital statistics
☐ Health
☐ Education
☐ Law enforcement
☐ Geography
☐ Parks and recreation
☐ Elections

☐ Federal Government
☐ State and local government
☐ National defense
☐ Social insurance
☐ Labor force
☐ Income
☐ Prices
☐ Banking

☐ Business
☐ Communications
☐ Energy
☐ Science
☐ Land transportation
☐ Air and water transportation
☐ Agriculture

☐ Forests and fisheries
☐ Mining
☐ Construction and housing
☐ Manufactures
☐ Domestic trade
☐ Foreign commerce
☐ Outlying areas
☐ International statistics

b. indicate topics for which you would like to see more coverage.

2. **To what degree do you find our current presentation of data in tables clear, meaningful, and easy to understand?** *Mark (X) one.*

☐ Very much   ☐ Somewhat   ☐ Not at all

3. **Which of the following do you feel currently interferes with the clarity of the tables presented?** *Mark (X) all that apply.*

☐ Hard to understand column headings and row indentations
☐ Too many numbers in the tables—too much data to absorb
☐ Not clear what numbers mean when they are rounded ("In thousands" or "in millions," for example)
☐ Too many notes in the tables

☐ Some concepts too difficult to understand
☐ Other — *Specify*

_____

_____

4. Please indicate which of these features you might like to see expanded or reduced in future editions.

Mark (X) the appropriate column for each feature.

	Expand	Reduce	OK as is
a. State rankings *(pp. xii–xxi)*			
b. Telephone contact list *(pp. xxii–xxiv)*			
c. Introductory text for sections			
d. Guide to Sources *(pp. 887–925)*			
e. Metropolitan Concepts and Components *(pp. 926–935)*			
f. Statistical Methodology (now called Limitations of the Data) *(pp. 936–950)*			
g. Index *(pp. 959–1011)*			
h. Charts and graphs			

5. **Indicate your level of satisfaction with the Abstract.** *Mark (x) one.*

☐ Very satisfied   ☐ Satisfied   ☐ Indifferent   ☐ Unsatisfied   ☐ Very unsatisfied

Figure 17.1 A Usability Survey

Figure 17.2
Defining a
Document's
Usability Criteria

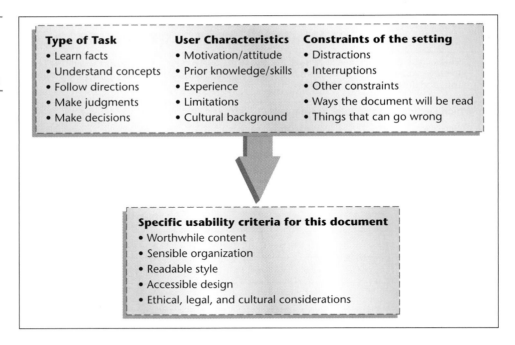

Type of Task
- Learn facts
- Understand concepts
- Follow directions
- Make judgments
- Make decisions

User Characteristics
- Motivation/attitude
- Prior knowledge/skills
- Experience
- Limitations
- Cultural background

Constraints of the setting
- Distractions
- Interruptions
- Other constraints
- Ways the document will be read
- Things that can go wrong

Specific usability criteria for this document
- Worthwhile content
- Sensible organization
- Readable style
- Accessible design
- Ethical, legal, and cultural considerations

The Checklist for Usability (page 342) identifies criteria shared by many technical documents. In addition, specific elements (visuals, page design) and document types (proposals, memos, instructions) have their own usability criteria. These are detailed in the individual usability checklists throughout this book (for example, pages 299, 321, 421).

USABILITY ISSUES IN ONLINE OR MULTIMEDIA DOCUMENTS

In contrast to printed documents, online or multimedia documents have unique usability issues (Holler 25; Humphreys 754–55):

- ◆ Online documents tend to focus more on "doing" than on detailed explanations. Workplace readers typically use online documents for reference or training rather than for study or memorizing.
- ◆ Users of online instructions rarely need persuading (say, to pay attention or follow instructions) because they are guided interactively through each step of the procedure.
- ◆ Visuals play an increasingly essential role in online instruction.
- ◆ Online documents are typically organized to be read interactively and selectively rather than in linear sequence. Users move from place to place, depend-

ing on their immediate needs. Organization is therefore flexible and modular, with small bits of easily accessible information that can be combined to suit a particular user's needs and interests.

◆ Online readers can easily lose their bearings. Unable to shuffle or flip through a stack of printed pages in linear order, readers need constant orientation (to retrieve some earlier bit of information or to compare something on one page with another). In the absence of page or chapter numbers, index, or table of contents, "Find," "Search," and "Help" options need to be plentiful and complete.

See Chapter 20 for usability considerations in Web pages and other electronic documents.

GUIDELINES

for Testing a Document's Usability

1. *Identify the Document's Purpose.* Determine how much *learning* versus *performing* is required in performing the task—and assess the level of difficulty (Mirel et al. 79; Wickens 232, 243, 250):

   ◆ *Simply learning facts.* "How rapidly is this virus spreading?"
   ◆ *Mastering concepts or theories.* "What biochemical mechanism enables this virus to mutate?"
   ◆ *Following directions.* "How do I inject the vaccine?"
   ◆ *Navigating a complex activity that requires decisions or judgments.* "Is this diagnosis accurate?" "Which treatment option should I select?"

2. *Identify the human factors.* Determine which characteristics of the user and the work setting (**human factors**) enhance or limit performance (Wickens 3):

   ◆ *User's abilities/limitations.* What do these users know already? How experienced are they in this task area? How educated? What cultural differences could create misunderstanding?
   ◆ *Users' attitudes.* How motivated or attentive are they? How anxious or defensive? Do users need persuading to pay attention or be careful?
   ◆ *Users' reading styles.* Will users be scanning the document, studying it, or memorizing it? Will they read it sequentially (page by page) or consult the document periodically and randomly? (For example, expert users often look only for key words and key concepts in titles, abstracts, or purpose lines, then determine what sections of the document they will read.)
   ◆ *Workplace constraints.* Under what conditions will the document be read? What distractions or interruptions does the work setting pose? (For example, procedures for treating a choking victim, posted in a

busy restaurant kitchen.) Will users always have the document in front of them while performing the task?

- *Possible failures.* How might the document be misinterpreted or misunderstood (Boiarsky 100)? Any potential "trouble spots" (material too complex for these users, too hard to follow or read, too loaded with information)?

For a closer look at human factors, review the Audience and Use Profile Sheet (page 29).

3. *Design the usability test.* Ask respondents to focus on specific problems (Hart 53–57; Daugherty 17–18):
   - *Content:* "Where is there too much or too little information?" "Does anything seem inaccurate?"
   - *Organization:* "Does anything seem out of order, or hard to find or follow?"
   - *Style:* "Is anything hard to understand?" "Are any words inexact or too complex?" "Do any expressions seem wordy?"
   - *Design:* "Are there any confusing headings, or too many or too few?" "Are any paragraphs, lists, or steps too long?" "Are there any misleading or overly complex visuals?" "Could any material be clarified by a visual?" "Is anything cramped and hard to read?"
   - *Ethical, legal, and cultural considerations:* "Does anything mislead?" "Could anything create potential legal liability or cross-cultural misunderstanding?"

   Figure 17.3 shows a sample usability survey (Carliner 258). Notice how the phrasing encourages users to respond with examples instead of just "yes" or "no."

4. *Administer the test.* Have respondents perform the task under controlled conditions. To ensure dependable responses, look for these qualities (Daugherty 19–20):
   - *A range of responses.* Select evaluators at various levels of expertise with this task.
   - *Independent responses.* Ask each evaluator to work alone when testing the document and recording the findings.
   - *Reliability.* Have evaluators perform the test twice.
   - *Group consensus.* After individual testing, arrange a meeting for evaluators to compare notes and to agree on needed revisions.
   - *Thoroughness.* Once the document is revised/corrected, repeat this entire testing procedure.

Basic Usability Survey

1. Briefly describe why this document is used. _____

2. Evaluate the *content:*
   - Identify any irrelevant information. _____
   - Indicate any gaps in the information. _____
   - Identify any information that seems inaccurate. _____
   - List other problems with the content. _____

3. Evaluate the *organization:*
   - Identify anything that is out of order or hard to locate or follow. _____
   - List other problems with the organization. _____

4. Evaluate the *style:*
   - Identify anything you misunderstood on first reading. _____
   - Identify anything you couldn't understand at all. _____
   - Identify expressions that seem wordy, inexact, or too complex. _____
   - List other problems with the style. _____

5. Evaluate the *design:*
   - Indicate any headings that are missing, confusing, or excessive. _____
   - Indicate any material that should be designed as a list. _____
   - Give examples of material that might be clarified by a visual. _____
   - Give examples of misleading or overly complex visuals. _____
   - List other problems with design. _____

6. Identify anything that seems misleading or that could create legal problems or cross-cultural misunderstanding. _____

7. Please suggest other ways of making this document easier to use. _____

Figure 17.3   A Basic Usability Survey

Checklist FOR USABILITY

(Numbers in parentheses refer to the first page of discussion.)

Content

- ☐ Is all material relevant to this user for this task? (30)
- ☐ Is all material technically accurate? (150)
- ☐ Is the level of technicality appropriate for this audience? (24)
- ☐ Are warnings and cautions inserted where needed? (470)
- ☐ Are claims, conclusions, and recommendations supported by evidence? (526)
- ☐ Is the material free of gaps, foggy areas, or needless details? (31)
- ☐ Are all key terms clearly defined? (424)
- ☐ Are all data sources documented? (578)

Organization

- ☐ Is the structure of the document visible at a glance? (193)
- ☐ Is there a clear line of reasoning that emphasizes what is most important? (195)
- ☐ Is material organized in the sequence users are expected to follow? (208)
- ☐ Is everything easy to locate? (193)
- ☐ Is the material "chunked" into easily digestible parts? (193)

Style

- ☐ Is each sentence understandable the first time it is read? (216)
- ☐ Is rich information expressed in the fewest words possible? (224)
- ☐ Are sentences put together with enough variety? (231)
- ☐ Are words chosen for exactness, and not for camouflage? (235)
- ☐ Is the tone appropriate for the situation and audience? (244)

Design

- ☐ Is page design inviting, accessible, and appropriate for the user's needs? (305)
- ☐ Are there adequate aids to navigation (heads, lists, typestyles)? (308)
- ☐ Are adequate visuals used to clarify, emphasize, or summarize? (257)
- ☐ Do supplements accommodate the needs of a diverse audience? (324)

Ethical, Legal, and Cultural Considerations

- ☐ Does the document indicate sound ethical judgment? (61)
- ☐ Does the document comply with copyright law and other legal standards? (73)
- ☐ Does the document respect users' cultural diversity? (48)

COLLABORATIVE PROJECT

Test the usability of a document prepared for this course:

a. As a basis for your "alpha" test, adapt the guidelines on page 339, the audience and use profile sheet on page 29, the general checklist above, and whichever specific checklist applies to this particular document (as on page 456, for example).

b. As a basis for your "beta" test, adapt the Usability Survey on page 341.

c. Revise the document based on your findings.

d. Appoint a group member to explain the usability testing procedure and the results to the class.

PART

V

Specific Documents and Applications

Memo Reports and Electronic Mail

PURPOSE OF MEMO REPORTS

Reports present ideas and facts to decision makers. Workplace audiences rely on short reports as a basis for informed decisions on matters as diverse as the most comfortable office chairs to buy or the best recruit to hire for management training. Such reports usually take the form of a memorandum.

Memos, the major form of internal written communication in organizations, leave a *paper trail* (of directives, inquiries, instructions, requests, and recommendations) for future reference.

NOTE

Email memos leave their own trail. Although generally less formal and more quickly written than paper documents, email messages are saved in both hard copy and on-line—and can inadvertently be forwarded to someone never intended to receive or read them.

Organizations rely on memos to trace decisions, track progress, recheck data, or assign responsibility. Therefore, any memo can have far-reaching ethical and legal implications. Be sure your memo includes the date and your initials for verification, and that its information is specific, accurate, and unambiguous.

ELEMENTS OF A USABLE MEMO

A usable memo is easy to scan, file, and retrieve. The memo has a heading that names the organization and identifies the sender, recipient, subject (often in caps or underlined for emphasis), and date. (Placement of these items may differ among firms.) Other memo elements are summarized in Figure 18.1.

INTERPERSONAL CONSIDERATIONS IN WRITING A MEMO

A form of "in-house" correspondence, memos circulate among colleagues, subordinates, and superiors to address questions like these:

What memo recipients want to know

- ◆ What are we doing right, and how can we do it better?
- ◆ What are we doing wrong, and how can we improve?
- ◆ Who's doing what, and when, and where?

Memo topics often involve evaluations or recommendations about policies, procedures, and, ultimately, the *people with whom we work.*

Because people are sensitive to criticism (even when it is merely implied) and often resistant to change, an ill-conceived or aggressive memo can spell disaster for its author. Consider, for instance, this evaluation of one company's training for new employees:

A hostile approach

No one tells new employees what it's *really* like to work here—how to survive politically: for example, never tell anyone what you *really* think; never observe how few

NAME OF ORGANIZATION

MEMORANDUM
Date: (also serves as a chronological record for future reference)
To: Name and title (the title also serves as a record for reference)
From: Your name and title (your initials for verification)
Subject: ELEMENTS OF A USABLE MEMO

Subject Line
Announce the memo's purpose and contents, to orient readers to the subject and help them assess its importance. An explicit title also makes filing by subject easier.

Introductory Paragraph
Unless you have reason for being indirect, state your main point immediately.

Topic Headings
When discussing a number of subtopics, include headings (as shown here). Headings help you organize and they help readers locate information quickly.

Visuals
To summarize numerical data, your memo may include one or more visuals (as on page 355).

Paragraph Spacing
Do not indent the first line of paragraphs. Single space within paragraphs and double space between them.

Second-Page Header
When the memo exceeds one page, include a header on each subsequent page. For example: Ms. Baxter, June 12, 20xx, page 2.

Memo Verification
Do not sign your memos. Initial the "From" line, after your name.

Copy Notation
When sending copies to people not listed on the "To" line, include a copy notation two spaces below the last line, and list, by rank, the names and titles of those recipients. For example,

Copies: J. Spring, V.P., Production
         H. Baxter, General Manager, Production

Figure 18.1   Elements of a Usable Memo

women are in management positions, or how disorganized things seem to be. New employees shouldn't have to learn these things the hard way. We need to demand clearer behavioral objectives.

Instead of emphasizing deficiencies, the following version focuses on the *benefits* of change:

A more reasonable approach

New employees would benefit from a concrete guide to the personal and professional traits expected in our company. Training sessions could be based on the attitudes, manners, and behavior appropriate in business settings.

Here are some common mistakes that can offend coworkers:

When memos go wrong

- *Griping or complaining.* Everyone has problems of their own. Never complain without suggesting a solution.
- *Being too critical or judgmental.* Making someone look bad means making an enemy.
- *Being too bossy.* The imperative mood is best reserved for instructions.
- *Neglecting to provide a copy to each person involved.* No one appreciates being left "out of the loop."

Before releasing any memo designed to influence people's thinking, review Chapter 4 carefully.

COMMON TYPES OF MEMO REPORTS

Memo reports cover every conceivable topic. Common types include recommendation, justification, progress, periodic, and survey reports.

Recommendations Reports

Recommendation reports interpret data, draw conclusions, and make recommendations, often in response to a specific request. The recommendation report on page 348 is addressed to the writer's boss. This is only one example of daily problem solving that is recorded in memo form.

A Problem-Solving Recommendation

Bruce Doakes is assistant manager of occupational health and safety for a major airline that employs over two hundred reservation and booking agents. Each agent spends eight hours daily seated at a workstation that has a computer, telephone, and other electronic equipment. Many agents have complained of chronic discomfort from their work: headache, eyestrain and irritation, blurred or double vision, backache, and stiff neck and joints.

Bruce's boss asked him to study the problem and recommend improvements in the work environment. Bruce surveyed employees and consulted ophthalmologists, chiropractors, orthopedic physicians, and the latest publications on ergonomics (tailoring work environments for employees' physical and psychological well-being). After completing his study, Bruce composed his report. ▲

Bruce's report had to be persuasive as well as informative.

RECOMMENDATION MEMO

Trans Globe Airlines

MEMORANDUM
To: R. Ames, Vice President, Personnel
From: B. Doakes, Health and Safety
Date: August 15, 20xx
Subject: **Recommendations for Reducing Agents' Discomfort**

Provides immediate orientation by giving brief background and main point

In our July 20 staff meeting, we discussed physical discomfort among reservation and booking agents, who spend eight hours daily at automated workstations. Our agents complain of headaches, eyestrain and irritation, blurred or double vision, backaches, and stiff joints. This report outlines the apparent causes and recommends ways of reducing discomfort.

Causes of Agents' Discomfort

Interprets findings and draws conclusions

For the time being, I have ruled out the computer display screens as a cause of headaches and eye problems for the following reasons:

1. Our new display screens have excellent contrast and no flicker.
2. Research findings about the effects of low-level radiation from computer screens are inconclusive.

The headaches and eye problems seem to be caused by the excessive glare on display screens from background lighting.

Other discomforts, such as backaches and stiffness, apparently result from the agents' sitting in one position for up to two hours between breaks.

Recommended Changes

Makes general recommendations

We can eliminate much discomfort by improving background lighting, workstation conditions, and work routines and habits.

Expands on each recommendation

Background Lighting. To reduce the glare on display screens, these are recommended changes in background lighting:

1. Decrease all overhead lighting by installing lower-wattage bulbs.
2. Keep all curtains and adjustable blinds on the south and west windows at least half-drawn, to block direct sunlight.

3. Install shades to direct the overhead light straight downward, so that it is not reflected by the screens.

Workstation Conditions. These are recommended changes in the workstations:

1. Reposition all screens so light sources are neither at front nor back.
2. Wash the surface of each screen weekly.
3. Adjust each screen so the top is slightly below the operator's eye level.
4. Adjust all keyboards so they are 27 inches from the floor.
5. Replace all fixed chairs with adjustable, armless, secretarial chairs.

Work Routines and Habits. These are recommended changes in agents' work routines and habits:

1. Allow frequent rest periods (10 minutes each hour instead of 30 minutes twice daily).
2. Provide yearly eye exams for all terminal operators, as part of our routine health-care program.
3. Train employees to adjust screen contrast and brightness whenever the background lighting changes.
4. Offer workshops on improving posture.

Discusses benefits of following the recommendations

These changes will give us time to consider more complex options such as installing hoods and antiglare filters on terminal screens, replacing fluorescent lighting with incandescent, covering surfaces with nonglare paint, or other disruptive procedures.

cc. J. Bush, Medical Director
    M. White, Manager of Physical Plant

For advice on thinking critically as you formulate, evaluate, and refine recommendations, see pages 526–28.

Justification Reports

Justification reports (a type of recommendation report) are in a unique class in that they are often initiated by the writer rather than requested by the recipients. As the name implies, such reports *justify* the writer's position on some issue. Justification reports therefore typically begin rather than end with the request or recommendation. Such reports answer this key question for recipients: *Why should we?*

Typically, justification reports follow a version of this arrangement:

How to organize a justification report

1. State the problem and your recommendations for solving it.
2. Point out the cost, savings, and benefits of your plan.
3. If needed, explain how your suggestion can be implemented.
4. Conclude by encouraging the reader to act.

The next writer uses a version of the preceding arrangement: she begins with the problem and recommended solution, spells out costs and benefits, and concludes by reemphasizing the major benefit. The tone is confident yet diplomatic—appropriate for an unsolicited recommendation to a superior.

JUSTIFICATION MEMO

Greentree Bionomics, Inc.

MEMORANDUM
To: D. Spring, Personnel Director
            Greentree Bionomics, Inc. (GBI)
From: M. Noll, Biology Division
Date: April 18, 20xx
Subject: **The Need to Hire Additional Personnel**

Introduction and Recommendation

Opens with the problem

With twenty-six active employees, GBI has been unable to keep up with scheduled contracts. As a result, we have a contract backlog of roughly $500,000. This backlog is caused by understaffing in the biology and chemistry divisions.

Recommends a solution

To increase production and ease the work load, I recommend that GBI hire three general laboratory assistants.

Expands on the recommendation

The lab assistants would be responsible for cleaning glassware and general equipment; feeding and monitoring fish stocks; preparing yeast, algae, and shrimp cultures; preparing stock solutions; and assisting scientists in various tests and procedures.

Cost and Benefits

Shows how benefits would offset costs

While costing $74,880 yearly (at $12.00/hour), three full-time lab assistants would have a positive effect on overall productivity:

1. Uncleaned glassware would no longer pile up, and the fish holding tanks could be cleaned daily (as they should be) instead of weekly.
2. Because other employees would no longer need to work more than forty hours weekly, morale would improve.
3. Research scientists would be freed from general maintenance work (cleaning glassware, feeding and monitoring the fish stock, etc.). With more time to perform client tests, the researchers could eliminate our backlog.
4. With our backlog eliminated, clients would no longer have cause for impatience.

Conclusion

Encourages acceptance of recommendation

Increased production at GBI is essential to maintaining good client relations. These additional personnel would allow us to continue a reputation of prompt and efficient service, thus ensuring our steady growth and development.

Progress Reports

Large organizations depend on progress (or status) reports to track activities, problems, and progress on various projects. Daily progress reports are vital in a business that assigns crews to many projects. Management uses progress reports to evaluate the project and its supervisor, and to decide how to allocate funds.

Often, a progress report is one of a series. Together, the project proposal (Chapter 24), progress reports (the number varies with the scope and length of the project), and the final report (Chapter 25) provide a record and history of the project.

To give management the answers it needs, progress reports must, at a minimum, answer these questions:

<table>
<tr><td>

What recipients of a progress report want to know

</td><td>

- *How much has been accomplished since the last report?*
- *Is the project on schedule?*
- *If not, what went wrong?*
- *How was the problem corrected?*
- *How long will it take to get back on schedule?*
- *What else needs to be done?*
- *What is the next step?*
- *Any unexpected developments?*
- *When do you anticipate completion? Or, on a long project, when do you anticipate completion of the next phase?*

</td></tr>
</table>

If the report is part of a series, you might also refer to prior problems or developments.

Many organizations have forms for organizing progress reports, so no one format is best. But each report in a series should use the same organization. The following memorandum illustrates how one writer organized her report.

PROGRESS REPORT (ON THE JOB)
Subject: Progress Report: Equipment for New Operations Building

Work Completed

Our training group has met twice since our May 12 report in order to answer the questions you posed in your May 16 memo. In our first meeting, we identified the types of training we anticipate.

Types of Training Anticipated

- Divisional Surveys
- Loan Officer Work Experience
- Divisional Systems Training
- Divisional Clerical Training (Continuing)
- Divisional Clerical Training (New Employees)
- Divisional Management Training (Seminars)
- Special/New Equipment Training

In our second meeting, we considered various areas for the training room.

Training Room

The frequency of training necessitates having a training room available daily. The large training room in the Corporate Education area (10th floor) would be ideal. Before submitting our next report, we need your confirmation that this room can be assigned to us.

To support the training programs, we purchased this equipment:

(marginal notes:)

Summarizes achievements to date

Details the achievements

- ◆ Audioviewer
- ◆ LCD monitor
- ◆ Videocassette recorder and monitor
- ◆ CRT
- ◆ Software for computer-assisted instruction
- ◆ Slide projector
- ◆ Tape recorder

This equipment will allow us to administer training in a variety of modes, ranging from programmed and learner-controlled instruction to group seminars and workshops.

Work Remaining

Describes what remains to be done

To support the training, we need to furnish the room appropriately. Because the types of training will vary, the furniture should provide a flexible environment. Outlined here are our anticipated furnishing needs.

- ◆ Tables and chairs that can be set up in many configurations. These would allow for individual or group training and large seminars.
- ◆ Portable room dividers. These would provide study space for training with programmed instruction, and allow for simultaneous training.
- ◆ Built-in storage space for audiovisual equipment and training supplies. Ideally, this storage space should be multipurpose, providing work or display surfaces.
- ◆ A flexible lighting system, important for audiovisual presentations and individualized study.
- ◆ Independent temperature control, to ensure that the training room remains comfortable regardless of group size and equipment used.

Gives a rough timetable

The project is on schedule. As soon as we receive your approval of these specifications, we will proceed to the next step: sending out bids for room dividers, and having plans drawn for the built-in storage space.

cc.   R. S. Pike, SVP
       G. T. Bailey, SVP

As you work on a longer report or term project, your instructor might require a progress report. In this next memo, Karen Granger documents her progress on her term project: an evaluation of the Environmental Protection Agency's effectiveness in cleaning a heavily contaminated harbor.

PROGRESS REPORT ON TERM PROJECT

Progress Report

To: Dr. John Lannon
From: Karen P. Granger
Date: April 17, 20xx
Subject: **Evaluation of the EPA's** *Remedial Action Master Plan*

Project Overview

As my term project, I have been evaluating the issues of politics, scheduling, and safety surrounding the EPA's published plan to remove PCB contaminants from New Bedford Harbor.

Work Completed

February 23:   Began general research on the PCB contamination of the New Bedford Harbor.

March 8:   Decided to analyze the *Remedial Action Master Plan* (RAMP) in order to determine whether residents are being "studied to death" by the EPA.

March 9–19:   Drew a map of the harbor to show areas of contamination. Obtained the RAMP from Pat Shay of the EPA.

Interviewed Representative Grimes briefly by phone; made an appointment to interview Grimes and Sharon Dean on April 13.

Interviewed Patricia Chase, President of the New England Sierra Club, briefly, by phone.

March 24:   Obtained *Public Comments on the New Bedford RAMP*, a collection of reactions to the plan.

April 13:   Interviewed Grimes and Dean; searched Grimes's files for information. Also searched the files of Raymond Soares, New Bedford Coordinator, EPA.

Work in Progress

Contacting by telephone the people who commented on the RAMP.

Work to Be Completed

April 25:   Finish contacting commentators on the RAMP.

April 26:   Interview an EPA representative about the complaints that the commentators raised on the RAMP.

Date for Completion: May 3, 20xx

Complications

The issue of PCB contamination is complicated and emotional. The more I uncovered, the more difficult I found it to remain impartial in my research and analysis. As a New Bedford resident, I expected to find that we are indeed being studied to death; because my research seems to support my initial impression, I am not sure I have remained impartial.

Lastly, the people I talk to do not always have the time to find answers for my questions. Everyone, however, has been interested and encouraging, if not always informative.

Margin notes:
- Summarizes achievements to date
- Describes work remaining, with timetable
- Describes problems encountered

Periodic Activity Reports

The periodic activity report resembles the progress report in that it summarizes activities over a specified period. But unlike progress reports, which summarize specific accomplishments on a given project, periodic reports summarize general activities during a given period. Manufacturers requiring periodic reports often have prepared forms, because most of their tasks are quantifiable (e.g., units produced).

But most white-collar jobs do not lend themselves to prepared-form reports. You may have to develop your own format, as the next writer does.

DeWitt's report answers her boss's primary question: *What did you accomplish last month?* Her response has to be detailed and informative.

PERIODIC ACTIVITY REPORT

Date: 6/18/xx
To: N. Morgan, Assistant Vice President
From: F. C. DeWitt
Subject: **Coordinating Meetings for Training**

Gives overview of recent activities, and their purpose

For the past month, I've been working on a cooperative project with the Banking Administration Institute, Computron Corporation, and several banks. My purpose has been to develop training programs, specific to banking, appropriate for computer-assisted instruction (CAI).

We have focused on three major areas: Proof/Encoding Training, Productivity Skills for Management, and Banking Principles.

Gives details

I hosted two meetings for this task force. On June 6, we discussed Proof/Encoding Training, and on June 7, Productivity Training. The objective for the Proof/Encoding meeting was to compare ideas, information, and current training packages available on this topic. We are now designing a training course.

The objective of the meeting on Productivity was to discuss skills that increase productivity in banking (specifically Banking Operations). Discussion included instances in which computer-assisted instruction is appropriate for teaching productivity skills. Computron also discussed computer applications used to teach productivity.

On June 10, I attended a meeting in Washington, D.C., to design a course in basic banking principles for high-level clerical/supervisory-level employees. We also discussed the feasibility of adapting this course to CAI. This type of training, not currently available through Corporate Education, would meet a definite supervisor/management need in the division.

Explains the benefits of these activities

My involvement in these meetings has two benefits. First, structured discussions with trainers in the banking industry provide an exchange of ideas, methods, and experiences. This involvement expedites development of our training programs because it saves me time on research. Second, microcomputers will continue to affect future training. With a working knowledge of these systems and their applications, I now am able to assist my group in designing programs specific to our needs.

Employees use periodic reports to inform management of what they are doing and how well they are doing it. Therefore, accuracy, clarity, and appropriate level of detail are important—as is the persuasive (and ethical) dimension. Make sure that recipients have all the necessary facts—and that they understand these facts as clearly as you do.

Survey Reports

Brief survey reports are often used to examine conditions that affect an organization (consumer preferences, available markets). The following memo, from the research director for a midwestern grain distributor, gives clear and specific information directly. To simplify interpretation, the data are arranged in a table.

SURVEY REPORT

Subject: Food-Grain Consumption, U.S., 1997–2000

Announces purpose of memo

Here are the data you requested on April 7, as part of your division's annual marketing survey.

Presents data in tabular form, for easy comparison

U.S. PER-CAPITA CONSUMPTION OF FOOD GRAINS IN POUNDS, 1997–2000				
	1997	**1998**	**1999**	**2000**
Corn Products				
Cornmeal and other	16.4	16.4	16.6	16.7
Corn syrup and sugar	22.3	22.7	23.3	23.8
Oat Products	3.8	4.1	4.9	4.9
Barley Products	1.2	1.2	1.2	1.2
Wheat				
Flour	116.2	116.4	117.1	117.8
Breakfast cereals	2.9	2.9	2.9	2.9
Rye, Flour	1.2	1.2	1.2	1.2
Rice, Milled	8.3	6.7	7.7	7.0
Source: U.S. Dept. of Agriculture, Economic Research Service				

Names the source

Discusses the conclusions to be drawn from the table

Food-grain consumption during the period remained stable or increased by less than 10 percent, with the notable exception of oat products (21 percent increase).

Of greater concern is the effect of the 17 percent decrease in milled-rice consumption on our long-term marketing objectives.

cc: C. B. Schulz, Vice President, Marketing

If the above data had originated from *primary* sources (interviews, questionnaire, company records) the report would also describe how these data were collected.

Minutes of Meetings

Many team or project meetings on the job require someone to record the proceedings. Minutes are the records of such meetings. Copies of minutes are distributed to all members and interested parties, to keep track of proceedings. The appointed secretary records the minutes. When you record minutes, answer these questions:

What recipients of minutes want to know

- ◆ *What group held the meeting? When, where, and why?*
- ◆ *Who chaired the meeting? Who else was present?*
- ◆ *Were the minutes of the last meeting approved (or disapproved)?*
- ◆ *Who said what?*
- ◆ *Was anything resolved?*
- ◆ *Who made which motions and what was the vote?*

Minutes are filed as part of an official record, and so must be precise, clear, highly informative, and free of the writer's personal commentary ("As usual, Ms. Jones disagreed with the committee") or judgmental words ("good," "poor," "irrelevant").

MINUTES OF A MEETING

Subject: **Minutes of Managers' Meeting, October 5, 20xx**

Members Present

Tells who attended

Harold Tweeksbury, Jeannine Boisvert, Sheila DaCruz, Ted Washington, Denise Walsh, Cora Parks, Cliff Walsh, Joyce Capizolo

Agenda

Summarizes discussion of each item

1. The meeting was called to order on Wednesday, October 5, at 10 A.M. by Cora Parks.
2. The minutes of the September meeting were approved unanimously.
3. The first order of new business was to approve the following policies for the Christmas season:

   a. Temporary employees should list their ID numbers in the upper-left corner of their receipt envelopes to help verification. Discount Clerical assistant managers will be responsible for seeing that this procedure is followed.

   b. When temporary employees turn in their envelopes, personnel from Discount Clerical should spot-check them for completeness and legibility. Incomplete or illegible envelopes should be corrected, completed, or rewritten. *Envelopes should not be sealed.*

Tells who said what

4. Jeannine Boisvert moved that we also hold one-day training workshops for temporary employees in order to teach them our policies and procedures. The motion was seconded. Joyce Capizolo disagreed, saying that on-the-job training (OJT) was

Tells what was voted

   enough. The motion for the training session carried 6–3. The first workshop, which Jeannine agreed to arrange, will be held October 25.
5. Joyce Capizolo requested that temporary employees be sent a memo explaining the

temporary employee discount procedure. The request was converted to a motion and seconded by Cliff Walsh. The motion passed by a 7–2 vote.

6. Cora Parks adjourned the meeting at 11:55 A.M.

ELECTRONIC MAIL

As early as 1998, roughly 75 million U.S. businesspeople were using email, the average worker sending and receiving from 60 to 190 messages daily (Reiss 6; Stepanek 74). By 2002, international email messages sent will reach 600 million daily ("No Need" 15).

Email Benefits

Compared to phone, fax, or conventional mail (or even face-to-face conversation, in some cases), email offers real advantages:

Email facilitates communication and collaboration

- *Email is rapid, convenient, efficient, and relatively nonintrusive.* Unlike conventional mail, email travels instantly. Also, even though a fax network transmits print copy rapidly, email eliminates paper shuffling, manual dialing, and many other steps. Besides eliminating "telephone tag," email is less intrusive than the telephone, offering recipients the choice of when to read a message and whether to respond.
- *Email can be democratic.* Email messages usually appear as plain print on a screen, with no special type styles, fancy letterheads, page design, or paper texture—allowing readers to focus on the message instead of the medium. (Multimedia enhancements such as sound or motion might be added, but these are inappropriate for routine correspondence.)

  Email also allows anyone at any level in an organization to contact anyone else. For instance, the mail clerk could conceivably email the company president directly, whereas a conventional memo or phone call would be routed through the chain of management or screened by assistants (D. Goodman 33–35). In addition, people who are ordinarily shy in face-to-face conversations may be more willing to express their views in an email conversation.

NOTE *Communication expert Stuart Selber offers this important caution: "The mere presence of email on the job does not mean that someone's work environment supports democratic uses of email. In fact, just the opposite could be true: a company can use email to monitor and [intimidate employees]."*

- *Email can foster creative thinking.* Email dialogues involve give-and-take, much like a conversation. Email users generally communicate spontaneously—without worrying about page design, paragraph structure, or perfect phrasing. This relatively free exchange of views can lead to new insights or ideas (Bruhn 43).
- *Email is an excellent tool for collaborative work and research.* Collaborative teams can keep in touch via email, and researchers can contact people who

have the answers they need. Documents or electronic files of any length can be attached and sent for the recipient to download.

NOTE   *While electronically mediated collaboration increases the quality and sharing of ideas, groups who meet only via network develop less trust than groups who have some face-to-face contact (Ross-Flanigan 57, 58).*

Email Privacy Issues

Gossip, personal messages, or complaints about the boss or a colleague—all might reach unintended recipients. While phone companies and other private carriers are governed by laws protecting privacy, no such legal protection has been developed for Internet communication (Peyser and Rhodes 82). The Electronic Privacy Act of 1986 offers limited protection against unauthorized reading of another person's email, but employers are exempt (Extejt 63). In fact, employers generally claim legal right to monitor any of their company's information:

Monitoring of email by an employer is legal

> In some instances it may be proper for an employee to monitor E-mail, if it has evidence of safety violations, illegal activity, racial discrimination, or sexual improprieties, for instance. Companies may also need access to business information, whether it is kept in an employee's drawer, file cabinet, or computer E-mail (Bjerklie, "E-Mail" 15).

Email privacy can be compromised in other ways as well:

How email can compromise privacy

♦ Everyone on a group mailing list—intended recipient or not—automatically receives a copy of the message.
♦ Even when deleted from the system, messages can live on, saved in a backup file.
♦ Anyone with access to your network and password can read your document, alter it, use parts of it out of context, pretend to be its author, forward it, plagiarize your ideas, or even author a document or conduct illegal activity in your name. (One partial safeguard is encryption software, which scrambles the message, enabling only those who possess the special code to unscramble it.)

Email Quality Issues

Free exchange via computer screen adds to the *quantity* of information exchanged, but not always to its *quality:*

How email can compromise information quality

♦ The ease of sending and exchanging messages generates overload and junk mail—a party announcement sent to three hundred employees on a group mailing list, or the indiscriminate mailing of a political statement to dozens of newsgroups (*spamming*).
♦ Some electronic messages are poorly edited and long-winded.
♦ Off-the-cuff messages or responses might offend recipients. Email users seem less restrained about rude remarks (*flaming*) than they would be in a face-to-face encounter.

◆ Recipients might misinterpret the tone. "Emoticons" or "smileys," punctuation cues that signify pleasure :-), displeasure :-(, sarcasm ;-), anger >:-< and other emotional states, offer some assistance but are not always adequate or appropriate substitutes for the subtle cues in spoken conversation. Also, common email abbreviations (FYI, BTW, HAND—which mean "for your information," "by the way," and "have a nice day") might strike recipients as too informal.

GUIDELINES

for Using
Electronic Mail*

1. *Check and answer your email daily.* Like an unreturned phone call, unanswered email is annoying. If you're really busy, at least acknowledge receipt and respond later.

2. *Check your distribution list before each mailing.* Verify that your message will reach all intended recipients—but no unintended ones.

3. *Assume that your email is permanent and could be read by at anyone any time.* Don't write anything you couldn't say face-to-face. Avoid *spamming* (sending junk mail) and *flaming* (making rude remarks).

4. *Don't use email for confidential information.* Avoid complaining, criticizing, or evaluating people, or saying anything private.

5. *Don't use the company email network for personal correspondence or for anything not work-related.* Employers increasingly monitor their email networks.

6. *Before you forward an incoming message to other recipients, obtain permission from the sender.* Assume that anything you receive is the private property of the sender. (See page 126 for email copyright issues.)

7. *Limit your message to a single topic.* Remain focused and concise.

8. *Limit your message to a single screen, if possible.* Don't force recipients to scroll needlessly.

9. *Use a clear subject line to identify your topic.* Instead of "Test Data" or "Data Request," announce your purpose clearly: "Request for Beta Test Data for Project #16"). Recipients scan subject lines in deciding which new mail to read immediately.

10. *Refer clearly to the message to which you are responding:* ("Here are the Project 16 Beta test data you requested on Oct. 10").

11. *Keep sentences and paragraphs short, for easy reading.*

12. *Don't write in FULL CAPS—unless you want to SCREAM at the recipient!*

13. *Close with a signature section.* Include the name of your company or department, phone and fax number, and any other contact information.

*Adapted from Bruhn 43; D. Goodman 33–35, 167; Kawasaki, "The Rules" 286; Munger; Nantz and Drexel 45–51; Peyser and Rhodes 82.

GUIDELINES

for Choosing Email versus Paper or the Telephone

Email is excellent for reaching a lot of people quickly with a relatively brief, informal message. But there are often good reasons to put something on paper—or to speak directly with the recipient instead. Consider the following guidelines.

1. *Don't use email when a more personal medium is preferable.* Sometimes an issue is best resolved by a phone call, or even by voice mail—whereas email might imply that the sender can't be bothered to speak with the recipient directly.

2. *Don't use email for a detailed discussion.* In contrast to a rapid-fire email message, preparing a paper document is generally more deliberate, giving you a chance to shape and clarify your thinking, to choose words carefully, and to revise. Also, a well-crafted paper document (say, a memo or a letter) is likely to be read more attentively. (For in-house recipients, consider transmitting your paper document attached to a brief, introductory email.)

3. *Don't use email to send formal correspondence.* Because of the volume and casual style of email, recipients might overlook the significance of a message. For instance, don't apply for a job or respond to clients or customers via email, unless recipients request or approve this method. (Chapter 19 treats letters—including electronic job hunting.)

Usability Checklist FOR MEMO REPORTS

Use this checklist as a guide to refining and revising your memo reports. (Numbers in parentheses refer to the first page of discussion.)

☐ Have you chosen the best report medium (paper, email, phone) for the situation? (360)

☐ Does the memorandum have a complete heading? (345)

☐ Does the subject line forecast the memo's contents? (346)

☐ Are recipients given enough information to make an *informed* decision? (31)

☐ Are the conclusions and recommendations clear? (349)

☐ Are paragraphs single-spaced within and double-spaced between? (346)

☐ Do headings, charts, or tables appear as needed? (346)

☐ If more than one reader is receiving copies, does the memo include a distribution notation (cc:) to identify other readers? (346)

☐ Is the writing style clear, concise, exact, fluent, and appropriate? (215)

☐ Does the document's appearance create a favorable impression? (305)

☐ Does the document create an appropriate interpersonal connection? (345)

EXERCISES

1. We would all like to see changes in our schools' policies or procedures, whether they are changes in our majors, school regulations, social activities, grading policies, or registration procedures. Find some area of your school that needs obvious changes, and write a justification report to the person who might initiate that change. Explain why the change is necessary and describe the benefits. Follow the format on page 350.

2. Think of an idea you would like to see implemented in your job (e.g., a way to increase productivity, improve service, increase business, or improve working conditions). Write a justification memo, persuading your audience that your idea is worthwhile.

3. Write a memo to your employer, justifying reimbursement for this course. *Note:* You might have written another version of this assignment for Exercise 3 in Chapter 1. If so, compare early and recent versions for content, organization, style, and format.

4. Identify a dangerous or inconvenient area or situation on campus or in your community (endless cafeteria lines, a poorly lit intersection, slippery stairs, a poorly adjusted traffic light). Observe the problem for several hours during a peak use period. Write a justification report to a *specifically identified* decision maker, describing the problem, listing your observations, making recommendations, and encouraging reader support or action.

5. Assume that you have received a $10,000 scholarship, $2,500 yearly. The only stipulation for receiving installments is that you send the scholarship committee a yearly progress report on your education, including courses, grades, school activities, and cumulative average. Write the report.

6. In a memo to your instructor, outline your progress on your term project. Describe your accomplishments, plans for further work, and any problems or setbacks. Conclude your memo with a specific completion date.

7. Keep accurate minutes for one class session (preferably one with debate or discussion). Submit the minutes in memo form to your instructor.

8. Conduct a brief survey (e.g., of comparative interest rates from various banks on a car loan, comparative tax and property evaluation rates in three local towns, or comparative prices among local retailers for an item). Arrange your data and report your findings to your instructor in a memo that closes with specific recommendations for the most economical choice.

9. Although the campus store (or cafeteria) is a convenient place to buy small items (or food), you believe students are overcharged for the convenience. To prove your point, you decide to do a comparative analysis of the cost of five items at the campus store (or cafeteria) and two local stores (or restaurants). Be sure to compare like sizes, weights, brands. Conduct your survey, analyze your findings, and submit the results in memo form to the student senate. Your instructor might want you to include recommendations.

10. Recommendation Report (choose one)

    a. You are legal consultant to the leadership of a large autoworkers' union. Before negotiating its next contract, the union needs to anticipate effects of robotics technology on assembly-line autoworkers within ten years. Do the research and write a report recommending a course of action.

    b. You are a consulting engineer to an island community of two hundred families suffering a severe shortage of fresh water. Some islanders have raised the possibility of producing drinking water from salt water (desalination). Write a report for the town council, summarizing the process and describing instances in which desalination has been used successfully or unsuccessfully.

Would desalination be economically feasible for a community this size? Recommend a course of action.

c. You are a health officer in a town less than one mile from a massive radar installation. Citizens are disturbed about the effects of microwave radiation. Do they need to worry? Should any precautions be taken? Find the facts and write your report.

d. You are an investment broker for a major firm. A longtime client calls to ask your opinion. She is thinking of investing in a company that is fast becoming a leader in fiber optics communication links. "Should I invest in this technology?" your client wants to know. Find out, and give her your recommendations in a short report.

e. The buildings in the condominium complex you manage have been invaded by carpenter ants. Can the ants be eliminated by any insecticide *proven* nontoxic to humans or pets? (Many dwellers have small children and pets.) Find out, and write a report making recommendations to the maintenance supervisor.

f. The "coffee generation" wants to know about the properties of caffeine and the chemicals used on coffee beans. What are the effects of these substances on the body? Write your report, making specific recommendations about precautions that coffee drinkers can take.

g. As a consulting dietitian to the school cafeteria in Blandville, you've been asked by the school to report on the most dangerous chemical additives in foods. Parents want to be sure that foods containing these additives are eliminated from school menus, insofar as possible. Write your report, making general recommendations about modifying school menus.

h. Dream up a scenario of your own in which information and recommendations would make a real difference. (Perhaps the question could be one you've always wanted answered.)

11. Individually or in small groups, decide whether each of the following documents would be appropriate for transmission via a company email network. Be prepared to explain your decisions.

Sarah Burnes' memo about benzene levels (page 15)

The "Rational Connection" memo (page 39)

The "better" memo to the maintenance director (page 43)

Tom Ewing's letter to a potential customer (page 53)

The medical report written for expert readers (page 24)

A memo reporting illegal or unethical activity in your company

A personal note to a colleague

A request for a raise or promotion

Minutes of a meeting

Announcement of a no-smoking policy

An evaluation or performance review of an employee

A reprimand to an employee

A notice of a meeting

Criticism of an employee or employer

A request for volunteers

A suggestion for change or improvement in company policy or practice

A gripe

A note of praise or thanks

A message you have received and have decided to forward to other recipients

COLLABORATIVE PROJECT

Organize into groups of four or five and choose a topic for which group members can agree on a position. Here are some possibilities:

- Should your college abolish core requirements?
- Should every student in your school pass a writing proficiency exam before graduating?
- Should courses outside one's major be graded pass/fail at the student's request?
- Should your school drop or institute student evaluation of teachers?
- Should all students be required to be computer literate before graduating?
- Should campus police carry guns?
- Should school security be improved?

- Should students with meal tickets be charged according to the type and amount of food they eat, instead of paying a flat fee?

As a group, decide your position on the issue, and brainstorm collectively to justify your recommendation to a stipulated primary audience in addition to your classmates and instructor. Complete an audience and use profile (page 29), and compose a justification report. Appoint one member to present the report in class.

Letters and Employment Correspondence

We often have good reason to correspond in more formal medium than a memo or email message. A well-crafted letter is appropriate in situations like these:

When to send a
letter instead of a
memo

- To personalize your correspondence, conveying the sense that this message is prepared exclusively for your particular recipient.
- To convey a dignified, professional impression.
- To act as a representative of your company or organization.
- To present a reasoned, carefully constructed case.
- To respond to clients, customers, or anyone outside your organization.
- To provide an official notice or record (as in a letter announcing legal action or confirming a verbal agreement).

Because a letter often has a persuasive purpose, proper tone is essential for connecting with the recipient. Because your signature indicates your approval—and responsibility—for the contents of the letter (which may serve as a legal document), precision is crucial.

This chapter covers three common letter types: inquiry letters, claim letters, and letters of application, along with résumés. Other types are discussed in Chapters 16 and 24.

ELEMENTS OF USABLE LETTERS

The conventional arrangement of workplace letters allows recipients to locate what they need immediately (Figure 19.1).

Basic Parts of Letters

A typical business letter has six parts:

HEADING AND DATE. If your stationery has a company letterhead, simply include the date two lines below the letterhead, at the right or left margin. On blank stationery, include your return address and the date (but not your name):

Street address
City, state, zip code
Month, day, year

| 154 Sea Lane
| Harwich, MA 02163
| July 15, 20xx

Use the Postal Service's two-letter state abbreviations (MA for Massachusetts, VT for Vermont) in your heading, inside address, and on the envelope.

INSIDE ADDRESS. Two to six line spaces below the heading and abutting the left margin is the inside address.

Heading

LEVERETT LAND & TIMBER COMPANY, INC.

creative land use
quality building materials
architectural construction

Date

January 17, 20xx

Inside address

Mr. Thomas E. Muffin
Clearwater Drive
Amherst, MA 01022

Salutation

Dear Mr. Muffin:

Letter text

I have examined the damage to your home caused by the ruptured water pipe and consider the following repairs to be necessary and of immediate concern:

> Exterior:
> Remove plywood soffit panels beneath overhangs
> Replace damaged insulation and plumbing
> Remove all built-up ice within floor framing
> Replace plywood panels and finish as required
>
> Northeast Bedroom—Lower Level:
> Remove and replace all sheetrock, including closet
> Remove and replace all door casings and baseboards
> Remove and repair windowsill extensions and moldings
> Remove and reinstall electric heaters
> Respray ceilings and repaint all surfaces

This appraisal of damage repair does not include repairs and/or replacements of carpets, tile work, or vinyl flooring. Also, this appraisal assumes that the plywood subflooring on the main level has not been severely damaged.

Leverett Land & Lumber Company, Inc. proposes to furnish the necessary materials and labor to perform the described damage repairs for the amount of six thousand one hundred and eighty dollars ($6,180).

Complimentary closing

Sincerely,

Signature

P.A. Jackson

Title

Gerald A. Jackson, President

Typist's initials

GAJ/ob

Enclosure notation

Enc. Itemized estimate

Figure 19.1   A Standard Design for a Workplace Letter

Name and position	Dr. Ann Mello, Dean of Students
Company name	Western University
Street address	30 Mogul Hill Road
City, state, zip code	Stowe, VT 51350

Whenever possible, address a specifically named recipient, and include his or her job title. Using "Mr." or "Ms." before the name is optional. (See page 250 for avoiding sexist usage in titles and salutations.)

NOTE   *Depending on the letter's length, adjust the horizontal placement of your return and inside address to achieve a balanced page.*

SALUTATION.   The salutation, two line spaces below the inside address, begins with *Dear* and ends with a colon (*Dear Ms. Smith:*). If you don't know the recipient's name, use the position title (*Dear Manager*) or, preferably, an attention line (page 368). Only address the recipient by first name if that is the way you would address this individual in person.

Salutations	Dear Ms. Smith:
	Dear Managing Editor:
	Dear Professor Lexington-Trudeau:

No satisfactory guidelines exist for addressing several people within an organization. *Gentlemen* or *Dear Sirs* implies bias. *Ladies and Gentlemen* sounds too much like the beginning of a speech. *Dear Sir or Madam* is old-fashioned. *To Whom It May Concern* is vague and impersonal. Your best bet is to eliminate the salutation by using an attention line.

LETTER TEXT.   The text of your letter begins two line spaces below the salutation. Workplace letters typically include (1) a brief *introduction* paragraph (five lines or fewer) that identifies you and your purpose; (2) one or more *body* paragraphs that contain details of your message; (3) a *conclusion* paragraph that sums up and encourages action.

The shape of workplace letters

Keep the paragraphs short (usually fewer than eight lines). If your body section is long, divide it into shorter paragraphs, as on page 376. Notice that the body section in that letter is broken down into four questions for easy answering.

COMPLIMENTARY CLOSING.   The closing, two line spaces below the last line of text, should parallel the level of formality used in the salutation and should reflect your relationship to the recipient (polite but not overly intimate). These possibilities are ranked in decreasing order of formality:

| Complimentary closings | Respectfully, |
| | Sincerely, *(most often used)* |

Cordially,

Best wishes,

Warmest regards,

Regards,

Best,

In a modified block letter (Figure 19.2), align the closing with your heading or the date. In a full block letter (Figure 19.3), position the closing flush with the left margin.

SIGNATURE.   Type your full name and title on the fourth and fifth lines below and aligned with the closing. Sign your name in the triple space between the two.

The signature block

Sincerely yours,

Martha S. Jones

Martha S. Jones
Personnel Manager

If you are representing your company or a group that bears legal responsibility for the correspondence, type the company's name in full caps two line spaces below your complimentary closing; place your typed name and title four line spaces below the company name, and sign in the triple space between.

Signature block
representing the
company

Yours truly,

HASBROUCK LABORATORIES

Lester Fong

L. H. Fong
Research Associate

Specialized Parts of Letters
Some letters require one or more of the following specialized parts.

ATTENTION LINE.   Use an attention line when you write to an organization and do not know your recipient's name but are directing the letter to a specific department or position.

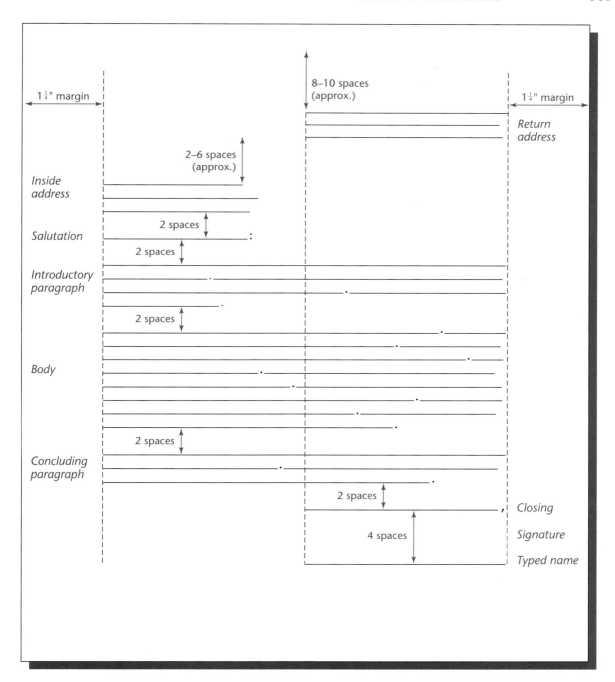

Figure 19.2   Modified Block Letter

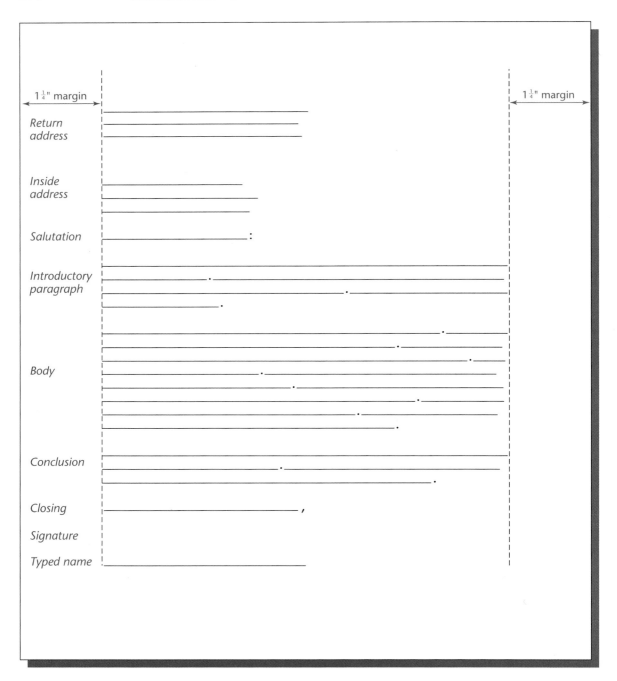

Figure 19.3 Block Letter

An attention line can replace your salutation

> Glaxol Industries, Inc.
> 232 Rogaline Circle
> Missoula, MT 61347
>
> ATTENTION: <u>Director of Research and Development</u>

Drop two line spaces below the inside address, and place the attention line either flush with the left margin or centered on the page.

SUBJECT LINE.   Because it announces the topic of your letter, the subject line is a good device for attracting a busy person's attention.

A subject line can attract attention

> SUBJECT: <u>Placement of the Subject Line</u>

Place the subject line below the inside address or attention line. Leave one line space above and below. Underline the subject or highlight it typographically, say, using boldface type.

TYPIST'S INITIALS.   If someone else types your letter, your initials (in caps) and your typist's (in lowercase letters) should appear below the typed signature, flush with the left margin.

Typist's initials

> JJ/pl

ENCLOSURE NOTATION.   When other documents accompany your letter, add an enclosure notation one line below the typist's initials, flush with the left margin. Indicate the number of enclosures.

Enclosure noted

> Enclosure
> Enclosures 2
> Encl. 3

If the enclosures are important documents such as legal certificates, checks, or specifications, name them in the notation.

Enclosure named

> Enclosures: 2 certified checks, 1 set of KBX plans.

DISTRIBUTION NOTATION.   If you distribute copies of your letter to other recipients, indicate this by inserting the notation *CC, cc, Cc* (for *carbon copy*) or *Copy* one line space below any enclosure notation.

Distribution notations

> cc: office file
>   Melvin Blount

c: S. Furlow
B. Smith

Most copies are distributed on an *FYI* (*For Your Information*) basis, but writers sometimes use the distribution notation to maintain a paper trail or to signal the primary recipient that this information is being shared with others (e.g., superiors, legal authorities).

POSTSCRIPT.   A postscript (typed or handwritten) draws attention to a point you wish to emphasize or adds a personal note. Do not use a postscript if you forget to mention a point in the body of the letter. Rewrite the letter instead.

A postscript

P. S. Because of its terminal position in your letter, a postscript can draw attention to a point that needs reemphasizing.

Place the postscript two line spaces below any other notation, and flush with the left margin. Because readers often regard postscripts as sales letter gimmicks, use them sparingly in professional correspondence.

Design Factors

Although several formats are acceptable, and your company may have its own, the two most popular formats for workplace letters are *modified block* and *block*.

In the modified block form, the first line of a paragraph is not indented. Paragraphs are separated by a line space. The return address, complimentary closing, and signature align at page center (Figure 19.2).

In the block form, every line begins at the left margin (Figure 19.3). This form is popular because it looks businesslike and eliminates the need to tab and center.

Additional design factors make workplace letters appear inviting, accessible, and professional:

QUALITY STATIONERY.   Use high-quality, 20-pound bond, $8^{1}/_{2}$" × 11" white stationery with a minimum fiber content of 25 percent.

UNIFORM MARGINS AND SPACING.   When using stationery without a letterhead, frame your letter with $1^{1}/_{2}$-inch top and side margins and bottom margins of 1 to $1^{1}/_{2}$ inches. Use single spacing within paragraphs and double spacing between. Vary these guidelines to suit your needs, but strive for a balanced look.

HEADERS FOR SUBSEQUENT PAGES.   Head each additional page with a notation identifying the recipient, date, and page number.

Subsequent-page
header

Adrianna Fonseca, June 25, 19xx, p. 2

Never use an additional page solely for your complimentary closing and signature. Instead, reformat the letter so that at least two lines of final text appear above the closing.

THE ENVELOPE.   Your envelope (usually a #10 envelope) should be of the same quality as your stationery. Place the reader's name and address at a central point on the envelope. Place your own name and address in the upper left corner. Single-space these elements.

INTERPERSONAL CONSIDERATIONS IN WORKPLACE LETTERS

In addition to presenting an accessible and inviting design, an effective letter enhances the relationship between sender and recipient. Interpersonal elements make a *human* connection.

FOCUS ON YOUR RECIPIENT'S INTERESTS: THE "YOU" PERSPECTIVE.   When speaking face-to-face, you unconsciously modify your statements and expression as you read the listener's signals: a smile, a frown, a raised eyebrow, a nod. In a telephone conversation, a voice provides cues that signal approval, dismay, anger, or confusion. In writing a letter, however, you might forget that a flesh-and-blood recipient will be reacting to what you are saying—or seem to be saying.

A letter displaying a "you" perspective focuses on content important to the recipient and conveys respect for his or her feelings and attitudes.

To achieve a "you" perspective, ask yourself how the recipient will react to what you have written. Even a single word, carelessly chosen, can offend. In writing to correct a billing error, for example, you might feel tempted to say this:

A needlessly offensive tone

> Our record keeping is very efficient, so this obviously is your error.

Such an accusatory tone might be appropriate after numerous failed attempts to achieve satisfaction on your part, but in your initial correspondence it would be offensive. Here is a more considerate version:

A tone that conveys the "you" perspective

> If my paperwork is wrong, please let me know and I will send you a corrected version immediately.

Instead of hastily judging the recipient, this second version invites a response.

USE PLAIN ENGLISH.   Workplace correspondence often suffers from *letterese,* those tired, stuffy, and overblown phrases some writers think they need to make their communications seem important. Here is a typically overwritten closing sentence:

Letterese

> Humbly thanking you in anticipation of your kind cooperation, I remain
> Faithfully yours,

Although no one *speaks* this way, some writers lean on such heavy prose instead of simply writing this:

Clear phrasing

> I will appreciate your cooperation.

Here are a few of the many old standards that are popular because they are easy to use but that make letters unimaginative and boring:

Plain-English translations

LETTERESE	PLAIN-ENGLISH
As per your request	As you requested
Contingent upon receipt of	As soon as we receive
I am desirous of	I want, I would like
Please be advised that I	I
This writer	I
In the immediate future	Soon
In accordance with your request	As you requested
Due to the fact that	Because
I wish to express my gratitude.	Thank you.

Be natural. Write as you would speak in a classroom or office.

FOCUS ON THE HUMAN CONNECTION.   As you plan and write, answer these questions:

Questions for connecting with the recipient

- ◆ *What do I want this person to do, think, or feel?* Offer a job, give advice or information, follow my instructions, grant a favor, accept bad news?
- ◆ *What details and emphasis does this person expect?* Measurements, dates, costs, model numbers, enclosures, other details?
- ◆ *To whom am I writing?* When possible, write to a person, not a title.
- ◆ *What is my relationship to this person?* Is this a potential employer, employee, a person doing a favor, a person whose products are disappointing, an acquaintance, an associate, a stranger?

ANTICIPATE THE RECIPIENT'S REACTION.   After you have written a draft, answer the following three questions, which pertain to the effect of your letter. Will readers be inclined to respond favorably?

Questions for anticipating the recipient's reaction

- ◆ *How will this person react?* With anger, hostility, pleasure, confusion, fear, guilt, resistance, satisfaction?
- ◆ *What impression of me does this letter convey?* Do you seem intelligent, courteous, friendly, articulate, pretentious, illiterate, confident?
- ◆ *Am I ready to sign my letter with confidence?* Think about it.

DECIDE ON A DIRECT OR INDIRECT PLAN.   The reaction you anticipate should determine the organizational plan of your letter: either *direct* or *indirect*.

◆ *Will the recipient feel pleased, angry, or neutral?*
◆ *Will the message cause resistance, resentment, or disappointment?*

The direct plan puts the main point in the first paragraph, followed by the explanation. Use the direct plan when you expect the recipient to react with approval or when you want to convey immediately the point of your letter (e.g., in good news, inquiry, or application letters—or other routine correspondence).

If you expect resistance, consider an indirect plan. Give the explanation before the main point (as in refusing a request, admitting a mistake, or requesting a pay raise). An indirect plan might make recipients more tolerant of bad news or more receptive to your argument.

Whenever you consider using an indirect plan, think carefully about its ethical implications. Never try to deceive the recipient—and never create an impression that you have something to hide.

INQUIRY LETTERS

Inquiry letters may be solicited or unsolicited. You often write the first type as a consumer requesting information about an advertised product. Such letters are welcomed because the recipient stands to benefit. You can be brief: "Please send me your software price list."

Other inquiries are unsolicited—asking an unsuspecting recipient to spend time reading your letter, considering your request, collecting the information, and responding. Always apologize for the imposition, express appreciation, and state a reasonable request clearly and briefly. Long, involved inquiries usually go unanswered.

Before you ask specific questions, do your homework. A vague request ("Please send me all your data on . . .") is likely to be ignored.

Sample Situation

You are preparing an analytical report on the feasibility of harnessing solar energy for home heating in northern climates. You learn that a private, nonprofit research group has been experimenting with solar energy systems. After deciding to write for details, you plan and compose your inquiry.

Tell the recipient who you are and why you want information. Maintain the "you" perspective with opening statements that spark interest and goodwill.

In the body of your letter, write specific and clearly worded questions that are easy to understand and answer. If you have several questions, arrange them in a list. (Lists help readers organize their answers, increasing your chances of getting all the information you want.) Number each question, and separate it from the others, perhaps leaving space for responses right on the page. If you have more than five questions, consider using an attached questionnaire.

Conclude by explaining how you plan to use the information and, if possible, how your reader might benefit. If you have not done so earlier, specify a date by which you need a response. Offer to send a copy of your finished report. Close with a statement of appreciation.

INQUIRY LETTER

States the purpose

As a student at Evergreen College, I am preparing a report (April 15 deadline) on the feasibility of solar energy as a viable source of home heating in northern climates.

Makes a reasonable request

While gathering data on home solar heating, I encountered references (in *Scientific American* and elsewhere) to your group's pioneering work in solar energy systems. Would you please allow me to benefit from your experience? I would appreciate answers to these four questions in particular:

Presents a list of specific questions

1. At this stage of development, have you found active or passive solar heating more practical?
2. Do you hope to surpass the 60 percent limit of heating needs supplied by the active system? If so, what level of efficiency do you expect to achieve, and how soon?
3. What is the estimated cost of building materials for your active system, per cubic foot of living space?
4. What metal do you use in collectors to obtain the highest thermal conductivity at the lowest maintenance costs?

Tells how the material will be used, and offers to share findings

Your answers, along with any recent findings you can share, will enrich a learning experience I hope to put into practice after graduation by building my own solar-heated home. I would be glad to send you a copy of my report, along with the house plans I have designed. Thank you.

Sometimes your questions will be too numerous or involved to be answered by letter or questionnaire. You might then request an informative interview (if the respondent is nearby). Karen Granger, the next writer, sought a state representative's "opinions on the EPA's progress" in cleaning up local contamination. Anticipating a complex answer, she requested an interview.

LETTER REQUESTING AN INFORMATIVE INTERVIEW

As a technical writing student at the University of Massachusetts, I am preparing a report evaluating the EPA's progress in cleaning up PCB contamination in New Bedford Harbor.

In my research, I have encountered your name repeatedly. Your dedicated work has had a definite influence on this situation, and I am hoping to benefit from your knowledge.

I was surprised to learn that, although this contamination is considered the most extensive anywhere, the EPA still has not moved beyond conducting studies. My own study questions the need for such extensive data gathering. Your opinion, as I can ascertain from *Standard Times* articles, is that the EPA is definitely moving too slowly.

The EPA refutes that argument by asserting they simply do not yet have the information necessary to begin a clean-up operation.

As both a writer and a New Bedford resident, I am very interested in your opinions on the EPA's progress. Could you find time in your busy schedule to grant me an interview? With your permission, I will phone in a few days to arrange an appointment.

I would deeply appreciate your assistance and I would gladly send you a copy of my completed report.

Whenever you seek a written response to an unsolicited inquiry, include a stamped, self-addressed envelope for the reply.

Telephone and Email Inquiries

Unsolicited inquiries via telephone or email can be efficient and productive but can also be unwelcome and intrusive. A traditional letter implies greater respect for the recipient's privacy and provides a certain distance from which that person can contemplate a response—or even decide not to respond.

Why inquiries via regular mail may be preferable

One alternative is to inquire by traditional letter and to invite a response by phone (collect) or email, if your reader prefers. Or, if you must inquire via telephone or email, consider this suggestion: Establish a brief initial contact in which you apologize for any intrusion and then ask about the respondent's willingness to answer your questions at a convenient time in the near future. In this or any communication, don't sacrifice goodwill in the interest of efficiency.

Etiquette for telephone or email inquiries

CLAIM LETTERS

Claim letters request adjustments for defective goods or poor services, or they complain about unfair treatment or the like.

Routine Claims

Routine claims follow the direct plan, because the claim is backed by a contract, guarantee, or the company's reputation. State the request or problem in your introductory paragraph; then explain in the body section. Close courteously, restating the action you request.

Make the tone courteous and reasonable. Your goal is not to express dissatisfaction but to achieve results: a refund, replacement, apology. Press your claim objectively yet firmly by explaining it clearly and by stipulating the *reasonable* action that will satisfy you.

Explain the problem in enough detail to clarify the basis for your claim. Explain that your new alarm clock never rings, instead of merely saying it's defective. Identify the faulty item clearly, giving serial and model numbers, and date and place of purchase. Then propose a fair adjustment. Conclude by expressing goodwill and confidence in the company's integrity.

The next writer asks directly how to return his defective skis for repair. The attention line directs his claim to the right department. The subject line, and its reemphasis in the first sentence, makes clear the nature of the claim.

ROUTINE CLAIM LETTER

Attention: <u>Consumer Affairs Department</u>

Subject: <u>Delaminated Skis</u>

States problem and action desired	This winter, my Tornado skis began to delaminate. I want to take advantage of your lifetime guarantee to have them relaminated.
Provides details	I bought the skis from the Ski House in Erving, Massachusetts, in November 1973. Although I no longer have the sales slip, I did register them with you. The registration number is P9965.
Explains basis for claim	I'm aware that you no longer make metal skis, but as I recall, your lifetime guarantee on the skis I bought was a major selling point. Only your company and one other were backing their skis so strongly.
Courteously states desired action	Would you please let me know how to go about returning my delaminated skis for repair?

Arguable Claims

When your request is in some way unusual or debatable, you must *persuade* the recipient to grant your claim.

Use the indirect plan for an arguable claim. People are more likely to respond favorably *after* reading your explanation. Begin with a neutral statement both parties can agree to—but that also serves as the basis for your request (e.g., "Customer goodwill is often an intangible quality, but a quality that brings tangible benefits").

Once you've established agreement, explain and support your claim. Include enough information for a fair evaluation: date and place of purchase, order or policy number, dates of previous letters or calls, and background.

Conclude by requesting a *specific action* (a credit to your account, a replacement, a rebate). Ask confidently.

The next writer employs a tactful, reasonable tone and an indirect plan.

ARGUABLE CLAIM LETTER

Establishes early agreement	Your company has an established reputation as a reliable wholesaler of office supplies. For eight years we have counted on that reliability, but a recent episode has left us annoyed and disappointed.
Presents facts to support claim	On January 29, 1999, we ordered five cartons of 2.0 megabyte KAO diskettes (#A74–866) and thirteen cartons of Epson MX 70/80 black ribbon cartridges (#A19–556).
Offers more support	On February 5, the order arrived. But instead of the 2.0 MB KAO diskettes ordered, we received 1.4 MB 3M diskettes. And the Epson ribbons were blue, not the black we had ordered. We returned the order the same day.
Includes all relevant information	Also on the 5th, we called John Fitzsimmons at your company to explain our problem. He promised delivery of a corrected order by the 12th. Finally, on the 22nd we did receive an order—the original incorrect one—with a note claiming that the packages had been water damaged while in our possession.
Sticks to the facts— accuses no one	Our warehouse manager insists the packages were in perfect condition when he released them to the shipper. Because we had the packages only five hours, and had no rain on the 5th, we are certain the damage did not occur here.
Requests a specific adjustment	Responsibility for damages therefore rests with either the shipper or your warehouse staff. What bothers us is our outstanding bill from Hightone ($1,049.50) for the faulty shipment. We insist that the bill be canceled and that we receive a corrected statement. Until this misunderstanding, our transactions with your company were excellent. We hope they can be again.

RÉSUMÉS AND JOB APPLICATIONS

Attractive jobs are highly competitive. Some large companies receive thousands of résumés yearly for a mere handful of openings. Whether you are applying for your first professional job or changing careers, you must market your skills effectively. Your résumé and letter of application must stand out among the competition.

Prospecting for Jobs

Begin your employment search by studying the job market to identify the careers and jobs for which you best qualify.

EVALUATING YOUR SKILLS AND APTITUDES.   Focus on openings that suit your skills, aptitudes, and goals:

Identify your assets

- What skills have you acquired in school, on the job, through hobbies or other interests?
- Do you have skills in leadership or in group projects (as demonstrated in employment, social organizations, or extracurricular activities)?
- Do you speak a foreign language? Have musical or artistic talent?
- Do you have communication skills? Are you a good listener?
- Do you perform well under pressure?

Besides helping focus your search, answers to these questions will be handy when you write your résumé and prepare for interviews.

LEARNING ABOUT THE JOB MARKET.   Launch your search early. Don't wait for the job to come to you. Take the initiative by observing these suggestions:

Explore all resources

- Scan the Help Wanted section in major Sunday newspapers for job descriptions, salaries, and qualifications. The World Wide Web offers electronic access to job listings nationwide. (For electronic job hunting see page 393).
- Ask a reference librarian to suggest occupational handbooks, government publications, newsletters, and magazines or journals in your field.
- Visit your college placement service. Openings are posted there, interviews are scheduled, and counselors can provide job-hunting advice.
- Sign up at your placement office for interviews with company representatives who visit the campus.
- Speak with people in your field to get an inside view and practical advice.
- Seek the advice of faculty in your major who do outside consulting or who have worked in business, industry, or government.
- Look for a summer job or internship in your field; this experience may count as much as your education.
- Many professional organizations invite student memberships (at reduced fees). Such affiliations can generate excellent contacts, and they look good on your résumé. Try to attend meetings of the local chapter.
- Do whatever additional *networking* you can. Notice that several job-search methods in Figure 19.4 include talking with people and exploring useful contacts.

Increased use of email and online job listings and résumé postings is changing the way people conduct job searches. (See page 393.)

Launch your search before senior year to learn whether certain courses make you more marketable. If you are changing jobs or careers, employers will be more interested in what you have accomplished *since* college. Be prepared to show how your experience is relevant to the new job.

Learn about career opportunities and the job market by consulting resources such as those on page 381 at your library or newsstand:

Figure 19.4
How People
Usually Find Jobs
Source: Tips for
Finding the Right
Job. *U.S.*
Department of
Labor.

Most Commonly Used Job-Search Methods		
Percent of Total Job-seekers Using the Method	Method	Effectiveness Rate*
66.0%	Applied directly to employer....................................	47.7%
50.8	Asked friends about jobs where they work....................	22.1
41.8	Asked friends about jobs elsewhere...........................	11.9
28.4	Asked relatives about jobs where they work.................	19.3
27.3	Asked relatives about jobs elsewhere	7.4
45.9	Answered local newspaper ads................................	23.9
21.0	Private employment agency	24.2
12.5	School placement office..	21.4
15.3	Civil Service test	12.5
10.4	Asked teacher or professor....................................	12.1
1.6	Placed ad In local newspaper	12.9
6.0	Union hiring hall...	22.2

* A percentage obtained by dividing the number of jobseekers who actually found work using the method, by the total number of jobseekers who tried to use that method, whether successfully or not.

Research potential
employers

- *Fortune, Business Week, Forbes,* the *Wall Street Journal,* or trade publications in your field—for the latest developments and the big picture on business, economy, technology, and the like. Articles about specific topics and companies can be searched in the *Business Index.*
- *Occupational Outlook Handbook* and its quarterly update, *Occupational Outlook Quarterly,* published by the U.S. Department of Labor—for descriptions of occupations, qualifications, employment prospects, and salaries.
- *Almanac of Jobs and Salaries, Dun's* ® *Employment Opportunities Directory,* or *Federal Career Opportunities*—for government and private-sector organizations that hire college graduates.
- *Moody's Industrial Index* or Standard and Poor's *Register of Corporations*—for data on plant locations, major subsidiaries, products, executive officers, and corporate assets.
- Annual reports, for data on a company's assets, innovations, performance, and prospects. (Many libraries collect annual reports. Ask your librarian.)

The above list is a mere sampling of the resources—increasingly available in electronic versions.

You might also register with an employment agency. Of course, a fee is payable after you are hired, but employers often pay this fee. Ask about fee arrangements *before* you sign up.

Once you have a clear picture of where you fit into the job market, you must answer the big question asked by all employers: *What do you have to offer?* Your answer must be a highly polished presentation of yourself, your education, work history, interests, and skills—in short, your résumé.

Preparing Your Résumé

The résumé is a summary of your experience and qualifications. Written before your application letter, the résumé provides background information to support your letter. This information supplies an employer with a one- or two-page reference. The letter will emphasize specific parts of your résumé and will discuss how your background is suited to a particular job.

Employers generally spend fifteen to forty-five seconds scanning a résumé initially. They look for an obvious and persuasive answer to this question: *What can you do for us?* Employers are impressed by a résumé that:

What employers expect in a résumé

1. looks good (conservative, tasteful, uncluttered, on quality paper);
2. reads easily (headings, typeface, spacing, and punctuation provide clear orientation); and
3. provides information the employer needs for making an interviewing decision.

Employers generally discard résumés that are mechanically flawed, cluttered, sketchy, or hard to follow. Make your résumé perfect.

Organize your information within these categories:

Résumé categories

- ◆ name and address
- ◆ job and career objectives
- ◆ educational background
- ◆ work experience
- ◆ personal data
- ◆ interests, activities, awards, and skills
- ◆ references

Select and organize material to emphasize what you can offer. Don't just list *everything;* be selective. Don't abbreviate, because some people might not know the referent. Use punctuation to clarify and emphasize, not to be "artsy." Try to limit your résumé to a single page, as most employers prefer. (Of course, if you are changing jobs or careers, or if your résumé looks cramped, you might need a second page.)

Begin your résumé well before your job search. Your final version can be duplicated for various similar targets—but each new type of job requires a new résumé that is tailored to fit the advertised demands of that job.

NOTE   *Never "invent" credentials. Your résumé should make you look as good as the facts allow. Distorting the facts is unethical and counterproductive. Companies routinely investigate claims made in a résumé, and people who have lied are fired.*

NAME AND ADDRESS. Include your full name, mailing and email address, and phone number (many interview invitations and job offers are made by phone). Include both school and summer addresses, indicating dates you can be reached at each.

JOB AND CAREER OBJECTIVES. From the information collected earlier (page 379), you should have a clear idea of the *specific* jobs for which you *realistically* qualify. Be prepared to have different statements of objectives to meet the requirements of different job descriptions.

The key to a successful résumé is the image of *you* it projects—disciplined and purposeful, yet flexible. State your immediate and long-range goals, including any plans to continue your education.

Statement of career objective | Intensive-care nursing in a teaching hospital, with the eventual goal of supervising and instructing.

Do not borrow a trite statement from some placement brochure. If applying to a company with various branches, express your willingness to relocate.

EDUCATIONAL BACKGROUND. If your education is more substantial than your work experience, place it first. Begin with your most recent school and work backward, listing degrees, diplomas, and schools attended *beyond* high school (unless the high school's prestige, its program, or your achievements warrant its inclusion). List your major, minor, and selected courses that have directly prepared you for the job you seek. If your class rank and grade point average are in the upper thirty percent, list them. Include schools or specialized training during military service. If you finance part or all of your education by working, say so, indicating the percentage of your contribution.

WORK EXPERIENCE. If you have experience that relates to the job you seek, list it before your education. Start with your most recent job and work backward. Provide dates and names of employers. Indicate whether the job was full-time, part-time (hours weekly), or seasonal. Describe your exact duties in each job, indicating promotions. If the job was major (and related to this one), describe it in detail, and give your reason for leaving. Include military experience. If you have no real experience, show your potential by emphasizing your preparation and by writing an enthusiastic letter.

Complete sentences are unnecessary in résumés. They take up room best left for emphasis on your other qualifications. Use action verbs (*supervised, developed, built, taught, installed, managed, trained, solved, planned, directed*) to stress your ability to produce results. If your résumé is likely to be scanned electronically, list key words as nouns (*leadership skills, software development, data processing, editing*) immediately below your name and address. (See page 395.)

PERSONAL DATA.   An employer cannot legally discriminate on the basis of sex, religion, race, age, national origin, disability, or marital status. Therefore, you aren't required to provide this information or a photograph. But if you believe that any of this information could advance your prospects, by all means include it.

PERSONAL INTERESTS, ACTIVITIES, AWARDS, AND SKILLS.   List hobbies, sports, and other pastimes; memberships in teams and organizations; offices held; and any recognition you have received. Include dates and types of volunteer work. Employers know that people with a well-rounded lifestyle are likely to take an active interest in their jobs. Be selective in this section. List only items that reflect the qualities employers seek.

REFERENCES.   Your list of references names at least three (five maximum) people *who have agreed* to write strong, positive assessments of your qualifications and personal qualities. Often a reference letter is the key to getting an employer to want to meet you; choose your references carefully.

Select references who can speak with authority about your ability and character. Avoid family members and close friends not in your field. Instead choose among former employers, professors, and respected community figures who know you well enough to write *concretely* on your behalf.

A mediocre letter of reference is more damaging than no letter at all. Don't simply ask, "Could you please act as one of my references?" This is hard to refuse, but a person who doesn't know you well or who doesn't think highly of your work might write a letter that does more harm than good. Instead, make an explicit request: "Do you feel you know me and my work well enough to write a strong letter of reference? If so, would you act as one of my references?" This second version gives your respondent the option of declining gracefully or it elicits a firm commitment to a positive recommendation.

Detailed recommendations are time-consuming to write. Your references have no time to write individual letters to every prospective employer. Ask for only one letter, with no salutation. Your reference keeps a file copy, you keep the original for your personal dossier (to reproduce as necessary); and a copy goes to the placement office for your placement dossier. Because the law permits you to read all material in your dossier, this arrangement provides you with your own copy of your credentials. (The dossier is discussed later in this chapter.)

NOTE

Under some circumstances you may—if you wish to—waive the right to examine your recommendations. Some applicants, especially those applying to professional schools, as in medicine and law, waive this right in concession to a general feeling that a letter writer who is assured of confidentiality is more likely to provide a balanced, objective, and reliable assessment of a candidate. Seek the advice of your major adviser or a career counselor.

If the people you select as references live elsewhere, you might make your request by letter:

LETTER REQUESTING A REFERENCE

From September 19xx to August 19xx, I worked at Teo's Restaurant as a waiter, cashier, and then assistant manager. Because I enjoyed my work, I decided to study for a career in the hospitality field.

In three months I will graduate from San Jose City College with an A.A. degree in Hotel and Restaurant Management. Next month I begin my job search. Do you remember me and my work enough to write a strong letter of recommendation? If so, would you kindly serve as one of my references?

To save your time, please omit the salutation from your letter. If you could send me the original, I will forward a copy to my college placement office.

To update you on my recent activities, I've enclosed my résumé. If I can answer any questions, please contact me by phone (collect) at 123–555–4321 or by email at JPURDY@VALNET.com.

Thank you for your help and support.

NOTE

Opinion is divided about whether names and addresses of references should appear in a résumé. If saving space is important, simply state, "References available on request," keeping your résumé only one page long, but if your résumé already takes up more than one page, you should probably include names and addresses of references. (A prospective employer might recognize a name, and thus notice your name among the crowd of applicants.) If you are changing careers, a full listing of references is extremely important.

Organizing Your Résumé

Emphasize your assets

Organize your résumé to convey the strongest impression of your qualifications, skills, and experience. Depending on your background, you can arrange your material in reverse chronological order, functional order, or a combination.

REVERSE CHRONOLOGICAL ORGANIZATION.   In a reverse chronological résumé you list your most recent experience first, moving backward through your earlier experiences. Use this arrangement to show a pattern of job experiences or progress along a specific career track (as in Figure 19.5).

FUNCTIONAL ORGANIZATION.   In a functional résumé, you emphasize skills, abilities, and achievements that relate specifically to the job for which you are applying. Use this arrangement if you have limited job experience, gaps in your job record, or if you are changing careers.

James David Purdy
203 Elmwood Avenue
San Jose, CA 95139
Tel.: 214-316-2419
Email: jpur@valnet.com

Career
Objective   Customer relations for a hospitality chain, leading to market
                     management.

Education
1997–2000   *San Jose College, San Jose, CA*
                     Associate of Arts Degree in Hotel/Restaurant Management, June 2000.
                     Grade point average: 3.25 of a possible 4.00. All college expenses
                     financed by scholarship and part-time job (20 hours weekly).

Employment
1997–2000   *Peek-a-Boo Lodge, San Jose, CA*
                     Began as desk clerk and am now desk manager (part-time) of this 200-unit
                     resort. Responsible for scheduling custodial and room service staff,
                     convention planning, and customer relations.

1996–1997   *Teo's Restaurant, Pensacola, FL*
                     Beginning as waiter, advanced to cashier and finally to assistant manager.
                     Responsible for weekly payroll, banquet arrangements, and supervising
                     dining room and lounge staff.

1994–1996   *Encyclopaedia Britannica, Inc., San Jose, CA*
                     Sales representative (part-time). Received top bonus twice.

1993–1994   *White's Family Inn, San Luis Obispo, CA*
                     Worked as busperson, then server (part-time).

Personal   *Awards*
                     Captain of basketball team, 1992; Lion's Club Scholarship, 1994.

                     Special Skills
                     Speak French fluently; expert skier.

                     Activities
                     High school basketball and track teams (3 years); college student senate
                     (2 years); Innkeepers' Club—prepared and served monthly dinners at the
                     college (2 years).

                     Interests
                     Skiing, cooking, sailing, oil painting, and backpacking.

References   Placement Office, San Jose City College, San Jose, CA 90462

Figure 19.5   Résumé for an Entry-Level Candidate (Reverse Chronological Arrangement)

COMBINED ORGANIZATION.   Most employers prefer résumés that are chronologically organized because they are easier to scan. However, electronic scanning of résumés (page 395) will call for a more functional pattern. One alternative is a modified-functional résumé which preserves the logical progression that employers prefer but also highlights your abilities and job skills (as in Figure 19.6).

A Sample Situation

Imagine that you are a twenty-four-year-old student about to graduate from a community college with an A.A. degree in Hotel and Restaurant Management. Before college, you worked at related jobs for more than three years. You now seek a junior management position while you continue your education part time.

You have spent two weeks compiling information for your résumé and obtaining commitments from four references. Figure 19.5 shows your résumé. Notice that this résumé mentions nothing about salary. Wait until this matter comes up in your interview, or later.

Your résumé should be attractive, neat, and free of mechanical or grammatical errors. Using a word processor and laser printer, you can make changes to suit various jobs and have a perfect document for each version.

NOTE

The importance of proofreading your résumé cannot be overstated. By relying solely on your computer's spelling checker you might end up expressing pride in receiving a "plague" instead of a "plaque," in receiving a "bogus award" instead of a "bonus award" or in "ruining" your own business instead of "running" it.

Preparing Your Job Application Letter

YOUR IMAGE.   Although it elaborates on your résumé, your application letter must emphasize personal qualities and qualifications convincingly. In your résumé you present raw facts. In your application letter you relate these facts to the company to which you are applying. The tone and insight you bring to your discussion suggest a good deal about who you are. The letter is your chance to explain how you see yourself fitting into the organization, to interpret your résumé, and to show how you can be an asset to this employer. Your letter's immediate purpose is to secure an interview.

TARGETS.   Unlike the résumé, the letter should never be photocopied. Although you can base letters to different employers on the same model—with appropriate changes—each letter should be prepared anew.

Sometimes you will apply for positions advertised in print or by word of mouth (solicited applications). At other times you will write prospecting letters to organizations that have not advertised but might need someone like you (unsolicited applications). In either case, tailor your letter to the situation.

THE SOLICITED LETTER.   Imagine that you are James Purdy (Figure 19.5). In *Innkeeper's Monthly,* you read the advertisement on page 389 and decide to apply.

Karen P. Granger
82 Mountain Street
New Bedford, MA 02740
Telephone (617) 864-9318
Email: kgrang@swis.net

Objective	A summer internship documenting microcomputer software.
Qualifications	Software and hardware documentation. Editing. Desktop publishing. Usability testing. Computer science. Internet research. World Wide Web collaboration. Networking technology. Instructor-led training. IBM PC, Macintosh, DEC 20 mainframe and VAX 11/780 systems. Pagemaker, Netscape, Powerpoint, Excel, and Lotus Notes software. Logo, Pascal, HTML, and C++ program languages.
Education	Attending University of Massachusetts at Dartmouth; B.A. expected January 2001. Major: English/Communication. Minor: Computer Science. Dean's List, all semesters. GPA: 3.54. Class rank 110 of 1792.
Experience Intern Technical Writer	**Conway Communications, Inc.,** 39 Wall Street, Marlboro, MA 02864. Learned local area network (LAN) technology and Conway's product line. Wrote, designed, and tested five hardware upgrade manuals. Produced a hardware installation/maintenance manual from another writer's work. Specified and approved all illustrations. Designed home page and hypertext links. Summers 1998, 1999.
Writing Tutor	**Writing/Reading Center,** UMD. Tutored writing and word processing for individuals and groups. Edited WRC student newsletter. Trained new tutors. Cowrote and acted in a video about the WRC. Designed WRC home page for World Wide Web. Fall 1997–present.
Managing Editor	**The Torch,** UMD Weekly Newspaper. Organized staff meetings, generated story ideas, edited articles, and supervised page layout and pasteup. Fall 1998–present.
Achievements	Two writing samples published in Dr. John M. Lannon's *Technical Communication,* 8th ed. (Longman, 2000); Massachusetts State Honors Scholarship, 1997–2000.
Activities	Student member, Society for Technical Communication and American Soc. for Training and Development; student representative, College Curriculum Committee; UMD Literary Soc.
References	Placement Office, University of Massachusetts, No. Dartmouth MA 02747

Figure 19.6   Résumé for a Summer Internship Candidate (Modified-Functional Arrangement)

RESORT MANAGEMENT OPENINGS

Liberty International, Inc., is accepting applications for several junior management positions at our new Lake Geneva resort. Applicants must have three years of practical experience, along with formal training in all areas of hotel/restaurant management. Please
apply by June 1, 20xx, to

Sara Costanza
Personnel Director
Liberty International, Inc.
Lansdowne, PA 24135

Now plan and compose your letter.

Introduction.   Begin by directly stating your reason for writing. Name the job,
and remember you are talking *to* someone; use the pronoun "you" instead of awkward or impersonal constructions such as "One can see from the enclosed résumé."
If you can, establish a connection by mentioning a mutual acquaintance—but only
with that person's permission. Finally, after referring to your enclosed résumé, discuss your qualifications.

Body.   Concentrate on the experience, skills, and aptitudes you can bring to this
specific job. Follow these suggestions.

How to compose a persuasive application

- *Don't come across as a jack-of-all-trades.* Relate your qualifications specifically
  to the job for which you are applying.
- *Avoid flattery:* "I am greatly impressed by your remarkable company."
- *Be specific.* Replace "much experience," "many courses," or "increased sales"
  with "three years of experience," "five courses," or "a 35 percent increase in sales
  between June and October 1996."
- *Support all claims with evidence, to show how your qualifications will benefit this
  employer.* Instead of saying, "I have leadership skills," say, "I was student senate
  president during my senior year and captain of the lacrosse team."
- *Create a dynamic tone by using active* voice *and action verbs:*

Weak	Management responsibilities were steadily given to me.
Strong	I steadily assumed management responsibilities.

- *Trim the fat from your sentences:*

Flabby	I have always been a person who enjoys a challenge.
Lean	I enjoy a challenge.

◆ *Express self-confidence:*

Unsure   It is my opinion that I have the potential to become a successful manager because. . . .

Confident   I will be a successful manager because. . . .

◆ *Never be vague:*

Vague   I am familiar with the 1022 interactive database management system, and RUNOFF, the text processing system.

Definite   As a lab grader for one semester, I kept grading records on the 1022 database management system, and composed lab procedures on the RUNOFF text processing system.

◆ *Avoid letterese.* Write in plain English.
◆ *Be enthusiastic.* An enthusiastic attitude can be as important as your background, in some instances.

Conclusion.   Restate your interest and emphasize your willingness to retrain or relocate (if necessary). If the recipient is nearby, request an interview; otherwise, request a phone call, stating times you can be reached. Your conclusion should leave the impression that you are someone worth knowing.

Revision.   *Never* settle for a first draft—or a second or third! This letter is your model for letters serving in various circumstances. Make it perfect.

After several revisions, James Purdy finally signed the letter shown below.

JOB APPLICATION LETTER

203 Elmwood Avenue
San Jose, CA 10462
April 22, 2000

Sara Costanza
Personnel Director
Liberty International, Inc.
Lansdowne, PA 24135

Dear Ms. Costanza:

Writer identifies
self and purpose

Establishes a
connection

Please consider my application for a junior management position at your Lake Geneva resort. I will graduate from San Jose City College on May 30 with an Associate of Arts degree in Hotel/Restaurant Management. Dr. H. V. Garlid, my nutrition professor, described his experience as a consultant for Liberty International and encouraged me to apply.

Relates specific
qualifications to the
job opening

For two years I worked as a part-time desk clerk, and I am now the desk manager at a 200-unit resort. This experience, combined with earlier customer relations work in a variety of situations, has given me a clear and practical understanding of customers' needs and expectations.

As an amateur chef, I know of the effort, attention, and patience required to prepare fine food. Moreover, my skiing and sailing background might be assets to your resort's recreation program.

Expresses
confidence and
enthusiasm
throughout

I have confidence in my hospitality management skills. My experience and education have prepared me to work well with others and to respond creatively to changes, crises, and added responsibilities.

Makes follow-up
easy for the reader

Should my background meet your needs, please phone me any weekday after 4 p.m. at 214-316-2419.

Sincerely,

James D. Purdy

James D. Purdy

Enclosure

Purdy wisely emphasizes practical experience because his background is varied and impressive. An applicant with less practical experience would emphasize education instead, discussing related courses, extracurricular activities, and aptitudes.

As an additional example, here is the letter composed by Karen Granger in her quest for a summer internship.

INTERNSHIP APPLICATION LETTER

Dear Mr. White:

Begins by stating
purpose

I read in *Internship 2000* that your company offers a summer documentation internship. Because of my education and previous technical writing employment, I am highly interested in such a position.

Identifies herself
and college
background

In January 2001, I will graduate from the University of Massachusetts with a B.A. in English/Writing. I have prepared specifically for a computer documentation career by taking computer science, mathematics, and technical writing courses.

Expands on
background

In one writing course, the Computer Documentation Seminar, I wrote three software manuals. One manual uses a tutorial to introduce beginners to the Apple Macintosh and Microsoft Word. The other manuals describe two IBM PC applications that arrived at the university's computer center with no documentation.

Describes work
experience

The enclosed résumé describes my work as the intern technical writer with Conway Communications, Inc. for two summers. I learned local area networking (LAN) by documenting Conway's LAN hardware and software. I was responsible for several projects simultaneously and spent much of my time talking with engineers and testing procedures. If you would like samples of my writing, please let me know.

Explains interest in
this job

Although Conway has invited me to return next summer and to work full time after graduation, I would like more varied experience before committing myself to permanent employment. I know I could make a positive contribution to Birchwood Group, Inc. May I telephone you next week to arrange a meeting?

Sincerely,

THE UNSOLICITED LETTER.   Ambitious jobseekers do not limit their search to advertised openings. (Studies indicate that fewer than twenty percent of all job openings are advertised.) The unsolicited, or prospecting, letter is a good way to uncover possibilities beyond the Help Wanted section. Such letters have advantages and disadvantages.

Disadvantages.   The unsolicited approach has two drawbacks: (1) you might waste time writing to organizations that have no openings; and (2) you cannot tailor your letter to specific requirements.

Advantages.   For an advertised opening you compete with legions of applicants, whereas your unsolicited letter might arrive just as an opening materializes. Even when no opening exists, most companies welcome unsolicited letters and they file impressive applications—often online—or pass them along to a company that has an opening.

Reader Interest.   Because your unsolicited letter is unexpected, attract attention immediately. Don't begin: "I am writing to inquire about the possibility of obtaining a position with your company. If you can't establish a connection through a mutual acquaintance, use a forceful opening:

Opens forcefully

Does your hotel chain have a place for a junior manager with a college degree in hospitality management, proven commitment to quality service, and customer relations experience that extends far beyond textbooks? If so, please consider my application for a position.

Address your letter to the person most likely in charge of hiring. (Consult company Web sites or the business directories listed on page 381 for names of company officers.)

In Brief HOW APPLICANTS ARE SCREENED FOR PERSONAL QUALITIES

As many as twenty-five percent of résumés contain falsified credentials, such as a nonexistent degree or a contrived affiliation with a prestigious school (Parrish 1+). A security director for one major employer estimates that fifteen to twenty percent of job applicants have something personal to hide: a conviction for drunk driving or some other felony, trouble with the IRS, bad credit, or the like (Robinson 285).

With yearly costs of employee dishonesty or bad judgment amounting to billions of dollars, companies use preemployment screening for integrity, emotional stability, and a host of other personal qualities (Hollwitz and Pawlowski 203, 209).

Screening often begins with a background check of education, employment history, and references. One corporation checks up to ten references (from peers, superiors, and subordinates) per candidate (Martin, "So" 78). In addition, roughly ninety-five percent of corporations check on the applicant's character, trustworthiness, and reputation: they may examine driving, credit, and criminal records, and interview neighbors and coworkers (Robinson 285).

The law affords some protection by requiring employers to notify the applicant before checking on character, reputation, and credit history and to provide a copy of any report that leads to a negative hiring decision (Robinson 285). Once an applicant is hired, however, the picture changes: more than fifty percent of companies provide personal information to credit agencies, banks, and landlords without informing employees, and forty percent don't inform employees about what kinds of records on them are being kept (Karaim 72).

Beyond screening for background, employers use aptitude and personality tests to pinpoint desirable qualities. A sampling of test questions (Garner 86; Kane 56; Martin, "So" 77, 78):

- *Ability to perform under pressure:* "Do you get nervous and confused at busy intersections?"
- *Emotional stability and even temper:* "Do you honk your horn often while driving?"
- *Sense of humor:* "Tell us a joke."
- *Ability to cope with people in stressful situations.* "Do you like to argue and debate?" "Are you good at taking control in a crisis?"
- *Persuasive skills:* "Write a brief memo to a client, explaining why X [stipulated on the test] can't be done on time."
- *Presentation skills:* "Prepare and give a five-minute speech on some aspect of the industry as it relates to this company."

Above all, most employers look for candidates who are *likable*. One employer checks with each person an applicant speaks with during the company visit—including the receptionist. Another employer has candidates join in a company softball game (Martin, "So" 77).

ELECTRONIC JOB HUNTING

The computer and the World Wide Web continue to improve the quality and speed of contact between jobseekers and employers. Beyond ease and efficiency, electronic job hunting offers real benefits:

Benefits of online job searches

- You can search for jobs worldwide.
- You can focus your search by region, industry, or job category.
- You can research companies comprehensively from many perspectives.

◆ You can create your own Web site, with hyperlinks to samples of your work, employment references, or other supporting material.
◆ You can search "passively" and discretely (say, while employed elsewhere) by specifying preferences for salary, region, types of industry, and then receive an email message when the service provider identifies an opening that matches your "profile" (Martin, "Changing Jobs?" 206).
◆ Your search can be ongoing, in that your résumé remains part of an active computer file, until you delete it.

Online Employment Resources

More than fifty percent of major companies now recruit online (Weingarten 17), spending upwards of $200 million yearly (Croal 81). They usually pay a fee for searches of résumés posted on online databases such as these:

A sampling of job sites on the Web

◆ *Career Mosaic* at <http://www.careermosaic.com> provides lists of "hot" jobs, direct access to employer Web sites, free posting of résumés, online job fairs (at which you might chat with an interviewer online), career advice, and domestic and international job listings.
◆ *CareerPath.com* at <http://www.careerpath.com> posts job listings from major newspapers nationwide, and enables jobseekers to post résumés free.
◆ A Web search via *Yahoo!* or *Netscape Navigator* of the linked categories "Business and Economy: Employment: Jobs" yields job listings in specific fields or locations, company profiles, employers who hire in particular specialties, sites on which résumés or onscreen application forms can be posted free, and templates for creating online résumés.

Electronic recruiting centers such as careerWEB match applicants with employers in engineering, marketing, data processing, and technical fields worldwide. Countless specialized sites include *Boston Job Bank, College Grad Job Hunter, Eurojobs On-line, HiTechCareers, Hospitality Industry Job Exchange, Jobs in Atomic and Plasma Physics*, and *Positions in Bioscience and Medicine*.

Major companies such as Johnson & Johnson, Boeing, John Deere, and IBM—to name a few—post openings on their own Web sites. Working from a computer, you can visit a company's site, learn about its history, products, and priorities, search its job listings, e-mail your résumé and application letter, and arrange an interview.

NOTE

For a company's real story, look beyond its Web site (for which information has been selected to paint the rosiest possible picture). To see what industry analysts think about the company's prospects, consult an impartial research site such as Hoover's Online at <http://www.hoovers.com>. Here, in addition to relatively objective financial and management profiles, you can access recent articles about the company (Martin, "Changing Jobs?" 208).

Web sites maintained by professional organizations offer additional job listings, along with career advice and industry prospects. Some sites also allow for posting of interactive, hyperlinked résumés.

Professional organizations on the Web

- Society for Technical Communication <http://www.stc-va.org>
- International Television and Video Association <http://www.itva.org>
- Institute for Electrical and Electronic Engineers <http://www.ieee.org>

NOTE

Record each date and site at which you post an online résumé (or fill out an on-screen application) so you can keep track of responses and edit or delete your material as needed (Curry 100).

Electronic Scanning of Résumés

Countless large and midsize companies are scanning résumés electronically. Electronic storage of online or hard copy résumés offers an efficient way to screen countless applicants, to compile a database of applicants (for later openings), and to evaluate all applicants fairly.

How scanning works

An optical scanner feeds in the printed page, stores it as a file, and searches the file for key words associated with the job opening (nouns instead of traditional "action verbs"). Those résumés containing the most key words ("hits") are selected and ranked (Pender 120).

How to Prepare Content for a Scannable Résumé

Using nouns as key words, list all your skills, credentials, and job titles. (Help Wanted ads are a source for key words.)

Key words to list on a résumé

- *List specialized skills:* marketing, C++ programming, database management, user documentation, Internet collaboration, software development, graphic design, hydraulics, fluid mechanics, editing.
- *List general skills:* teamwork coordination, conflict management, oral communication, report and proposal writing, problem solving, troubleshooting, Spanish speaker.
- *List credentials:* student member, Institute of Electrical and Electronic Engineers, board-certified, B.S. Electrical Engineering, top 5 percent of class.
- *List job titles:* manager, director, supervisor, intern, coordinator, project leader.
- *List synonymous versions of key terms (to increase the chances of a hit):* procurement *and* purchasing multimedia *and* hypermedia, Web page design, *and* HTML, management *and* supervision (McNair 14). If you lack skills or experience, emphasize your personal qualifications: *analytical skills, energy, efficiency, flexibility, imagination, motivation.* In any résumé for the global marketplace, include willingness to relocate and travel—if you are truly willing to do so.

NOTE

Don't hesitate to make your scannable résumé longer than your hard copy version (but no more than three pages total). The longer the résumé being scanned, the more hits are possible (McNair 13).

How to Design a Scannable Résumé

Design your scannable résumé to provide effective computer access:

How to make a
scannable résumé
computer-accessible

- *Keep it simple.* Use a plain typeface (such as Helvetica for headings and Times Roman for body text), 10- to 14-point type, and white paper and black ink (McNair 13).
- *Avoid fancy highlighting.* Use **boldface** or FULL CAPS for emphasis. Avoid fancy fonts, italics, underlining, bullets, slashes, dashes, parentheses, or ruled lines (Pender 120).
- *Avoid a two-column format.* Multiple columns can be jumbled by scanners that read across the page.
- *Do not staple or fold your pages.* Submit the résumé in a large envelope so it will not be folded.

You might submit two versions of your résumé: one traditionally designed and one scannable—or include a keyword section, as in Figure 19.6 on page 388. Or you might submit a version via email and bring hard copies to the interview. Figure 19.7 shows a scannable version of Figure 19.6.

Electronic Résumés

Online versions of hard copy résumés can be submitted as email or hypertext.

EMAIL RÉSUMÉS. You might submit your résumé directly as email text or attach a file containing a version formatted for hard copy. Check with the recipient beforehand. Not all systems can receive or decode attached files—so when in doubt, send the résumé as email text (Skarzenski 18).

One way to ensure that your résumé can be read by any computer is to create an ASCII version (or "text file"). Select "Save As Text Only" from your desktop menu, and reformat your ASCII page as needed (Robart 14).

> **NOTE**
>
> *When submitting a résumé in electronic form, include a cover letter, either as an email document or, preferably, as an attractively formatted attachment. Also, indicate the software (and version) used in composing the résumé, or ask the employer to specify the desired software (Robart 13).*

HYPERLINKED RÉSUMÉS. Consider placing a hyperlinked résumé, which might include an online portfolio containing a personal statement, blueprints, writing samples, or indicators of your talents, on your own Web site or that of your school or professional society (Krause 159–60). Figure 19.8 provides a sample of the first page of a hyperlinked résumé.

You then include the address for your hyperlinked résumé on your hard copy or scannable résumé. Some employers refuse to track down a résumé on a Web page, so be sure to provide other delivery options as well (Robart 13–14).

KAREN P. GRANGER
82 Mountain Street
New Bedford, MA 02720
Telephone (617) 864-9318
Email: kgrang@swis.net

OBJECTIVE
A summer internship in software documentation.

QUALIFICATIONS
Software and hardware documentation. Editing. Desktop publishing. Usability testing. Computer science. Internet research. World Wide Web collaboration. Networking technology. Instructor-led training. IBM PC, Macintosh, DEC 20 mainframe and VAX 11/780 systems. Pagemaker, Netscape, Powerpoint, Excel, and Lotus Notes software. Logo, Pascal, HTML, and C++ program languages.

EDUCATION
UNIVERSITY OF MASSACHUSETTS AT DARTMOUTH: B.A. expected January 2001. English and Communications major. Computer Science minor. GPA: 3.54. Class rank: top 7 percent.

EXPERIENCE
CONWAY COMMUNICATIONS, INC., 39 WALL STREET, MARLBORO, MA 02864: Intern Technical Writer. LAN technology. Writing, designing, and testing hardware upgrade manuals. Desktop publishing of installation and maintenance manual. Specifying art and illustrations. Designing home page and hypertext links. Summers 1998 and 1999.

WRITING AND READING CENTER, UMD: Tutor. Individual and group instruction in writing and word processing. Training new tutors. Newsletter editing. Scriptwriting and acting in a training video. Designing home page. Fall 1994–present.

THE TORCH, UMD WEEKLY NEWSPAPER: Managing Editor. Responsible for conducting staff meetings, generating story ideas, supervising page layout, pasteup, and copyediting. Fall 1998–present.

ACHIEVEMENTS AND AWARDS
Writing samples published in Dr. John M. Lannon's TECHNICAL COMMUNICATION, 8th ed., Longman, 2000. Massachusetts State Honors Scholarship, 1997–2000. Dean's list, each semester.

ACTIVITIES
Student member, Society for Technical Communication and American Society for Training and Development. Student representative, College Curriculum Committee. UMD Literary Society.

REFERENCES
Placement Office, University of Massachusetts, No. Dartmouth MA 02747.

Figure 19.7 A Computer-Scannable Résumé

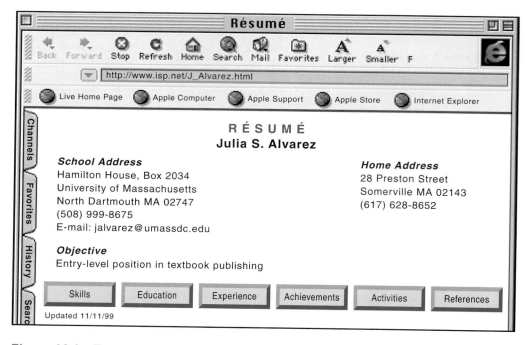

Figure 19.8   The First Page of a Hyperlinked Résumé

NOTE   *Be sure your hyperlinked résumé can download quickly. Keep in mind that complex graphics and multimedia download very slowly.*

SUPPORT FOR THE APPLICATION

Your Dossier

Your dossier contains your credentials: college transcript, recommendation letters, and any other items (such as a scholarship award or commendation letter) that document your achievements. An employer impressed by what you say about yourself will want to read what others think and will request your dossier. By collecting recommendations in one folder, you spare your references from writing the same letter repeatedly.

Your college placement office will keep your dossier on file and send copies to employers. Always keep your own copy as well. Then, if an employer requests your dossier, you can make a photocopy and mail it, advising your reader that the placement copy is on the way. This is not needless repetition! Most employers establish a specific timetable for (1) advertising an opening, (2) reading letters and résumés, (3) requesting and reviewing dossiers, (4) holding interviews, and (5) making an offer. Timing is crucial. Too often, dossier requests may sit on someone's desk and may even get lost in a busy placement office. Weeks can pass before your dossier is mailed. In these situations the only loser is you.

Employment Interviews

An employer impressed by your credentials will arrange an interview. The interview's purpose is to confirm the employer's impressions from your application.

You might meet with one interviewer, a group, or several groups in succession. You might be interviewed alone or with several other candidates at once. Interviews can last one hour or less, a full day, or several days. The interview can range from a pleasant chat to a grueling interrogation. Some interviewers may antagonize you deliberately to observe your reaction. Unprepared interviewees make mistakes like the following (Dumont and Lannon 620):

How people fail job interviews

- know little or nothing about the company
- have inflated ideas about their own worth
- have little idea of how their education prepares them for work
- dress inappropriately
- exhibit no self-confidence
- have only vague ideas of how they could benefit the employer
- inquire only about salary and benefits
- speak negatively of former employers or coworkers

Careful preparation is the key to a productive interview.

Prepare for the interview by learning about the company (its products or services, history, prospects, branch locations) in trade journals and industrial indexes. Request company literature and annual reports. Speak with people who know about the company. Prepare specific answers to the obvious questions:

Questions to expect in a job interview

- Why do you wish to work here?
- What do you know about our company: its goals, products, future plans?
- What do you know about our core values (say, informal management structure, commitment to diversity or the environment)?
- What do you know about the expectations and demands of this job?
- What are the major issues affecting this industry?
- What do you see as your biggest weakness? Your biggest strength?
- Can you describe an instance in which you came up with a new and better way of doing something?
- What would you like to be doing in ten years?

Plan informative and direct answers to questions about your background, training, experience, and salary requirements.

Prepare your own list of *well-researched* questions about the job and the organization; you will be invited to ask questions, and what you ask can be as revealing as any answers you give.

GUIDELINES

for Surviving a Job Interview

1. *Get your timing right.* Confirm the interview's exact time and location. Arrive at least a few minutes early.

2. *Bring support materials.* Bring extra copies of your résumé and a portfolio (if appropriate) with examples of your work.

3. *Make a positive first impression:*
   ◆ Come dressed as if you already work for the company.
   ◆ Extend a firm handshake
   ◆ Relax but do not slouch.
   ◆ Keep your hands in your lap.
   ◆ Do not fiddle with your face, hair, or other body parts.

4. *Do not worry about having all the answers.* If you cannot answer a question, say so, and relax. Interviewers typically do about seventy percent of the talking (Kane 56).

5. *Avoid abrupt yes or no answers—as well as life stories.* Keep answers short and sweet.

6. *Never criticize a previous employer.* Above all, interviewers like nice people.

7. *Don't be afraid to allow silence.* An interviewer simply may stop talking, just to observe your reaction to silence. If you have nothing more to say or ask, don't feel compelled to speak. Let the interviewer make the next move.

8. *Take a hint.* When your interviewer hints that the meeting is ending (perhaps by checking a watch), restate your interest, ask when you might expect further word, thank the interviewer, and leave.

9. *Display etiquette and restraint.* If you are invited to lunch, don't order the most expensive dish on the menu; don't order an alcoholic beverage; don't smoke; don't salt your food before tasting it; don't eat too quickly; don't put your elbows on the table; don't speak with your mouth full; and don't order a huge dessert. In short, remember this adage: "There's no such thing as a free lunch." And try to order last.

10. *Send a follow-up message.* Send a brief thank-you note via email or conventional mail.

The Follow-Up Letter

Within a few days after the interview, refresh the employer's memory with a letter or email restating your interest. James Purdy sent Sara Costanza this follow-up memo:

Refresh the
employer's memory

> Thank you again for your hospitality during my visit to your Lansdowne offices. The tour of your resort facilities was informative and enjoyable.
>
> After meeting you and your colleagues and touring the resort, I believe I could be a productive member of your staff. I would welcome the opportunity to prove my abilities.

Letters of Acceptance or Refusal

You may receive a job offer by phone or letter. If by phone, request a written offer, and respond with a formal letter of acceptance. This letter may serve as part of your contract; spell out the terms you are accepting. Here is Purdy's letter of acceptance:

Accept an offer
with enthusiasm

> I am delighted to accept your offer of a position as assistant recreation supervisor at Liberty International's Lake Geneva Resort, with a starting salary of $34,500.
>
> As you requested, I will phone Bambi Druid in your personnel office for instructions on reporting date, physical, exam, and employee orientation.
>
> I look forward to a long and satisfying career with Liberty International.

You may have to refuse offers. Even if you refuse by phone, write a prompt and cordial letter of refusal, explaining your reasons, and allowing for future possibilities. Purdy handled one refusal this way:

Decline an offer
diplomatically

> Although I was impressed by the challenge and efficiency of your company's operations, I am unable to accept your offer of a position as assistant desk manager of your London hotel.
>
> I have accepted a position with Liberty International because Liberty has offered me the chance to participate in its manager-trainee program. Also, Liberty will pay tuition for the courses I take in completing my B.S. degree in hospitality management.
>
> If any later openings should materialize at your Aspen resort, however, I would again appreciate your considering me as a candidate.
>
> Thank you for your interest and courtesy.

In Brief HOW TO EVALUATE A JOB OFFER

Fortunately, most organizations will not expect you to accept or reject an offer on the spot. You will probably be given at least a week to make up your mind. Although there is no way to remove all risks from this career decision, you will increase your chances of making the right choice by thoroughly evaluating each offer—weighing all the advantages against all the disadvantages of taking the job.

The Organization

Background information on the organization—be it a company, government agency, or nonprofit concern—can help you decide whether it is a good place for you to work.

Is the organization's business or activity in keeping with your own interests and beliefs? It will be easier to apply yourself to the work if you are enthusiastic about what the organization does.

How will the size of the organization affect you? Large firms generally offer a greater variety of training programs and career paths, more managerial levels for advancement, and better employee benefits than small firms. Large employers also have more advanced technologies in their laboratories, offices, and factories. However, jobs in large firms tend to be highly specialized—workers are assigned relatively narrow responsibilities. On the other hand, jobs in small firms may offer broader authority and responsibility, a closer working relationship with top management, and a chance to clearly see your contribution to the success of the organization.

Should you work for a fledgling organization or one that is well established? New businesses have a high failure rate, but for many people, the excitement of helping create a company and the potential for sharing in its success more than offset the risk of job loss. It may be almost as exciting and rewarding, however, to work for a young firm that already has a foothold on success.

Does it make any difference to you whether the company is private or public? A private company may be controlled by an individual or a family, which can mean that key jobs are reserved for relatives and friends. A public company is controlled by a board of directors responsible to the stockholders. Key jobs are open to anyone with talent.

Is the organization in an industry with favorable long-term prospects? The most successful firms tend to be in industries that are growing rapidly.

Where is the job located? If it is in another city, you need to consider the cost of living, the availability of housing and transportation, and the quality of educational and recreational facilities in the new location. Even if the place of work is in your area, consider the time and expense of commuting and whether you can use public transportation.

Where are the firm's headquarters and branches located? Although a move may not be required now, future opportunities could depend on your willingness to move to these places.

It is usually easy to get background information on an organization simply by telephoning its public relations office. A public company's annual report to the stockholders tells about its corporate philosophy, history, products or services, goals, and financial status. Most government agencies can furnish reports that describe their programs and missions. Press releases, company newsletters or magazines, and recruitment brochures can also be useful. Ask the organization for any other items that might interest a prospective employee.

Background information on the organization may also be available at your public or school library. If you cannot get an annual report, check the library for reference directories that provide basic facts about the company, such as earnings, products and services, and number of employees.

Stories about an organization in magazines and newspapers can tell a great deal about its successes, failures, and plans for the future. You can identify articles on a company by looking under its name in periodical or computerized indexes such as the *Business Periodicals Index, Reader's Guide to Periodical Literature, Newspaper Index, Wall Street Journal Index,* and *New York Times Index.* It will probably not be useful to look back more than two or three years.

The library may also have government publications that present projections of growth for the industry in which the organization is classified. Long-term projections of employment and output for more than two hundred industries, covering the entire economy, are developed by the Bureau of Labor Statistics and revised every other year—consult the current *Monthly Labor Review* for the most recent projections. The *U.S. Industrial Outlook,* published annually by the U.S. Department of Commerce, presents detailed analysis of growth prospects for a large number of industries. Trade magazines also have frequent articles on the trends for specific industries.

Career centers at colleges and universities often have information on employers that is not available in libraries. Ask the career center librarian how to find out about a particular organization.

The Nature of the Work

Even if everything else about the job is good, you will be unhappy if you dislike the day-to-day work. Determining in advance whether you will like the work may be difficult. However, the more you find out about it before accepting or rejecting the job offer, the more likely you are to make the right choice. Ask yourself questions like the following.

Does the work match your interests and make good use of your skills? The duties and responsibilities of the job should be explained in enough detail to answer this question.

How important is the job in this company? An explanation of where you fit in the organization and how you are supposed to contribute to its overall objectives should give an idea of the job's importance.

Are you comfortable with the supervisor?
Do employees seem friendly and cooperative?
Does the work require travel?
Does the job call for irregular hours?

How long do most people who enter this job stay with the company? High turnover can mean dissatisfaction with the nature of the work or something else about the job.

The Opportunities

A good job offers you opportunities to grow and move up. It gives you chances to learn new skills, increase your earnings, and rise to positions of greater authority, responsibility, and prestige.

The company should have a training plan for you. You know what your abilities are now. What valuable new skills does the company plan to teach you?

The employer should give you some idea of promotion possibilities within the organization. What is the next step on the career ladder? If you have to wait for a job to become vacant before you can be promoted, how long does this usually take? Employers differ on their policies regarding promotion from within the organization. When opportunities for advancement do arise, will you compete with applicants from outside the company? Can you apply for jobs for which you qualify elsewhere within the organization or is mobility within the firm limited?

The Salary and Benefits

Wait for the employer to introduce these subjects. Most companies will not talk about pay until they have decided to hire you. In order to know if their offer is reasonable, you need a rough estimate of what the job should pay. You may have to go to several sources for this information. Talk to friends who were recently hired in similar jobs. Ask your teachers and the staff in the college placement office about starting pay for graduates with your qualifications. Scan the Help Wanted ads in newspapers. Check the library or your school's career center for salary surveys, such as the College Placement Council Salary Survey and Bureau of Labor Statistics occupational wage surveys. If you are considering the salary and benefits for a job in another geographic area, make allowances for differences in the cost of living, which may be significantly higher in a large metropolitan area than in a smaller city, town, or rural area. Use the research to come up with a base salary range for yourself, the top being the best you can hope to get and the bottom being the least you will take. An employer cannot be specific about the amount of pay if it includes commissions and bonuses. The way the plan works, however, should be explained. The employer also should be able to tell you what most people in the job earn.

Also take into account that the starting salary is just that, the start. Your salary should be reviewed on a regular basis—many organizations do it every twelve months. If the employer is pleased with your performance, how much can you expect to earn after one, two, three, or more years?

Don't think of your salary as the only compensation you will receive—consider benefits.

Benefits can add a lot to your base pay. Health insurance and pension plans are among the most important benefits. Other common benefits include life insurance, paid vacations and holidays, and sick leave. Benefits vary widely among smaller and larger firms, among full-time and part-time workers, and between the public and private sectors. Find out exactly what the benefit package includes and how much of the costs you must bear.

Asking yourself these kinds of questions won't guarantee that you make the best career decision—only hindsight could do that—but it will probably help you make a better choice than if you act on impulse.

Source: Excerpted from U.S. Department of Labor. To-morrow's Jobs. *Washington, DC. GPO, 1995.*

Usability Checklist FOR LETTERS

(Numbers in parentheses refer to the first page of discussion.)

Content
- [ ] Is the letter addressed to a specifically named person? (367)
- [ ] Does the letter contain all of the standard parts? (365)
- [ ] Does the letter have all needed specialized parts? (368)
- [ ] Have you given the reader all necessary information? (367)
- [ ] Have you identified the name and position of your reader? (367)

Arrangement
- [ ] Does the introduction immediately engage the reader and lead naturally to the body? (367)
- [ ] Are transitions between letter parts clear and logical? (646)

- [ ] Does the conclusion encourage the reader to act? (367)
- [ ] Is the format correct? (372)
- [ ] Is the design acceptable? (372)

Style
- [ ] Is the letter in conversational language (free of letterese)? (373)
- [ ] Does the letter reflect a "you" perspective throughout? (373)
- [ ] Does the tone reflect your relationship with your reader? (245)
- [ ] Is the reader likely to react favorably to this letter? (374)
- [ ] Is the style clear, concise, and fluent throughout? (215)
- [ ] Is the letter grammatical? (Appendix C)
- [ ] Does the letter's appearance enhance your image? (372)

EXERCISES

1. Bring to class a copy of a business letter addressed to you or a friend. Compare letters. Choose the most and least effective.

2. Write and mail an unsolicited letter of inquiry about the topic you are investigating for an analytical report or research assignment. In your letter you might request brochures, pamphlets, or other informative literature, or you might ask specific questions. Submit a copy of your letter and the response to your instructor.

3. a. As a student in a state college, you learn that your governor and legislature have cut next year's operating budget for all state colleges by twenty percent. This cut will cause the firing of young and popular faculty members, drastically reduce admissions, financial aid, new programs, and wreck college morale. Write a claim letter to your governor or representative, expressing your strong disapproval and justifying a major adjustment in the proposed budget.

   b. Write a claim letter to a politician about some issue affecting your school or community.

   c. Write a claim letter to an appropriate school official to recommend action on a campus problem.

4. Write a five hundred to seven hundred word personal statement applying to a college for transfer or for graduate or professional school admission. Cover two areas: (1) what you can bring to this school by way of attitude, background, and talent; and (2) what you expect to gain in personal and professional growth.

5. Write a letter applying for a part-time or summer job, in response to a specific ad. Choose an organization related to your career goal. Identify the exact hours and calendar period during which you are free to work.

6. a. Identify a job you hope to have. Using newspaper ads, library (see your reference librarian for assistance), placement office, and personal sources, write your own description of the job: duties, responsibilities, work hours, salary range, requirements for promotion, highest promotion possible, unemployment rate in the field, employment outlook during the next decade, need for further education (advanced degrees, special training), employment rate in terms of geography, optimum age bracket (as in football, is one "over the hill" after thirty-five?).

   b. Using the sources listed above, construct a profile of the ideal employee for this job. If you were the personnel director screening applicants, what specific qualifications would you require (education, experience, age, physical ability, appearance, special skills, personality traits, attitude, outside interests, and so on)? Try to locate an actual newspaper ad describing job responsibilities and qualifications. Or assume that you're a personnel officer, and compose an ad for the job.

   c. Assess your own credentials against the ideal employee profile. Review the plans you have made to prepare for this job: specific courses, special training, work experience. Assume that you have completed your preparation. How do you measure up to the requirements in part b? Are your goals realistic? If not, why not? What alternative plan should you create?

   d. Using your list from part c, compose a perfect résumé.

   e. Write a letter applying for the job described in part a.

   f. Write a follow-up letter thanking your fictional employer for your interview and restating your interest.

   g. Write a letter accepting a job offer from this same employer.

   h. Write a letter graciously declining this job offer.

   i. Submit each of these items, in order, to your instructor.

Note: Use the sample letters in this chapter for guidance, but don't borrow specific expressions.

7. Most of these sentences need to be overhauled before being included in a letter. Identify the weakness in each statement, and revise as needed.

    a. Pursuant to your ad, I am writing to apply for the internship.

    b. I need all the information you have about methane-powered engines.

    c. You idiots have sent me a faulty disk drive!

    d. It is imperative that you let me know of your decision by January 15.

    e. You are bound to be impressed by my credentials.

    f. I could do wonders for your company.

    g. I humbly request your kind consideration of my application for the position of junior engineer.

    h. If you are looking for a winner, your search is over!

    i. I have become cognizant of your experiments and wish to ask your advice about the following procedure.

    j. You will find the following instructions easy enough for an ape to follow.

    k. I would love to work for your wonderful company.

    l. As per your request I am sending the country map.

    m. I am in hopes that you will call soon.

    n. We beg to differ with your interpretation of this leasing clause.

    o. I am impressed by the high salaries paid by your company.

8. Write a complaint letter about a problem you've had with goods or service. State your case clearly and objectively, and request a specific adjustment.

9. *For Class Discussion:* Under what circumstances might it be acceptable to contact a potential inquiry respondent by email? When should you just leave the person alone?

COLLABORATIVE PROJECTS

1. Assume that this ad has appeared in your school newspaper.

STUDENT CONSULTANT WANTED

The Dean of Students invites applications for the position of student consultant to the Dean for the next academic year. Duties: (1) meeting with students as individuals and groups to discuss issues, opinions, questions, complaints, and recommendations about all areas of college policy; (2) presenting oral and written reports to the Dean of Students; and (3) attending college planning sessions as student spokesperson. Time commitment: 15 hours weekly. Salary: $4,000.

Candidates should be full-time students with at least one year of student experience at this college. The ideal applicant will be skilled in report writing and oral communication, will work well in groups, and will demonstrate a firm commitment to our college. Application deadline: May 15.

    a. Compose a résumé and a letter of application for this position.

    b. Divide your class into screening, interview, and hiring committees.

    c. Exchange your group's letters and résumés with those of another group.

    d. As an individual committee member, read and evaluate each of the applications you have received. Rank each application privately on paper, according to the criteria in this chapter, before discussing them with your group. *Note:* While screening applicants, you will be competing for selection by another committee, which is reviewing your own application.

    e. As a committee, select the three strongest applications, and interview each finalist for ten minutes, after you have prepared a list of standardized questions.

f. On the basis of these interviews, rank your preferences privately, on paper, giving specific reasons for your final choice.

g. Compare your conclusions with those of your colleagues and choose the winning candidate.

h. As a committee, compose a memo to your instructor, justifying your final recommendations. (See pages 349–50 for more on justification reports.

2. Form groups according to college majors. Prepare a set of instructions for entry-level job-seekers in your major, telling them how to launch their search. Base at least part of your advice on your analysis of Figure 19.4. Limit your document to one double-sided page, using an inviting and accessible design and any visuals you consider appropriate. Appoint one group member to present your final document to the class.

Web Pages and Other Electronic Documents

Except for manuals and reference books, paper documents (letters, memos, reports, proposals) are usually structured for reading in linear sequence: front to back. Information builds on the information that precedes it. Email documents (page 357), often printed out at their destination, usually display this linear structure as well.

But some electronic documents allow users to design their own sequence in searching for specific chunks of information or merely browsing through parts of a topic in no particular order, as they might read a newspaper or encyclopedia (Grice and Ridgway 35–43). To allow readers to design their own information sequence, these documents are composed in small, discrete modules, that can stand alone in meaning.

This chapter introduces types of electronic documents essential to workplace communication.

ONLINE DOCUMENTATION

People who use computers in their jobs need instructions and training for operating the systems and understanding their equipment's many features. Online documentation is designed to support specific tasks and provide answers to specific questions.

Although computers come with printed manuals, the computer itself is often the preferred training medium. In computer-based training, on-screen documentation explains how the system works and how to use it. Examples of online help:

Types of online documentation

- ◆ error messages and troubleshooting advice
- ◆ reference guides to additional information or instructions
- ◆ tutorial lessons that include interactive exercises with immediate feedback
- ◆ help and review options to accommodate different learning styles
- ◆ link to software manufacturer's Web site

Instead of leafing through a printed manual, users find what they need by typing a simple command, clicking a mouse button, using a help menu, or following an electronic prompt.

The documentation itself might appear (a) in dialog boxes that ask the user to input a response or click on an option, or (b) pop-up or balloon help that appears when the user clicks on an icon or points to an item on the screen for more information. (Explore, for example, the online help resources on your own computer.)

HYPERTEXT

One extension of online documentation increasingly used in tutorial and training software is *hypertext*. Unlike printed text, designed to be read front to back, a hypertext document is nonlinear, offering multiple informational paths through the material, the particular path (or paths) determined by what and how much a user wants to know. *Hypermedia* expands the applications of hypertext by adding graphics, sound, video, and animation.

In addition to online documentation, hypertext serves as a research and reference tool for navigating complex cross-references and retrieving information electronically. In a hypertext system, a topic can be explored from any angle, at any level of detail. Assume, for instance, that you are using a hypertext database to research the AIDS epidemic. The database contains chunks of related topics organized in a network (or web) of files linked electronically (Horton, "Is Hypertext" 22), as in Figure 20.1.

After accessing the initial file ("The AIDS Epidemic"), you can navigate the network in any direction, choosing which file to open and where to go next, thereby customizing the direction of your search. The files themselves might be printed words, graphics, sound, video, or animation.

Because hypertext offers multiple layers of information, from general to specific (Figure 20.2), you also customize the depth of your search.

Communication specialist William Horton explains the instructional power achieved through hypertext layering:

> Paper . . . lacks depth. All information must be on the same level or layer. With hypertext, however, the screen can have deeper reserves of information. These deeper layers do not clutter the screen, but are available if needed. ("Is Hypertext" 25)

Hypertext accommodates inquiry in various directions

Hypertext accommodates inquiry at various depths

Figure 20.1
Topics (or Files) in a Hypertext Network are Linked Electronically

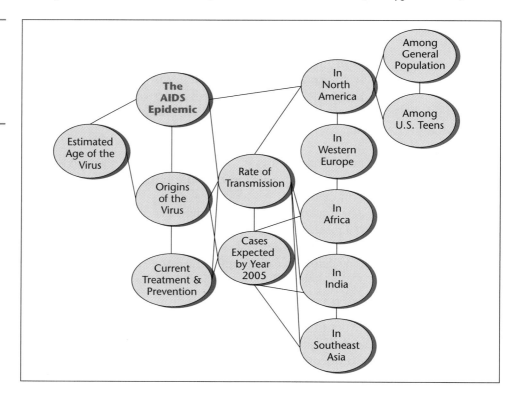

Figure 20.2
Hypertext Topics
Can Be Layered,
to Serve
Different
Audiences.
*Source: Visual
adapted from
Horton, William. "Is
Hypertext the Best
Way to Document
Your Product?"*
Technical
Communication
*39.1 (1991):25.
Used with
permission from the
Society for Technical
Communication,
Arlington, VA.*

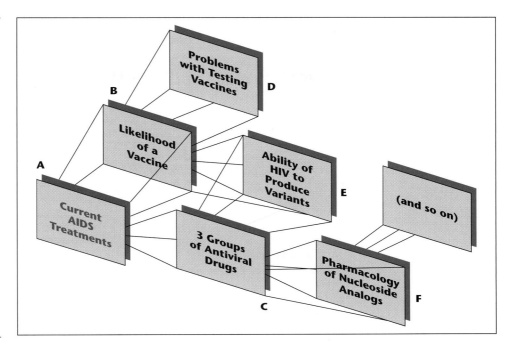

People navigating a hypertext document invent their own "text" and can therefore discover combinations, relationships, and chains of knowledge that are impossible to express in the sequential pages of a printed document (Bernstein 42).

Following is an application of hypertext instruction in the auto industry:

A hypertext
application

> A mechanic can zoom from a picture of a car engine to a video of a specific malfunctioning part, and then move directly to the text that tells how to fix it. (Morse 7)

For all their potential as research and instructional tools, hypertext documents do have limitations:

Limitations of
hypertext
documents

- ◆ Confronting countless possible paths through the material, readers can get lost in "hyperspace."

  | Where do I go next? How do I get out? How do I organize?

- ◆ Ease of navigation depends on how effectively individual chunks (or "nodes") of information are segmented and linked.
- ◆ Readers may resist a document that leaves them responsible for organizing their own learning. They generally rely on the writer to help organize their thinking (Horton, "Is Hypertext" 26).
- ◆ Instead of always enhancing learning, hypertext documents can in fact interfere with understanding and impede performance (Barfield, Haselkorn, and Weatbrook 22, 27). Users in one study took longer to read a hypertext document than the equivalent paper document (Rubens, "Reading" 36).

◆ Some hypertexts can be too highly structured, leaving users with *fewer* navigational choices than they would have with a printed document (Selber).

Communicators can address these problems by making the right decisions (Horton, "Is Hypertext" 27; Nickels-Shirk 191):

DECISIONS IN CREATING HYPERTEXT DOCUMENTS

◆ How much information should be included on one screen?
◆ What level of detail should be presented?
◆ How can each chunk be written so it can be read in any order and still make sense?
◆ How can the material best be linked for easiest navigation of various possible paths?
◆ Which combination of media should be employed (printed words, speech, sound effects, animation, music)?

Creating hypertext documents is usually not a one-person job; it often requires writers, graphic artists, computer specialists, and animators, among others. This is one example of how communication technology makes collaboration not only possible but also necessary.

The most rapidly expanding application of hypertext occurs via the Internet, on the World Wide Web. (Chapter 8 covers Internet research.)

THE WEB

Like a CD-ROM or electronic database, the Web offers a collection of electronic documents and multimedia. But hypertext enables the Web to link information in nonlinear patterns, providing countless routes to be explored—worldwide— according to the user's particular needs (Hunt 377). Some unique characteristics of the Web (December 371–72):

How the Web differs from other media

◆ *The Web is interactive.* Each user constructs his/her own hypertextual path through the material and often can respond/add to the message.
◆ *The Web allows reciprocal use.* Besides getting information, users can also provide it.
◆ *The Web is porous.* A web document can be entered at various points because a Web site usually offers multiple files, which are linked.
◆ *The Web is ever changing.* A Web page or site is "a work in progress"—not only in its content but also in the technology itself (software, hardware, modems, servers). Unlike paper, software, or non-rewritable CD-ROM, the Web has no "final state."

These features allow Web users to discover and create their own connections among an endless array of ideas.

In Brief HOW WEB SITES ENHANCE WORKPLACE TRANSACTIONS

The Web is a tool for advertising, learning about new products or companies, updating product information, or ordering products (Teague 236, 238). Each organization advertises its services and products via its own *home page,* a type of electronic billboard that introduces the organization and provides links to additional pages.

Specific Benefits

◆ *Visibility.* A Web site attracts business by establishing a presence in markets worldwide.

◆ *Access:* A Web site is accessible twenty-four hours a day (Dulude 47).

◆ *Customer Relations.* Through enhanced customer service and support and rapid response, a Web site increases customer satisfaction and enhances a company's caring persona (Hoger, Cappel, and Myerscough 41).

◆ *Efficiency.* Two-way, real-time communication allows sudden problems, errors, or areas of danger to be broadcast rapidly. The audience can control the viewing of messages and respond immediately. Onscreen instructions (say, for installing a modem) can be enhanced with high-resolution, 3-D graphics; parts can be color coded for assembly; and material can be updated instantly (Dulude 49–60).

◆ *Economy.* The cost of an Internet/Web bank transaction drops from over $1 to roughly 1 cent; the cost of processing an airline ticket drops from $8 to $1 (IBM 13). By radically reduced printing, mailing, and distribution costs, a site can facilitate mass publishing. Also, an advertiser can embed deeper and deeper levels of product details, without using extra page space. Ultimately, as the cost of business transactions drops, so does the number of required employees.

◆ *Data Gathering.* Tracking software provides customer data by recording who uses the Web site, how often they use it, and exactly where they go. Employees access reference materials from journal and trade magazines, as well as addresses of researchers, and the latest information about legal issues and government regulations (Ritzenthaler and Ostroff 17–18).

◆ *Information Sharing.* Intranets and extranets (Chapter 8) increase the flow of ideas up and down, and from outside to inside the company and vice versa. Knowledge audits identify who knows what and this information is then listed in the company intranet directory ("yellow pages").

◆ *Collaboration.* Company sites help reduce the length of and need for face-to-face meetings. And people who do meet are well prepared because they have shared information beforehand.

Web Applications In Major Companies

◆ For training employees in rapidly changing job skills, companies rely on distance-learning programs offered by colleges and universities. Such programs include email correspondence with faculty, online discussion groups, assignments and examples downloaded from the school Web site, and searches of virtual libraries (for special training or MBA work, etc.).

◆ General Electric expects to save $500 million over three years by purchasing via the Internet (McWilliams and Stepanik 172).

◆ NASA posts requests for proposals (RFPs) on an engineering Web site, to which bidders and contractors can respond via email— thereby speeding the whole process, eliminating fax, phone, and copying time (Machlis 45).

◆ The Volvo Corporation is connecting all its branches, warehouses, and truck dealerships in Sweden and the United States via a global network to keep track of parts, specifications, and product updates, and to allow authorized employees worldwide access to company databases so that "all data will be available anywhere" (Hamblen 51+).

NOTE

Keep in mind that Web pages, like all online screens, take at least twenty-five percent longer to read than paper documents. One possible solution: high-resolution screens, as readable as paper copy, should be available and affordable within a decade (Neilsen, "Be Succinct").

ELEMENTS OF A USABLE WEB SITE

Although more diverse than typical users of paper documents, Web users share common expectations. Following are basic usability requirements for a Web site.

Accessibility

Users expect a site to be easy to enter, navigate, and exit. Instead of reading word for word, they tend to skim, looking for key material without having to scroll through pages of text. They look for chances to interact, and they want to download material at a reasonable speed.

In Brief HOW SITE NEEDS AND EXPECTATIONS DIFFER ACROSS CULTURES

Despite its U.S. origin, the Internet has rapidly become international and cross-cultural. Yet countries vary greatly in their level of "Internet maturity," with much of the world years behind the United States (Neilsen, "Global"). A useful international site therefore reflects careful regard for cost, clarity, and cultural sensitivity.

Cost

High phone rates in many countries affect Internet costs. In Japan, for example—whose Internet use ranks second to that in the United States—monthly cost for one hour daily online is more than double the U.S. cost for unlimited access (Neilsen, "Global"). A usable site therefore omits graphics that are slow to load.

Clarity

To facilitate access and avoid misunderstandings, international communication via the Internet incorporates measures like these:

◆ Sites often provide home page versions in various languages (or links to a translation package).
◆ Time zones, currencies, and other units of measurement differ (10 A.M. in San Francisco

equals 6 P.M. in London, 7 P.M. in Stockholm or 3 A.M. in Tokyo). In arranging real-time interaction (say, an online conference), the host specifies the recipient's time as well as the home time (Neilsen, "International").

◆ Because the value of a "dollar" in countries such as Australia, Canada, Singapore, or Zimbabwe differs from the value of the U.S. dollar, businesses specify "US $12.50," and so on. Also, offering payment options in the culture's own currency helps avoid currency-exchange ambiguities (Hodges 28–29).
◆ A date listed as "6/10/98" might be confusing in other cultures. It would be preferable to say "10 June 1988" or "June 10, 1998."
◆ Temperature measurements are specified as "Fahrenheit" or "Celsius."

Cultural Sensitivity

A site that is truly "international" in ambiance—and not only "American"—enables anyone anywhere to feel at home (Neilsen, "Global"). For example, it avoids sarcasm or irreverence (which some cultures consider highly offensive), and exclusive references or colloquialisms such as "bear markets," "the Wild West," and "Super Bowl."

Worthwhile Content

Users expect the site to contain all the explanations they need (help screens, links, and so on). They want material that is accurate and constantly up-to-date (say, product and price updates). They expect clear error messages that spell out appropriate corrective action. They look for links to other, high-quality sites as indicators of credibility. They look for an email address and other contact information to be prominently displayed.

Sensible Arrangement

Users want to know where they are, and where they are going. They expect a recognizable design and layout, with links easily navigated forward or backward, back links to the home page, and no dead ends. They look for navigation bars and hot buttons to be explicitly labeled ("Company Information" "Ordering" "Job Openings," and so on).

Instead of a traditional introduction, discussion, and conclusion, users expect the punch line right up front. Because they hate to scroll, users often read only what is on the first screen.

Good Writing and Page Design

Users expect a writing style that is easy to read and error-free. They look for concise pages that are quick to scan, with short sentences and paragraphs, headings and bulleted lists. Instead of having to wade through overstatement and exaggeration to "get at the facts," users expect restrained, impartial language (Neilsen, "Be Succinct" 2). Figure 20.3 illustrates the effect of good writing on usability.

Good Graphics and Special Effects

Some users look for images or multimedia special effects—as long as they are neither excessive nor gratuitous. Since other users often disable their browser's visual capability (to save memory and downloading time) they look for a prose equivalent of each visual (*visual/prose redundancy*). They expect to recognize each icon and screen element—hot buttons, links, help options, and the like.

SCRIPTING A WEB DOCUMENT WITH HTML

HTML (HyperText Markup Language) is a computer scripting language that can be understood by all Web browsers and that specifies the positioning of each element on a Web page: text, art, headings, lists, hot buttons, and so on. For any type of computer or operating system, HTML provides a common "interface" enabling all users on a network to create, access, exchange, or edit information (Culshaw 34).

An HTML document is coded by use of *tags:* the command, enclosed in angle brackets, appears on both ends of the content to be acted on. The tagged command plus the related text are known as an *HTML element.*

A typical HTML element

| \<TITLE>English 266 Technical Writing\</TITLE>

Figure 20.4 shows basic HTML commands.

Site Version	Sample Paragraph	Usability Improvement (relative to control condition)
Promotional writing (control condition) using the "marketese" found on many commercial Web sites	Nebraska is filled with internationally recognized attractions that draw large crowds of people every year, without fail. In 1996, some of the most popular places were Fort Robinson State Park (355,000 visitors), Scotts Bluff National Monument (132,166), Arbor Lodge State Historical Park & Museum (100,000), **Carhenge** (86,598), Stuhr Museum of the Prairie Pioneer (60,002), and Buffalo Bill Ranch State Historical Park (28,446).	0% (by definition)
Concise text with about half the word count as the control condition	In 1996, six of the best-attended attractions in Nebraska were Fort Robinson State Park, Scotts Bluff National Monument, Arbor Lodge State Historical Park & Museum, **Carhenge,** Stuhr Museum of the Prairie Pioneer, and Buffalo Bill Ranch State Historical Park.	58%
Scannable layout using the same text as the control condition in a layout that facilitates scanning	Nebraska is filled with internationally recognized attractions that draw large crowds of people every year, without fail. In 1996, some of the most popular places were: • Fort Robinson State Park (355,000 visitors) • Scotts Bluff National Monument (132,166) • Arbor Lodge State Historical Park & Museum (100,000) • **Carhenge** (86,598) • Stuhr Museum of the Prairie Pioneer (60,002) • Buffalo Bill Ranch State Historical Park (28,446)	47%
Objective language using neutral rather than subjective, boastful, or exaggerated language (otherwise the same as the control condition)	Nebraska has several attractions. In 1996, some of the most-visited places were Fort Robinson State Park (355,000 visitors), Scotts Bluff National Monument (132,166), Arbor Lodge State Historical Park & Museum (100,000), **Carhenge** (86,598), Stuhr Museum of the Prairie Pioneer (60,002), and Buffalo Bill Ranch State Historical Park (28,446).	27%
Combined version using all three improvements in writing style together: concise, scannable, and objective	In 1996, six of the most-visited places in Nebraska were: • Fort Robinson State Park • Scotts Bluff National Monument • Arbor Lodge State Historical Park & Museum • **Carhenge** • Stuhr Museum of the Prairie Pioneer • Buffalo Bill Ranch State Historical Park	124%

Figure 20.3   The Effect of Good Writing on Usability   *Source: Neilsen, Jakob. "Reading on the Web." Oct. 1997. Alertbox. 8 Aug. 1999. <http://www.useit.com/ALERTBOX/9710a.html>*

- To denote a Web page (so browsers can identify this page as an HTML file):

    \<HTML\> entire HTML file \</HTML\> *(slash [/] denotes an ending tag)*

- To denote the main parts of an HTML file:

    \<HEAD\> . . . . \</HEAD\> *(sets off prefatory material, such as title)*
    \<BODY\> . . . . \</BODY\> *(sets off all material browsers will display)*

- To signify a break in the text*:

    \<BR\>   *(line break; begins next line at left margin)*
    \<P\>   *(paragraph)*
    \<HR\>   *(displays horizontal rule, across text width of page)*

*No closing tag is needed for a mere insertion into the page, in which no specific content is modified. These are called "empty tags."

- To specify headings:

    \<H1\> . . . . \</H1\> *(for highest level head)*
    \<H2\> . . . . \</H2\> *(and so on, down to as far as sixth-level heads)*

- To align elements elsewhere than at left margin *(center, right, justify):*

    \<H2 ALIGN=right\> . . . . \</H2\> *(aligns head on right margin)*
    \<P ALIGN=center\> . . . . \</P\> *(centers the paragraph)*
    \<BLOCKQUOTE\> . . . . \</BLOCKQUOTE\> *(sets off quoted material)*

- To display a list:

    \<UL\>   *(unnumbered list, with bullets displayed before each item)*
     \<LI\>   *(1st item in list)*
     \<LI\>   *(2nd item, and so on)*
    \</UL\>   *(end of list)*
    \<OL\>   *(numbered list, with a number displayed before each item)*
     \<LI\>   *(1st item in list)*
     \<LI\>   *(2nd item, and so on)*
    \</OL\>   *(end of list)*

- To specify typestyles:

\<Q\> . . . . \</Q\>	"Show enclosed content in quotation marks."
\<B\> . . . . \</B\>	**Boldface**
\<I\> . . . . \</I\>	*Italics*
\<BIG\> . . . . \</BIG\>	Larger than current font
\<SMALL\> . . . . \</SMALL\>	Smaller than current font

- To insert a figure in the text:

    \<FIG SRC="filename or URL"\> . . . \</FIG\>

- To create links (HREF stands for Hypertext Reference):

    \<A HREF="some URL"\>Go to some URL \</A\> *(to other Web sites)*
    \<A HREF="#price list"\>price list\</A\> *(to another part of document)*

- To receive email from site users:

    \<A HREF\>= "mail to:your_address@your_domain"\>Your email address\</A\>

Figure 20.4 A Sampling of HTML Commands

NOTE

To visit a Web site offering HTML advice, instruction, and useful links, go to <http://www.utoronto/websites/HTMLdocs/NewHTML/into.html>. Also, keep in mind that WYSIWYG ("what you see is what you get") editing programs, such as Adobe PageMill,™ Microsoft FrontPage,™ or Symantec Visual Page™ largely eliminate the need for HTML coding by hand. These authoring tools provide step-by-step instructions and templates for creating Web pages, positioning graphics and other page elements, adding clip art, and so on.

Java enhances
HTML documents

HTML produces only static pages, with limited possibilities for data display. However, a programming language called *Java* can be embedded in an HTML script to provide "dynamic" content (graphics, motion, and sound). Java-enhanced Web pages allow more sophisticated, interactive applications, such as simulations and computer-based training.

GUIDELINES

for Creating a Web Site

NOTE

Organizational Web sites are generally developed by a Web team: content developers, graphic designers, programmers, and managers. Whether or not you are an actual team member, expect a collaborative role in your organization's site development and maintenance.

Planning Your Site

1. *Identify the site's intended audience.* Are they potential customers seeking information, people purchasing a product or service, customers seeking product support or updates or troubleshooting advice (Wilkinson 33)? Will different audiences be seeking different material?

2. *Decide on the site's purpose.* Is the purpose to publish information, sell a product, promote an idea, solicit customer feedback, advertise talents, create goodwill? Should the site convey the image of a "cool" cutting-edge company (or individual), displaying skill with the latest Web technologies (animation, interaction, fancy design)? For specific ideas, find and examine other sites that display the features you are considering.

3. *Decide on what the site will contain.* Will it display only print documents or graphics, audio, and video as well? Will links be provided and, if so, how many and to where? Will user feedback be solicited and, if so, in what form: survey questions, email comments, or the like?

4. *Decide on the level of user interaction.* Will this be a document-only site, offering no interaction beyond downloading and printing? Will it offer dynamic marketing (Dulude 69): online questions and answers, technical support, downloadable software, online catalogs? Will users be able

to download documents, software, or documentation? Will an email button be included?

Laying Out Your Pages

1. *Design your pages (Chapter 15) to guide the user.* Highlight important material with headings, lists, typestyles, color, and white space. Remember that too much white space causes excessive scrolling. Prefer sans serif fonts (page 314). Use storyboards (page 201) to sketch the basic elements of each page. Limit page size to 30K, to speed downloading.

2. *Use graphics that download quickly.* Avoid excessive complexity and color, especially in screen background. Keep maximum image size below 30K. Create an individual file for each graphic and use thumbnail sketches on the home page, with links to the larger images, each in its own file.

3. *Include text-only versions of all visual information.* Roughly twenty percent of users turn off the graphics function on their browsers (Gannon 22–23).

4. *Provide orientation.* Structure the content to reflect its relative importance and frequency of use. Place the material that is most important to users right up front and create links to more detailed information. Date each page to announce the exact time of each update—or include a "What's New" head, so readers can keep track of changes.

5. *Provide navigational aids.* Keep links logical and always link back to the home page. Don't overwhelm the user with excessive choices. Label each link explicitly (for example, use "Product Updates," instead of merely "Click here").

6. *Define and shape the content.* Use hypertext to chunk information into subtopics, each in a digestible node, and link the nodes—but remember that hypertext takes longer to download and print (Neilsen, "Be Succinct"). Structure each hypertext node as an "inverse pyramid," in which you begin with the conclusion (Neilsen, "How Users Read" 1). The inverted pyramid works like a newspaper article, in which the major news/conclusion appears first (say, "The jury deliberated only two hours before returning a guilty verdict"), followed by the details (Neilsen, "Inverted Pyramids"). Last but not least, use restraint: give users the opportunity to receive *less* information (Outing). Think hard about what users need and give them only that.

7. *Sharpen the style.* Make the online text at least fifty percent shorter than its hard copy equivalent. Try to summarize (Chapter 11). Use short sen-

tences and paragraphs. Avoid "marketese" or promotional language that exaggerates ("breakthrough," "revolutionary," "cutting edge").

8. *Check your site.* Double-check the accuracy of numbers, dates, data, and such; check for broken links; and check for correct spelling, grammar, and so on.

9. *Attend to legal considerations.* Have your legal department approve all material before you post it (Wilkinson 33). Obtain written permission before linking to other Web pages or borrowing any graphic element from another site. Display a privacy notice that explains how each transaction is being recorded, collected, and used. To protect your own intellectual property, display a copyright notice on every page of the site (Evans, "Whose" 48, 50).

10. *Test your site for usability.* Test for usability with unfamiliar users (beta testing) and keep track of their problems and questions. What do users like and dislike? Can they navigate effectively to get to what they need? Are the icons recognizable? Is the site free of needless complexity or interactivity? Test your document with various browsers to be sure it can be downloaded.

11. *Maintain your site.* Review the site regularly, update often, and redesign as needed. If the site accommodates email queries, respond within one business day (Dulude 117).

PRIVACY ISSUES ON THE WEB

Information sharing between computers is what makes the Internet and World Wide Web possible. For instance, when a user visits a site, the host computer needs to know which browser is being used. Also, for improved client service, a host site can track the links visitors follow, the files they open or download, and the pages they visit most often (Reichard 106).

Often, however, more information gets "shared" than the user intended. For instance, the host computer can record the user's domain name, place of origin, and usage patterns (James-Catalano 32).

Some servers and sites display privacy notices explaining how usage patterns or transactions are being recorded, collected, and used. But this offers only limited protection. Any Internet transaction is routed through various browsers and servers and can be intercepted anywhere along the way.

Checklist FOR WEB SITE USABILITY

(Numbers in parentheses refer to first page of discussion.)

Accessibility

☐ Is the site easy to enter, navigate, and exit? (414)

☐ Is required scrolling kept to a minimum? (414)

☐ Is downloading speed reasonable? (419)

☐ If interaction is offered, is it useful—not superfluous? (418)

☐ Does the site avoid overwhelming the user with excessive choices? (420)

Content

☐ Are all needed explanations, error messages, and help screens provided? (415)

☐ Is the time of each update clearly indicated? (419)

☐ Is everything accurate and up-to-date? (420)

☐ Are links connected to high-quality sites? (415)

☐ Does everything belong (nothing excessive or superfluous or needlessly complex)? (419)

☐ Is an email button or other contact method prominently displayed? (415)

☐ Does the content accommodate international users? (414)

Arrangement

☐ Is the key part of the message on the first page? (419)

☐ Are navigation bars, hot buttons, and help options clearly displayed and explicitly labeled? (415)

☐ Are links easily navigated—backward and forward—with back links to the home page? (415)

Writing and Page Design

☐ Is the text easy to scan, with short sentences and paragraphs, and do headings, lists, typestyles, and color highlight important material? (415)

☐ Is overall word count roughly one-half of the hard copy equivalent? (419)

☐ Is the tone reasonable and restrained—free of overstatement and "marketese." (420)

Graphics and Special Effects

☐ Is each graphic easy to download? (415)

☐ Is each graphic backed up by a text-only version? (419)

☐ Is each graphic or special effect necessary? (415)

Legal Considerations

☐ Does the site display a privacy notice that explains how the transaction is being recorded, collected, and used? (420)

☐ Does each page of the site display a copyright notice? (420)

☐ Has written permission been obtained for each link to other sites and for each graphic element borrowed from another site? (420)

☐ Has all posted material received prior legal approval? (420)

INDIVIDUAL OR COLLABORATIVE PROJECTS

1. Consult the previous checklist and evaluate a Web site for usability. You might select a site at your school or place of employment. You might begin by deciding on specific information you seek (such as "internship opportunities" or "special programs" or "campus crime statistics" or "average SAT scores of admitted students") and use this as a basis for assessing the site's accessibility, content, arrangement, and so on.

Complete your evaluation and report any

problems, or suggest improvements in a memo to a designated decision maker. (Your instructor might ask different class groups to evaluate the same site and to compare their findings in class.)

2. Download and print pages from a Web site. Edit these pages to improve their layout and writing style. Submit copies to your instructor.

3. Examine Web sites from three or four competing companies (say, computer makers IBM™, Apple™, Gateway™, Dell™, and Compaq™— or automakers, and so on). Which site do you think is the most effective; the least effective; why? Report your findings in a memo to your classmates.

Technical Definitions

PURPOSE OF DEFINITIONS

LEVELS OF DETAIL IN A DEFINITION

EXPANSION METHODS

SITUATIONS REQUIRING DEFINITIONS

PLACEMENT OF DEFINITIONS

Usability Checklist **for Definitions**

All useful writing shares one feature—*clarity*. Clear writing begins with clear thinking; clear thinking begins with an understanding of what all the terms mean. Therefore, clear writing depends on definitions that both audience and writer understand. Unless you are sure that your audience knows the exact or special meaning you intend, always define a term the first time you use it.

PURPOSE OF DEFINITIONS

Definitions either explain a term that is specialized or unfamiliar to an audience or convey your exact definition of a word that has more than one meaning.

Most fields have specialized terms. Engineers talk about "prestressed concrete," "torque," or "tolerances"; attorneys discuss "easements," "liens," and "escrow accounts." For people outside the field, these terms must be defined. Some people are unaware that even certain familiar terms such as "disability," "guarantee," "tenant," "lease," and "mortgage" take on very specialized meanings in some contexts. What "guarantee" means in one situation is not necessarily what it means in another.

Contracts are detailed (and legal) definitions of the specific terms of an agreement. Because we are legally responsible for the documents we prepare, we rely on the technique of clear definition.

Definitions have ethical requirements, too. Chapter 5, for instance, explains how the shuttle *Challenger* explosion resulted from faulty definition: namely, the definition of "acceptable risk" was finally determined not by the technical facts, but by social pressures to launch on schedule. This social definition of acceptable risk was passed forward and influenced the top decision makers to approve the disastrous launch. Agreeing on meaning in such cases may be hard, but we are ethically bound to communicate an accurate interpretation of the facts as we understand them.

*Definitions have
legal implications*

*Definitions have
ethical implications*

LEVELS OF DETAIL IN A DEFINITION

How much detail is necessary for your audience to understand a term or a concept? Can you use a synonym (a term with a similar meaning)? Will you need a sentence, a paragraph, or multiple pages?

Parenthetical Definition

Often, you can clarify the meaning of a word by using a more familiar synonym or a clarifying phrase:

*Parenthetical
definitions*

> The *leaching field* (sievelike drainage area) requires crushed stone.
>
> The trees on the site are mostly *deciduous* (shedding foliage at season's end).

A parenthetical definition of "leaching field" might be adequate in a progress report to a client whose house you are building. But a public health report titled

"Groundwater Contamination from Leaching Fields" would call for an expanded definition.

Be sure that the synonym or explanatory phrase will clarify your meaning, not obscure it. Don't say:

A tumor is a neoplasm.

A solenoid is an inductance coil that serves as a tractive electromagnet. *(Appropriate for an engineering manual, but too specialized for general readers.)*

Do say:

A tumor is a growth of cells that occurs independently of surrounding tissue and serves no useful function.

A solenoid is a coil that converts electrical energy to magnetic energy capable of performing mechanical functions.

Sentence Definition

More complex terms may require a *sentence definition* (which may be stated in more than one sentence). These definitions follow a fixed pattern: (1) the name of the item to be defined, (2) the class to which the item belongs, and (3) the features that differentiate the item from all others in its class.

Elements of sentence definitions

Term	Class	Distinguishing Features
carburetor	a mixing device	in gasoline engines that blends air and fuel into a vapor for combustion within the cylinders
diabetes	a metabolic disease	caused by a disorder of pituitary gland or pancreas and characterized by excessive urination, persistent thirst, and inability to metabolize sugar
brief	a legal document	containing all the facts and points of law pertinent to a specific case, and filed by an attorney before the case is argued in court
stress	an applied force	that strains or deforms a body
fiber optics	a technology	that uses light energy to transmit voices, video images, and data via hair-thin glass fibers

These elements are combined into one or more complete sentences:

A complete sentence definition

Diabetes is a metabolic disease caused by a disorder of the pituitary gland or pancreas. This disease is characterized by excessive urination, persistent thirst, and inability to metabolize sugar.

Sentence definition is especially useful if you need to stipulate your precise working meaning of a term that has several possible meanings. State your working definitions at the beginning of your report:

A working
definition

| Throughout this report, the term "disadvantaged student" means. . . .

CLASSIFYING THE ITEM.   The narrower your class, the clearer your meaning. "Transit" is classified as a "surveying instrument," not as a "thing" or an "instrument." "Stress" is classified as "an applied force"; to say that stress "takes place when . . ." or "is something that . . ." fails to reflect a specific classification. "Diabetes" is precisely classified as "a metabolic disease," not as "a medical term."

DIFFERENTIATING THE TERM.   If the distinguishing features are too broad, they will apply to more than this one item. A definition of "brief" as "a legal document used in a courtroom" fails to differentiate "brief" from all other legal documents (wills, affidavits, and so on). Conversely, differentiating "carburetor" as "a mixing device used in lawn mower engines" is too narrow because it ignores the carburetor's use in other gasoline engines.

Also, avoid circular definitions (repeating the word you are defining as part of the distinguishing feature). "Stress is an applied force that places stress on a body" is a circular definition.

Expanded Definition

The sentence definition of "solenoid" on page 425 is good for a layperson who simply needs to know what a solenoid is. An instruction manual for mechanics, however, would define solenoid in much greater detail (pages 432–33); mechanics need to know how a solenoid works and how to use it. Your choice of parenthetical definition, sentence definition, or expanded definition depends on the amount of information people need.

The problem with defining an abstract and general word, such as "condominium" or "loan," is different. "Condominium" is a vaguer term than "solenoid" (solenoid A is pretty much like solenoid B) because the former refers to many types of ownership agreements, and so requires expanded definition for almost any audience.

An expanded definition may be a short paragraph (as for a simple tool) or may extend to scores of pages (as for a digital dosimeter—a device for measuring radiation exposure); sometimes the definition itself is the whole report.

The following expanded definition, from an auto insurance policy, defines "coverage for bodily injury to others," a phrase that could have many possible meanings.

An expanded
definition

PART I. BODILY INJURY TO OTHERS

Under this Part, we will pay damages to people injured or killed by your auto in Massachusetts accidents. Damages are the amounts an injured person is legally entitled to collect for bodily injury through a court judgment or settlement. We will pay only if you or someone else using your auto with your consent is legally responsible for the accident. The most we will pay for injuries to any one person as a result of any one accident is $5,000. The most we will pay for injuries to two or more people as a result of any one accident is a total of $10,000. This is the most we will pay as the result of a single accident no matter how many autos or premiums are shown on the Coverage Selections page. We will *not* pay:

1. For injuries to guest occupants of your auto.
2. For accidents outside of Massachusetts or in places in Massachusetts where the public has no right of access.
3. For injuries to any employees of the legally responsible person if they are entitled to Massachusetts workers' compensation benefits.

EXPANSION METHODS

How you expand a definition depends on the audience's questions as shown in Figure 21.1. Begin with a sentence definition, and then select from the following expansion strategies.

Etymology

A word's origin (its development and changing meanings) can clarify its definition. *Biological control* of insects is derived from the Greek "bio," meaning *life* or *living organism,* and the Latin "contra," meaning *against* or *opposite.* Biological control, then, is the use of living organisms against insects. College dictionaries contain etymological information, but your best bets are *The Oxford English Dictionary* (or its Web site) and encyclopedic dictionaries of science, technology, and business.

Figure 21.1
Directions In
Which a
Definition Can
Be Expanded

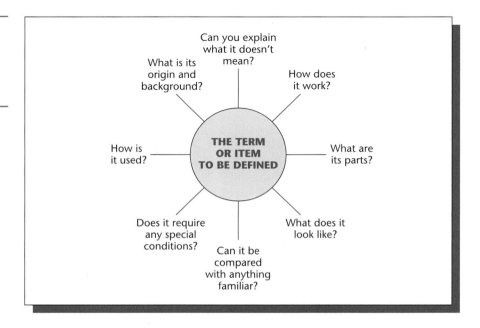

Some technical terms are acronyms, derived from the first letters or parts of several words. *Laser* is an acronym for *light amplification by stimulated emission of radiation.*

Sometimes a term's origin can be colorful as well as informative. *Bug* (jargon for *programming error*) derives from an early computer at Harvard that malfunctioned because of a dead bug blocking the contacts of an electrical relay. Because programmers were reluctant to acknowledge error, the term became a euphemism for *error.* Correspondingly, *debugging* is the elimination of errors in a program.

History and Background

The meaning of specialized terms such as "radar," "bacteriophage," "silicon chips," or "x-ray" can often be clarified through a background discussion: discovery or history of the concept, development, method of production, applications, and so on. Specialized encyclopedias are a good background source.

"Where did it come from?"

The idea of lasers . . . dates back as far as 212 B.C., when Archimedes used a [magnifying] glass to set fire to Roman ships during the siege of Syracuse. (Gartaganis 22)

"How was it perfected?"

The early researchers in fiber optic communications were hampered by two principal difficulties—the lack of a sufficiently intense source of light and the absence of a medium which could transmit this light free from interference and with a minimum signal loss. Lasers emit a narrow beam of intense light, so their invention in 1960 solved the first problem. The development of a means to convey this signal was longer in coming, but scientists succeeded in developing the first communications-grade optical fiber of almost pure silica glass in 1970. (Stanton 28)

Negation

Some definitions can be clarified by an explanation of what the term *does not* mean. The policy excerpt on page 426 defines coverage for "bodily injury to others" partly by using negation: "We will *not* pay: 1. For injuries to guest occupants of your auto."

Operating Principle

Most items work according to an operating principle, whose explanation should be part of your definition:

"How does it work?"

A clinical thermometer works on the principle of heat expansion: As the temperature of the bulb increases, the mercury inside expands, forcing a mercury thread up into the hollow stem.

Air-to-air solar heating involves circulating cool air, from inside the home, across a collector plate (heated by sunlight) on the roof. This warmed air is then circulated back into the home.

> Basically, a laser [uses electrical energy to produce] coherent light, light in which all the waves are in phase with each other, making the light hotter and more intense. (Gartaganis 23)

> [A fiber optics] system works as follows: An electrical charge activates the laser . . . , and the resulting light . . . energy passes through the optical fiber. At the other end of the fiber, a . . . receiver . . . converts this light signal back into electrical impulses. (Stanton 28)

Even abstract concepts or processes can be explained on the basis of their operating principle:

> Economic inflation is governed by the principle of supply and demand: If an item or service is in short supply, its price increases in proportion to its demand.

Analysis of Parts
When your subject can be divided into parts, identify and explain them:

"What are its parts?"

> The standard frame of a pitched-roof wooden dwelling consists of floor joists, wall studs, roof rafters, and collar ties.

> Psychoanalysis is an analytic and therapeutic technique consisting of four parts: (1) free association, (2) dream interpretation, (3) analysis of repression and resistance, and (4) analysis of transference.

In discussing each part, of course, you would further define specialized terms such as "floor joists" and "repression."

Analysis of parts is particularly useful for helping laypersons understand a technical subject. This next analysis helps explain the physics of lasing by dividing the process into three discrete parts:

> 1. [Lasers require] a source of energy, [such as] electric currents or even other lasers.

> 2. A resonant circuit . . . contains the lasing medium and has one fully reflecting end and one partially reflecting end. The medium—which can be a solid, liquid, or gas—absorbs the energy and releases it as a stream of photons [electromagnetic particles that emit light]. The photons . . . vibrate between the fully and partially reflecting ends of the resonant circuit, constantly accumulating energy—that is, they are amplified. After attaining a prescribed level of energy, the photons can pass through the partially reflecting surface as a beam of coherent light and encounter the optical elements.

> 3. Optical elements—lenses, prisms, and mirrors—modify size, shape, and other characteristics of the laser beam and direct it to its target. (Gartaganis 23)

Figure 1 (page 430) shows the three parts of a laser.

Visuals
Well-labeled visuals (such as the laser drawing) help clarify definitions. Always introduce your visual and explain it. If your visual is borrowed, credit the source. Un-

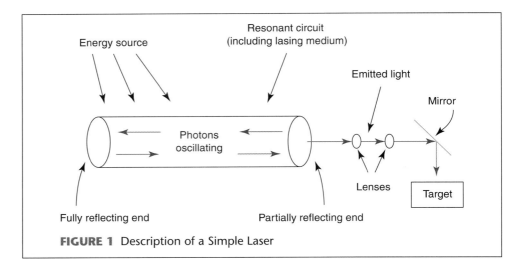

FIGURE 1 Description of a Simple Laser

less the visual takes up one whole page or more, do not place it on a separate page. Include the visual near its discussion.

Comparison and Contrast

Comparisons and contrasts help readers understand. Analogies (a type of comparison) to something familiar can help explain the unfamiliar:

"Does it resemble anything familiar?"

> To visualize how a simplified earthquake starts, imagine an enormous block of gelatin with a vertical knife slit through the middle of its lower half. Gigantic hands are slowly pushing the right side forward and pulling the left side back along the slit, creating a strain on the upper half of the block that eventually splits it. When the split reaches the upper surface, the two halves of the block spring apart and jiggle back and forth before settling into a new alignment. Inhabitants on the upper surface would interpret the shaking as an earthquake. ("Earthquake Hazard Analysis" 8)
>
> The average diameter of an optical cable is around two-thousandths of an inch, making it about as fine as a hair on a baby's head (Stanton 29–30).

Here is a contrast between optical fiber and conventional copper cable:

"How does it differ from comparable things?"

> Beams of laser light coursing through optical fibers of the purest glass can transmit many times more information than the present communications systems. . . . A pair of optical fibers has the capacity to carry more than 10,000 times as many signals as conventional copper cable. A $\frac{1}{2}$-inch optical cable can carry as much information as a copper cable as thick as a person's arm. . . .
>
>   Not only does fiber optics produce a better signal, [but] the signal travels farther as well. All communications signals experience a loss of power, or attenuation, as they move along a cable. This power loss necessitates placement of repeaters at one- or two-

mile intervals of copper cable in order to regenerate the signal. With fiber, repeaters are necessary about every thirty or forty miles, and this distance is increasing with every generation of fiber. (Stanton 27–28)

Here is a combined comparison and contrast:

"How is it both similar *and* different?"

Fiber optics technology results from the superior capacity of lightwaves to carry a communications signal. Sound waves, radio waves, and light waves can all carry signals; their capacity increases with their frequency. Voice frequencies carried by telephone operate at 1000 cycles per second, or hertz. Television signals transmit at about 50 million hertz. Light waves, however, operate at frequencies in the hundreds of trillions of hertz. (Stanton 28)

Required Materials or Conditions
Some items or processes need special materials and handling, or they may have other requirements or restrictions. An expanded definition should include this important information.

"What is needed to make it work (or occur)?"

Besides training in engineering, physics, or chemistry, careers in laser technology require a strong background in optics (study of the generation, transmission, and manipulation of light).

Abstract concepts might also be defined in terms of special conditions:

To be held guilty of libel, a person must have defamed someone's character through written or pictorial statements.

Example
Familiar examples showing types or uses of an item can help clarify your definition. This example shows how laser light is used as a heat-generating device:

"How is it used or applied?"

Lasers are increasingly used to treat health problems. Thousands of eye operations involving cataracts and detached retinas are performed every year by ophthalmologists . . . . Dermatologists treat skin problems. . . . Gynecologists treat problems of the reproductive system, and neurosurgeons even perform brain surgery—all using lasers transmitted through optical fibers. (Gartaganis 24–25)

The next example shows how laser light is used to carry information:

The use of lasers in the calculating and memory units of computers, for example, permits storage and rapid manipulation of large amounts of data. And audiodisc players use lasers to improve the quality of the sound they reproduce. The use of optical cable to transmit data also relies on lasers. (Gartaganis 25)

And this final example shows how optical fiber can relay a video signal:

Acting, in essence, as tiny cameras, optical fibers can be inserted into the body and relay an image to an outside screen. (Stanton 28)

Examples are a powerful communication tool—as long as you tailor the examples to your audience's level of understanding.

Whichever expansion strategies you use, be sure to document your information sources.

SITUATIONS REQUIRING DEFINITIONS

The following definitions employ expansion strategies appropriate to their audiences' needs (and labeled in the margin). Each definition, like a good essay, is unified and coherent: Each paragraph is developed around one main idea and logically connected to other paragraphs. Visuals are incorporated. Transitions emphasize the connection between ideas. Each definition is at a level of technicality that connects with the intended audience.

This example is preceded by an audience and use profile based on the worksheet on page 29.

An Expanded Definition for Semitechnical Readers

Audience and Use Profile.   The intended users of this material are beginning student mechanics. Before they can repair a solenoid, they will need to know where the term *solenoid* comes from, what a solenoid looks like, how it works, how its parts operate, and how it is used. This definition is designed as an *introduction,* so it offers only a general (but comprehensive) view of the mechanism.

Because the users are not engineering students, they do *not* need details about electromagnetic or mechanical theory (e.g., equations or graphs illustrating voltage magnitudes, joules, lines of force). ▲

EXPANDED DEFINITION: SOLENOID

Formal sentence definition

Etymology

A solenoid is an electrically energized coil that forms an electromagnet capable of performing mechanical functions. The term "solenoid" is derived from the word "sole," which in reference to electrical equipment means "a part of," or "contained inside, or with, other electrical equipment." The Greek word *solenoides* means "channel," or "shaped like a pipe."

Description and analysis of parts

A simple plunger-type solenoid consists of a coil of wire attached to an electrical source, and an iron rod, or plunger, that passes in and out of the coil along the axis of the spiral. A return spring holds the rod outside the coil when the current is deenergized, as shown in Figure 1.

Special conditions and operating principle

When the coil receives electric current, it becomes a magnet and thus draws the iron rod inside, along the length of its cylindrical center. With a lever attached to its end, the rod can transform electrical energy into mechanical force. The amount of mechanical force produced is the product of the number of turns in the coil, the strength of the current, and the magnetic conductivity of the rod.

Example and analysis of parts

The plunger-type solenoid in Figure 1 is commonly used in the starter motor of an automobile engine. This type is $4\frac{1}{2}$ inches long and 2 inches in diameter, with a steel cas-

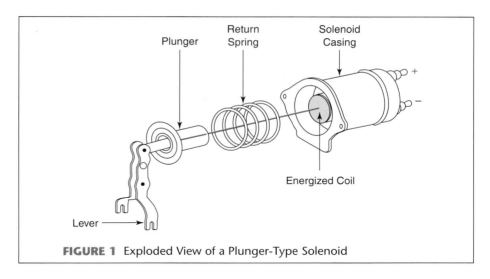

FIGURE 1 Exploded View of a Plunger-Type Solenoid

Explanation of
visual

ing attached to the casing of the starter-motor. A linkage (pivoting lever) is attached at one end to the iron rod of the solenoid, and at the other end to the drive gear of the starter, as shown in Figure 2. When the ignition key is turned, current from the battery is supplied to the solenoid coil, and the iron rod is drawn inside the coil, thereby shifting the attached linkage. The linkage, in turn, engages the drive gear, activated by the starter motor, with the flywheel (the main rotating gear of the engine).

Comparison of
sizes and
applications

Because of the solenoid's many uses, its size varies according to the work it must do. A small solenoid will have a small wire coil, hence a weak magnetic field. The larger the coil, the stronger the magnetic field; in this case, the rod in the solenoid can do harder work. An electronic lock for a standard door would, for instance, require a much smaller solenoid than one for a bank vault.

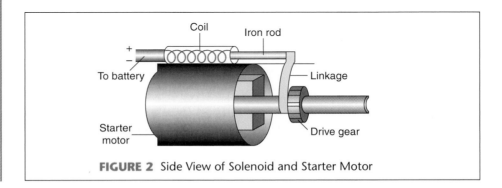

FIGURE 2 Side View of Solenoid and Starter Motor

The audience for the following definition (an entire community) is too diverse to define precisely, so the writer wisely addresses the lowest level of technicality— to ensure that all readers will understand.

An Expanded Definition for Nontechnical Readers

Audience and Use Profile.   The following definition is written for members of a community whose water supply (all obtained from wells, because the town has no reservoir) is doubly threatened: (1) by chemical seepage from a recently discovered toxic dump site, and (2) by a two-year drought that has severely depleted the water table. This definition forms part of a report that analyzes the severity of the problems and explores possible solutions.

To understand the problems, these users first need to know what a water table is, how it is formed, what conditions affect its level and quality, and how it figures into town planning decisions. The concepts of *recharge* and *permeability* are vital to understanding the problem here, so these terms are defined parenthetically. This audience has no interest in geological or hydrological (study of water resources) theory. They simply need a broad picture. ▲

EXPANDED DEFINITION: WATER TABLE

Formal sentence definition

Example

Operating principle

The water table is the level below the earth's surface at which the ground is saturated with water. Figure 1 shows a typical water table that might be found in the East. Wells driven into such a formation will have a water level identical to that of the water table.

The world's fresh-water supply comes almost entirely as precipitation that originates with the evaporation of sea and lake water. This precipitation falls to earth and follows one of three courses: It may fall directly onto bodies of water, such as rivers or lakes, where it is directly used by humans; it may fall onto land, and either evaporate or run over the ground to the rivers or other bodies of water; or it may fall onto land, be contained, and seep into the earth. The latter precipitation makes up the water table.

Comparison

Similar in contour to the earth's surface above, the water table generally has a level that reflects such features as hills and valleys. Where the water table intersects the ground surface, a stream or pond results.

Operating principle

A water table's level, however, will vary, depending on the rate of recharge (replacement of water). The recharge rate is affected by rainfall or soil permeability (the

FIGURE 1 A Typical Water Table (Eastern United States)

Example

Special conditions
and examples

Special conditions

ease with which water flows through the soil). A water table then is never static; rather, it is the surface of a body of water striving to maintain a balance between the forces which deplete it and those which replenish it. In areas of Florida and some western states where the water table is depleted, the earth caves in, leaving sinkholes.

The water table's depth below ground is vital in water resources engineering and planning. It determines an area's suitability for wastewater disposal, or a building lot's ability to handle sewage. A high water table could become contaminated by a septic system. Also, bacteria and chemicals seeping into a water table can pollute an entire town's water supply. Another consideration in water-table depth is the cost of drilling wells. These conditions obviously affect an industry's or homeowner's decision on where to locate.

The rising and falling of the water table give an indication of the pumping rate's effect on a water supply (drawn from wells) and of the sufficiency of the recharge rate in meeting demand. This kind of information helps water resources planners decide when new sources of water must be made available.

PLACEMENT OF DEFINITIONS

Poorly placed definitions interrupt the information flow. If you have only a few parenthetical definitions, place them after the terms. Any more than a few definitions per page will be disruptive. Rewrite them as sentence definitions and place them in a "Definitions" section of your introduction, or in a glossary. Definitions of terms in the report's title belong in your introduction.

Place expanded definitions in one of three locations:

Where to place
expanded
definitions

♦ If the definition is essential to the user's understanding of the *entire* document, place it in the introduction. A report titled "The Effects of Aerosol on the Earth's Ozone Shield" would require expanded definitions of "aerosol" and "ozone shield" early.

♦ When the definition clarifies a major part of your discussion, place it in that section of your report. In a report titled "How Advertising Influences Consumer Habits," "operant conditioning" might be defined early in the appropriate section. Too many expanded definitions *within* a report, however, can be disruptive.

♦ If the definition serves only as a reference, place it in an appendix. A report on fire safety in a public building might have an expanded definition of "carbon monoxide detectors" in an appendix.

Electronic documents pose special problems for placement of definitions. In a hypertext document, for instance, each reader explores the material differently. One option for making definitions available when they are needed is the "pop-up note": The term to be defined is highlighted in the text, to indicate that its definition can be called up and displayed in a small window on the actual text screen (Horton, "Is Hypertext" 25).

Usability Checklist FOR DEFINITIONS

(Numbers in parentheses refer to the first page of discussion.)

Content

☐ Is the type of definition (parenthetical, sentence, expanded) suited to its purpose and users' needs? (424)
☐ Does the definition adequately classify the item? (426)
☐ Does the definition clarify, rather than obscure, the meaning? (425)
☐ Is the expanded definition adequately developed? (426)
☐ Are all data sources documented? (432)
☐ Are visuals employed adequately and appropriately? (429)
☐ Does the sentence definition describe features that distinguish the item from all other items in the same class? (426)

Arrangement

☐ Is the expanded definition unified and coherent (like an essay)? (432)
☐ Are transitions adequate? (646)
☐ Does the definition appear in the appropriate location? (435)

Style and Page Design

☐ Is the definition in plain English? (425)
☐ Will the level of technicality connect with the audience? (24)
☐ Are sentences clear, concise, and fluent? (215)
☐ Is word choice precise? (240)
☐ Is the definition grammatical? (Appendix C)
☐ Is the definition ethically acceptable? (424)
☐ Is the page design inviting and accessible? (305)

EXERCISES

1. Sentence definitions require precise classification and differentiation. Is each of these definitions adequate for a layperson? Rewrite those that seem inadequate. Consult dictionaries and encyclopedias as needed.

   a. A bicycle is a vehicle with two wheels.
   b. A transistor is a device used in transistorized electronic equipment.
   c. Surfing is when one rides a wave to shore while standing on a board specifically designed for buoyancy and balance.
   d. Bubonic plague is caused by an organism known as *pasteurella pestis.*
   e. Mace is a chemical aerosol spray used by the police.
   f. A Geiger counter measures radioactivity.
   g. A cactus is a succulent.
   h. In law, an indictment is a criminal charge against a defendant.
   i. A prune is a kind of plum.
   j. Friction is a force between two bodies.
   k. Luffing is what happens when one sails into the wind.
   l. A frame is an important part of a bicycle.
   m. Hypoglycemia is a medical term.
   n. An hourglass is a device used for measuring intervals of time.
   o. A computer is a machine that handles information with amazing speed.
   p. A Ferrari is the best car in the world.
   q. To meditate is to exercise mental faculties in thought.

2. Standard dictionaries define for the layperson, whereas specialized reference books define for the specialist. Choose an item in your field and copy the definition (1) from a standard dictionary and (2) from a technical reference book. For the technical definition, label each expansion strategy. Rewrite the specialized definition for a layperson.

3. Using reference books as necessary, write sentence definitions for these terms or for terms from your field.

generator	gyroscope
dewpoint	coronary bypass
microprocessor	oil shale
capitalism	chemotherapy
local area network	estuary
marsh	Boolean logic
artificial intelligence	classical conditioning
economic inflation	hypothermia
anorexia nervosa	thermistor
low-impact camping	aquaculture
hemodialysis	nuclear fission
modem	

4. Select an item from the list in Exercise 3 or from an area of interest. Identify an audience and purpose. Complete an audience and use profile sheet (page 29). Begin with a sentence definition of the term. Then write an expanded definition for a first-year student in that field. Next, write the same definition for a layperson (client, patient, or other interested party). Leave a margin at the left side of your page to list expansion strategies. Use at least four expansion strategies in each version, including at least one visual or an art brief (page 282) and a rough diagram. In preparing each version, consult no fewer than four outside references. Cite and document each source, using one of the documentation styles discussed in Appendix A. Submit, with your two versions, an explanation of your changes from the first version to the second.

5. Figure 21.2 shows a page from a brochure titled *Cogeneration*. The brochure provides an expanded definition for potential users of fuel conservation systems engineered and packaged by Ewing Power Systems. The intended users are plant engineers and other technical experts unfamiliar with cogeneration.

Another page of the brochure is designed in a question-and-answer format. Figure 21.3 shows parts of that page.

Identify each expansion strategy in Figures 21.2 and 21.3. Is the definition appropriate for a technical audience? Why, or why not? Be prepared to discuss your analysis and evaluation in class.

COLLABORATIVE PROJECTS

1. Divide into small groups by majors or interests. Appoint one person as group manager. Decide on an item, concept, or process that would require expanded definition for laypersons.

EXAMPLES

From computer science: an algorithm, an applications program, artificial intelligence, binary coding, top-down procedural thinking, or systems analysis.

From nursing: a pacemaker, coronary bypass surgery, or natural childbirth.

Complete an audience and use profile (page 434).

Once your group has decided on the appropriate expansion strategies (etymology, negation, etc.), the group manager will assign each member to work on one or two specific strategies as part of the definition. As a group, edit and incorporate the collected material into an expanded definition, revising as often as needed.

The group manager will assign one member to present the definition in class, using either opaque or overhead projection, a large-screen monitor, or photocopies.

2. Beyond informative definition, your group's goal in this project is to develop a *persuasive* definition.

Assume that you work for an organization that has recently formulated a policy to eliminate sexual harassment. Your charge as the communications group is to develop printed material that will publicize this new policy. Your specific task is to develop an expanded

TECHNICAL CONSIDERATIONS

Turbine generator sets make electricity by converting a steam pressure drop into mechanical power to spin the generator. Conceptually, steam turbines work much the same way as water turbines. Just as water turbines take the energy from water as it flows from a high elevation to a lower elevation, steam turbines take the energy from steam as it flows from high pressure to low pressure. The amount of energy that can be converted to electricity is determined by the difference between the inlet pressure and the exhaust pressure (pressure drop) and the volume of steam flowing through the turbine.

Steam turbines have been used in industry in a variety of applications for decades and are the most common way utilities generate electricity. Exactly how a steam turbine generator can be used in your plant depends upon your circumstances.

IF YOU USE WASTE AS A BOILER FUEL

If you use wood waste or incinerator waste as a boiler fuel you can afford to condense turbine exhaust steam in a condenser. This allows you to convert waste fuel into electricity.

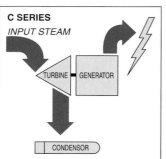

The simplest form of a condensing turbine generator set is the Ewing Power Systems C Series. All surplus steam enters the turbine at high pressure and exhausts to a condenser at a very low pressure, usually a vacuum. Because of the very low exhaust pressure, the pressure drop through the turbine is greater and more energy is extracted from each pound of steam. This is the same basic design as utilities use to produce power. The condenser can be either air or water cooled. In water cooled systems the "cooling" water can be hot enough for use as process hot water or for space heat.

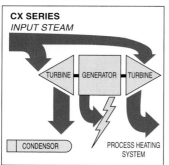

In situations where there is surplus fuel and also a need for low pressure process steam, the Ewing Power Systems CX Series is the system of choice. This arrangement includes a back pressure turbine and a condensing turbine connected to a common generator. Low pressure process or space heating loads are met with the back pressure turbine while surplus steam is directed to the condensing turbine to maximize power production.

IF YOU PURCHASE BOILER FUEL SUCH AS OIL OR GAS

If oil or gas is used as boiler fuel the best use of a turbine is as a replacement for a steam pressure reducing valve. Many plants produce steam at high pressure and then use some or all of the steam at low pressure after passing it through a pressure reducing valve. Other plants have high pressure boilers but run them at low pressure because they do not need high pressure steam for their process. In either case, a turbine generator can turn the pressure drop energy potential into electricity.

The Ewing Power Systems BP Series turbine generator sets are designed for pressure reducing (back pressure) applications. Very little energy is consumed by the turbine, so most of the inlet steam is available for process. The turbine generator uses about 3631 BTU's per hour for each kilowatt-hour produced. At 40 cents per gallon for No. 6 fuel oil and 85% boiler efficiency, it will cost about 1.1 cents per kilowatt-hour to generate your own power with a BP Series turbine. Generating costs for gas-fired boilers are similar.

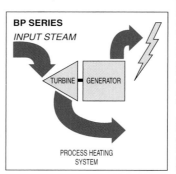

To generate power at this very low cost, all the exhaust steam must be used productively. Generator output is therefore completely governed by process steam demand. For example, if steam is used for space heating you will make more electricity on cold days than on warmer days because more steam will flow through the turbine.

ELECTRICAL CONSIDERATIONS

In most cases the generator will be connected to your plant electrical system and to your utility. This means that you will not give up the security of utility power. It also means that you do not have to generate *all* your own power; most cogeneration systems provide only part of the plant load. The more power the generator is making, the less you buy from the utility. If you make more power than you use, you will be able to sell the excess to the utility. If your generator is off-line for any reason, you will be able to buy power from the utility, just as you do now.

There are two primary generator designs: induction and synchronous. Induction generators are similar to induction motors and are much simpler than synchronous generators. Synchronous generators require more elaborate controls and are usually more expensive but offer the advantage of stand-alone capability. Whereas induction generators cannot operate unless they are connected to a utility grid, synchronous sets can be operated in isolation as emergency units or when it is economically advantageous to avoid interconnection with the utility.

All turbine generator sets from Ewing Power Systems include a complete electrical control panel. Our standard panels meet most utility interconnection requirements and we will customize the panel to meet unusual requirements. Synchronous panels can be built for full utility paralleling, stand-alone capability, or both.

Figure 21.2   Expanded Definition in a Technical Brochure   *Source: Courtesy of Ewing Power Systems, So. Deerfield, MA 01373.*

definition of *sexual harassment* to be published in the company newsletter.

At this stage, the company is plagued by confusion, misunderstanding, resentment, and paranoia about the harassment issue. Therefore, beyond compliance with legal requirements, your organization seeks to improve gender relations between coworkers, in the hope of boosting productivity.

Thus, you face an informative, persuasive, and ethical challenge: to move beyond the usual matters of clarity so that your definition promotes real understanding and reconciliation. In short, you need to amplify the legal definition so that employees are able to recog-nize harassment and to understand clearly what does and what does not constitute harassment. But unless you also do something to change their us-against-them *attitude*, you will only create greater tension and overreaction.

In other words, you want your definition not merely to *inhibit* behavior, but to *enlighten* the readers. Insensitive readers, of course, are likely to change their behavior only if they feel coerced. But appeals to fear almost always have limited success, and people generally tend to be reasonable in the long run. You want your definition to have rational appeal, to cause people to *internalize* the values that underlie the issue. No sermons, please.

Q. What is cogeneration?

A. It is the simultaneous production of electricity and useful thermal energy. This means that you can generate electricity with the same steam you are now using for heating or process. *You can use the same steam twice.* In modern usage cogeneration has also come to mean using waste fuel for in-plant electricity generation.

Q. How does it save money?

A. Cogeneration saves money by allowing you to produce your own electricity for a fraction of the cost of utility power. Cogenerated power is cheaper because cogeneration systems are much more efficient than central utility plants. By using the same steam twice, cogeneration systems can achieve efficiencies of up to 80%, whereas the best utilities can do is about 40%.

Q. Are there other benefits to cogeneration?

A. Yes. For companies using waste fuel, elimination of waste disposal can be a very important benefit. Depending upon design, a cogeneration system can provide emergency standby power and can smooth out boiler load swings.

Q. Is cogeneration new?

A. No. It's been done ever since the beginning of the electrification of industrial America. Originally, most electric power was cogenerated by individual manufacturers, not the utilities. In the 1920s and '30s, as cheaper utility electricity became available, cogeneration waned. With cheap oil available, power rates continued to decline through the 1950s and '60s. Then came the 1973–74 Arab Oil Embargo. Everything changed abruptly. Since then electricity prices have risen. Further upward pressure was produced by some utilities' nuclear power plant building programs. Now many companies are getting back to their original source of power: cogeneration.

Figure 21.3   Expanded Definition in a Technical Brochure   *Source: Courtesy of Ewing Power Systems, So. Deerfield, MA 01373.*

CHAPTER 22

Technical Descriptions and Specifications

PURPOSE OF TECHNICAL DESCRIPTION

ELEMENTS OF USABLE DESCRIPTION

A GENERAL MODEL FOR TECHNICAL DESCRIPTION

A SITUATION REQUIRING TECHNICAL DESCRIPTION

SPECIFICATIONS

Usability Checklist **for Descriptions**

Description creates a word picture. Because it helps an audience *visualize,* description is a common denominator in all writing.

Technical description conveys information about a product or mechanism to someone who will use it, buy it, operate it, assemble it, or manufacture it, or who needs to know more about it. Any item can be visualized from countless perspectives. Therefore, *how* you describe—your perspective—depends on your purpose and the user's needs.

PURPOSE OF TECHNICAL DESCRIPTION

Manufacturers use technical descriptions to sell products; banks require detailed descriptions of any business or construction venture before approving a loan; medical personnel maintain daily or hourly descriptions of a patient's condition and treatment.

What an audience expects to learn from a technical description

No matter what the subject of description, it should answer questions like these: *What is it? What does it do? What does it look like? What is it made of? How does it work? How is it put together?* The description in Figure 22.1, part of an installation and operation manual, answers applicable questions for do-it-yourself homeowners.

ELEMENTS OF USABLE DESCRIPTION

Clear and Limiting Title

Give an immediate forecast

The title "A Description of a Velo Ten-Speed Racing Bicycle" promises a comprehensive description. If you intend to describe the braking mechanism only, be sure your title indicates this: "A Description of the Velo's Center-Pull Caliper Braking Mechanism."

Overall Appearance and Component Parts

Show the whole item before describing each part.

Begin with the big picture

> The standard stethoscope is roughly 24 inches long and weighs about 5 ounces. The instrument consists of a sensitive sound-detecting and amplifying device whose flat surface is pressed against a bodily area. This amplifying device is attached to rubber and metal tubing that transmits the body sound to a listening device inserted in the ear.
>
> The stethoscope's Y-shaped structure contains seven interlocking pieces: (1) diaphragm contact piece, (2) lower tubing, (3) Y-shaped metal piece, (4) upper tubing, (5) U-shaped metal strip, (6) curved metal tubing, and (7) hollow ear plugs. These parts form a continuous unit (Figure 1).

Visuals

Use drawings, diagrams, or photographs generously. Note how the description of the stethoscope is greatly clarified by Figure 1.

Visuals generated by computers (for example, 3-D drawing or architectural drafting programs) are particularly appropriate for technical descriptions. Other

Let the visual
repeat, restate, or
reinforce the prose

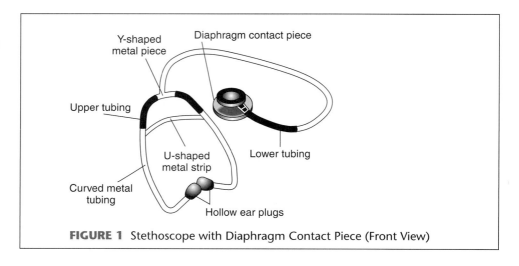

FIGURE 1 Stethoscope with Diaphragm Contact Piece (Front View)

sources for graphics in technical descriptions include clip art, electronic scanning, and downloading from the Internet. (See page 294 for a sampling of useful Web sites and discussion of legal issues regarding the use of computer graphics.)

Function of Each Part
Explain what each part does and how it relates to the whole.

Describe each part

> Body sounds cause the diaphragm contact piece to vibrate. The diaphragm contact piece receives, amplifies, and transmits the sound impulse.

In Figure 22.1 notice how steps 1–6 spell out the role of each part in heating the water and storing it for use.

Appropriate Level of Technicality
Give enough detail to create a clear picture, but omit unnecessary information. Identify your audience and its reasons for using your description.

Give users exactly
and only what they
need

    Assume that you want to describe a specific bicycle model. The picture you create will depend on the details you select. How will your audience use this description? What is the audience's level of technical understanding? Is this a customer interested in the bike's appearance—its flashy looks and racy style? Is it a repair technician who needs to know how parts operate? Is it a helper in your bicycle shop who needs to know how to assemble this bike? Or is it the bike's manufacturer who needs precise specifications (page 452)?

    The description of the hot water maker in Figure 22.1 focuses on *what this model looks like, what it's made of,* and *how it works.* Its intended audience of do-it-yourselfers will already know what a hot water maker is and what it does. That audience needs no background. A description of *how it was put together* appears with the installation and maintenance instructions later in the manual.

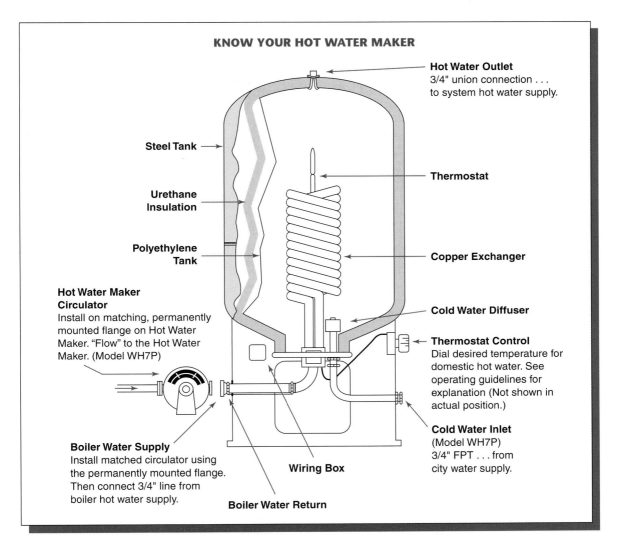

Figure 22.1   A Mechanism Description.   *Source: Courtesy of AMTROL, Inc.*

In contrast, *specifications* for manufacturing the hot water maker would describe each part in exacting detail (e.g., the steel tank's required thickness and pressure rating as well as required percentages of iron, carbon, and other constituents in the steel alloy).

Objectivity

Objective description filters out personal impressions to whatever extent is appropriate, focusing instead on observable details ("The drilling crew worked in freezing rain and gale-force winds"). Subjective description colors objective details with

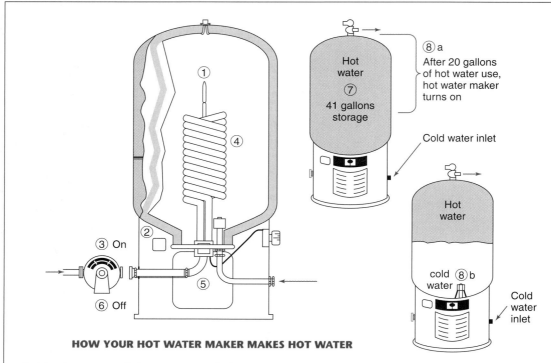

HOW YOUR HOT WATER MAKER MAKES HOT WATER

1. The thermostat calls for energy to make hot water in your Hot Water Maker.

2. The built-in relay signals your boiler/burner to generate energy by heating boiler water.

3. The Hot Water Maker circulator comes on and circulates hot boiler water through the inside of the Hot Water Maker heat exchanger.

4. Heat energy is transferred, or "exchanged" from the boiler water inside the exchanger to the water surrounding it in the Hot Water Maker.

5. The boiler water after the maximum of heat energy is taken out of it, is returned to the boiler so it can be reheated.

6. When enough heat has been exchanged to raise the temperature in your Hot Water Maker to the desired temperature, the thermostat will de-energize the relay and turn off the Hot Water Maker circulator and your boiler/burner. This will take approximately 23 minutes—when you first

start up the Hot Water Maker. During use, reheating will be approximately 9–12 minutes.

7. You now have 41 gallons of hot water in storage . . . ready for use in washing machines, showers, sinks, etc. This 41 gallons of hot water will stay hot up to 10 hours, if you don't use it, without causing your boiler/burner to come on. (Unless, of course, you need it for heating your home in the winter.)

8. When you do use hot water, you will be able to use approximately 20 gallons, before the Hot Water Maker turns on. Then you will still have 21 gallons of hot water left for use, as your Hot Water Maker "recoups" 20 gallons of cold water. This means, during normal use (3 1/2 GPM Flow), you will never run out of hot water.

9. You can expect substantial energy savings with your Hot Water Maker, as its ability to store hot water and efficiently transfer energy to make more hot water will keep your boiler off for longer periods of time.

Figure 22.1   A Mechanism Description   *Continued*

personal impressions and metaphors. It usually strives to create a mood or share a feeling about the subject ("The weather was miserable").

Except for promotional writing, technical descriptions should be impartial, if they are to be ethical. Pure objectivity is, of course, humanly impossible. Each observer has a unique perspective on the facts and their meaning, and therefore chooses what to put in and what to leave out. Nonetheless, we are expected to communicate the facts as we know and understand them. Even positive claims made in promotional writing (for example, "reliable," "rugged," and so on in Figure 22.5) should be based on objective and verifiable evidence.

NOTE   *Remaining "objective" does not mean forsaking personal evaluation in cases in which a product may be unsafe or unsound. An ethical communicator "is obligated to express her or his opinions of products, as long as these opinions are based on objective and responsible research and observation" (Mackenzie 3).*

The following suggestions will help you to achieve objectivity in writing technical descriptions.

SELECT DETAILS THAT ARE CONCRETE AND SPECIFIC ENOUGH TO CONVEY AN UNMISTAKABLE PICTURE.

Focus on details any observer could recognize, details a camera would record. Avoid details that provide no distinct visual image, as in the italicized words below:

Subjective

His office has an *awful* view, *terrible* furniture, and a *depressing* atmosphere.

This next version provides a clear and exact picture:

Objective

His office has broken windows looking out on a brick wall, a rug with a six-inch hole in the center, missing floorboards, broken chairs, and a ceiling with chunks of plaster missing.

USE PRECISE AND INFORMATIVE LANGUAGE.

Instead of "large," "long," and "near," give exact measurements, weights, dimensions, and ingredients. Specify location and spatial relationships: "above," "oblique," "behind," "tangential," "adjacent," "interlocking," "abutting," and "overlapping." Specify position: "horizontal," "vertical," "lateral," "longitudinal," "in cross-section," "parallel." Avoid judgmental words ("impressive," "poor"), unless your judgment is requested and can be supported by facts.

Notice how the precise words below help users to *visualize:*

Indefinite	Precise
a late-model car	a 1999 Lexus ES 300
an inside view	a cross-sectional, cutaway, or exploded view

next to the foundation   adjacent to the right side
a small red thing   a red activator button with a 1-inch diameter and a
                           concave surface

Never confuse precision with overly complicated technical terms or needless jargon. Don't say "phlebotomy specimen" instead of "blood," or "thermal attenuation" instead of "insulation," or "proactive neutralization" instead of "damage control." The clearest writing aimed at laypersons uses precise but plain language—as long as the simpler words do the job.

Clearest Descriptive Sequence

Organize for the user's understanding

Any item usually has its own logic of organization, based on (1) the way it appears as a static object, (2) the way its parts operate in order, or (3) the way its parts are assembled. These relationships—spatial, functional, and chronological—are discussed below.

A spatial sequence parallels the user's angle of vision in viewing the item

SPATIAL SEQUENCE.   Part of all physical descriptions, a spatial sequence answers these questions: *What is it? What does it do? What does it look like? What parts and material is it made of?* Use this sequence when you want users to visualize the item as a static item or mechanism at rest (an office interior, a document, the Statue of Liberty, a plot of land, a chainsaw, or a computer keyboard). Can this item best be visualized from front to rear, left to right, top to bottom? (What logical path do the parts create?) A retractable pen would logically be viewed from outside to inside. The specifications in Figure 22.2 on page 454 proceed from the ground upward.

A functional sequence parallels the order in which parts operate

FUNCTIONAL SEQUENCE.   The functional sequence answers: *How does it work?* It is best used in describing a mechanism in action, such as a 35-millimeter camera, a nuclear warhead, a smoke detector, or a car's cruise-control system. The logic of the item is reflected by the order in which its parts function. Like the hot water maker in Figure 20.1, a mechanism usually has only one functional sequence. The stethoscope description on page 441 follows the sequence of parts through which sound travels.

When describing a solar home-heating system, you would begin with the heat collectors on the roof, moving through the pipes, pumping system, and tanks for the heated water, to the heating vents in the floors and walls—from source to outlet. After this functional sequence of operating parts, you could describe each part in a spatial sequence.

A chronological sequence parallels the order in which parts are assembled

CHRONOLOGICAL SEQUENCE.   A chronological sequence answers: *How is it put together?* Use the chronological sequence for an item that is best visualized in terms of its order of assembly (such as a piece of furniture, an umbrella tent, or a prehung window or door unit). Architects might find a spatial sequence best for describing a proposed beach house to clients; however, they would use a chronological sequence (of blueprints) for specifying for the builder the prescribed dimensions, materials, and construction methods at each stage.

COMBINED SEQUENCES. The description of a bumper jack on page 450 alternates among all three sequences; a spatial sequence (bottom to top) for describing the overall mechanism at rest, a chronological sequence for explaining the order in which the parts are assembled, and a functional sequence for describing the order in which the parts operate.

A GENERAL MODEL FOR TECHNICAL DESCRIPTION

Description of a complex mechanism almost invariably calls for an outline. This model is adaptable to any description.

   I. Introduction: General Description[1]
     A. Definition, Function, and Background of the Item
     B. Purpose (and Audience—for classroom only)
     C. Overall Description (with general visuals, if applicable)
     D. Principle of Operation (if applicable)
     E. List of Major Parts

   II. Description and Function of Parts
     A. Part One in Your Descriptive Sequence
       1. Definition
       2. Shape, dimensions, material (with specific visuals)
       3. Subparts (if applicable)
       4. Function
       5. Relation to adjoining parts
       6. Mode of attachment (if applicable)
     B. Part Two in Your Descriptive Sequence (and so on)

   III. Summary and Operating Description
     A. Summary (used only in a long, complex description)
     B. Interrelation of Parts
     C. One Complete Operating Cycle

You might modify, delete, or combine certain parts of this outline to suit your subject, purpose, and audience.

Introduction: General Description

Give users only as much background as they need to grasp the picture.

---

[1]In most descriptions, the subdivisions in the introduction can be combined and need not appear as individual headings in the document.

A DESCRIPTION OF THE STANDARD STETHOSCOPE
Introduction

Definition and function

History and background

Purpose and audience

The stethoscope is a listening device that amplifies and transmits body sounds to aid in detecting physical abnormalities.

This instrument has evolved from the original wooden, funnel-shaped instrument invented by a French physician, R. T. Lennaec, in 1819. Because of his female patients' modesty, he found it necessary to develop a device, other than his ear, for auscultation (listening to body sounds).

This report explains to the beginning paramedical or nursing student the structure, assembly, and operating principle of the stethoscope. [*Omit this section if you submit an audience and use profile to your instructor or if you write for a workplace audience.*]

To complete the introduction, give a brief, overall description of the item, discuss its principle of operation, and list its major parts, as in the overall stethoscope description on page 441.

Description and Function of Parts

The body of your text describes each major part. After arranging the parts in sequence, follow the logic of each part. Provide only as much detail as users need.

Users of this description will use a stethoscope daily, so they need to know how it works, how to take it apart for cleaning, and how to replace worn or broken parts. (Specifications for the manufacturer would require many more technical details—dimensions, alloys, curvatures, tolerances, and so on.)

Definition, size, shape, and material

Subparts

Diaphragm Contact Piece.   The diaphragm contact piece is a shallow metal bowl, about the size of a silver dollar (and twice its thickness). Various body sounds cause it to vibrate.

Three separate parts make up the piece: hollow steel bowl, plastic diaphragm, and metal frame, as shown in Figure 2.

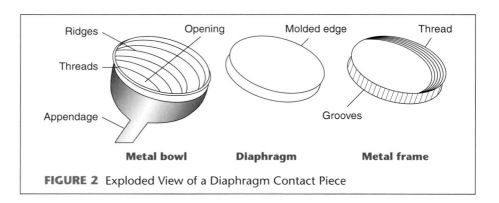

FIGURE 2 Exploded View of a Diaphragm Contact Piece

Function and relation to adjoining parts

Mode of attachment

The stainless steel metal bowl has a concave inner surface, with concentric ridges that funnel sound toward an opening in the tapered base, then out through the hollow appendage. Lateral threads ring the outer circumference of the bowl to accommodate the interlocking metal frame. A fitted diaphragm covers the bowl's upper opening.

The diaphragm is a plastic disk, 2 millimeters thick, 4 inches in circumference, with a molded lip around the edge. It fits flush over the metal bowl and vibrates sound toward the ridges. A metal frame that screws onto the bowl holds the diaphragm in place.

The stainless steel frame fits over the disk and metal bowl. A $\frac{1}{4}$-inch ridge between the inner and outer edge accommodates threads for screwing the frame to the bowl. The frame's outside circumference is notched with equally spaced, perpendicular grooves—like those on the edge of a dime—to provide a gripping surface.

The diaphragm contact piece receives, amplifies, and transmits sound through the system of attached tubing. The piece attaches to the lower tubing by an appendage on its apex (narrow end), which fits inside the tubing.

Each part of the stethoscope, in turn, is described according to its own logic of organization.

Summary and Operating Description
Conclude by explaining how the parts work together to make the whole item function.

How parts interrelate

One complete operating cycle

Summary and Operating Description
The seven major parts of the stethoscope provide support for the instrument, flexibility of movement for the operator, and ease in use.

In an operating cycle, the diaphragm contact piece, placed against the skin, picks up sound impulses from the body's surface. These impulses cause the plastic diaphragm to vibrate. The amplified vibrations, in turn, are carried through a tube to a dividing point. From here, the amplified sound is carried through two separate but identical series of tubes to hollow ear plugs.

A SITUATION REQUIRING TECHNICAL DESCRIPTION

The following description of an automobile jack, aimed toward a general audience, follows the outline model.

A Mechanism Description for a Nontechnical Audience
Audience and Use Profile.   Some users of this description (written for an owner's manual) will have no mechanical background. Before they can follow instructions for *using* the jack safely, they will have to learn what it is, what it looks like, what its parts

are, and how, generally, it works. They will *not* need precise dimensions (e.g., "The rectangular base is 8 inches long and $6\frac{1}{2}$ inches wide, sloping upward $1\frac{1}{2}$ inches from the front outer edge to form a secondary platform 1 inch high and 3 inches square"). The engineer who designed the jack might include such data in specifications for the manufacturer. Laypersons, however, need only the dimensions that will help them recognize specific parts and understand their function, for safe use and assembly.

Also, this audience will need only the broadest explanation of how the leverage mechanism operates. Although the physical principles (*torque, fulcrum*) would interest engineers, they would be of little use to people who simply need to operate the jack safely. ▲

DESCRIPTION OF A STANDARD BUMPER JACK
Introduction—General Description

Definition, purpose, and function

The standard bumper jack is a portable mechanism for raising the front or rear of a car through force applied with a lever. This jack enables even a frail person to lift one corner of a 2-ton automobile.

Overall description (spatial sequence)

The jack consists of a molded steel base supporting a free-standing, perpendicular, notched shaft (Figure 1). Attached to the shaft are a leverage mechanism, a bumper catch, and a cylinder for insertion of the jack handle. Except for the main shaft and leverage mechanism, the jack is made to be dismantled. All its parts fit neatly in the car's trunk.

Operating principle

The jack operates on a leverage principle, with the operator's hand traveling 18 inches and the car only $\frac{3}{8}$ of an inch during a normal jacking stroke. Such a device requires many strokes to raise the car off the ground but may prove a lifesaver to a motorist on some deserted road.

List of major parts (chronological sequence)

Five main parts make up the jack: base, notched shaft, leverage mechanism, bumper catch, and handle.

Description of Parts and Their Function

First major part
Definition, shape, and material

Function and mode of attachment

Base.   The rectangular base is a molded steel plate that provides support and a point of insertion for the shaft (Figure 2). The base slopes upward to form a platform containing a 1-inch depression that provides a stabilizing well for the shaft. Stability is increased by a 1-inch cuff around the well. As the base rests on its flat surface, the bottom end of the shaft is inserted into its stabilizing well.

Second major part, etc.

Shaft.   The notched shaft is a steel bar (32 inches long) that provides a vertical track for the leverage mechanism. The notches, which hold the mechanism in position on the shaft, face the operator.

The shaft vertically supports the raised automobile, and attached to it is the leverage mechanism, which rests on individual notches.

Leverage Mechanism.   The leverage mechanism provides the mechanical advantage needed for the operator to raise the car. It is made to slide up and down the notched shaft. The main body of this pressed-steel mechanism contains two units: one for transferring the leverage and one for holding the bumper catch.

The leverage unit has four major parts: the cylinder, connecting the handle and a pivot point; a lower pawl (a device that fits into the notches to allow forward and prevent backward motion), connected directly to the cylinder; an upper pawl, connected at

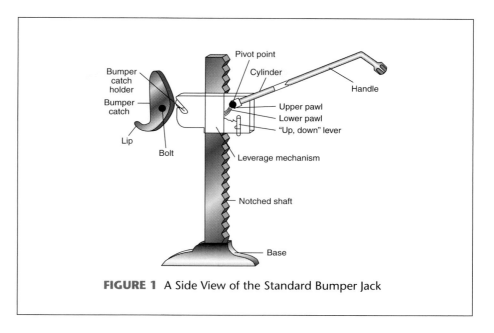

FIGURE 1 A Side View of the Standard Bumper Jack

the pivot point; and an "up-down" lever, which applies or releases pressure on the upper pawl by means of a spring (Figure 1). Moving the cylinder up and down with the handle causes the alternate release of the pawls, and thus movement up or down the shaft—depending on the setting of the "up-down" lever. The movement is transferred by the metal body of the unit to the bumper catch holder.

The holder consists of a downsloping groove, partially blocked by a wire spring (Figure 1). The spring is mounted in such a way as to keep the bumper catch in place during operation.

Bumper Catch.   The bumper catch is a 9-inch molded plate that attaches the leverage mechanism to the bumper and is bent to fit the shape of the bumper. Its outer $\frac{1}{2}$-inch is bent up to form a lip (Figure 1), which hooks behind the bumper to hold the

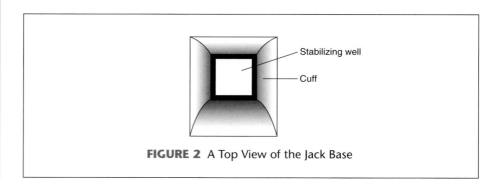

FIGURE 2 A Top View of the Jack Base

catch in place. The two sides of the plate are bent back 90 degrees to leave a 2-inch bumper contact surface, and a bolt is riveted between them. This bolt slips into the groove in the leverage mechanism and provides the attachment between the leverage unit and the car.

Jack Handle.   The jack handle is a steel bar that serves both as lever and lug bolt (or lugnut) remover. This round bar is 22 inches long, $\frac{5}{8}$ inch in diameter, and is bent 135 degrees roughly 5 inches from its outer end. Its outer end is a socket wrench made to fit the wheel's lug bolts. Its inner end is beveled to form a bladelike point for prying the wheel covers and for insertion into the cylinder on the leverage mechanism.

Conclusion and Operating Description

Assembly

One quickly assembles the jack by inserting the bottom of the notched shaft into the stabilizing well in the base, the bumper catch into the groove on the leverage mechanism, and the beveled end of the jack handle into the cylinder. The bumper catch is then attached to the bumper, with the lever set in the "up" position.

One complete operating cycle (functional sequence)

As the operator exerts an up-down pumping motion on the jack handle, the leverage mechanism gradually climbs the vertical notched shaft until the car's wheel is raised above the ground. When the lever is in the "down" position, the same pumping motion causes the leverage mechanism to descend the shaft.

SPECIFICATIONS

Airplanes, bridges, smoke detectors, and countless other items are produced according to certain *specifications*. A particularly exacting type of description, specifications (or "specs") prescribe standards for performance, safety, and quality. For virtually any product, specifications describe these features:

Descriptions contained in a typical set of specifications

- ◆ the methods for manufacturing, building, or installing the product
- ◆ the materials and equipment to be used
- ◆ the size, shape, and weight of the product

While technical descriptions are used to sell, instruct, record observations, or persuade to action, specifications seek to insure compliance with an authority's rules.

Specifications have ethical and legal implications

Because specifications define an "acceptable" level of quality, any product that does not meet these specs provides grounds for a lawsuit. When injury or death results (as in a bridge collapse caused by inferior reinforcement), the contractor, subcontractor, or supplier who cut corners is criminally liable.

How specifications originate

Federal and state regulatory agencies routinely issue specifications to ensure safety. The Consumer Product Safety Commission specifies that power lawn mowers be equipped with a "kill switch" on the handle, a blade guard to prevent foot injuries, and a grass thrower that aims downward to prevent eye and facial injury. This same agency issues specifications for baby products, as in governing the fire retardancy of pajama fabric. Passenger airline specifications for aisle width, seat belt configurations, and emergency equipment are issued by the Federal Aviation Administration. State and local agencies issue specifications in the form of building codes, fire codes, and other standards for safety and reliability.

Government departments (Defense, Interior, etc.) issue specifications for all types of military hardware and other equipment. A set of NASA specifications for spacecraft parts can be hundreds of pages long, prescribing the standards for even the smallest nuts and bolts, down to screw thread depth and width in millimeters.

The private sector issues specifications for countless products or projects, to help ensure that customers get exactly what they want. Figure 22.2 shows partial specifications drawn up by an architect for a medical-clinic building. This section of the specs covers only the structure's "shell." Other sections detail the requirements for plumbing, wiring, and interior finish work.

Specifications like those in Figure 22.2 must be clear enough for identical interpretation by a broad audience (Glidden 258–59):

Specifications address a diverse audience

- *the customer,* who has the big picture of what is needed and who wants the best product at the best price
- *the designer* (architect, engineer, computer scientist, etc.), who must translate the customer's wishes into the actual specification
- *the contractor or manufacturer,* who won the job by making the lowest bid, and so must preserve profit by doing only what is prescribed
- *the supplier,* who must provide the exact materials and equipment
- *the workforce,* who will do the actual assembly, construction, or installation (managers, supervisors, subcontractors, and workers—some working on only one part of the product, such as plumbing or electrical)
- *the inspectors* (such as building, plumbing, or electrical inspectors), who evaluate how well the product conforms to the specifications

Each of these parties needs to understand and agree on exactly *what* is to be done and *how* it is to be done. In the event of a lawsuit over failure to meet specifications, the readership broadens to include judges, lawyers, and jury. Figure 22.3 depicts how a set of clear specifications unifies all users (their various viewpoints, motives, and levels of expertise) in a shared understanding.

In addition to guiding a product's design and construction, specifications can facilitate the product's use and maintenance. For instance, specifications in a computer manual include the product's performance limits, or *ratings:* its power requirements; its work, processing, or storage capacity; operating environment requirements; the makeup of key parts; and so on. Product support literature for appliances, power tools, and other items routinely contains ratings to help customers select a good operating environment or replace worn or defective parts (Riney 186). The ratings in Figure 22.4 are taken from the owner's manual for an ink jet printer.

Specifications play a persuasive role in technical marketing by spelling out the product and service quality that potential customers can expect in return for their

Ruger, Filstone, and Grant Architects

SPECIFICATIONS FOR THE POWNAL CLINIC BUILDING

Foundation
    footings: 8" x 16" concrete (load-bearing capacity: 3,000 lbs. per sq. in.)
    frost walls: 8" x 4' @ 3,000 psi
    slab: 4" @ 3,000 psi, reinforced with wire mesh over vapor barrier

Exterior Walls
    frame: eastern pine #2 timber frame with exterior partitions set inside posts
    exterior partitions: 2" x 4" kiln-dried spruce set at 16" on center
    sheathing: 1/4" exterior-grade plywood
    siding: #1 red cedar with a 1/2" x 6' bevel
    trim: finished pine boards ranging from 1" x 4" to 1" x 10"
    painting: 2 coats of Clear Wood Finish on siding; trim primed and finished with one
            coat of bone white, oil base paint

Roof system
    framing: 2" x 12" kiln-dried spruce set at 24" on center
    sheathing: 5/8" exterior-grade plywood
    finish: 240 Celotex 20-year fiberglass shingles over #15 impregnated felt roofing paper
    flashing: copper

Windows
    Anderson casement and fixed-over-awning models, with white exterior cladding,
       insulating glass and screens, and wood interior frames

Landscape
    driveway: gravel base, with 3" traprock surface
    walks: timber defined, with traprock surface
    cleared areas: to be rough graded and covered with wood chips
    plantings: 10 assorted lawn plants along the road side of the building

Figure 22.2   Specifications for a Building Project (Partial)

Figure 22.3
Users and
Potential Users
of Specifications

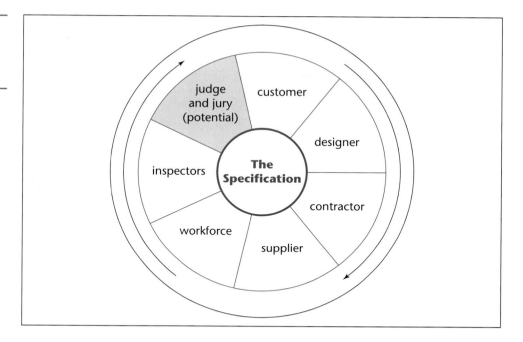

Figure 22.4
Specifications for
the Color
StyleWriter™
2400
*Source: Reprinted by
permission of Apple
Computer, Inc.*

General specifications

Marking engine
- Thermal ink jet engine

Resolution
- 360 dots per inch (dpi) for text and graphics (180 dpi for draft quality)

Engine speed
- Printing speed depends on the images printed and on the Macintosh computer used.

Connector cable
- Apple System/Peripheral-8 cable

Interface
- High-speed serial (RS-422)
- Optional LocalTalk

Paper feed in pounds (lb.) and grams/meter² (g/m²)
- Sheet feeder holds up to 100 sheets of 20-lb. (75-g/m²) paper or 15 envelopes.

investment. Figure 22.5 (pages 459–60), displays marketing literature that accompanies the expanded definition in Figures 21.2 and 21.3 (pages 438–39). After a brief introduction to the C Series Cogeneration System, the description focuses on the system's major components and specifications as well as the support services offered by the vendor. Representational diagrams and system schematics depict the system's general appearance and operating principles.

Usability Checklist FOR DESCRIPTIONS

Use this list to refine the content, arrangement, and style of your description. (Numbers in parentheses refer to first page of discussion.)

Content

☐ Does the title promise exactly what the description delivers? (441)
☐ Are the item's overall features described, as well as each part? (441)
☐ Is each part defined before it is discussed? (447)
☐ Is the function of each part explained? (442)
☐ Do visuals appear whenever they can provide clarification? (441)
☐ Will users be able to visualize the item? (445)
☐ Are any details missing, needless, or confusing for this audience? (442)
☐ Is the description ethically acceptable? (445)

Arrangement

☐ Does the description follow the clearest possible sequence? (446)
☐ Are relationships among the parts clearly explained? (448)

Style and Page Design

☐ Is the description sufficiently impartial? (443)
☐ Is the language informative and precise? (445)
☐ Is the level of technicality appropriate for the audience? (442)
☐ Is the description written in plain English? (446)
☐ Is each sentence clear, concise, and fluent? (215)
☐ Is the description written in grammatical English? (Appendix C)
☐ Is the page design inviting and accessible? (305)

EXERCISES

1. Select an item from this list or a device used in your major field. Using the general outline as a model, develop an objective description. Include (a) all necessary visuals or (b) a brief (page 282) and a rough diagram for each visual or (c) a "reference visual" (a copy of a visual published elsewhere) with instructions for adapting your visual from that one. (If you borrow visuals from other sources, provide full documentation.) Write for a specific use by a specified audience. Attach your written audience and use profile (based on the worksheet, page 29) to your document.

    soda-acid fire extinguisher
    breathalyzer
    sphygmomanometer
    transit
    Skinner box
    distilling apparatus

saber saw
hazardous waste site
brand of woodstove
photovoltaic panel
catalytic converter
radio

Remember, you are simply describing the item, its parts, and its function: *do not* provide directions for its assembly or operation.

As an optional assignment, describe a place you know well. You are trying to convey a visual image, not a mood; therefore, your description should be impartial, discussing only observable details.

2. The bumper jack description in this chapter is aimed toward a general audience. Evaluate it using the revision checklist. In one or two paragraphs, discuss your evaluation, and suggest revisions.

3. Figure 22.5 is designed to promote as well as describe a technical product. Answer the following questions about the document:

- When read in conjunction with Figures 21.2 and 21.3, is 22.5 an effective introduction to this particular product for its intended audience of engineers and other technical experts? Why or why not?
- Is the overall page design effective? Why or why not? Be specific. (See Chapter 15 and "Using Color," pages 290–93.)
- Are the visuals adequate and appropriate? Why or why not? (See Chapter 14). Why aren't the diagrams on page one labeled more extensively?
- Is this a sufficiently impartial description? Why or why not? Given its purpose as a marketing document, is the description ethically appropriate? (See Chapter 5.)

Be prepared to discuss your analysis and evaluation in class.

4. Locate a description and specifications for a particular brand of automobile or some other consumer product. Evaluate this material for promotional and descriptive value and ethical appropriateness.

COLLABORATIVE PROJECTS

1. Divide into small groups. Assume that your group works in the product-development division of a large and diversified manufacturing company.

Your division has just thought of an idea for an inexpensive consumer item with a potentially vast market. (Choose one from the following list.)
flashlight
nail clippers
retractable ballpoint pen
scissors
stapler
any simple mechanism—as long as it has moving parts

Your group's assignment is to prepare three descriptions of this invention:
a. for company executives who will decide whether to produce and market the item
b. for the engineers, machinists, and so on who will design and manufacture the item
c. for the customers who might purchase and use the item

Before writing for each audience, be sure to collectively complete audience and use profiles (page 449).

Appoint a group manager, who will assign tasks (visuals, typing, etc.) to members. When the descriptions are fully prepared, the group manager will appoint one member to present them in class. The presentation should include explanations of how the descriptions differ for the different audiences.

2. Assume that your group is an architectural firm designing buildings at your college. Develop a set of specifications for duplicating the interior of the classroom in which this course is held. Focus only on materials, dimensions, and equipment (whiteboard, desk, etc.) and

use visuals as appropriate. Your audience includes the firm that will construct the classroom, teachers, and school administrators. Use the same format as in Figure 22.2, or design a better one. Appoint one member to present the completed specifications in class. Compare versions from each group for accuracy and clarity.

3. As a group, select a particular product (sound system, laptop computer, video game, or the like) for which descriptions and specifications are available (in product manuals, brochures, and so on). Using Figure 22.5 as a model, design a marketing document that describes and promotes this product in a one-page, double-sided format. Include (a) all necessary visuals or (b) a brief (page 282) and a rough diagram for each visual or (c) a "reference visual" with instructions for adapting your visual from that one. (If you borrow visuals from other sources, provide full documentation.)

C Series System Description

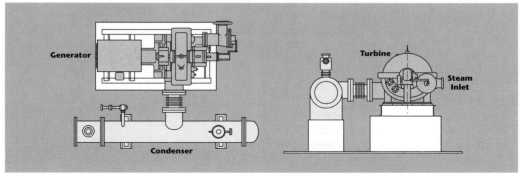

The Ewing Power Systems C Series is a complete single-stage condensing turbine generator package designed for use where maximum electricity production is desired and surplus steam is available. High-pressure steam passes through the turbine and exits at a vacuum pressure to a close-coupled condenser. The condenser may be either water cooled or air cooled. In water-cooled systems the cooling water can be used for heating or the heat can be dissipated in a cooling tower. This series is ideal for converting waste fuel into valuable electricity. It is generally not suited for applications where oil or gas is the primary boiler fuel unless the condenser cooling water will be used for heating.

Features and Specifications

Turbine Features
- Coppus RLHA turbine, fully proven in world wide applications
- Integrated steam control system including all transmitters, actuators and controllers
- Dual electronic and mechanical overspeed trip mechanisms
- Handvalves for maximum operating efficiency
- Very low maintenance
- Rugged, reliable design, 20 year minimum service life
- Meets all applicable NEMA and API specifications

Steam Specifications
- Recommended Inlet Pressure
  - Maximum: 700 psig
  - Minimum: 14 psig
- Recommended Exhaust Pressure
  - Maximum: 0 psig (14.7 psia)
  - Minimum: –10 psig (4.7 psia)
- Steam Flow
  2,500 pounds of steam per hour (75 boiler horsepower) or greater

Generator Features
- Louis Allis induction generator, renowned for high efficiency and dependability*
- Models from 55 kW to 800 kW continuous duty at 480 volts
- Models to 2,000 kW at higher voltages
- Extra high efficiency design is standard

*Synchronous generators also available

Standard Prewired Electrical Controls
- Shunt trip, 3-pole, motor operated circuit breaker with stored energy trip mechanism
- Utility Grade Protective Relays
  - Over/under Voltage
  - Over/under Frequency
  - Ground overcurrent
- Stator thermostats
- Time delay relay to disconnect on motoring
- Pilot lights for operating and trip status
- Ammeter and voltmeter
- Digital tachometer
- Kilowatt meter
- Synchronous panels available

NOTE: *We will customize our control panels to meet the interconnection requirements of any utility.*

Condenser Features
- Air cooled models
  - Specially designed to minimize power consumption
  - Freeze protection system
  - Standard design is for 97°F ambient, higher temperatures available
- Water cooled models
  - Includes cooling tower
  - Steel shell and tubesheet and admiralty brass tubes
  - Integral hot-well
- Both models include:
  - Steam ejector to remove non condensables
  - Condensate pump

(EWING POWER SYSTEMS)

Figure 22.5　Specifications in Marketing Literature　*Source: Courtesy of Erving Power Systems, So. Deerfield, MA. 01373*

Typical System Schematic

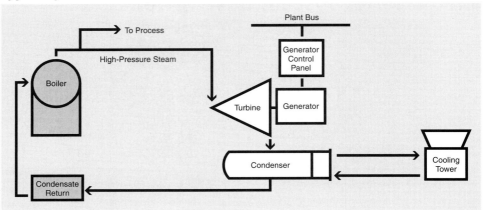

C Series with water cooled condenser and cooling tower.

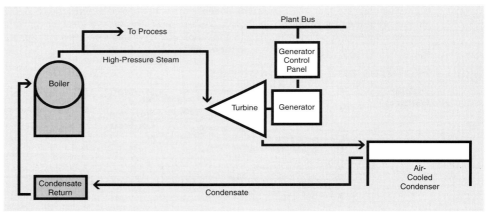

C Series with air-cooled condenser.

Services

In addition to supplying the finest available equipment, Ewing Power Systems also provides complete engineering services. We recognize that you may not be familiar with many of the engineering details of cogeneration so we go to extra lengths to assure our systems will meet your specific requirements for steam and electricity and will provide detailed prints and installation supervision. We will also work with your local utility to assure that your system is designed to take maximum advantage of their rate structure and will meet all their technical requirements. These engineering services, together with our complete economic analysis, assure that your cogeneration system will provide the shortest possible payback.

Figure 22.5   Specifications in Marketing Literature   *Continued*

Procedures
and Processes

A *procedure* is a sequence of steps required to complete a task (say, in creating a Web page). A *process* is a sequence of events leading to a product or result (say, in the production of global warming).

PURPOSE OF PROCESS-RELATED EXPLANATION

In the workplace, you might instruct a colleague or customer in accessing a database or shipping radioactive waste. The new employee needs instructions for operating the office machines; the employee going on vacation writes instructions for the person filling in. The computer user consults the manuals (or *documentation*) for instructions on connecting a printer or running a program.

What an audience expects to learn from a set of instructions

- ◆ How do I do it?
- ◆ Why do I do it?
- ◆ What materials and equipment will I need?
- ◆ What do I do first, next, and so on?
- ◆ Are there any precautions?

Instructions emphasize the user's role, explaining each step in enough detail for the user to complete the task safely and efficiently.

Besides showing how to do something, you often have to explain how things occur, as in showing coworkers how the budget for various departments is determined, or how the system for employee evaluations in your company works.

What an audience expects to learn from a process analysis

- ◆ How does it happen, or how is it produced?
- ◆ What happens first, next, and so on?
- ◆ What is the result?

Process analysis emphasizes the process itself—instead of the user's role.

FAULTY INSTRUCTIONS AND LEGAL LIABILITY

As many as ten percent of workers are injured yearly on the job (Clement 149). Certain medications produce depression that can lead to suicide (Caher 5). Countless injuries result from misuse of consumer products such as power tools, car jacks, or household cleaners—misuse often caused by defective instructions.

Anyone injured because of unclear, inaccurate, or incomplete instructions can sue the writer. Courts have ruled that a defect in product support literature carries the same type of liability as a defect in the product itself (Girill, "Technical Communication and Law" 37).

Those who prepare instructions are potentially liable for damage or injury resulting from omissions such as the following (Caher 5–7; Manning 13; Nordenberg 7):

Examples of faulty instructions that create legal liability

- *Failure to instruct users in the proper use of a product:* for example, a medication's proper dosage or possible interaction with other drugs.
- *Failure to warn against hazards from proper use of a product:* for example, the risk of repetitive stress injury resulting from extended use of a computer keyboard.
- *Failure to warn against the possible misuses of a product:* for example, the childhood danger of suffocation posed by plastic bags.
- *Failure to explain a product's benefits and risks in language that average consumers can understand.*
- *Failure to convey the extent of risk with forceful language.*
- *Failure to display warnings prominently.*

Some legal experts argue that defects in the instructions carry even greater liability than defects in the product because they are more easily demonstrated to a nontechnical jury (Bedford and Stearns 128).

ELEMENTS OF USABLE INSTRUCTION

Clear and Limiting Title

Give an immediate forecast

The title "Instructions for Cleaning the Drive Head of a Laptop Computer" tells users what to expect: instructions for a specific procedure on a selected part. But the title "The Laptop Computer" gives no forecast; a document with this title might contain a history of the laptop, a description of each part, or a wide range of related information.

Informed Content

Know the procedure

Ignorance on your part makes you no less liable for faulty or inaccurate instructions:

Ignorance provides no legal excuse

> If the author of [a car repair] manual had no experience with cars, yet provided faulty instructions on the repair of the car's brakes, the home mechanic who was injured when the brakes failed may recover [damages] from the author. (Walter and Marsteller 165)

Do not write instructions for any task with which you are not very familiar.

Visuals

Instructions also often include a persuasive dimension: to promote interest, commitment, or action. In addition to showing what to do, visuals attract the user's attention and help keep words to a minimum.

Types of visuals especially suited to instructions include icons, representational and schematic diagrams, flowcharts, photographs, and prose tables.

Visuals generated by computers are especially useful for instructions. Other sources for instructional graphics include clip art, electronic scanning, and down-

how to locate something

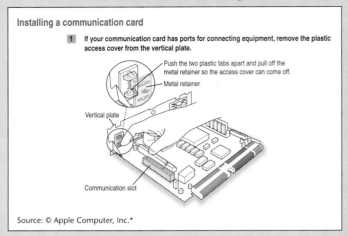

Installing a communication card

1 If your communication card has ports for connecting equipment, remove the plastic access cover from the vertical plate.

Push the two plastic tabs apart and pull off the metal retainer so the access cover can come off.

Metal retainer

Vertical plate

Communication slot

Source: © Apple Computer, Inc.*

how to operate something

Source: *Superstock*

how to handle something

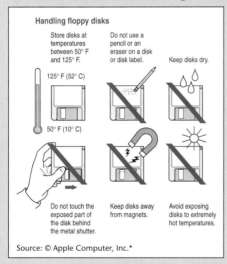

Handling floppy disks

Store disks at temperatures between 50° F and 125° F.

Do not use a pencil or an eraser on a disk or disk label.

Keep disks dry.

125° F (52° C)

50° F (10° C)

Do not touch the exposed part of the disk behind the metal shutter.

Keep disks away from magnets.

Avoid exposing disks to extremely hot temperatures.

Source: © Apple Computer, Inc.*

how to assemble something

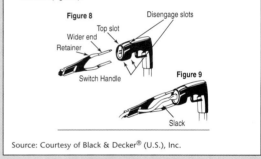

Extension Cord Retainer

1. Look into the end of the Switch Handle and you will see 2 slots. The WIDER end of the Retainer goes into the TOP slot (Figure 8).
2. Plug extension cord into Switch Handle and weave cord into Retainer, leaving a little slack (Figure 9).

Figure 8
Disengage slots
Top slot
Wider end
Retainer
Switch Handle
Figure 9
Slack

Source: Courtesy of Black & Decker® (U.S.), Inc.

how to avoid damage or injury

△ **Important:** The fixing assembly in the printer operates at very high temperatures. When you need to open the printer, be careful not to touch the fixing assembly. △

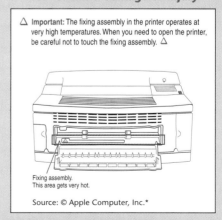

Fixing assembly. This area gets very hot.

Source: © Apple Computer, Inc.*

how to position something

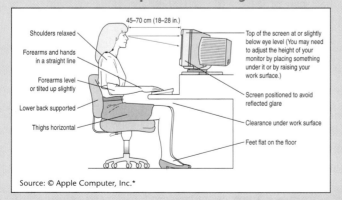

45–70 cm (18–28 in.)

Shoulders relaxed

Forearms and hands in a straight line

Forearms level or tilted up slightly

Lower back supported

Thighs horizontal

Top of the screen at or slightly below eye level (You may need to adjust the height of your monitor by placing something under it or by raising your work surface.)

Screen positioned to avoid reflected glare

Clearance under work surface

Feet flat on the floor

Source: © Apple Computer, Inc.*

Figure 23.1 Common Types of Instructional Visuals and Their Functions   *Illustrations © Apple Computer, Inc. 1993. Used with permission. Apple, the Apple logo and Power Macintosh are registered trademarks™ of Apple Computer, Inc. All rights reserved.

how to diagnose and solve problems

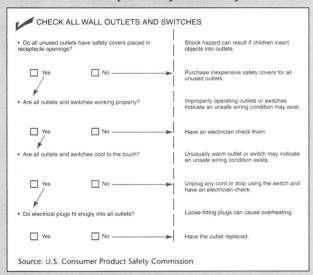

GENERAL TROUBLESHOOTING CHART

If the amplifier is otherwise operating satisfactorily the more common causes of trouble may generally be attributed to the following:
1. Incorrect connections or loose terminal contacts. Check the speakers, record player, tape deck, antenna and line cord.
2. Improper operation. Before operating any audio component, be sure to read the instructions.

3. Improper location of audio components. The proper positioning of components, such as speakers and turntable, is vital to stereo.
4. Defective audio components.

Following are some other common causes of malfunction and what to do about them.

PROGRAM	SYMPTOM	PROBABLE CAUSE	WHAT TO DO
AM, FM or MPX reception	a. Constant or intermittent noise heard at certain times or in a certain area	* Discharge or oscillation caused by electrical appliances, such as fluorescent lamps, TV sets, D.C. motors, rectifier and oscillator * Natural phenomena, such as atmospherics, static, and thunderbolt * Insufficient antenna input due to reinforced concrete walls or long distance from the station * Wave interference from other electrical appliances	* Attach a noise limiter to the electrical appliance that causes the noise, or attach it to the power source of the amplifier. * Install an outdoor antenna and ground the amplifier to raise the signal-to-noise ratio. * Reverse the power cord plug-receptacle connections. * If the noise occurs at a certain frequency, attach a wave trap to the ANT. input. * Place the set away from other electrical appliances.

Source: Courtesy of Sansui Electronic Co. Ltd.

how to proceed systematically

☑ CHECK ALL WALL OUTLETS AND SWITCHES

- Do all unused outlets have safety covers placed in receptacle openings?

☐ Yes ☐ No —— Shock hazard can result if children insert objects into outlets.

Purchase inexpensive safety covers for all unused outlets.

- Are all outlets and switches working properly?

☐ Yes ☐ No —— Improperly operating outlets or switches indicate an unsafe wiring condition may exist.

Have an electrician check them.

- Are all outlets and switches cool to the touch?

☐ Yes ☐ No —— Unusually warm outlet or switch may indicate an unsafe wiring condition exists.

Unplug any cord or stop using the switch and have an electrician check.

- Do electrical plugs fit snugly into all outlets?

☐ Yes ☐ No —— Loose-fitting plugs can cause overheating.

Have the outlet replaced.

Source: U.S. Consumer Product Safety Commission

how to make the right decisions

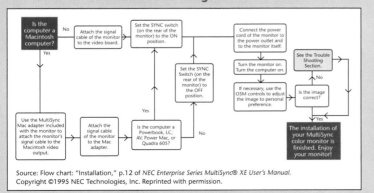

Source: Flow chart: "Installation," p.12 of *NEC Enterprise Series MultiSync® XE User's Manual.* Copyright ©1995 NEC Technologies, Inc. Reprinted with permission.

how to identify safe or acceptable limits

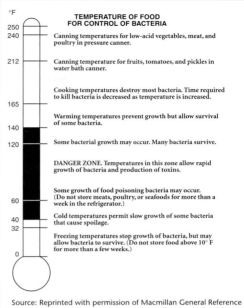

TEMPERATURE OF FOOD FOR CONTROL OF BACTERIA

°F

250 240 — Canning temperatures for low-acid vegetables, meat, and poultry in pressure canner.

212 — Canning temperature for fruits, tomatoes, and pickles in water bath canner.

Cooking temperatures destroy most bacteria. Time required to kill bacteria is decreased as temperature is increased.

165 — Warming temperatures prevent growth but allow survival of some bacteria.

140
120 — Some bacterial growth may occur. Many bacteria survive.

DANGER ZONE. Temperatures in this zone allow rapid growth of bacteria and production of toxins.

60 — Some growth of food poisoning bacteria may occur. (Do not store meats, poultry, or seafoods for more than a week in the refrigerator.)

40 — Cold temperatures permit slow growth of some bacteria that cause spoilage.

32

Freezing temperatures stop growth of bacteria, but may allow bacteria to survive. (Do not store food above 10° F for more than a few weeks.)

0

Source: Reprinted with permission of Macmillan General Reference USA, a Division of Simon & Schuster, from *The New York Public Library Desk Reference.* Copyright ©1989, 1993, 1998 by the New York Public Library and the Stonesong Press, Inc.

why action is important

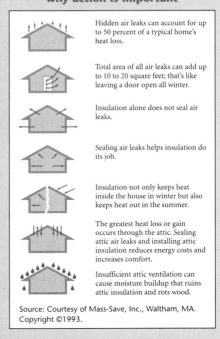

Hidden air leaks can account for up to 50 percent of a typical home's heat loss.

Total area of all air leaks can add up to 10 to 20 square feet; that's like leaving a door open all winter.

Insulation alone does not seal air leaks.

Sealing air leaks helps insulation do its job.

Insulation not only keeps heat inside the house in winter but also keeps heat out in the summer.

The greatest heat loss or gain occurs through the attic. Sealing attic air leaks and installing attic insulation reduces energy costs and increases comfort.

Insufficient attic ventilation can cause moisture buildup that ruins attic insulation and rots wood.

Source: Courtesy of Mass-Save, Inc., Waltham, MA. Copyright ©1993.

Figure 23.1 Common Types of Instructional Visuals and Their Functions *Continued* loading from the Internet. (Page 294 describes useful Web sites and discusses legal issues in the use of computer graphics.)

Use visuals generously

Illustrate any step that might be hard for users to visualize. Parallel the user's angle of vision in performing the activity or operating the equipment—and name the angle (side view, top view) if you think people will have trouble figuring it out for themselves. The less specialized your audience, the more visuals they are likely to need. But do not illustrate any action simple enough for users to visualize on their own, such as "Press RETURN" for anyone familiar with a keyboard.

Figure 23.1 presents an array of visuals and their specific instructional functions. Each of these visuals is easily constructed and some could be further enhanced, depending on your production budget and graphics capability. Writers and editors often provide a brief (page 282) and a rough sketch describing the visual and its purpose for the graphic designer or art department.

Appropriate Level of Technicality

Unless you know your users have the relevant background and skills, write for laypersons, and do three things:

Provide exactly and only what users need

1. Give them enough background to understand why they need your instructions.
2. Give them enough detail to understand what to do.
3. Give them enough examples to visualize the procedure clearly.

These next examples show how you might adapt instructions titled "How to Create a Floppy Disk Backup to a Hard Disk" for novice Macintosh users.

PROVIDE BACKGROUND.   Begin by explaining the purpose of the procedure.

Tell users why they are doing this

> You might easily lose information stored on a hard disk if:
>
> ♦ the disk is damaged by repeated use, jarring, moisture, or extreme temperature;
> ♦ the disk is erased by a power surge, a computer malfunction, or a user error; or
> ♦ the stored information is scrambled by a nearby magnet (telephone, computer terminal, or the like).
>
> Always make a backup copy of any disk that contains important material.

Also, state your assumptions about the user's level of technical understanding.

Spell out what users should already know

> To follow these instructions, you should be able to identify these parts of a Macintosh system: computer, monitor, keyboard, mouse, floppy disk drive, and 3.5-inch floppy disk.

Define any specialized terms that appear in your instructions.

Tell users what each key term means

> *Initialize:* Before you can store or retrieve information on a disk, you must initialize the blank disk. Initializing creates a format the computer can understand—a directory of

specific memory spaces (like post office boxes) on the disk, where you can store information and retrieve it as needed.

When the user understands *what* and *why,* you are ready to explain *how* to carry out the procedure.

Make instructions complete but not excessive

PROVIDE ADEQUATE DETAIL.   Include enough detail for users to understand and perform the task successfully, but omit general information that users probably know.

Inadequate detail for laypersons

FIRST AID FOR ELECTRICAL SHOCK
1. Check vital signs.
2. Establish an airway.
3. Administer external cardiac massage as needed.
4. Ventilate, if cyanosed.
5. Treat for shock.

Not only are the above details inadequate, but terms such as "vital signs" and "cyanosed" are too technical for laypersons. Such instructions posted for workers in a high-voltage area would be useless. Illustrations and explanations are needed, as in these instructions for item 2 above, establishing an airway:

Adequate detail for laypersons

MOUTH-TO-MOUTH BREATHING

If there are no signs of breathing, place one hand under the victim's neck and gently lift. At the same time, push with the other hand on the victim's forehead. This will move the tongue away from the back of the throat to open the airway.

Source: Reprinted with permission of Macmillan General Reference USA, a Division of Simon & Schuster, from *The New York Public Library Desk Reference.* Copyright ©1989, 1993, 1998 by the New York Public Library and the Stonesong Press, Inc.

Don't overestimate what people know, especially about a procedure you can perform almost automatically. (Think about when you were learning to drive a car—or a time when you have tried to teach someone else.) Always assume that the user knows less than you. A colleague will know at least a little less; a layperson will know a good deal less—maybe nothing—about this procedure.

Exactly how much information is enough? Consider these suggestions:

How to provide adequate but not excessive detail

◆ Give everything users need, so the instructions can stand alone.
◆ Give only what users need. Don't tell them how to build a computer when they only need to know how to copy a disk.

* Instead of focusing on the *product* ("How does it work?"), focus on the *task* ("How do I use it?" or "How do I do it?") (Grice, "Focus" 132).
* Omit steps *(Seat yourself at the computer)* that are obvious to users.
* Adjust the *information rate* ("the amount of information presented in a given page," Meyer 17) to the user's background and the difficulty of the task. For complex or sensitive steps, slow the information rate. Don't make users do too much too fast.
* Reinforce the prose with visuals. Don't be afraid to repeat information if it saves users from flipping pages.
* When writing instructions for consumer products, assume "a barely literate reader" (Clement 151). Simplify.
* Recognize the persuasive dimension of the instructions. Users may need persuading that this procedure is necessary or beneficial, or that they can complete this procedure easily and competently.

OFFER EXAMPLES. Procedures require specific examples (how to load a program, how to order a part):

Give plenty of examples

To load your program, type this command:

> Load "Style Editor"

Then press RETURN.

Like visuals, examples *show* users what to do.

INCLUDE TROUBLESHOOTING ADVICE. Anticipate things that commonly go wrong in this procedure—the paper jams in the printer, the tray of the CD-ROM drive won't open, or some other malfunction. Explain the probable cause(s) and offer solutions.

Explain what to do when things go wrong

Note: IF X doesn't work, first check Y and then do Z.

In the sample procedure that follows, careful use of background, detailed explanation, visual examples, and troubleshooting advice create a user-friendly level of technicality for computer novices.

Background

FIRST STEP: HOW TO INITIALIZE YOUR FLOPPY DISK
Before you can copy or store information on a blank disk, you must initialize the disk. Follow this procedure:

Instructional details

1. Switch on the computer.
2. Insert your floppy disk in the floppy disk drive. Unable to recognize this new disk, the computer will ask if you wish to initialize the disk (Figure 1).

Visual example
reinforces the
verbal message

Figure 1   The "Initialize" Message

3. Using your mouse, place the on-screen pointer (small arrow) inside the "Initialize" box.
4. Press and quickly release the mouse button. Within 15–20 seconds, the initializing will be completed, and a message will appear, asking you to name your disk (Figure 2).

Visual appears close
to the related step

Figure 2   The "Disk naming" Message

Troubleshooting
tips

NOTE: If your disk is rejected, it might be improperly seated in the drive or damaged, or the drive itself might be damaged.

(a) Eject and reinsert the disk, and repeat steps 2–4.
(b) If (a) fails, insert a different disk. If this doesn't work, have your disk drive checked.

Make verbal and
visual information
redundant

In the previous sample, visuals and prose are *redundant* (Weiss 100). Chapter 13 warns against *style redundancy* (extra words that give no extra information). Effective instructions, however, often exhibit *content redundancy,* giving the same information in verbal and visual form. When you can't be sure how much is enough, risk overexplaining rather than underexplaining.

Logically Ordered Steps

Organize for the
user's understanding

Instructions are almost always arranged in chronological order, with warnings and precautions inserted for specific steps.

Show how the
steps are connected

You can't splice two wires to make an electrical connection until you have removed the insulation. To remove the insulation, you will need. . . .

Notes and Hazard Notices

Alert users to special considerations and hazards

Here are the only items that should interrupt the steps in a set of instructions (Van Pelt 3):

- ◆ A *note* clarifies a point, emphasizes vital information, or describes options or alternatives.

> NOTE: If you don't name a newly initialized disk, the computer automatically names it "Untitled."

While a note is designed to enhance performance and prevent error, the following hazard notices—ranked in order of severity—are designed to prevent damage, injury, or death.

- ◆ A *caution* prevents possible mistakes that could result in injury or equipment damage:

The least forceful notice

> CAUTION: A momentary electrical surge or power failure will erase the contents of internal memory. To avoid losing your work, every few minutes save on disk what you have just typed into the computer.

- ◆ A *warning* alerts users to potential hazards to life or limb:

A moderately forceful notice

> WARNING: To prevent electrical shock, always disconnect your printer from its power source before cleaning internal parts.

- ◆ A *danger* notice identifies an immediate hazard to life or limb:

The most forceful notice

> DANGER: The red canister contains DEADLY radioactive material. **Do not break the safety seal** under any circumstances.

Content requirements for hazard notices

Inadequate notices of warning, caution, or danger are a common cause of lawsuits (page 463). Each hazard notice is legally required to (1) describe the specific hazard, (2) spell out the consequences of ignoring the hazard, and (3) offer instruction for avoiding the hazard (Manning 15).

Visual requirements for hazard notices

Even the most emphatic verbal notice might be overlooked by an impatient or inattentive user. Direct the user's attention with symbols, or icons, as a visual signal (Bedford and Stearns 128):

Use hazard symbols

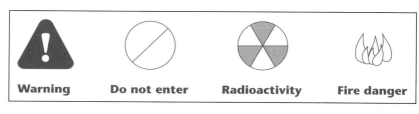

Warning   **Do not enter**   **Radioactivity**   **Fire danger**

Visibility
requirements for
hazard notices

Keep the hazards prominent in the user's awareness: Preview the hazards in your introduction, and place each notice, clearly highlighted (by a ruled box, a distinct typeface, larger typesize, or color) immediately before the related step.

NOTE

Use hazard notices only when needed; overuse will dull their effect, and readers may overlook their importance.

Readability

Make instructions
immediately
readable

Instructions must be understood at the first reading because readers usually take *immediate* action.

Like descriptions (page 445), instructions name parts, use location and position words, and state exact measurements, weights, and dimensions. Instructions also require strict attention to phrasing, sentence structure, and paragraph structure.

USE DIRECT ADDRESS, ACTIVE VOICE, AND IMPERATIVE MOOD.

To emphasize the user's role, write instructions in the second person, as direct address.

Begin all steps with action verbs, using the *active voice* and *imperative mood* ("Insert the disk" instead of "The disk should be inserted" or "You should insert the disk").

Indirect or
confusing

- The user types in his or her access code.
- You should type in your access code.
- It is important to type in the access code.
- The access code is typed in.

In the next version, the opening verb immediately signals the specific action to be taken.

Direct and clear

Type in your access code.

NOTE

Certain cultures consider the direct imperative bossy and offensive. For cross-cultural audiences, you might rephrase an instruction declaratively: from "Type in your access code" to "The access code should be typed in." Or you might use an indirect imperative such as "Be sure to type in your access code" (Coe, "Writing" 18).

USE SHORT AND LOGICALLY SHAPED SENTENCES.

Use shorter sentences than usual, but don't "telegraph" your message by omitting articles *(a, an, the)*, as on page 217. Use one sentence for one step, so users can perform one step at a time.

If a single step covers two related actions, describe these actions in their required sequence:

Confusing

Before switching on the computer, insert the disk in the drive.

Logical

Insert the disk in the drive; then switch on the computer.

Simplify explanations by using a familiar-to-unfamiliar sequence:

Hard	You must initialize a blank disk before you can store information on it.
Easier	Before you can store information on a blank disk, you must initialize the disk.

USE PARALLEL PHRASING.   Parallelism is important in all writing but especially in instructions, because repeating grammatical forms emphasizes the step-by-step organization.

Not parallel

> To log on to the VAX 20, follow these steps:
>
> 1. Switch the terminal to "on."
> 2. The CONTROL key and C key are pressed simultaneously.
> 3. Typing LOGON, and pressing the ESCAPE key.
> 4. Type your user number, and then press the ESCAPE key.

All steps should be in identical grammatical form:

Parallel

> To log on to the VAX 20, follow these steps:
>
> 1. Switch the terminal to "on."
> 2. Press the CONTROL key and C key simultaneously.
> 3. Type LOGON, and then press the ESCAPE key.
> 4. Type your user number, and then press the ESCAPE key.

PHRASE INSTRUCTIONS AFFIRMATIVELY.   Research shows that users respond more quickly and efficiently to instructions phrased affirmatively rather than negatively (Spyridakis and Wenger 205).

Negative	Verify that your disk is not contaminated with dust.
Affirmative	Examine your disk for dust contamination.

USE TRANSITIONS TO MARK TIME AND SEQUENCE.   Transitional expressions bridge related ideas. Some transitions ("in addition," "next," "meanwhile," "finally," "ten minutes later," "the next day," "before") mark time and sequence. They help users understand the step-by-step process:

Transitions enhance continuity

> **PREPARING THE GROUND FOR A TENT**
> Begin by clearing and smoothing the area that will be under the tent. This step will prevent damage to the tent floor and eliminate the discomfort of sleeping on uneven ground. *First,* remove all large stones, branches, or other debris within a level 10 × 13-foot area. Use your camping shovel to remove half-buried rocks that cannot easily be moved by hand. *Next,* fill in any large holes with soil or leaves. *Finally,* make several light surface passes with the shovel or a large, leafy branch to smooth the area.

Effective Design

An effective instructional design conveys the sense that the task is within a qualified user's range of abilities.

To help users find, recognize, and remember what they need, follow these suggestions:

How to design
instructions

- ◆ *Provide informative headings.* A heading such as "How to Initialize Your Blank Disk" is more informative than "Disk Initializing."
- ◆ *Arrange steps in a numbered list.* Unless the procedure consists of simple steps (as in "Preparing the Ground for a Tent," above), list and number each step.
- ◆ *Separate steps visually.* Single-space within steps and double-space between.
- ◆ *Make warning, caution, and danger notices highly visible.* Use ruled boxes or highlighting, and plenty of white space.
- ◆ *Keep the visual and the step close together.* If room allows, place the visual right beside the step; if not, right after it. Set off the visual with plenty of white space.
- ◆ *Consider a multicolumn design.* If steps are brief and straightforward and require back-and-forth reference from prose to visuals, consider multiple columns. Figure 23.2 shows how to connect peripheral devices (scanners, extra

---

Connecting cables

▲ **Warning:** When making SCSI connections, *always* turn off power to all devices in the chain. Failure to do so can cause the loss of information and damage to your equipment. ▲

1. **Shut down your PowerBook and all SCSI devices in the chain.**

2. **To connect the first device, use an Apple HDI-30 SCSI System Cable.**

   Attach the smaller end of the cable to your computer's SCSI port (marked with the icon ◆) and the larger end of the cable to either SCSI port on the device.

3. **To connect the next device, use a SCSI peripheral interface cable.**

   Both cable connectors are the same. Attach one connector to the available SCSI port on the first device, and the other connector to either SCSI port on the next device.

4. **Repeat step 3 for each additional device you want to connect.**

The illustration shows where to add cable terminators.

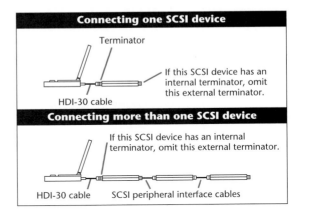

△ **Important:** The total length of an SCSI chain should not exceed 20 feet (6 meters). Apple SCSI cables are designed to meet this restriction. If you are using SCSI cables from another vendor, check the length of the chain. △

Once your SCSI devices are connected, always turn them on before turning on your PowerBook. If you turn the computer on first, it may not be able to start up, or it may not recognize the SCSI devices.

Figure 23.2   A Multicolumn Design   *Source: Reprinted by permission of Apple Computer, Inc.*

drives, and so on) to the computer using SCSI (Small Computer System Interface) cables.

◆ *Keep it simple.* Users can be overwhelmed by a page with excessive or inconsistent designs.

Consult Chapter 15 for additional design considerations.

NOTE   *Like any material displayed on a computer screen, online instructions have their own design requirements, which are discussed in Chapters 15 and 20. Also, despite online documentation's increasing popularity, many information technology professionals continue to find printed manuals more convenient and easier to navigate (Foster 10).*

A GENERAL MODEL FOR INSTRUCTIONS

You can adapt the following outline to any instructions. Here are the possible sections to include:

I. Introduction
   A. Definition, Benefits, and Purpose of the Procedure
   B. Intended Audience (usually omitted for workplace audiences)
   C. Prior Knowledge and Skills Needed by the Audience
   D. Brief Overall Description of the Procedure
   E. Principle of Operation
   F. Materials, Equipment (in order of use), and Special Conditions
   G. Working Definitions (always in the introduction)
   H. Warnings, Cautions, Dangers (previewed here and spelled out at steps)
   I. List of Major Steps

II. Required Steps
   A. First Major Step
      1. Definition and purpose
      2. Materials, equipment, and special conditions for this step
      3. Substeps (if applicable)
         a.
         b.
   B. Second Major Step (and so on)

III. Conclusion
   A. Review of Major Steps (for a complex procedure only)
   B. Interrelation of Steps
   C. Troubleshooting or Follow-up Advice (as needed)

This outline is only tentative; you might modify, delete, or combine some elements, depending on your subject, purpose, and audience.

Introduction

The introduction should help users to begin "doing" as soon as they are able to proceed safely, effectively, and confidently (van der Meij and Carroll 245–46). Most users are interested primarily in "how to use it or fix it," and will require only a general understanding of "how it works." You don't want to bury users in a long introduction, nor do you want to set them loose on the procedure without adequate preparation. Know your audience—what they need and don't need.

Following is an introduction from instructions for people using a college library. Some users will be computer experts; some will be novices—but all will require a detailed introduction to computerized literature searches before they conduct a search on their own.

Clear and limiting title	**HOW TO USE THE OCLC TERMINAL TO SEARCH FOR A BOOK**
	Introduction
Definition of the procedure	Our library's OCLC (Online Computer Library Center) terminal offers an efficient way to search for books, journals, government publications, and other printed materials. This terminal is connected to the OCLC database in Columbus, Ohio.
Definition (continued)	The Ohio database contains more than 8 million records of books and other published materials in libraries nationwide. Each record lists the information found on a catalog card, and identifies the libraries holding the work.
Purpose	These instructions will enable you to determine whether a book you seek can be found in our library or in other libraries throughout the country.
Audience and required skills	Any library patron can use the OCLC terminal to search for a book. To operate the terminal, you only need to be familiar with the keyboard (Figure 1) and to have an operator's manual handy for general reference.
General description of the procedure	After logging on the system, you can search for a book by TITLE, AUTHOR, or AUTHOR and TITLE. Once you have viewed the catalog entry for the book you seek, you can use the terminal to determine the libraries from which you might borrow the book. These instructions show a search by TITLE only.
Materials and equipment	The only additional equipment you need is a copy of the *Manual of OCLC Participating Institutions,* to find a listing of library names according to their OCLC symbol displayed on your terminal screen.
Working definitions	Specialized terms that appear in these instructions are defined here:

Cursor: a small, blinking rectangle that indicates the screen position of the next character you will type.

HOME Position: the cursor location at the extreme top left of your screen.

HOME position is where you will type most of your messages to the computer, and where the computer will display its messages to you.

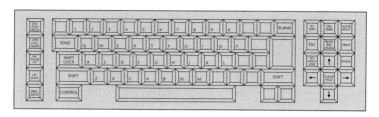

Figure I   The OCLC Terminal Keyboard

Cautions and notes

Pay close attention to the NOTES that accompany the logging-on step and to the CAUTION preceding the logging-off step.

List of major steps

The major steps in using OCLC to search for a book by its TITLE are (1) logging on the system, (2) initiating the search, (3) viewing the information on your title, (4) locating the book, and (5) logging off.

Body: Required Steps

In the body section (labeled Required Steps), give each step and substep in order. Insert warnings, cautions, and notes as needed. Begin each step with its definition or purpose or both. Users who understand the reasons for a step will do a better job. A numbered list (like the following) is an excellent way to segment the steps.

Required Steps

Heading previews the step

1. *Logging on the OCLC System*
   To activate the system, you must first log on.

   a. Flick the red power switch (right edge of keyboard).

Capitals and spacing set off note

   NOTE: *Press HOME before typing, to place the cursor at HOME position.*

   b. Type the authorization number (07-34-6991).

   NOTE: *If you make a typing error, place the cursor over the error and retype. Move the cursor by using the arrow keys.*

   c. Press SEND and wait for the HELLO response.

Headings 1–5 offer an overview of the process

2. *Initiating the Search*
   To initiate a computer search, enter your title's exact code name.
   a. Type in the code for your desired title, excluding articles (*a, an, the*) when they are the first word of the title. Type your entry exactly as in this example:

Example shows
what to do

Suppose your title is *The Logic of Failure;* you would type **log, of, fai**

b. Press DISPLAY/RECD and then SEND.

The computer will display a title that matches the code you have entered (Figure 2).

Visual reinforces
the prose

```
The Logic of Failure: recognizing and avoiding error in
complex situations/Dorner, Dietrich. Metropolitan Books.
Distributed in the U.S. and Canada by Addison-Wesley,
c 1996.
```

Figure 2   The Title That Matches the Code You Typed

3. *Viewing the Information on Your Title*
   [Steps for viewing detailed information on the title, locating the book, and logging off continue.]

Conclusion

The conclusion of a set of instructions has several possible functions:

- You might summarize the major steps in a long and complex procedure, to help users review their performance.
- You might describe the results of the procedure.
- You might offer follow-up advice about what could be done next or refer the user to further sources of documentation.
- You might give troubleshooting advice about what to do if anything goes wrong.

You might do all these things—or none of them. If your procedural section has given users all they need, omit the conclusion altogether.

In the case of the OCLC instructions, these concluding remarks are useful and appropriate:

Follow-up advice

Conclusion

Once you locate the title you seek, ask a reference librarian for help requesting the book via Interlibrary Loan. (Books usually arrive within two days.)

Troubleshooting
advice

In case the first library is unable to supply the book requested, you should initially choose several libraries from your printout. The Interlibrary Loan system will automatically forward your request to each lender you have specified, until one lender indicates it can supply the book.

A SITUATION REQUIRING INSTRUCTIONS

The instructions for felling a tree in Figure 23.3 on pages 479–82 are patterned after the general outline, shown earlier. They will appear in a series of brochures for forestry students who are about to begin summer jobs with the Idaho Forestry Service.

Instructions for a Semitechnical Audience

Audience and Use Profile.   These instructions are aimed at a partially informed audience who knows how to use chainsaws, axes, and wedges but who is approaching this dangerous procedure for the first time. Therefore, no visuals of cutting equipment (chainsaws and so on) are included because the audience knows what these items look like. Basic information (such as what happens when a tree binds a chainsaw) is omitted because the audience has this knowledge also. Likewise, these users need no definition of general forestry terms such as *culling* and *thinning,* but they *do* need definitions of terms that relate specifically to tree felling (*undercut, holding wood,* and so on).

For clarity, visuals illustrate the final three steps. The conclusion, for these users, will be short and to the point—a simple summary of major steps with emphasis on safety. ▲

INSTRUCTIONS FOR FELLING A TREE 1

INTRODUCTION

Forestry Service personnel fell (cut down) trees to cull or thin a forested plot, to eliminate the hazard of dead trees standing near power lines, to clear an area for construction, and the like.

These instructions explain how to remove sizable trees for personnel who know how to use a chainsaw, axe, and wedge safely.

When you set out to fell a tree, expect to spend most of your time planning the operation and preparing the area around the tree. To fell the tree, you will make two chainsaw cuts, severing the stem from the stump. Depending on the direction of the cuts, the weather, and the terrain, the severed tree will fall into a predetermined clearing.

> **WARNING:** Although these instructions cover the basic procedure, felling is very dangerous. Trees, felling equipment, and terrain vary greatly. Even professionals are sometimes killed or injured because of judgment errors or misuse of tools.
>
> Your main concern is safety. Be sure to have an expert demonstrate this procedure before you try it. Also, pay attention to warnings in steps 1 and 4.

To fell sizable trees, you will need this equipment:
—a 3- to 5-horsepower chainsaw with a 20-inch blade
—a single-blade splitting axe
—two or more 12-inch steel wedges

The major steps in felling a tree are (1) choosing the lay, (2) providing an escape path, (3) making the undercut, and (4) making the backcut.

REQUIRED STEPS

1. *How to Choose the Lay*

   The "lay" is where you want the tree to fall. On level ground in an open field, which way you direct the fall makes little difference. But such ideal conditions are rare.

   Consider ground obstacles and topography, surrounding trees, and the condition of the tree to be felled. Plan your escape path and the location of your cuts depending on surrounding houses, electrical wires, and trees. Then follow these steps:

   a. Make sure the tree is still alive.

Figure 23.3 A Set of Instructions

b. Determine the direction and amount of lean.

2

> **WARNING:** If the tree is dead and leaning substantially, do not try to fell it without professional help. Many dead, leaning trees have a tendency to split along their length, causing a massive slab to fall spontaneously.

c. Find an opening into which the tree can fall in the direction of lean or as close to the direction of lean as possible.

Because most trees lean downhill, try to direct the fall downhill. If the tree leans slightly away from your desired direction, use wedges to direct the fall.

2. *How to Provide an Escape Path*
Falling trees are unpredictable. Avoid injury by planning a definite escape path. Follow these steps:

a. Locate the path in the direction opposite of the fall (Figure 1).

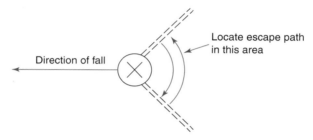

Figure 1   Escape Path Location

b. Clear a path 2 feet wide extending beyond where the top of the tree could land.

3. *How to Make the Undercut*
The undercut is a triangular slab of wood cut from the trunk on the side toward which you want the tree to fall. Follow these steps:

a. Start the chainsaw.

b. Holding the saw with blade parallel to the ground, make a first cut 2 to 3 feet above ground. Cut horizontally, to no more than 1/3 the tree's diameter (Figure 2).

Figure 23.3   **A Set of Instructions**   *Continued*

3

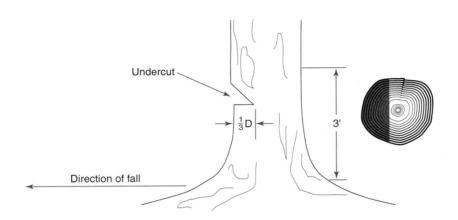

Figure 2 Making the Undercut

    c. Make a downward-sloping cut, starting 4 to 6 inches above the first so that the cuts intersect at 1/3 the diameter (Figure 2).

4. *How to Make the Backcut*
After completing the undercut, you make the backcut to sever the stem from the stump. This step requires good reflexes and absolute concentration.

> **WARNING:** Observe tree movement closely during the backcut. If the tree shows any sign of falling in your direction, drop everything and move out of its way.
>   Also, do not cut completely through to the undercut; instead, leave a narrow strip of "holding wood" as a hinge, to help prevent the butt end of the falling tree from jumping back at you.

To make your backcut, follow these steps:

    a. Holding the saw with the blade parallel to the ground, start your cut about 3 inches above the undercut, on the opposite side of the trunk. Leave a narrow strip of "holding wood" (Figure 3).

Figure 23.3 A Set of Instructions *Continued*

4

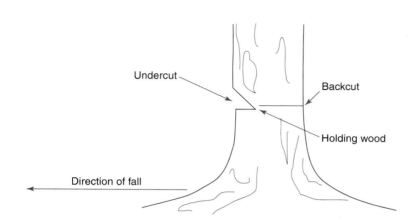

Figure 3   Making the Backcut

NOTE: If the tree begins to bind the chainsaw, hammer a wedge into the backcut with the blunt side of the axe head. Then continue cutting.

  b. As soon as the tree begins to fall, turn off the chainsaw, withdraw it, and step back immediately—the butt end of the tree could jump back toward you.

  c. Move rapidly down the escape path.

CONCLUSION

Felling is a complex and dangerous procedure. Choosing the lay, providing an escape path, making the undercut, and making the backcut are the basic steps—but trees, terrain, and other circumstances vary greatly.

For the safest operation, seek professional advice and help whenever you foresee *any* complications whatsoever.

Figure 23.3   A Set of Instructions   *Continued*

A GENERAL MODEL FOR PROCESS ANALYSIS

An explanation of how things work or happen divides the process into its parts or principles. Colleagues and clients need to know how stock and bond prices are governed, how your bank reviews a mortgage application, how an optical fiber conducts an impulse, and so on. Process analysis emphasizes *the process itself* so that the audience can follow it as "observers." (For example, the instructions in Figure 23.3 for *how to fell a tree* might be translated into an analysis of *how trees are felled.*)

Like instructions, a process analysis must be detailed enough to allow users to follow the process step by step. Because it emphasizes the process itself, rather than the audience's role, process analysis is written in the third person. Instead of leading to immediate action, process analysis helps readers gain understanding. Format and paragraph structure, therefore, resemble those in a traditional essay.

Much of your college writing explains how things happen. Your audience is the professor, who will evaluate what you have learned. Because this person knows *more* than you do about the subject, you often discuss only the main points, omitting details.

But your real challenge comes in analyzing a process for audiences who know *less* than you, and who are neither willing nor able to fill in the blank spots; you then become the teacher, and the users become your students.

For an uninformed audience, introduce your analysis by telling what the process is, and why, when, and where it happens. In the body, tell how it happens, analyzing each stage in sequence. In the conclusion, summarize the stages, and describe one full cycle of the process.

Sections from the following general outline can be adapted to any process analysis:

I. Introduction
   A. Definition, Background, and Purpose of the Process
   B. Intended Audience (usually omitted for workplace audiences)
   C. Prior Knowledge Needed to Understand the Process
   D. Brief Description of the Process
   E. Principle of Operation
   F. Special Conditions Needed for the Process to Occur
   G. Definitions of Special Terms
   H. Preview of Stages

II. Stages in the Process
   A. First Major Stage
      1. Definition and purpose
      2. Special conditions needed for the specific stage
      3. Substages (if applicable)
         a.
         b.
   B. Second Stage (and so on)

III. Conclusion
    A. Summary of Major Stages
    B. One Complete Process Cycle

Adapt this outline to the structure of the process you are explaining.

A SITUATION REQUIRING PROCESS ANALYSIS

Following is a document patterned after the sample outline. The author, Bill Kelly, belongs to an environmental group studying the problem of acid rain in its Massachusetts community. (Massachusetts is among the states most affected by acid rain.) To gain community support, the environmentalists must educate citizens about the problem. Bill's group is publishing and mailing a series of brochures. The first brochure explains how acid rain is formed.

A Process Analysis for a Nontechnical Audience

Audience and Use Profile.   Some will already be interested in the problem; others will have no awareness (or interest). Therefore, explanation is given at the lowest level of technicality (no chemical formulas, equations). But the explanation needs to be vivid enough to appeal to less aware or less interested readers. Visuals create interest and illustrate simply. To give an explanation thorough enough for broad understanding, the process is divided into three chronological steps: how acid rain develops, spreads, and destroys. ▲

HOW ACID RAIN DEVELOPS, SPREADS, AND DESTROYS

Introduction

Definition

Acid rain is environmentally damaging rainfall that occurs after fossil fuels burn, releasing nitrogen and sulfur oxides into the atmosphere. Acid rain, simply stated, increases the acidity level of waterways because these nitrogen and sulfur oxides combine with the air's normal moisture. The resulting rainfall is far more acidic than normal rainfall. Acid rain is a silent threat because its effects, although slow, are cumulative. This analysis explains the cause, the distribution cycle, and the effects of acid rain.

Purpose

Brief description of the process

Most research shows that power plants burning oil or coal are the primary causes of acid rain. The burnt fuel is not completely expended, and some residue enters the atmosphere. Although this residue contains several potentially toxic elements, sulfur oxide and, to a lesser extent, nitrogen oxide are the major problem, because they are transformed when they combine with moisture. This chemical reaction forms sulfur dioxide and nitric acid, which then rain down to earth.

Preview of stages

The major steps explained here are (1) how acid rain develops, (2) how acid rain spreads, and (3) how acid rain destroys.

The Process

First stage

How Acid Rain Develops.   Once fossil fuels have been burned, their usefulness ends. Unfortunately, it is here that the acid rain problem begins.

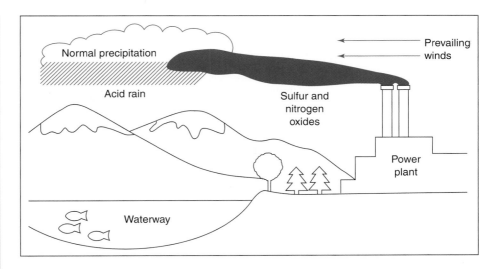

Figure 1   How Acid Rain Develops

Fossil fuels contain a number of elements that are released during combustion. Two of these, sulfur oxide and nitrogen oxide, combine with normal moisture to produce sulfuric acid and nitric acid. (Figure 1 illustrates how acid rain develops.) The released gases undergo a chemical change as they combine with atmospheric ozone and water vapor. The resulting rain or snowfall is more acid than normal precipitation.

Definition

Acid level is measured by pH readings. The pH scale runs from 0 through 14—a pH of 7 is considered neutral. (Distilled water has a pH of 7.) Numbers above 7 indicate increasing degrees of alkalinity. (Household ammonia has a pH of 11.) Numbers below 7 indicate increasing acidity. Movement in either direction on the pH scale, however, means multiplying by 10. Lemon juice, which has a pH value of 2, is 10 times more acidic than apples, which have a pH of 3, and is 1,000 times more acidic than carrots, which have a pH of 5.

Because of carbon dioxide (an acid substance) normally present in air, unaffected rainfall has a pH of 5.6. At this time, the pH of precipitation in the northeastern United States and Canada is between 4.5 and 4. In Massachusetts, rain and snowfall have an average pH reading of 4.1. A pH reading below 5 is considered to be abnormally acidic, and therefore a threat to aquatic populations.

Second stage

How Acid Rain Spreads.   Although it might seem that areas containing power plants would be most severely affected, acid rain can in fact travel thousands of miles from its source. Stack gases escape and drift with the wind currents. The sulfur and nitrogen oxides are thus able to travel great distances before they return to earth as acid rain.

For an average of two to five days after emission, the gases follow the prevailing winds far from the point of origin. Estimates show that about 50 percent of the acid rain that affects Canada originates in the United States; at the same time, 15 to 25 percent of the U.S. acid rain problem originates in Canada.

The tendency of stack gases to drift makes acid rain a widespread menace. More than 200 lakes in the Adirondacks, hundreds of miles from any industrial center, are unable to support life because their water has become so acidic.

Third stage

How Acid Rain Destroys.   Acid rain causes damage wherever it falls. It erodes various types of building rock such as limestone, marble, and mortar, which are gradually eaten away by the constant bathing in acid. Damage to buildings, houses, monuments, statues, and cars is widespread. Some priceless monuments and carvings have already been destroyed, and even trees of some varieties are dying in large numbers.

Substage

More important, however, is acid rain damage to waterways in the affected areas. (Figure 2 illustrates how a typical waterway is infiltrated.) Because of its high acidity, acid rain dramatically lowers the pH in lakes and streams. Although its effect is not immediate, acid rain can eventually make a waterway so acidic that it dies. In areas with natural acid-buffering elements such as limestone, the dilute acid has less effect. The northeastern United States and Canada, however, lack this natural protection, and so are continually vulnerable.

Substage

The pH level in an affected waterway drops so low that some species cease to reproduce. In fact, a pH level of 5.1 to 5.4 means that fisheries are threatened; once a waterway reaches a pH level of 4.5, no fish reproduction occurs. Because each creature is part of the overall food chain, loss of one element in the chain disrupts the whole cycle.

In the northeastern United States and Canada, the acidity problem is compounded by the runoff from acid snow. During the cold winter months, acid snow sits with little melting, so that by spring thaw, the acid released is greatly concentrated. Aluminum and other heavy metals normally present in soil are also released by acid rain and runoff.

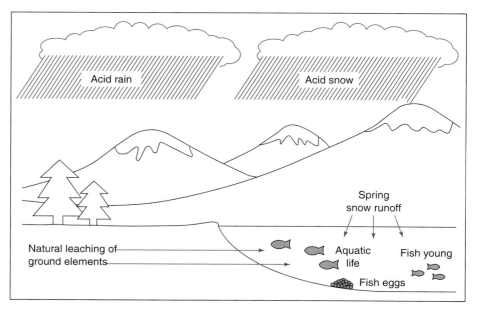

Figure 2   How Acid Rain Destroys

These toxic substances leach into waterways in heavy concentrations, affecting fish in all stages of development.

Summary

One complete cycle

Acid rain develops from nitrogen and sulfur oxides emitted by industrial and power plants burning fossil fuels. In the atmosphere, these oxides combine with ozone and water to form acid rain: precipitation with a lower-than-average pH. This acid precipitation returns to earth many miles from its source, severely damaging waterways that lack natural buffering agents. The northeastern United States and Canada are the most severely affected areas in North America.

Usability Checklist FOR INSTRUCTIONS

Use this checklist to evaluate the usability of instructions. (Numbers in parentheses refer to the first page of discussion.)

Content

☐ Does the title promise exactly what the instructions deliver? (463)
☐ Is the background adequate for the intended audience? (466)
☐ Do explanations enable users to understand what to do? (467)
☐ Do examples enable users to see how to do it correctly? (468)
☐ Are the definition and purpose of each step given as needed? (466)
☐ Is all needless information omitted? (467)
☐ Are all obvious steps omitted? (468)
☐ Do notes, cautions, or warnings appear whenever needed, before the step? (470)
☐ Is the information rate appropriate for the user's abilities and the difficulty of this procedure? (468)
☐ Are visuals adequate for clarifying the steps? (463)
☐ Do visuals repeat prose information whenever necessary? (468)
☐ Is everything accurate? (463)

Organization

☐ Is the introduction adequate without being excessive? (475)
☐ Do the instructions follow the exact sequence of steps? (476)

☐ Is all the information for a particular step close together? (473)
☐ For a complex step, does each sentence begin on a new line? (471)
☐ Is the conclusion necessary and, if necessary, adequate? (477)

Style

☐ Do introductory sentences have enough variety to maintain interest? (233)
☐ Does the familiar material appear *first* in each sentence? (472)
☐ Do steps generally have short sentences? (471)
☐ Does each step begin with the action verb? (471)
☐ Are all steps in the active voice and imperative mood? (471)
☐ Do all steps have parallel phrasing? (472)
☐ Are transitions adequate for marking time and sequence? (472)

Page Design

☐ Does each heading clearly tell users what to expect? (473)
☐ On a typed page, are steps single-spaced within, and double-spaced between? (473)
☐ Do white space and highlights set off discussion from steps? (473)
☐ Are notes, cautions, or warnings set off or highlighted? (473)
☐ Are visuals beside or near the step, and set off by white space? (473)

EXERCISES

1. Improve readability by revising the diction, voice, and design of these instructions.

What to Do Before Jacking Up Your Car

Whenever the misfortune of a flat tire occurs, some basic procedures should be followed before the car is jacked up. If possible, your car should be positioned on as firm and level a surface as is available. The engine has to be turned off; the parking brake should be set; and the automatic transmission shift lever must be placed in "park," or the manual transmission lever in "reverse." The wheel diagonally opposite the one to be removed should have a piece of wood placed beneath it to prevent the wheel from rolling. The spare wheel, jack, and lug wrench should be removed from the luggage compartment.

2. Select part of a technical manual in your field or instructions for a general audience and make a copy of the material. Using the checklist above, evaluate the sample's usability. In a memo to your instructor, discuss the strong and weak points of the instructions. Or be prepared to explain to the class why the sample is effective or ineffective.

3. Select a specialized process that you understand well and that has several distinct steps. Using the process analysis on pages 484–87 as a model, explain this process to classmates who are unfamiliar with it. Begin by completing an audience and use profile (page 484). Some possible topics: how the body metabolizes alcohol, how economic inflation occurs, how the federal deficit affects our future, how a lake or pond becomes a swamp, how a volcanic eruption occurs, the Greenhouse Effect.

4. Choose a topic from this list, your major, or an area of interest. Using the general outline in this chapter as a model, outline instructions for a procedure that requires at least three major steps. Address a general audience, and begin by completing an audience and use profile. Include (a) all necessary visuals or (b) a brief (page 282) and a rough diagram for each visual or (c) a "reference visual" (a copy of a visual published elsewhere) with instructions for adapting your visual from that one. (If you borrow visuals from other sources, provide full documentation.)

planting a tree	hitting a golf ball
hot-waxing skis	removing the rear
hanging wallpaper	wheel of a bicycle
filleting a fish	avoiding hypothermia

5. Assume that you are assistant to the communications manager for a manufacturer of outdoor products. Among the company's best-selling items are its various models of gas grills. Because of fire and explosion hazards, all grills must be accompanied by detailed instructions for safe assembly, use, and maintenance.

   One of the first procedures in the manual is the "leak test," to ensure that the gas supply-and-transport apparatus is leak free. One of the engineers has prepared the instructions in Figure 23.4. Before being published in the manual, they must be approved by communications management. Your boss directs you to evaluate the instructions for accuracy, completeness, clarity, and appropriateness, and to report your findings in a memo. Because of the legal implications, your evaluation must spell out all positive and negative details of content, organization, style, and design. (Use the checklist on page 487 as a guide.) The boss is busy and impatient, and expects your report to be no longer than two pages. Do the evaluation and write the memo.

6. Select any one of the instructional visuals in Figure 23.1 and write a prose version of those instructions—without using visual illustrations or special page design. Bring your version to class and be prepared to discuss the conclusions you've derived from this exercise.

7. Locate examples of five or more visuals from the list on page 490.

Tank must be filled prior to this step (See "Propane Tank" section for details about filling this tank)

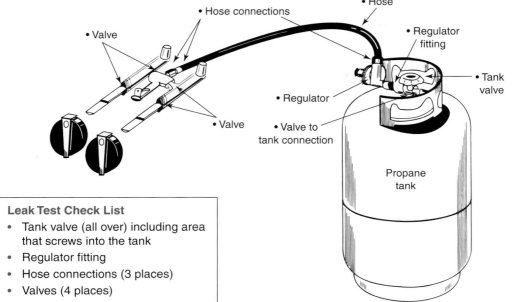

Leak Test Check List

- Tank valve (all over) including area that screws into the tank
- Regulator fitting
- Hose connections (3 places)
- Valves (4 places)

SAFETY! **All models must be Leak Tested! Take the hose, valve, and regulator assembly outdoors in a well ventilated area. Keep away from open flames or sparks. Do Not smoke during this test. Use only a soap and water solution to test for leaks.**

1. Have propane tank filled with propane gas only by a reputable propane gas dealer.
2. Attach regulator fitting to tank valve.
   - Regulator fitting has left-hand threads. Turn counterclockwise to attach.
3. Tighten regulator fitting securely with wrench.
4. Place the two matching control knobs onto the valve stems.
5. Turn the control knobs to the right, clockwise. This is the **"Off"** position.
6. Make a solution of half liquid detergent and half water.

7. Turn gas supply "ON" at the tank valve (counterclockwise).
8. Brush soapy mixture on all connections listed in the **Leak Test Check List.**
9. Observe each place for bubbles caused by leaks.
10. Tighten any leaking connections, if possible.
    - If leak cannot be stopped. **Do Not Use Those Parts!** Order new parts.
11. Turn gas supply **"Off"** at the tank valve.
12. Push in and turn control knobs to the left ("HI" position) to release pressure in hose.
13. Disconnect the regulator fitting from the tank valve.
    - Regulator fitting has left-hand threads. Turn fitting clockwise to disconnect.
14. Leave tank outdoors and return to the grill assembly with valve, hose, and regulator.

Figure 23.4 Instructions for Leak Testing a Grill   *Source: Instructions reprinted by permission of Thermos® Division.*

A visual that shows:

- how to locate something
- how to operate something
- how to handle something
- how to assemble something
- how to position something
- how to avoid damage or injury
- how to diagnose and solve a problem
- how to identify safe or acceptable limits
- how to proceed systematically
- how to make the right decision
- why an action or procedure is important

Bring your examples to class for discussion, evaluation, and comparison.

COLLABORATIVE PROJECTS

1. Select a specialized process you understand well (how gum disease develops, how an earthquake occurs, how steel is made, how a computer compiles and executes a program). Write a brief explanation of the process. Include visuals, as stipulated in Exercise 4 above. Exchange your analysis with a classmate in another major. Study your classmate's explanation for fifteen minutes and then write the same explanation *in your own words,* referring to your classmate's paper as needed. Now, evaluate your classmate's version of your original explanation for accuracy. Does it show that your explanation was understood? If not, why not? Discuss your conclusions in a memo to your instructor, submitted with all samples.

2. Draw a map of the route from your classroom to your dorm, apartment, or home—whichever is closest. Be sure to include identifying landmarks. When your map is completed, write instructions for a classmate who will try to duplicate your map from the information given in your written instructions. Be sure your classmate does not see your map! Exchange your instructions and try to duplicate your classmate's map. Compare your results with the original map. Discuss your conclusions about the usability of these instructions.

3. Divide into small groups and visit your computer center, library, or any place on campus or at work where you can find operating manuals for computers, lab or office equipment, or the like. (Or look through the documentation for your own computer hardware or software.) Locate fairly brief instructions that could use revision for improved content, organization, style, or format. Choose instructions for a procedure you are able to carry out. Make a copy of the instructions, test them for usability, and revise as needed. Submit all materials to your instructor, along with a memo explaining the improvements. Or be prepared to discuss your revision in class.

4. Visit your computer center and ask to borrow a software package that arrived without documentation or with very limited instructions. (Or perhaps you've received such programs as a member of a software club.) Run the program, and then prepare instructions for the next users. Test the usability of your instructions by having a classmate use them to run the program. Revise as needed. Remember, you are writing for a user with no experience.

5. Using a common word processor or desktop publishing software, design a form to be used for advisee course scheduling, course evaluations, or some other school function. Conduct a usability study for this document and redesign it as needed. Then write a report analyzing and evaluating the form before and after the usability study.

6. Working in small groups, revise Figure 23.5 for improved usability. Appoint one member to present your group's version to the class, explaining the specific criteria used for revision.

Proper Care Gives Safer Wear

• Follow, and save, the directions that come with your lenses. If you didn't get a patient information booklet about your lenses, request it from your eye-care practitioner.

• Use only the types of lens-care enzyme cleaners and saline solutions your practitioner okays.

• Be exact in following the directions that come with each lens-care product. If you have questions, ask your practitioner or pharmacist.

• Wash and rinse your hands before handling lenses. Fragrance-free soap is best.

• Clean, rinse, and disinfect reusable lenses *each* time they're removed, even if this is several times a day.

• Clean, rinse, and disinfect again if storage lasts longer than allowed by your disinfecting solution.

• Clean, rinse, and air-dry the lens case each time you remove the lenses. Then put in fresh solution. Replace the case every six months.

• Get your practitioner's okay before taking medicines or using topical eye products, even those you buy without a prescription.

• Remove your lenses and call your practitioner right away if you have vision changes, redness of the eye, eye discomfort or pain, or excessive tearing.

• Visit your practitioner every six months (more often if needed) to catch possible problems early.

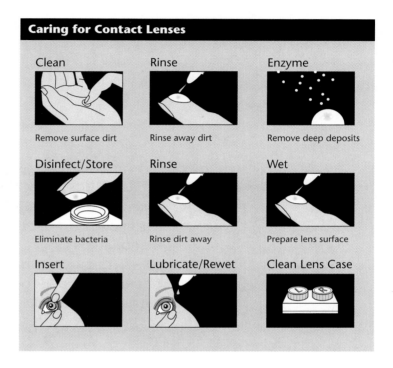

Caring for Contact Lenses

Clean	Rinse	Enzyme
Remove surface dirt	Rinse away dirt	Remove deep deposits
Disinfect/Store	Rinse	Wet
Eliminate bacteria	Rinse dirt away	Prepare lens surface
Insert	Lubricate/Rewet	Clean Lens Case

Watch Out:

• Never use saliva to wet your lenses.

• Never use tap water, distilled water, or saline solution made at home with salt tablets for any part of your lens care. Use *only* commercial sterile saline solution.

• Never mix different brands of cleaner or solution.

• Never change your lens-care regimen or products without your practitioner's okay.

• Never let cosmetic lotions, creams, or sprays touch your lenses.

• Never wear lenses when swimming or in a hot tub.

• Never wear daily-wear lenses during sleep, not even a nap.

• Never wear your lenses longer than prescribed by your eye-care practitioner. ∎

Figure 23.5 Instructions for Caring for Contact Lenses *Source: Farley, Dixie. "Keeping an Eye on Contact Lenses."* FDA Consumer *Mar.–Apr. 1998: 17–21.*

Proposals attempt to *persuade* an audience to take some form of direct action: to authorize a project, purchase a service or product, or otherwise support a specific plan for solving a problem.

Your own proposal might consist of a letter to your school board to suggest changes in the English curriculum; it may be a memo to your firm's vice president to request funding for a training program for new employees; or it may be a one thousand page document to the Defense Department to bid for a missile contract (competing with proposals from other firms). You might work alone or collaboratively, as part of a team. Developing and writing the proposal might take hours or months.

PURPOSE OF PROPOSALS

In science, business, industry, government, or education, proposals are written for decision makers: managers, executives, directors, clients, board members, or community leaders. Inside or outside your organization, these people review various proposals and then decide whether the plan is worthwhile, whether the project will materialize, whether the service or product is useful.

Before people can be induced to act, they need persuasive answers to questions like these:

What reviewers expect to learn from a proposal

- ◆ What exactly is the problem?
- ◆ Why is this important?
- ◆ What exactly is your plan?
- ◆ What are the benefits of your plan?
- ◆ What are your qualifications?
- ◆ How much will this cost?
- ◆ When will it be completed?

If your job depends on funding from outside sources, proposals might be the most vital documents you produce.

THE PROPOSAL PROCESS

The basic proposal process can be summarized like this: someone offers a plan for something that needs to be done. In business and government, this process has three stages:

Stages in the proposal process

1. Client *X* needs a service or product.
2. Firms *A, B,* and *C* propose a plan for meeting the need.
3. Client *X* awards the job to the firm offering the best proposal.

The complexity of each phase will, of course, depend on the situation. Here is a typical scenario:

Submitting a Competitive Proposal

You manage a mining engineering firm in Tulsa, Oklahoma. You regularly read the *Commerce Business Daily,* an essential reference tool for anyone whose firm seeks government contracts. This publication lists the government's latest needs for *services* (salvage, engineering, maintenance) and for *supplies, equipment,* and *materials* (guided missiles, engine parts, and so on). On Wednesday, February 19, you spot this announcement:

> **Development of Alternative Solutions to Acid Mine Water Contamination from Abandoned Lead and Zinc Mines** near Tar Creek, Neosho River, Ground Lake, and the Boone and Roubidoux aquifers in northeastern Oklahoma. This will include assessment of environmental effects of mine drainage followed by development and evaluation of alternate solutions to alleviate acid mine drainage in receiving streams. An optional portion of the contract, to be bid on as an add-on and awarded at the discretion of the OWRB, will be to prepare an Environmental Impact Assessment for each of three alternative solutions as selected by the OWRB. The project is expected to take six months to accomplish, with anticipated completion date of September 30, 20XX. The projected effort for the required task is thirty person-months. Requests for proposals will be issued and copies provided to interested sources upon receipt of a written request. Proposals are due March 1. (044)
> Oklahoma Water Resources Board
> P. O. Box 53585
> 1000 Northeast 10th Street
> Oklahoma City, OK 73151
> (405)555–2541

Your firm has the personnel, experience, and time to do the job, so you decide to compete for the contract. Because the March 1 deadline is fast approaching, you write immediately for a copy of the request for proposal (RFP). The RFP will give you the guidelines for developing the proposal—guidelines for spelling out your plan to solve the problem (methods, timetables, costs).

You receive a copy of the RFP on February 21—only one week before deadline. You get right to work with the two staff engineers you have appointed to your proposal team. Because the credentials of your staff could affect the client's acceptance of the proposal, you ask team members to update their résumés for inclusion in an appendix to the proposal. ▲

In scenarios like this, the client will award the contract to the firm submitting the best proposal, based on the following criteria (and perhaps others):

Criteria by which reviewers evaluate proposals

- understanding of the client's needs, as described in the RFP
- soundness of the plan being offered
- quality of the project's organization and management
- ability to complete the job by deadline

- ability to control costs
- firm's experience on similar projects
- qualifications of staff to be assigned to the project
- firm's record for similar projects

A client's specific evaluation criteria are often listed (in order of importance or on a point scale) in the RFP. Although these criteria may vary, every client expects a proposal that is *clear, informative,* and *realistic.*

Some clients hold a preproposal conference for the competing firms. During this briefing, the firms are informed of the client's needs, expectations, specific start-up and completion dates, criteria for evaluation, and any other details that might help guide proposal development.

In most cases, the merits of your proposed plan are judged *solely* by what you have *on paper.* This reality is summed up in the following advice, from a government publication that solicits proposals:

Give reviewers everything they need

> [Proposal] reviewers will base their conclusions only on information contained in the proposal. It cannot be assumed that readers are acquainted with the firm or key individuals or any referred-to experiments.[1]

Any proposal should stand alone in meaning and persuasiveness.

PROPOSAL TYPES

Proposals are classified according to *origin, audience,* or *intention.* Based on its origin, a proposal is either *solicited* or *unsolicited*—that is, requested by an employer or potential client or initiated on your own because you have recognized a need. Business and government proposals are usually solicited, and they originate from a customer's request (as in the situation on page 496).

Based on its audience, a proposal may be *internal* or *external*—written for members of your organization or for clients and funding agencies. (The situation on page 509 calls for an external proposal.)

Based on its intention, a proposal may be a *planning, research,* or *sales* proposal. These last categories by no means account for all variations among proposals. Some proposals fall within all three categories.

Planning Proposal

A planning proposal offers solutions to a problem or suggestions for improvement. It might be a request for funding to expand the campus newspaper, an architectural plan for new facilities at a ski area, or a plan to develop energy alternatives to fossil fuels.

---

[1]*Small Business Innovation Research Program* (Washington: U.S. Department of Defense, 1996): 9.

The planning proposal that follows is external and solicited. The XYZ Corporation has contracted a team of communication consultants to design in-house writing workshops. The consultants must persuade the client that their methods are likely to succeed. In their proposal, addressed to the company's education officer, the consultants offer concrete and specific solutions to clearly identified problems.

After a brief introduction summarizing the problem, the authors develop their proposal under two headings, "Assessment of Needs" and "Proposed Plan." Under "Proposed Plan," subheadings offer an even more specific forecast. The "Limitations" section shows that the proposal authors are careful to promise no more than they can deliver. Because this proposal is external, it is cast in letter form.

Planning Proposal

States purpose

Dear Mary:

Thanks for sending the writing samples from your technical support staff. Here is what we're doing to design a realistic approach.

Identifies problem

Assessment of Needs

After conferring with technicians in both Jack's and Terry's groups and analyzing their writing samples, we identified this hierarchy of needs:

- improving readability
- achieving precise diction
- summarizing information
- organizing a set of procedures
- formulating various memo reports
- analyzing audiences for upward communication
- writing persuasive bids for transfer or promotion
- writing persuasive suggestions

Proposes solution

Proposed Plan

Based on the needs listed above, we have limited our instruction package to eight carefully selected and readily achievable goals.

Course Outline.   Our eight, two-hour sessions are structured as follows:

Details what will be done

1. achieving sentence clarity
2. achieving sentence conciseness
3. achieving fluency and precise diction
4. writing summaries and abstracts
5. outlining manuals and procedures
6. editing manuals and procedures
7. designing various reports for various purposes
8. analyzing the audience and writing persuasively

Details how it will be done

Classroom Format.   The first three meetings will be lecture-intensive with weekly exercises to be done at home and edited collectively in class. The remaining five weeks will combine lecture and exercises with group editing of work-related documents. We plan to remain flexible so we can respond to needs that arise.

Limitations

Sets realistic expectations

Given our limited contact time, we cannot realistically expect to turn out a batch of polished communicators. By the end of the course, however, our students will have begun to appreciate writing as a deliberate process.

Encourages reader response

If you have any suggestions for refining this plan, please let us know.

Notice that the word choice ("thanks," "what we're doing," "Jack and Terry") creates an informal, familiar tone—appropriate in this external document only because the consultants and client have spent many hours in conferences, luncheons, and phone conversations.

Research Proposal

Research (or grant) proposals request approval (and often funding) for a research project. A chemistry professor might address a research proposal to the Environmental Protection Agency for funds to identify toxic contaminants in local groundwater. Research proposals are solicited by many government and private agencies: National Science Foundation, National Institutes of Health, and others. Each granting agency has its own requirements for proposal format and content, but any successful research proposal spells out the goals of the project, the plan for achieving these goals, and the writer's qualifications.

In college, you might submit proposals for independent study, field study, or a thesis project. A technical writing student usually submits a relatively brief research proposal that will lead to the term project (such as the long proposal that begins on page 505 or the analytical report that begins on page 530).

In the following research proposal, Tom Dewoody requests his instructor's authorization to do a feasibility study (Chapter 25) that will produce an analytical report for an audience of potential investors. Dewoody's proposal is persuasive because it clearly answers the questions about *what, why, how, when,* and *where.* Because this proposal is internal, it is cast informally as a memo.

Research Proposal

To:	Dr. John Lannon
From:	T. Sorrells Dewoody
Date:	March 16, 20XX
Subject:	*Proposal for Determining the Feasibility of Marketing Dead Western White Pine*

Introduction

Over the past four decades, huge losses of western white pine have occurred in the northern Rockies, primarily attributable to white pine blister rust and the attack of the mountain pine beetle. Estimated annual mortality is 318 million board feet. Because of the low natural resistance of white pine to blister rust, this high mortality rate is expected to continue indefinitely.

If white pine is not harvested while the tree is dying or soon after death, the wood begins to dry and check (warp and crack). The sapwood is discolored by blue stain, a fungus carried by the mountain pine beetle. If the white pine continues to stand after death, heart cracks develop. These factors work together to cause degradation of the lumber and consequent loss in value.

Statement of Problem

White pine mortality reduces the value of white pine stumpage because the commercial lumber market will not accept it. The major implications of this problem are two: first, in the face of rising demand for wood, vast amounts of timber lie unused; second, dead trees are left to accumulate in the woods, where they are rapidly becoming a major fire hazard here in northern Idaho and elsewhere.

Proposed Solution

One possible solution to the problem of white pine mortality and waste is to search for markets other than the conventional lumber market. The last few years have seen a burst of popularity and growing demand for weathered barn boards and wormy pine for interior paneling. Some firms around the country are marketing defective wood as specialty products. (These firms call the wood from which their products come "distressed," a term I will use hereafter to refer to dead and defective white pine.) Distressed white pine quite possibly will find a place in such a market.

Scope

To assess the feasibility of developing a market for distressed white pine, I plan to pursue six areas of inquiry:

1. What products presently are being produced from dead wood, and what are the approximate costs of production?
2. How large is the demand for distressed-wood products?
3. Can distressed white pine meet this demand as well as other species meet it?
4. Does the market contain room for distressed white pine?
5. What are the costs of retrieving and milling distressed white pine?
6. What prices for the products can the market bear?

Methods

My primary data sources will include consultations with Dr. James Hill, Professor of Wood Utilization, and Dr. Sven Bergman, Forest Economist—both members of the College of Forestry, Wildlife, and Range. I will also inspect decks of dead white pine at several locations, and visit a processing mill to evaluate it as a possible base of operations. I will round out my primary research with a letter and telephone survey of processors and wholesalers of distressed material.

Opens with background and causes of the problem

Describes problem

Describes one possible solution

Defines scope of the proposed study

Describes how study will be done

Secondary sources will include publications on the uses of dead timber, and a review of a study by Dr. Hill on the uses of dead white pine.

My Qualifications

Gives the writer's qualifications for this project

I have been following Dr. Hill's study on dead white pine for two years. In June of this year I will receive my B.S. in forest management. I am familiar with wood milling processes and have firsthand experience at logging. My association with Drs. Hill and Bergman gives me the opportunity for an in-depth feasibility study.

Conclusion

Encourages reader acceptance

Clearly, action is needed to reduce the vast accumulations of dead white pine in our forests. The land on which they stand is among the most productive forests in northern Idaho. By addressing the six areas of inquiry mentioned earlier, I can determine the feasibility of directing capital and labor to the production of distressed white pine products. With your approval I will begin research at once.

Sales Proposal

The sales proposal is a major marketing tool that offers a service or product. Sales proposals may be solicited or unsolicited. If they are solicited, several firms may compete with proposals of their own. Because sales proposals are addressed to readers outside your organization, they are cast as letters if they are brief. Long sales proposals, like long reports, are formal documents with supplements (cover letter, title page, table of contents).

A successful sales proposal persuades customers that your product or service surpasses those of any competitors. In the following solicited proposal, the writer explains *why* his machinery is best for the job, *how* the job can be done efficiently, *what* qualifications *he offers,* and *how much* the job will cost. He will be legally bound by his estimate. As a protective measure, he points out possible causes of increased costs.

Sales Proposal

Subject: *Proposal to Dig a Trench and Move Boulders at Bliss Site*

Dear Mr. Haver:

Offers to undertake the job

I've inspected your property and would be happy to undertake the landscaping project necessary for the development of your farm.

Gives the writer's qualifications

The backhoe I use cuts a span 3 feet wide and can dig as deep as 18 feet—more than an adequate depth for the mainline pipe you wish to lay. Because this backhoe is on tracks rather than tires, and is hydraulically operated, it is particularly efficient in moving rocks. I have more than twelve years of experience with backhoe work and have completed many jobs similar to this one.

Explains how the job will be done

After examining the huge boulders that block access to your property, I am convinced they can be moved only if I dig out underneath and exert upward pressure with the hydraulic ram while you push forward on the boulders with your D-9 Caterpillar. With this method, we can move enough rock to enable you to farm that now inaccessible tract. Because of its power, my larger backhoe will save you both time and money in the long run.

Gives a qualified cost estimate

This job should take 12 to 15 hours, unless we encounter subsurface ledge formations. My fee is $200 per hour. The fact that I provide my own dynamiting crew at no extra charge should be an advantage to you because you have so much rock to be moved.

Encourages reader acceptance

Please phone me anytime for more information. I'm sure we can do the job economically and efficiently.

NOTE

Never underestimate costs by failing to account for all variables—a sure way to lose money or clients.

The introduction describes the subject and purpose of the proposal. The conclusion reinforces the confident tone throughout and encourages readers' acceptance by ending with—and thus emphasizing—two vital words: "economically" and "efficiently."

The proposal categories (planning, research, and sales) discussed in this section are neither exhaustive nor mutually exclusive. A research proposal, for example, may request funds for a study that will lead to a planning proposal. The Vista proposal partially shown below combines planning and sales features; if clients accept the preliminary plan, they will hire the firm to install the automated system.

ELEMENTS OF A PERSUASIVE PROPOSAL

Reviewers will evaluate your proposal on the basis of the following quality indicators. (See also the criteria listed on pages 494–95.)

Understanding of the Client's Needs

Focus on the problem and the objective

Clients want specific suggestions for meeting specific needs. Their biggest question is "What can this do for me?". Demonstrate a clear understanding of their problem and their expectations, and offer an appropriate solution.

In his proposal for automating office procedures at Vista, Inc., Gerald Beaulieu focuses on the company's inefficiencies and then outlines specific solutions.

Gives background

Statement of the Problem
Vista provides two services. (1) It locates freight carriers for its clients. The carriers, in turn, pay vista a 6 percent commission for each referral. (2) Vista handles all shipping paperwork for its clients. For this auditing service, clients pay Vista a monthly retainer.

Describes problem and its effects

Although Vista's business has increased steadily for the past three years, record keeping, accounting, and other paperwork are still done manually. These inefficient procedures have caused a number of problems, including late billings, lost commissions, and poor account maintenance. Updated office procedures seem crucial to competitiveness and continued growth.

Enables readers to visualize results

Objective
This proposal offers a realistic and effective plan for streamlining Vista's office procedures. We first identify the burden imposed on your staff by the current system, and then we show how to reduce inefficiency, eliminate client complaints, and improve your cash flow by automating most office procedures.

A Focus on Benefits

Spell out the benefits

Show clients what they (or their organization) stand to gain by adopting your plan. (Each benefit is forecasted here in the introduction, and will be described at length later in the body section.)

Relates benefits directly to client's needs

Once your automated system is operational, you will be able to

- identify cost-effective carriers
- coordinate shipments (which will ensure substantial client discounts)
- print commission bills
- track shipments by weight, miles, fuel costs, and destination
- send clients weekly audit reports on their shipments
- bill clients on a 25-day cycle
- produce weekly or monthly reports

Additional benefits include eliminating repetitive tasks, improving cash flow, and increasing productivity.

Honest and Supportable Claims

Promise only what you can deliver

Because they typically involve large sums of money as well as contractual obligations, proposals require a solid ethical and legal foundation. False promises not only demolish a company's reputation but also invite lawsuits.

Here is how the Vista proposal qualifies its promises:

Anticipates the unexpected

As countless firms have learned, imposing automated procedures on a staff can create severe morale problems—particularly among senior staff who feel coerced. To diminish employee resistance, we suggest that your entire staff be invited to comment on this proposal. To help avoid hardware and software problems once the system is operational, we have included recommendations and a budget for staff training. (Inadequate training is counterproductive to the automation process.)

If the best available solutions have limitations, say so. Notice how the above solutions are qualified ("diminish" and "help avoid" instead of "eliminate") so as not to promise more than the plan can achieve.

A proposal can be judged fraudulent if it misleads potential clients by

Ethical and legal violations in a proposal

- ◆ making unsupported claims,
- ◆ ignoring anticipated technical problems, or
- ◆ knowingly underestimating costs or time requirements.

See page 46 for advice on supporting your claims.

Appropriate Detail

Provide adequate but not excessive detail

Vagueness is a fatal flaw in a proposal. Be sure to spell things out. Instead of writing, "We will install state-of-the-art equipment," enumerate the products or services to be provided.

Spells out what will be provided

To meet your automation requirements, we will install twelve Power Macintosh G3 computers with 9-Gigabyte hard drives. The system will be networked for rapid file transfer between offices. The plan also includes interconnection with four Hewlett-Packard 5 MP printers, and one HP Desk Jet 1600 CM color printer.

To avoid misunderstandings that could produce legal complications, a proposal must elicit *one* interpretation only.

Place support material (maps, blueprints, specifications, calculations) in an appendix so as not to interrupt the discussion.

NOTE   *While concrete and specific detail is vital, never overburden reviewers with needless material. A precise audience analysis (Chapter 3) can pinpoint specific information needs.*

Readability

Make the proposal inviting and easy to understand

A readable proposal is straightforward, easy to follow and understandable. Its tone is confident and encouraging, not bossy and critical. Review Chapter 13 for style strategies.

Visuals

Emphasize key points in your proposal with relevant tables, flowcharts, and other visuals (Chapter 14), properly introduced and discussed.

As the flowchart (Figure 1) illustrates, Vista's routing and billing system creates redundant work for your staff. The routing sheet alone is handled at least six times. Such extensive handling leads to errors, misplaced paperwork, and late billing.

Gives a framework for interpreting the visual

Visual repeats, restates, or reinforces the prose

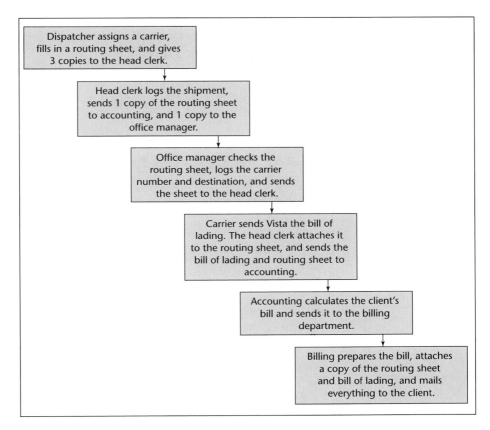

Figure 1   Flowchart of Vista's Manual Routing and Billing System

Accessible Page Design

Make the audience's job easy

Yours might be one of several proposals being reviewed. Help the audience to get in quickly, find what they need, and get out. Review Chapter 15 for design strategies.

Supplements Tailored for a Diverse Audience

Analyze the specific needs and interests of each major reviewer

A single proposal often addresses a diverse audience: executives, managers, technical experts, attorneys, politicians, and so on. Various reviewers are interested in various parts of your proposal. Experts look for the technical details. Others might be interested in the recommendations, costs, timetable, or expected results, but they will need an explanation of technical details as well.

How to give each
major reviewer
what he or she
expects

If the primary audience is expert or informed, keep the proposal text itself technical. For uninformed secondary reviewers (if any), provide an informative abstract, a glossary, and appendixes explaining specialized information. If the primary audience has no expertise and the secondary audience does, write the proposal itself for laypersons, and provide appendixes with the technical details (formulas, specifications, calculations) that experts will use to evaluate your plan. Review Chapter 16 for specific supplements.

If you are unsure which supplements to include in an internal proposal, ask the intended audience or study other proposals. For a solicited proposal (to an outside agency), follow the agency's instructions exactly.

A GENERAL MODEL FOR PROPOSALS

Depending on a proposal's complexity, each section contains some or all of the subsections listed in the following general outline:

Possible headings to
include in a
proposal

I. **Introduction**
   A. Statement of Problem and Objective
   B. Background
   C. Need
   D. Benefits
   E. Qualifications of Personnel
   F. Data Sources
   G. Limitations and Contingencies
   H. Scope

II. **Body**
   A. Methods
   B. Timetable
   C. Materials and Equipment
   D. Personnel
   E. Available Facilities
   F. Needed Facilities
   G. Cost
   H. Expected Results
   I. Feasibility

III. **Conclusion**
   A. Summary of Key Points
   B. Request for Action

The above headings can be rearranged, combined, divided, or deleted as needed. Not every proposal will contain all subsections; however, each major section must persuasively address specific information needs as illustrated in the sample proposal that begins on page 505.

Introduction

From the beginning, your goal is *to sell your idea*—to demonstrate the need for the project, your qualifications for tackling the project, and your clear understanding of how to proceed. Readers quickly lose interest in a wordy, evasive, or vague introduction.

Following is the introduction for a planning proposal titled "Proposal for Solving the Noise Problem in the University Library." Jill Sanders, a library work-study student, addresses her proposal to the chief librarian and the administrative staff. Because this proposal is unsolicited, it must first make the problem vivid through details that arouse concern and interest. This introduction is longer than it would be in a solicited proposal, whose audience would already agree on the severity of the problem.

NOTE *Title page, informative abstract, table of contents, and other supplements that ordinarily accompany long proposals of this type are omitted here to save space. See Chapter 16 for discussion and examples of each type of document supplement.*

INTRODUCTION
Statement of Problem

> Concise descriptions of problem and objective immediately alert the readers

During the October 20XX Convocation at Margate University, students and faculty members complained about noise in the library. Soon afterward, areas were designated for "quiet study," but complaints about noise continue. To create a scholarly atmosphere, the library should take immediate action to decrease noise.

Objective

This proposal examines the noise problem from the viewpoint of students, faculty, and library staff. It then offers a plan to make areas of the library quiet enough for serious study and research.

Sources

> This section comes early because it is referred to in the next section

My data come from a university-wide questionnaire; interviews with students, faculty, and library staff; inquiry letters to other college libraries; and my own observations for three years on the library staff.

Details of the Problem

> Details help readers to understand the problem

This subsection examines the severity and causes of the noise.

Severity.   Since the 20XX Convocation, the library's fourth and fifth floors have been reserved for quiet study, but students hold group study sessions at the large tables and disturb others working alone. The constant use of computer terminals on both floors adds to the noise, especially when students converse. Moreover, people often chat as they enter or leave study areas.

On the second and third floors, designed for reference, staff help patrons locate materials, causing constant shuffling of people and books, as well as loud conversation. At the computer service desk on the third floor, conferences between students and instructors create more noise.

The most frequently voiced complaint from the faculty members interviewed was about the second floor, where people using the Reference and Government Documents services converse loudly. Students complain about the lack of a quiet spot to study, especially in the evening, when even the "quiet" floors are as noisy as the dorms.

More than 80 percent of respondents (530 undergraduates, 30 faculty, 22 graduate students) to a university-wide questionnaire (Appendix A) insisted that excessive noise discourages them from using the library as often as they would prefer. Of the student respondents, 430 cited quiet study as their primary reason for wishing to use the library.

The library staff recognizes the problem but has insufficient personnel. Because all staff members have assigned tasks, they have no time to monitor noise in their sections.

Causes.   Respondents complained specifically about these causes of noise (in descending order of frequency):

1. Loud study groups that often lapse into social discussions.
2. General disrespect for the library, with some students' attitudes characterized as "rude," "inconsiderate," or "immature."
3. The constant clicking of computer terminals on all five floors, and of typewriters on the first three.
4. Vacuuming by the evening custodians.

All complaints converged on lack of enforcement by library staff.

Because the day staff works on the first three floors, quiet-study rules are not enforced on the fourth and fifth floors. Work-study students on these floors have no authority to enforce rules not enforced by the regular staff. Small, black-and-white "Quiet Please" signs posted on all floors go unnoticed, and the evening security guard provides no deterrent.

Needs

Excessive noise in the library is keeping patrons away. By addressing this problem immediately, we can help restore the library's credibility and utility as a campus resource. We must reduce noise on the lower floors and eliminate it from the quiet-study floors.

Scope

The proposed plan includes a detailed assessment of methods, costs and materials, personnel requirements, feasibility, and expected results.

Body

The body (or plan) section of your proposal will receive the most audience attention. The main goal of this section is to prove your plan will work. Here you spell out your plan in enough detail for the audience to evaluate its soundness. If this section is vague, your proposal stands no chance of being accepted. Be sure your plan is realistic and promise no more than you can deliver.

PROPOSED PLAN

This plan takes into account the needs and wishes of our campus community, as well as the available facilities in our library.

Phases of the Plan

Noise in the library can be reduced in three complementary phases: (1) improving publicity, (2) shutting down and modifying our facilities, and (3) enforcing the quiet rules.

Improving Publicity.   First, the library must publicize the noise problem. This assertive move will demonstrate the staff's interest. Publicity could include articles by staff members in the campus newspaper, leaflets distributed on campus, and a freshman library orientation acknowledging the noise problem and asking cooperation from new students. All forms of publicity should detail the steps being taken by the library to solve the problem.

Shutting Down and Modifying Facilities.   After notifying campus and local newspapers, you should close the library for one week. To minimize disruption, the shutdown should occur between the end of summer school and the beginning of the fall term.

During this period, you can convert the fixed tables on the fourth and fifth floors to cubicles with temporary partitions (six cubicles per table). You could later convert the cubicles to shelves as the need increases.

Then you can take all unfixed tables from the upper floors to the first floor, and set up a space for group study. Plans are already under way for removing the computer terminals from the fourth and fifth floors.

Enforcing the Quiet Rules.   Enforcement is the essential long-term element in this plan. No one of any age is likely to follow all the rules all the time—unless the rules are enforced.

First, you can make new "Quiet" posters to replace the present, innocuous notices. A visual-design student can be hired to draw up large, colorful posters that attract attention. Either the design student or the university print shop can take charge of poster production.

Next, through publicity, library patrons can be encouraged to demand quiet from noisy people. To support such patron demands, the library staff can begin monitoring the fourth and fifth floors, asking study groups to move to the first floor, and revoking library privileges of those who refuse. Patrons on the second and third floors can be asked to speak in whispers. Staff members should set an example by regulating their own voices.

Costs and Materials

- ◆ The major cost would be for salaries of new staff members who would help monitor. Next year's library budget, however, will include an allocation for four new staff members.
- ◆ A design student has offered to make up four different posters for $200. The university printing office can reproduce as many posters as needed at no additional cost.
- ◆ Prefabricated cubicles for 26 tables sell for $150 apiece, for a total cost of $3,900.
- ◆ Rearrangement on various floors can be handled by the library's custodians.

The Student Fee Allocations Committee and the Student Senate routinely reserve funds for improving student facilities. A request to these organizations presumably would yield at least partial funding for the plan.

Personnel

Describes personnel needed

The success of this plan ultimately depends on the willingness of the library administration to implement it. You can run the program itself by committees made up of students, staff, and faculty. This is yet another area where publicity is essential to persuade people that the problem is severe and that you need their help. To recruit committee members from among students, you can offer Contract Learning credits.

The proposed committees include an Antinoise Committee overseeing the program, a Public Relations Committee, a Poster Committee, and an Enforcement Committee.

Feasibility

Assesses probability of success

On March 15, 20XX, I mailed survey letters to twenty-five New England colleges, inquiring about their methods for coping with noise in the library. Among the respondents, sixteen stated that publicity and the administration's attitude toward enforcement were main elements in their success.

Improved publicity and enforcement could work for us as well. And slight modifications in our facilities, to concentrate group study on the busiest floors, would automatically lighten the burden of enforcement.

Benefits

Offers a realistic and persuasive forecast of benefits

Publicity will improve communication between the library and the campus. An assertive approach will show that the library is aware of its patrons' needs and is willing to meet those needs. Offering the program for public inspection will draw the entire community into improvement efforts. Publicity, begun now, will pave the way for the formation of committees.

The library shutdown will have a dual effect: it will dramatize the problem to the community, and it will provide time for the physical changes. (An antinoise program begun with carpentry noise in the quiet areas would hardly be effective.) The shutdown will be both a symbolic and a concrete measure, leading to reopening of the library with a new philosophy and a new image.

Continued strict enforcement will be the backbone of the program. It will prove that staff members care enough about the atmosphere to jeopardize their friendly image in the eyes of some users, and that the library is not afraid to enforce its rules.

Conclusion

The conclusion reaffirms the need for the project and induces the audience to act. End on a strong note, with a conclusion that is assertive, confident, and encouraging—and keep it short.

CONCLUSION AND RECOMMENDATION

Reemphasizes need and feasibility and encourages action

The noise in Margate University library has become embarrassing and annoying to the whole campus. Forceful steps are needed to restore the academic atmosphere.

Aside from the intangible question of image, close inspection of the proposed plan will show that it will work if you take the recommended steps and—most important—if daily enforcement of quiet rules becomes a part of the library's services.

In long proposals, especially those beginning with a comprehensive abstract, the conclusion can be omitted.

A SITUATION REQUIRING A PROPOSAL

This proposal in Figure 24.1 is aimed at one audience only: the citizens of Amherst and surrounding towns, who will be voting on whether to grant the funding for renovation of the regional high school.

Audience and Use Profile for a Formal Proposal

Audience Identity and Needs.   Most readers are broadly familiar with this issue, having seen it debated for nearly two years. But they still need an item-by-item explanation of the problems created by ever-increasing enrollment. Since the project would be funded partially by increased property taxes, voters are asking tough questions such as these:

◆ Just how big a problem is this?
◆ Says who?
◆ Why, exactly, do we need this project?
◆ What will it cost?
◆ Why can't we wait?
◆ Suppose we do nothing?
◆ What will we get for our money?

Attitude and Personality.   Nearly everyone (parents, taxpayers without school-age children, people on fixed incomes) has a real stake in this topic. Many voters tend to resist any request for more money by arguing that everyone has to economize during these difficult economic times. In short, audience attitude ranges from supportive to receptive to hesitant to defiant. This proposal is aimed at the hesitant population, since there is little chance of influencing those who reject the project outright.

To have a positive effect on its audience, this proposal needs to make a case for the project and spell out a realistic plan, showing that the various committees are sincere in their intention to eliminate nonessential building costs.

Expectations about the Document.   Voters will scrutinize and evaluate this proposal for its soundness. Especially in a budget request, the audience expects no shortcuts. Each element of the project will have to be justified. All the relevant numbers (enrollment figures, projected tax increase, costs of delaying the project, and so on) will have to be given and explained in ways readers can understand.

The proposal will be designed as an $8\frac{1}{2} \times 11$-inch, foldout brochure, to be mailed to each resident in the regional district. The document will be organized in a sequence that (1) identifies the problem, (2) lists costs to taxpayers, (3) justifies immediate need, (4) spells out relevant basic data and project background, (5) explains consequences of inaction, and (6) gives a detailed view of the additions and renovations. ▲

AMHERST REGIONAL HIGH SCHOOL BUILDING PROJECT

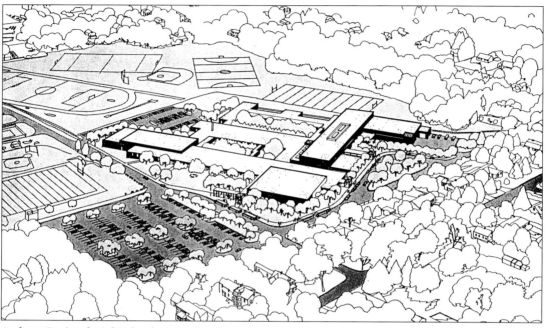

Amherst Regional High School   AERIAL VIEW

D•R•A   **Drummery**   Colby Hall   **Kuhn • Riddle Architects**
             Rosane   141 Herrick Road   Associate Architects
             Anderson   P.O. Box 299   7 North Pleasant Street
             Inc.   Newton Center, MA   Amherst, MA 01002
             Architecture   02159-0299   413-259-1630
             Interior Design 617-964-1700

Figure 24.1   A Planning Proposal   *Reprinted with permission. Copyright © 1994 Amherst/Pelham Regional School Board; Drummery, Rosane, and Anderson, Inc.; Kuhn Riddle Architects.*

1

WHY IS THE ADDITION/RENOVATION NEEDED?

Within one year both the junior and senior high schools will be above capacity and enrollments will continue to grow (Figure 1). Both buildings will require additions if the 7–9 and 10–12 split is maintained. By moving the 9th grade from the junior high school to the high school, we will eliminate the need for extensive additions to both buildings and create better opportunities for early adolescent and high school programming.

The students who will attend grades 9–12 over the next nine years are already enrolled in the elementary school of the four regional towns or in grades 7 and 8 in the present junior high. The number of students expected in the high school by the year 2000–1 nearly doubles from the current 759 (10–12) to 1,450 (9–12).

The high school was built in 1956 with an addition in 1964. All its systems are now between 30 and 40 years old and in need of upgrading. The state will only reimburse a school district for renovation if it is part of a large addition or expansion project. This project gives us the opportunity to renovate the existing building and bring it into compliance with current codes.

Figure 1   No Growth vs. Projected Growth in Grades 9–12 for 1993–2004

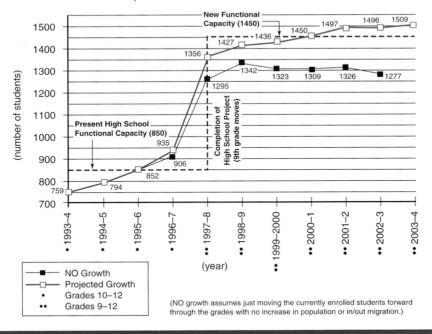

Figure 24.1   A Planning Proposal   *Continued*

WHAT WILL THE PROJECT COST?

The total construction project cost will not exceed $23,100,000, which includes the cost for construction, equipment, furnishings, and all fees. With the state reimbursing 67% of the project cost, the towns will be responsible for 33%, or approximately $7,700,000. This $7.7 million will be divided as follows:

Amherst	$5,953,600	(77.32%)
Leverett	$646,800	(8.40%)
Pelham	$495,800	(6.44%)
Shutesbury	$603,800	(7.84%)

The construction cost includes $13,475,000 for construction of the new addition and related site work plus $5,000,000 for renovation of the existing building. An additional $4,625,000 is budgeted for other project-related costs, including furnishings, equipment, surveys, testing, architectural/engineering fees, and contingencies.

WHY NOW?

Enrollments in the 1994–95 school year (Table 1) will effectively fill our secondary schools. Approval at this time by the four communities will allow ample time for preparation of construction documents necessary for state approval of the expansion. The project can then be bid in Spring 1995, with construction beginning in summer. Even under these circumstances, construction will not be complete until September 1997.

The School Building Committee is seeking approval at this time in order to:

1. allow construction to begin as soon as possible, providing additional space at the earliest possible date
2. provide ample time for design and final construction documents
3. minimize cost of temporary space arrangements
4. take advantage of favorable interest rates
5. avoid potential changes to the reimbursement formula which could increase the cost to the towns. Under the current formula the Amherst-Pelham Regional School District is eligible for 67% reimbursement from the state.

Table 1   October 1, 1993 Actual Enrollments

GRADE	K	1	2	3	4	5	6	7	8	9	10	11	12
AMHERST	222	265	252	239	252	253	272	273	231	216	227	198	183
PELHAM	24	17	15	17	22	17	16	12	11	20	11	11	11
LEVERETT	20	28	35	31	28	23	26	23	28	22	18	12	29
SHUTESBURY	26	27	29	30	39	27	31	28	24	18	26	15	18
TOTALS	292	337	331	317	341	320	345	336	294	276	282	236	241

Figure 24.1   A Planning Proposal   *Continued*

3

THE BASICS:

School Growth: General increase in the school population plus the addition of ninth
                  grade to the high school will require housing 1,450 students by the
                  year 2000. Current high school enrollment is 759.

The Scope: New construction of 118,500 square feet to include:
                      * Enlarged cafeteria
                      * Enlarged library
                      * New labs for science and technology education
                      * 22 multipurpose classrooms
                      * Large group instruction room
                      * 2 exercise rooms for physical education
                      * Expanded space for:
                          special education
                          T.B.E. (transitional bilingual education)
                          E.S.L. (English as a second language)
                          art instruction
                          music instruction
                          teacher resource areas
                      * Improved parking and bus access

                  Refurbishing/renovation of 118,750 of existing square feet to include:
                      * Heating   * Lighting
                      * Ventilating   * Electrical
                      * Plumbing   * Fire protection
                      * Reroofing   * Energy conservation

The Cost: $23.1 million for the total construction project including furnishings,
                  equipment and fees, of which 67% ($15.4 million) is reimbursable by
                  the state. The state will also reimburse 67% of the interest costs.

The Tax Impact: The tax impact will be spread over the twenty-year bonding period. In
                  general it will be higher in the earliest years and will decline over the
                  remaining time, with the state reimbursement being constant over this
                  period. The tax impact will vary among the four towns.

The Time Table: With a positive vote in all four towns this spring, ground could be
                  broken by June 1995. Phased construction will allow the school to
                  remain open during construction, with completion and full occupancy
                  by September 1997.

Figure 24.1   A Planning Proposal   *Continued*

4

BACKGROUND:

* Regional School Building Needs Committee, over 3 years, studied:
  a. Student population growth
  b. Space alternatives
  c. Possibility of third secondary school
  d. Feasibility of building additions/renovations to both junior and senior high school.

* Regional School Committee voted to move 9th grade from junior high to high school because the change:
  a. Unites the 9th grade with 9-12 curriculum in one location
  b. Requires only one building project.

* Specifications Committee, using State Department of Education criteria, determined:
  a. Use of existing space
  b. Total space requirements for a school with an enrollment of 1,450, based on continuation of existing school programs
  c. Flexibility needed for the future.

* School Committee, Building Committee, and State Department of Education have approved the Educational Specifications which reflect the current and projected needs for space by departments, programs and other building functions.

* Regional School Building Committee:
  a. Selected an architect for the project
  b. Selected a project design from several alternatives
  c. Reviewed thoroughly the conditions of the existing building
  d. Consulted with the state's School Building Assistance Bureau
  e. Reviewed and revised the scope and costs of the project.

WHAT WILL HAPPEN IF NO ACTION IS TAKEN?

If the high school is not expanded, both the junior and senior high schools will experience crippling overcrowding in the very near future. Both schools would likely require staggered sessions, and/or temporary modular classrooms, with early and late arrival/dismissal times. This will lead to a general erosion of the quality of education because of severely limited program and extracurricular opportunities for students. School operating expenses and transportation costs will increase. The cost of temporary classrooms is not reimbursable by the state. Additionally, the cost for ADA compliance and future repairs to major building systems would have to be borne 100% by local taxpayers.

Figure 24.1 A Planning Proposal   *Continued*

5

WHAT IS THE PROPOSED DESIGN?

The Building Committee established a program based on the educational specifications and a desire to:

* Avoid a sprawling floor plan
* Produce a functional and economical design
* Centralize locations of core facilities, such as the library
* Consolidate and centralize circulation of students.

Working together, the architects and Building Committee reviewed many options for adding on to the existing school. These options included adding over the existing building, filling in courtyards, adding many small additions and consolidating all the space in a one-, two-, or three-story addition. The final design, as described below, is a functional, compact, and economical response to the program requirements (Figures 2a and 2b).

The recommended addition/renovation project has two components:

(1) It adds 118,500 square feet of new construction to house the additional square footage needed due to increased enrollments and,

(2) It renovates 118,750 square feet of the existing building to upgrade the 40-year-old structure and bring it into the twenty-first century.

The central, organizing element of the project will be a new, wide, ramped corridor, or "Main Street," that will run the length of the school between the existing building on the west and the three-story addition on the east. The west side of the corridor will tie into all three of the main corridors of the existing school. Along the east side of the corridor will be a new library, new cafeteria and new entry into the school from the east side of the addition. At the north end of the "Street" on the first floor will be the technical education wing. The second floor will house the new science labs, the art classrooms and several multipurpose classrooms. Finally, the third floor will be devoted to multipurpose classroom space. To accomplish the meshing of old and new, and to achieve an efficient and compact plan, several existing science labs and shops will be razed.

The addition will house those core facilities that need to be expanded due to increased enrollments (i.e., library and cafeteria) and those spaces where substantial technological improvements were needed (i.e., art, tech. education, and science labs). Finally, the third floor's multipurpose classrooms give the entire building flexibility to adapt to a variety of educational philosophies, such as a "house" concept, should this approach be used in the future.

Figure 24.1   A Planning Proposal   *Continued*

6

The addition will be clad in masonry and massed in a way so that it blends in with the existing school. An elevator and three stairways will link the three floors in an efficient, functional manner. The addition will be energy-efficient, handicapped accessible, technologically up-to-date, and sprinklered for fire safety.

In order for the addition and existing school to blend together as a cohesive unit (and to obtain state approval for our reimbursement), the existing building must be updated and renovated. The existing cafeteria will be converted into a large group instruction area, expanded bathrooms for the auditorium, and special education spaces. The existing library will be converted into small classroom spaces. The present art classroom will be converted to an additional music room. The auditorium will be updated with new lighting and sound systems. For the most part, the existing classroom layout will remain the same and will be upgraded.

A concerted effort has been made to use existing systems where possible. All HVAC, electrical, plumbing, and fire safety systems will be upgraded as needed for energy efficiency and code compliance. A new roof, new lockers, insulation, and windows will be installed and all interior spaces will be updated for compliance with ADA and state accessibility guidelines.

A small addition will be constructed on the west side of the gym to house two new 2,500-square-foot exercise rooms to augment the need for additional physical education teaching stations. It will also house a training room and additional storage space.

Finally, the site will be improved in several ways (Figure 3). The existing parking lot will be expanded north for an additional 54 parking spaces, and an additional 58-car parking lot will be added on the west side of the gym. The road and intersection immediately south of the school will be rearranged to allow for safer and more efficient traffic patterns. A new entrance canopy on the south side will help focus attention to the existing main entrance. The Junior High connector road will be altered to allow for an automobile drop-off area adjacent to the new east entrance to the addition.

In summary, all the new square footage (except for the exercise rooms), is consolidated into one new addition, linked to the existing school by a new major circulation corridor. Every effort has been made to tie together an updated forty-year-old school with a new addition so they work as a whole and will serve our four regional communities well into the twenty-first century.

Figure 24.1   A Planning Proposal   *Continued*

7

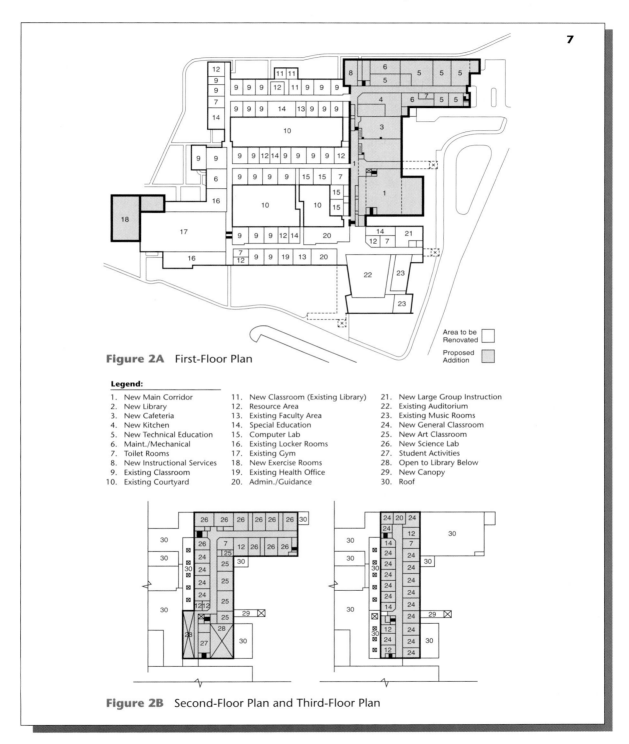

Figure 2A First-Floor Plan

Legend:

1. New Main Corridor
2. New Library
3. New Cafeteria
4. New Kitchen
5. New Technical Education
6. Maint./Mechanical
7. Toilet Rooms
8. New Instructional Services
9. Existing Classroom
10. Existing Courtyard

11. New Classroom (Existing Library)
12. Resource Area
13. Existing Faculty Area
14. Special Education
15. Computer Lab
16. Existing Locker Rooms
17. Existing Gym
18. New Exercise Rooms
19. Existing Health Office
20. Admin./Guidance

21. New Large Group Instruction
22. Existing Auditorium
23. Existing Music Rooms
24. New General Classroom
25. New Art Classroom
26. New Science Lab
27. Student Activities
28. Open to Library Below
29. New Canopy
30. Roof

Figure 2B Second-Floor Plan and Third-Floor Plan

Figure 24.1 A Planning Proposal *Continued*

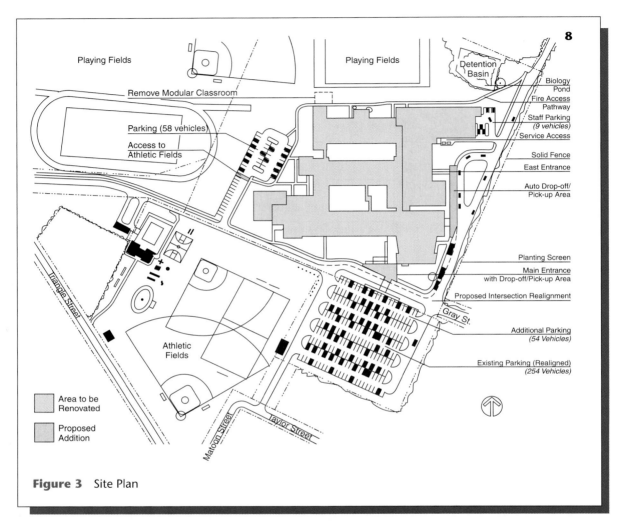

Figure 3   Site Plan

Figure 24.1   A Planning Proposal   *Continued*

Usability Checklist FOR PROPOSALS

(Numbers in parentheses refer to the first page of discussion.)

Format

☐ Is the short internal proposal in memo form and the short external proposal in letter form? (496)

☐ Does the long proposal have adequate supplements to serve the needs of different readers? (503)

☐ Is the format professional in appearance? (503)

☐ Are headings logical and adequate? (316)

☐ Does the title forecast the proposal's subject and purpose? (325)

Content

☐ Is the problem clearly identified? (505)

☐ Is the objective clearly identified? (505)

☐ Does everything in the proposal support its objective? (502)

☐ Does the proposal *show* as well as *tell?* (502)

☐ Is the proposed plan, service, or product beneficial? (501)

☐ Are the proposed methods practical and realistic? (506)

☐ Are all foreseeable limitations and contingencies identified? (501)

☐ Is the proposal free of overstatement? (501)

☐ Are visuals used effectively and whenever appropriate? (502)

☐ Is the proposal's length appropriate to the subject? (493)

☐ Is the proposal ethically acceptable? (501)

Arrangement

☐ Is there a recognizable introduction, body, and conclusion? (504)

☐ Are all *relevant* headings from the general outline included? (504)

☐ Does the introduction provide sufficient orientation to the problem and the plan? (505)

☐ Does the body explain *how, where,* and *how much?* (506)

☐ Does the conclusion encourage acceptance of the proposal? (508)

☐ Are there clear transitions between related ideas? (Appendix C)

Style and Page Design

☐ Is the level of technicality appropriate for primary readers? (504)

☐ Do supplements follow the appropriate style guidelines? (324)

☐ Will the informative abstract be understood by laypersons? (330)

☐ Is the tone appropriate? (502)

☐ Is the writing style clear, concise, and fluent throughout? (215)

☐ Is the language convincing and precise? (235)

☐ Is the proposal written in grammatical English? (Appendix C)

☐ Is the page design inviting and accessible? (503)

EXERCISES

1. Assume that the head of your high school English Department has asked you, as a recent graduate, for suggestions about revising the English curriculum to prepare students for writing. Write a proposal, based on your experience since high school. (Primary audience: the English Department head and faculty; secondary audience: the school committee.) In your external, solicited proposal, identify problems, needs, and benefits and spell out a realistic plan. Review the outline on page 504 before selecting specific headings.

2. After identifying your primary and secondary audience, compose a short planning proposal for improving an unsatisfactory situation in the classroom, on the job, or in your dorm or apartment (e.g., poor lighting, drab atmosphere, health hazards, poor seating arrangements). Choose a problem or situation whose resolution is more a matter of common sense and lucid observation than of intensive research. Be sure to (a) identify the problem clearly, give brief background, and stimulate interest; (b) clearly state the methods proposed to solve the problem; and (c) conclude with a statement designed to gain audience support for your proposal.

3. Write a research proposal to your instructor (or an interested third party) requesting approval for the final term project (an analytical report or formal proposal). Identify the subject, background, purpose, and benefits of your planned inquiry, as well as the intended audience, scope of inquiry, data sources, methods of inquiry, and a task timetable. Be certain that adequate primary and secondary sources are available. Convince your audience of the soundness and usefulness of the project.

4. As an alternate term project to the formal analytical report (Chapter 25), develop a long proposal for solving a problem, improving a situation, or satisfying a need in your school, community, or job. Choose a subject sufficiently complex to justify a formal proposal, a

topic requiring research (mostly primary). Identify an audience (other than your instructor) who will use your proposal for a specific purpose. Compose an audience and use profile, using the sample on page 509 as a model. Here are possible subjects for your proposal:

- improving living conditions in your dorm or fraternity/sorority
- creating a student-oriented advertising agency on campus
- creating a day care center on campus
- creating a new business or expanding a business
- saving labor, materials, or money on the job
- improving working conditions
- improving campus facilities for the disabled
- supplying a product or service to clients or customers
- increasing tourism in your town
- eliminating traffic hazards in your neighborhood or on campus
- reducing energy expenditures on the job
- improving security in dorms or in the college library
- improving in-house training or job orientation programs
- creating a one-credit course in job hunting or stress management for students
- improving tutoring in the learning center
- making the course content in your major more relevant to student needs
- creating a new student government organization
- finding ways for an organization to raise money
- improving faculty advising for students
- purchasing new equipment
- improving food service on campus
- easing first-year students through the transition to college
- changing the grading system at your school
- establishing more equitable computer terminal use
- designing a Web site for your employer or an organization to which you belong

Analytical Reports

PURPOSE OF ANALYSIS

TYPICAL ANALYTICAL PROBLEMS

ELEMENTS OF A USABLE ANALYSIS

A GENERAL MODEL FOR ANALYTICAL
REPORTS

A SITUATION REQUIRING AN ANALYTICAL
REPORT

Usability Checklist for Analytical Reports

The formal analytical report, like the short recommendation reports discussed in Chapter 18, leads to recommendations. The formal report replaces the memo when the topic requires lengthy discussion.

Analysis, a form of critical thinking in which we separate a whole into its parts, is essential to workplace problem solving. Assume, for example, that you receive this assignment from your supervisor:

A typical analytical problem

> Recommend the best method for removing the heavy-metal contamination from our company dump site.

To make the best recommendation, you will have to analyze the situation to learn everything you can about the nature and extent of the problem and the feasibility of various solutions.

The above situation calls for skills in critical thinking (Chapter 2) and in research (Chapters 7–10). Besides interviewing legal and environmental experts, you might search the literature and the World Wide Web. From these sources you can discover whether anyone has been able to solve a problem like yours, or learn about the newest technologies for toxic cleanup. Then you will have to decide how much, if any, of what others have done applies to your situation.

PURPOSE OF ANALYSIS

You may be assigned to evaluate a new assembly technique on the production line, or to locate and purchase the best equipment at the best price. You might have to identify the cause behind a monthly drop in sales, the reasons for low employee morale, the causes of an accident, or the reasons for equipment failure. You might need to assess the feasibility of a proposal for a company's expansion or investment.

The list is endless, but the procedure is always the same: (1) asking the right questions, (2) searching the best sources, (3) evaluating and interpreting your findings, and (4) drawing conclusions and making recommendations.

TYPICAL ANALYTICAL PROBLEMS

Far more than an encyclopedia presentation of information, the analytical report traces your inquiry, your evidence, and your reasoning to show exactly how you arrived at your conclusions and recommendations.

Workplace problem solving calls for skills in three broad categories: *causal analysis, comparative analysis,* and *feasibility analysis.* Each approach calls for its own type of reasoning.

Causal Analysis: "Why Does X Happen?"

Designed to attack a problem at its source, the causal analysis answers questions like this: *Why do so many apparently healthy people have sudden heart attacks?* Here is how causal reasoning proceeds:

Identify the problem

Medical researchers at the world-renowned Hanford Health Institute recently found that 20 to 30 percent of deaths from sudden heart attacks occur in people who have none of the established risk factors (weight gain, smoking, diabetes, lack of exercise, high blood pressure, or family history).

Examine possible causes

To better identify people at risk, researchers are now tracking down new and powerful risk factors such as bacteria, viruses, genes, stress, anger, and depression.

Recommend solutions

Once researchers identify these factors and their mechanisms, they can recommend preventive steps such as careful monitoring, lifestyle and diet changes, drug treatment, or psychotherapy (H. Lewis 39–43). ▲

A different version of causal analysis employs reasoning from effect to cause, to answer questions like this: *What are the health effects of exposure to electromagnetic radiation?*

NOTE   *Keep in mind that faulty causal reasoning is extremely common, especially when we ignore other possible causes or we confuse correlation with causation (page 158).*

Comparative Analysis: "Is *X* or *Y* Better for Our Purpose?"

Designed to rate competing items on the basis of specific criteria, the comparative analysis answers questions like this: *Which type of security (firewall/encryption) program should we install on our company's computer system?* Here is how reasoning proceeds in a comparative analysis:

Identify the criteria

XYZ Corporation needs to identify exactly what information (personnel files, financial records) or functions (in-house communication, file transfer) it wants to protect from whom. Does it need both virus and tamper protection? Does it wish to restrict network access or encrypt (scramble) email and computer files so they become unreadable to unauthorized persons? Does it wish to restrict access to or from the Web? In addition to the level of protection, how important are ease of maintenance and user-friendliness?

Rank the criteria

After identifying their specific criteria, XYZ decision makers need to rank them in order of importance (for example, 1. tamper protection, 2. user-friendliness, 3. secure financial records, and so on).

Compare items according to the criteria, and recommend the best one

On the basis of these ranked criteria, XYZ will assess relative strengths and weaknesses of competing security programs and recommend the best one (Schafer 93–94). ▲

Feasibility Analysis: "Is This a Good Idea?"

Designed to assess the practicality of an idea or plan, the feasibility analysis answers questions like this: *Will increased business justify the cost of an interactive, multimedia Web site?* Here is the reasoning process in a feasibility analysis:

Bigbyte, Inc., a retailer of discounted computer hardware and software, relies on catalog orders by telephone for its high-volume business. The company is now deciding whether to reallocate the bulk of its marketing resources to establish a high-profile presence on the Web.

Consider the
strength of
supporting reasons

Research uncovers numerous supporting reasons, including these: A Web site is an excellent medium for enhancing a company's name recognition and for announcing new products and catalog updates. Site visits can be tailored to individual customer preferences. "Cybershoppers" value the convenience of researching and comparing products from home, the absence of pressure from telephone salespersons, and the ability to customize their visits to a retailer's Web site.

Consider the
strength of
opposing reasons

Obstacles to Web-based sales include consumer reluctance to reveal credit card numbers and personal information online or to change their traditional shopping patterns, along with frustration with slow downloading. Also, creating and maintaining an attractive, navigable, and up-to-date site is costly, and few retail sites—even the most heavily visited—have shown actual profit.

Weigh the pros and
cons and
recommend a
course of action

After assessing the benefits and drawbacks of Web-based marketing in *this* situation, Bigbyte decision makers can make the appropriate recommendations (Hodges 22+). ▲

Combining Types of Analysis

Analytical categories overlap considerably. Any one study may in fact require answers to two or more of the previous questions. The sample report on pages 542–49 is both a feasibility analysis and a comparative analysis. It is designed to answer these questions: *Is technical marketing the right career for me? If so, which is my best option for entering the field?*

ELEMENTS OF A USABLE ANALYSIS

The analytical report incorporates many elements from assignments in earlier chapters, along with the suggestions that follow.

Clearly Identified Problem or Goal

Define your goal

To solve any problem or achieve any goal, you must first identify it precisely. Always begin by defining the main questions and thinking through any subordinate questions they may imply. Only then can you determine what to look for, where to look, and how much information you will need.

Your employer, for example, might pose this question: *Will a low-impact aerobics program significantly reduce stress among my employees?* The aerobics question obviously requires answers to three other questions: *What are the therapeutic claims for aerobics? Are they valid? Will aerobics work in this situation?* With the main questions identified, you can formulate a definite goal statement:

Goal statement

My goal is to examine and evaluate claims about the therapeutic benefits of low-impact aerobic exercise.

Words such as *examine* and *evaluate* (or *compare, identify, determine, measure, describe,* and so on) help readers understand the specific analytical activity that forms the subject of the report.

Adequate but Not Excessive Data

Decide how much is enough

A usable analysis filters material for the audience's understanding: Do decision makers in this situation need a closer look or are we presenting excessive detail when only general information is needed? Is it possible to have too much information? In some cases, yes—as behavioral expert Dietrich Dorner explains:

Excessive information hampers decision making

> The more we know, the more clearly we realize what we don't know. This probably explains why . . . organizations tend to [separate] their information-gathering and decision-making branches. A business executive has an office manager; presidents have . . . advisers; military commanders have chiefs of staff. The point of this separation may well be to provide decision makers with only the bare outlines of all the available information so they will not be hobbled by excessive detail when they are obliged to render decisions. Anyone who is fully informed will see much more than the bare outlines and will therefore find it extremely difficult to reach a clear decision. (99)

Confusing the issue with excessive information is no better than recommending hasty action on the basis of inadequate information (Dorner 104).

Accurate Data

Give readers all they need to make an informed judgment

Avoid stacking the evidence to support a preconceived point of view. Assume, for example, that you are asked to recommend the best chainsaw brand for a logging company. Reviewing test reports, you come across this information:

> Of all six brands tested, the Bomarc chainsaw proved easiest to operate. However, this brand also offers the fewest safety features.

In citing these equivocal findings, you need to present both of them accurately, and not simply the first—even though the Bomarc brand may be your favorite. Then argue for the feature you think should receive priority. (Refer to Chapter 10, pages 149–50 for presenting balanced and reasonable evidence.)

Fully Interpreted Data

Explain the significance of your data

Interpretation shows the audience "what is important and what unimportant, what belongs together and what does not" (Dorner 44). For example, you might interpret the above chainsaw data this way:

Explain the meaning of your evidence

> Our logging crews often work suspended by harness, high above the ground. Also, much work is in remote areas. Safety features therefore should be our first requirement in a chainsaw. Despite its ease of operation, the Bomarc saw does not meet our safety needs.

By saying "therefore" you engage in analysis—not just information sharing. Simply listing your findings is not enough. Explain what these findings mean. (Refer to pages 151–54.)

Subordination of Personal Bias

Evaluate and interpret evidence impartially

Each stage of analysis requires decisions about what to record, what to exclude, and where to go next. You must evaluate your data (*Is this reliable and important?*), interpret your evidence (*What does it mean?*), and make recommendations (*What action is needed?*). An ethically sound analysis presents a balanced and reasonable assessment of the evidence. Do not force viewpoints on the material that are not supported by dependable evidence. (Refer to page 153.)

Appropriate Visuals

Use visuals generously

Graphs are especially useful in an analysis of trends (rising or falling sales, radiation levels). Tables, charts, photographs, and diagrams work well in comparative analyses. Be sure to accompany each visual with a fully interpreted "story."

Valid Conclusions and Recommendations

Be clear about what the audience should think and do

Along with the informative abstract (page 330), conclusions and recommendations are the sections of a long report that receive the most audience attention. The goal of analysis is to reach a valid conclusion—an overall judgment about what all the material means (that *X* is better than *Y*, that *B* failed because of *C*, that *A* is a good plan of action). Here is the conclusion of a report on the feasibility of installing an active solar heating system in a large building:

Offer a final judgment

1. Active solar space heating for our new research building is technically feasible because the site orientation will allow for a sloping roof facing due south, with plenty of unshaded space.
2. It is legally feasible because we are able to obtain an access easement on the adjoining property, to ensure that no buildings or trees will be permitted to shade the solar collectors once they are installed.
3. It is economically feasible because our sunny, cold climate means high fuel savings and faster payback (fifteen years maximum) with solar heating. The long-term fuel savings justify our short-term installation costs (already minimal because the solar system can be incorporated during the building's construction—without renovations).

Conclusions are valid when they are logically derived from accurate interpretations.

Having explained *what it all means*, you then recommend *what should be done*. Taking all possible alternatives into account, your recommendations urge specific action (to invest in *A* instead of *B*, to replace *C* immediately, to follow plan *A*, or the like). Here are the recommendations based on the previous interpretations:

Tell what should be done

1. I recommend that we install an active solar heating system in our new research building.
2. We should arrange an immediate meeting with our architect, building contractor, and solar heating contractor. In this way, we can make all necessary design changes before construction begins in two weeks.

3. We should instruct our legal department to obtain the appropriate permits and easements immediately.

Recommendations are valid when they propose an appropriate response to the problem or question.

Because they culminate your research and analysis, recommendations challenge your imagination, your creativity, and—above all—your critical thinking skills. What strikes one person as a brilliant suggestion might be seen by others as irresponsible, offensive, or dangerous. (Figure 25.1 depicts the kinds of decisions writers encounter in formulating, evaluating, and refining their recommendations.)

NOTE *Keep in mind that solving one problem might create new and worse problems. For example, to prevent crop damage by rodents, an agriculture specialist might recommend trapping and poisoning. While rodent eradication may increase crop yield temporarily, it also increases the insects these rodents feed on—leading eventually to even greater crop damage. In short, before settling on any recommendation, try to anticipate its "side effects and long-term repercussions" (Dorner 15).*

When you do achieve definite conclusions and recommendations, express them with assurance and authority. Unless you have reason to be unsure, avoid noncommittal statements ("It would seem that" or "It looks as if"). Be direct and assertive ("The earthquake danger at the reactor site is acute," or "I recommend an immediate investment"). Announce where you stand.

If, however, your analysis yields nothing definite, do not force a simplistic conclusion on your material (pages 102–03). Instead, explain your position ("The contradictory responses to our consumer survey prevent me from reaching a definite conclusion. Before we make any decision about this product, I recommend a full-scale market analysis"). The wrong recommendation is far worse than no recommendation at all. (Refer to Chapter 10, pages 168–70 for helpful guidelines.)

Self-Assessment

Assess your analysis continuously

The more we are involved in a project, the larger our stake in its outcome—making self-criticism less likely just when it is needed most! For example, it is hard to admit that we might need to backtrack, or even start over, in instances like these (Dorner 46):

Things that might go wrong with your analysis

- During research you find that your goal isn't clear enough to indicate exactly what information you need.
- As you review your findings, you discover that the information you have is not the information you need.

It is even harder to test our recommendations and admit they are not working:

◆ After making a recommendation, you discover that what seemed like the right course of action turns out to be the wrong one.

If you meet such obstacles, acknowledge them immediately, and revise your approach as needed.

Figure 25.1
How to Think Critically About Your Recommendations
Source: Questions adapted from Ruggiero, Vincent R. The Art of Thinking, 5th ed. New York: Longman, 1998: 70–73.

Consider All the Details

- What exactly should be done—if anything at all?
- How exactly should it be done?
- When should it begin and be completed?
- Who will do it, and how willing are they?
- What equipment, material, or resources are needed?
- Are any special conditions required?
- What will this cost, and where will the money come from?
- What consequences are possible?
- Whom do I have to persuade?
- How should I order my list (priority, urgency, etc.)?

Locate the Weak Spots

- Is anything unclear or hard to follow?
- Is this course of action unrealistic?
- Is it risky or dangerous?
- Is it too complicated or confusing?
- Is anything about it illegal or unethical?
- Will it cost too much?
- Will it take too long?
- Could anything go wrong?
- Who might object or be offended?
- What objections might be raised?

Make Improvements

- Can I rephrase anything?
- Can I change anything?
- Should I consider alternatives?
- Should I reorder my list?
- Can I overcome objections?
- Should I get advice or feedback before I submit this?

A GENERAL MODEL FOR ANALYTICAL REPORTS

Whether you outline earlier or later, the finished report depends on a good outline. This model outline can be adapted to most analytical reports:

I. Introduction
    A. Definition, Description, and Background
    B. Purpose of the Report, and Intended Audience
    C. Method of Inquiry
    D. Working Definitions (here or in a glossary)
    E. Limitations of the Study
    F. Scope of the Inquiry (topics listed in logical order)

II. Collected Data
    A. First Topic for Investigation
        1. Definition
        2. Findings
        3. Interpretation of findings
    B. Second Topic for Investigation
        1. First subtopic
            a. Definition
            b. Findings
            c. Interpretation of findings
        2. Second subtopic (and so on)

III. Conclusion
    A. Summary of Findings
    B. Overall Interpretation of Findings (as needed)
    C. Recommendations (as needed and appropriate)

(This outline is only tentative. Modify as necessary.)

Two sample reports in this text follow the model outline. The first report, "Children Exposed to Electromagnetic Radiation: A Risk Assessment" (minus the document supplements that ordinarily accompany a long report), begins on page 530. The second report, "The Feasibility of a Technical Marketing Career," appears in Figure 25.2. Supplements for this report appear in Chapter 16.

Each report responds to slightly different questions. The first tackles these questions: *What are the effects of* X *and what should we do about them?* The second tackles two questions: *Is* X *feasible, and which version of* X *is better for my purposes?* At least one of these reports should serve as a model for your own analysis.

Introduction

The introduction engages and orients the audience and provides background—as briefly as possible. Identify the origin and significance of the topic, define or describe the problem or issue, and explain the report's purpose.

Identify your audience only in a report for your instructor, and only if you don't attach an audience and use profile. Briefly identify your methods of inquiry (interviews, surveys, literature searches, and so on), and explain any omissions of data (e.g., a person unavailable for an interview, or research still in progress). List working definitions, unless you have so many that you need a glossary. If you do include a glossary and appendixes, notify the audience early. Finally, spell out your research objectives by listing the major topics you will discuss in the body section.

NOTE   *Not all reports require every element. Give readers only what they need and expect.*

As you read the following introduction, think about the elements designed to engage and orient the audience (i.e., local citizens), and evaluate their effectiveness. (Review pages 98–99 for the situation that gave rise to this report.)

Children Exposed to Electromagnetic Radiation: A Risk Assessment

LAURIE A. SIMONEAU

INTRODUCTION

Definition and background

Wherever electricity flows—through the largest transmission line or the smallest appliance—it emits varying intensities of charged waves: an *electromagnetic field* (EMF). Some medical studies have linked human exposure to EMFs with definite physiologic changes and possible illness, including cancer, miscarriage, and depression.

Experts disagree over the health risk, if any, from EMFs. Some question whether EMF risk is greater from high-voltage transmission lines, the smaller distribution lines strung on utility poles, or household appliances. Conclusive research may take years; meanwhile, concerned citizens worry about avoiding potential risks.

Description of problem

In Bocaville, four sets of transmission lines—two at 115 Kilovolts (kV) and two at 500 kV—cross residential neighborhoods and public property. The Adams elementary school is less than 100 feet from this power line corridor. EMF risks—whatever they may be—are thought to increase with proximity.

Purpose and methods of inquiry

Based on examination of recent research and interviews with local authorities, this report assesses whether potential health risks from EMFs seem significant enough for Bocaville to (a) increase public awareness, (b) divert the transmission lines that run adjacent to the elementary school, and (c) implement widespread precautions in the transmission and distribution of electrical power throughout Bocaville.

Scope of inquiry

This report covers five major topics: what we know about various EMF sources, what research indicates about physiologic and health effects, how experts differ in evaluating the research, what the power industry and the public have to say, and what actions are being taken locally and nationwide to avoid risk.

Body

The body section describes and explains your findings. Present a clear and detailed picture of the evidence, interpretations, and reasoning on which you will base your conclusion. Divide topics into subtopics, and use informative headings as aids to navigation.

NOTE

Remember your ethical responsibility for presenting a fair *and* balanced *treatment of the material, instead of "loading" the report with only those findings that support your viewpoint.*

As you read the following section, evaluate how effectively it informs readers, keeps them on track, reveals a clear line of reasoning, and presents an impartial analysis.

First topic
Definition

DATA SECTION
Sources of EMF Exposure

Electromagnetic intensity is measured in *milligauss* (mG), a unit of electrical measurement. The higher the mG reading, the stronger the field. Consistent exposure above 1–2 mG is thought to increase cancer risk significantly (White 15), but no scientific evidence concludes that exposure even below 2.5 mG is safe.

Findings

Table 1 gives the EMF intensities from electric power lines at varying distances during average and peak usage.

Table 1 EMF Emissions from Power Lines (in milligauss)					
Types of Transmission Lines	Maximum on Right-of-Way	Distance from lines			
		50'	100'	200'	300'
115 Kilovolts (kV)					
Average usage	30	7	2	0.4	0.2
Peak usage	63	14	4	0.9	0.4
230 Kilovolts (kV)					
Average usage	58	20	7	1.8	0.8
Peak usage	118	40	15	3.6	1.6
500 Kilovolts (kV)					
Average usage	87	29	13	3.2	1.4
Peak usage	183	62	27	6.7	3.0

Source: United States Environmental Protection Agency, EMF in Your Environment. *Washington: GPO, 1992. Data from Bonneville Power Administration.*

Interpretation

As Table 1 indicates, EMF intensity drops substantially as distance from the power lines increases.

Although the EMF controversy has focused on 2 million miles of power lines crisscrossing the country, potentially harmful waves are also emitted by household wiring, appliances, computer terminals—and even from the earth's natural magnetic field. The background magnetic field (at a safe distance from any electrical appliance) in the average American home varies from 0.5 to 4.0 mG (U.S. EPA 10). Table 2 compares intensities of various sources.

Table 2 EMF Emissions from Selected Sources (in milligauss)	
Source	*Range*[a,b]
Earth's magnetic field	0.1–2.5
Blowdryer	60–1400
Four in. from TV screen	40–100
Four ft. from TV screen	0.7–9
Flourescent lights	10–12
Electric razor	1200–1600
Electric blanket	2–25
Computer terminal (12 inches away)	3–15
Toaster	10–60

[a]Data from Brodeur, Paul. "Annals of Radiation: The Cancer at Slater School." *The New Yorker* 7 Dec. 1992: 88; Miltane, John. Interview 5 Apr. 1999; National Institute of Environmental Health. *Questions and Answers about EMF.* Washington: GPO, 1995:3.

[b]Readings are made with a gaussmeter, and vary with technique, proximity of gaussmeter to source, its direction of aim, and other random factors.

Interpretation

Definitions

Finding

EMF intensity from certain appliances tends to be higher than from transmission lines because of the amount of current involved.

Voltage measures the speed and pressure of electricity in wires, but *current* measures the volume of electricity passing through wires. Current (measured in *amperage*) is what produces electromagnetic fields. The current flowing through a transmission line typically ranges from 200 to 400 amps. Most homes have a 200-amp service. This means that if every electrical item in the house were turned on at the same time, the house could run about 200 amps—almost as high as the transmission line. Consumers then have the ability to put 200 amps of current-flow into their homes, while transmission lines carrying 200 to 400 amps are at least 50 feet away (Miltane).

Proximity and duration of exposure, however, are other risk factors. People are exposed to EMFs from home appliances at close proximity, but appliances run only periodically: exposure is therefore sporadic, and intensity diminishes sharply within a few feet (Figure 1).

As Figure 1 indicates, EMF intensity drops dramatically over very short distances from the typical appliance.

Finding

Power line exposure, on the other hand, is at a greater distance (usually 50 feet or more), but it is constant. Moreover, its intensity can remain strong well beyond 100 feet (Miltane).

Interpretation

Research has yet to determine which type of exposure might be more harmful: briefly, to higher intensities or constantly, to lower intensities. In either case, proximity seems most significant because EMF intensity drops rapidly with distance.

Second topic

Physiologic Effects and Health Risks from EMF Exposure

Research on EMF exposure falls into two categories: epidemiologic studies and laboratory studies. The findings are sometimes controversial and inconclusive, but also disturbing.

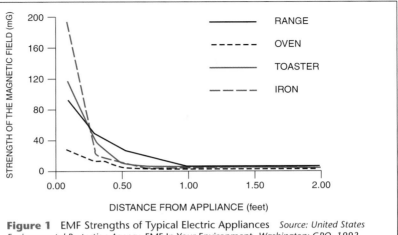

Figure 1 EMF Strengths of Typical Electric Appliances   *Source: United States Environmental Protection Agency.* EMF In Your Environment. *Washington: GPO, 1992.*

First subtopic
Definition
General Findings

Epidemiologic Studies.   Epidemiologic studies look for statistical correlations between EMF exposure and human illness or disorders. Of 77 such studies in recent decades, over 70 percent suggest that EMF exposure increases the incidence of the following conditions (Lee et al. 12–28; Pinsky 155–215):

◆ cancer, especially leukemia and brain tumors
◆ miscarriage
◆ stress and depression
◆ learning disabilities

In one study, for example, pregnant women working at computer terminals at least 20 hours weekly had a risk of miscarriage twice that of other pregnant clerical workers (Pinsky 212–13).

Following are summaries of four noted epidemiologic studies implicating EMFs in cancer.

Detailed findings

A Landmark Study of the EMF/Cancer Connection.   A 1979 Denver study by Wertheimer and Leeper was the first to implicate EMFs as a cause of cancer. Researchers compared hundreds of homes in which children had developed cancer with similar homes in which children were cancer free. Victims were two to three times as likely to live in "high-current homes" (within 130 feet of a transmission line or 50 feet of a distribution line).

This study has been criticized because (1) it was not "blind" (researchers knew which homes cancer victims were living in), and (2) they never took gaussmeter readings to verify their designation of "high current" homes (Pinsky 160–62; Taubes 96).

Detailed findings

Follow-up Studies.   Several major studies of the EMF/cancer connection have confirmed Wertheimer's findings:

- In 1988, Savitz studied hundreds of Denver houses and found that children with cancer were 1.7 times as likely to live in high-current homes. Unlike his predecessors, Savitz did not know whether a cancer victim lived in the home being measured, and he took gaussmeter readings to verify that houses could be designated "high-current" (Pinsky 162–63).
- In 1990, London and Peters found that Los Angeles children had 2.5 times more risk of leukemia if they lived near power lines (Brodeur 115).
- In 1992, a massive Swedish study found that children in houses with average intensities greater than 1 mG had twice the normal leukemia risk; at greater than 2 mG, the risk nearly tripled; at greater than 3 mG, it nearly quadrupled (Brodeur 115).
- In 1995, University of Colorado researchers evaluated findings from eleven separate studies of power line EMF effects (a *meta-analysis*). Among their conclusions: (1) prolonged exposure to crisscrossed power lines significantly increases risk for cancers in general; (2) prolonged exposure to any power line increases leukemia risk (Miller et al. 281–87).

Detailed findings

Workplace Studies.   More than 80 percent of 51 studies from 1981 to 1994—most notably a 1992 Swedish study—concluded that electricians, electrical engineers, and power line workers constantly exposed to an average of 1.5 to 4.0 mG had a significantly elevated cancer risk (Brodeur, 115; Pinsky 177–209).

Notable Recent Studies.   Two recent workplace studies seem to support or even amplify the above findings:

- A 1994 University of Southern California study indicates that high workplace exposure to EMFs triples the risk of Alzheimer's disease (Des Marteau 38; Maugh 3).
- A 1995 University of North Carolina study of 138,905 electric utility workers concluded that occupational EMF exposure roughly doubles brain cancer risk. This study, however, found no increased leukemia risk (Cavanaugh 8; Moore 16).

Interpretation

None of the above studies can be said to "prove" a direct cause-effect relationship, but their strikingly similar results suggest a conceivable link between prolonged EMF exposure and illness.

Second subtopic

Laboratory Studies.   Laboratory studies assess cellular, metabolic, and behavioral effects of EMFs on humans and animals. EMFs directly cause the following physiologic changes (Brodeur 88; Pinsky 24–29; Raloff, "EMFs'" 30):

General findings

- reduced heart rate
- altered brain waves
- impaired immune system
- interference with the synthesis of genetic material
- disrupted regulation of cell growth
- interaction with the biochemistry of cancer cells
- altered hormonal activity
- disrupted sleep patterns

These changes are documented in the following summaries of several significant laboratory studies.

EMF Effects on Cell Chemistry.   Recent studies have demonstrated previously unrecognized effects on cell growth and division.

Detailed findings

- A 1995 study at the School of Veterinary Medicine, Hannover, Germany, found marked increases of ornithine decarboxylase (a key enzyme in tumor growth) in tissues of rats exposed to electromagnetic fields for six weeks (Mevissen, Keitzmann, and Loscher 207–14).
- A 1995 University of Wisconsin study showed that cell metabolism is influenced by electromagnetic fields—the extent of effect depending on a cell's age and health. While this type of cellular stress does not appear to initiate cancer, it might help promote growth of an existing tumor (Goodman, Greenebaum, and Marron 279–338).

EMF Effects on Hormones.   Several studies have found that EMF exposure (say, from an electric blanket) inhibits production of melatonin, a hormone that fights cancer and depression, stimulates the immune system, and regulates bodily rhythms.

Detailed findings

- A 1997 study at the Lawrence National Laboratory found that EMF exposure can suppress both melatonin and the hormonelike, anticancer drug Tamoxifen (Raloff, "EMFs'" 30).
- In 1996, physiologist Charles Graham found that EMFs also elevate female estrogen levels and depress male testosterone levels—hormone alterations associated, respectively, with risk of breast or testicular cancer (Raloff, "EMFs'" 30)

Detailed findings

EMF Effects on Life Expectancy.   A 1994 South African study at the University of the Orange Free State measured the life span of mice exposed to EMFs. Both first and second generations of exposed mice showed significantly shortened life expectancy (de Jager and deBruyn 221–24).

Detailed findings

EMF Effects on Behavior.   Recent studies indicate that people who live adjacent to power lines have roughly twice the normal rate of depression (Pinsky 31–32). Also, rats exposed to EMFs exhibit slower rates of learning (Beardsley 20).

Interpretation

Although laboratory studies seem more conclusive than the epidemiologic studies, what these findings *mean* is debatable.

Third topic

Debate over Quality, Cost, and Status of EMF Research
Experts differ over the meaning of EMF research findings largely because of the following limitations attributed to various studies.

First subtopic

Findings

Limitations of Various EMF Studies
- Epidemiologic studies are criticized for overstating the evidence. For example, after reviewing the research, the American Physical Society reports "no consistent link between cancer and power line fields" (Broad 19). Some critics claim that so-called EMF-cancer links are produced by "data dredging" (making countless comparisons

Counterargument

between types of cancers and EMF sources until certain random correlations appear) (Taubes 99). Some studies are also accused of mistaking *coincidence* for *correlation*, without exploring "confounding factors" (e.g., exposure to toxic substances or to other adverse conditions—including the earth's natural magnetic field) (Moore 16).

Supporters of EMF research respond that the sheer volume of evidence seems overwhelming (Kirkpatrick 81, 83; White 15). Moreover, the Swedish studies cited earlier seem to invalidate the above criticisms (Brodeur 115).

Findings

◆ Laboratory studies are criticized—even by scientists who conduct them—because effects on an isolated culture of cells are not always equal to effects on the total human or animal organism. Also, effects in experimental animals are not always equivalent to effects in humans (Jauchem 190–94). Moreover, research has focused on EMF *hazards* only, virtually ignoring the possible health *benefits*, such as a potential role in heart attack prevention (Stix 33).

Breakthrough finding

◆ Until recently critics argued that no scientist had offered a reasonable hypothesis to explain the possible health effects (Palfreman 26). However, a 1997 study at Minnesota's Hughes Institute identified the precise biological mechanism by which EMF exposure activates tyrosine kinase enzymes that produce DNA damage (Raloff, "Electromagnetic" 119).

Second subtopic
Findings

Costs of EMF Research.   Critics claim that research and publicity about EMFs are becoming a profit venture, spawning "a new growth industry among researchers, as well as marketers of EMF monitors" ("Electrophobia" 1). Environmental expert Keith Florig identifies adverse economic effects of the EMF debate that include decreased property values, frivolous lawsuits, expensive but needless "low field" consumer appliances, and costly modifications to schools and public buildings (Monmonier 190).

Conflicting findings

Present Status of EMF Research.   In July 1998, an editor at the *New England Journal of Medicine* called for ending EMF/cancer research. He cited studies from the National Cancer Institute and other respected sources that showed "little evidence" of any causal connection. In a parallel development, federal and industry funding for EMF research has been reduced drastically (Stix 33). However, in August 1998, experts from the Energy Department and the National Institute of Environmental Health Sciences announced that EMFs should be officially designated a "possible human carcinogen" (Raloff, "Jury" 127).

Interpretation

After more than twenty years of study, the EMF/illness debate continues, even among respected experts. While most scientists agree that EMFs exert measurable effects on the human body, they disagree about whether a real hazard exists. Given the drastic cuts in research funding, definite answers are unlikely to appear anytime soon.

Fourth topic

Views from the Power Industry and the Public
While the experts continue their debate, other viewpoints are worth considering as well.

Findings

The Power Industry's Views.   The Electrical Power Research Institute (EPRI), the research arm of the nation's electric utilities, claims that recent EMF studies have provided valuable but inconclusive data that warrant further study (Moore 17).

What does our local power company think about the alleged EMF risk? Marianne Halloran-Barney, Energy Service Advisor for County Electric, expressed this view in an email correspondence:

> There are definitely some links, but we don't know, really, what the effects are or what to do about them.... There are so many variables in EMF research that it's a question of whether the studies were even done correctly.... Maybe in a few years there will be really definite answers.

Echoing Halloran-Barney's views, John Miltane, Chief Engineer for County Electric, added this political insight:

> The public needs and demands electricity, but in regard to the negative effects of generation and transmission, the pervasive attitude seems to be "not in my back yard!" Utilities in general are scared to death of the EMF issue, but at County Electric we're trying to do the best job we can while providing reliable electricity to 24,000 customers.

Public Perception.   Industry views seem to parallel the national perspective among the broader population: informed people are genuinely concerned, but remain unsure about what level of anxiety is warranted or what exactly should be done.

A 1997 survey by the Edison Electric Institute did reveal that EMFs are considered a serious health threat by 33 percent of the American public (Stix 33).

Risk-Avoidance Measures Being Taken

Although conclusive answers may require decades of research, concerned citizens are already taking action against potential EMF hazards.

Risk Avoidance Nationwide.   Here are steps taken by various communities to protect schoolchildren from EMF exposure:

- ◆ Hundreds of individuals and community groups have taken legal action to block proposed construction of new power lines. A single Washington law firm has defended roughly 140 utilities in cases related to EMFs (Dana and Turner 32).
- ◆ Houston schools "forced a utility company to remove a transmission line that ran within 300 feet of three schools. Cost: $8 million" (Kirkpatrick 85).
- ◆ California parents and teachers are pressuring reluctant school and public health officials to investigate cancer rates in the roughly 1,000 schools located within 300 feet of transmission lines, and to close at least one school (within 100 feet) in which cancer rates far exceed normal (Brodeur 118).

Given the expense of modifying power lines to reduce EMFs (an estimated $1 billion to $3 billion yearly), critics argue that the questionable risks fail to justify the costs of measures like those above (Broad 19). Nonetheless, widespread expressions of concern about EMF exposure continue to grow.

Risk Avoidance Locally.   Local awareness of the EMF issue seems low. The main public concern seems to be with property values. According to Halloran-Barney, County Electric receives one or two calls monthly from concerned customers, including people buying homes near power lines. The lack of public awareness adds another dimension to the EMF problem: People can't avoid a health threat that they don't know exists.

John Miltane stresses that County Electric takes the EMF issue very seriously: Whenever possible, new distribution lines are run underground and configured to diminish EMF intensity:

> Although EMFs are impossible to eliminate altogether, we design anything we build to emit minimal intensities. . . . Also, we are considering underground cable (at $1,200 per ft.) to replace 8,000 feet of transmission lines, but the bill would ring up to nearly $10 million, for the cable alone, without labor. You don't get environmental stuff for free, which is one of the problems.

Before risk avoidance can be considered on a broader community level, the public must first be informed about EMFs and the associated risks of exposure.

Conclusion

The conclusion is likely to interest readers most because it answers the original questions that sparked the analysis.

> **NOTE**   *Many workplace reports are submitted with the conclusion preceding the introduction and body sections.*

In the conclusion, you summarize, interpret, and recommend. Although you have interpreted evidence at each stage of your analysis, your conclusion presents a broad interpretation and suggests a course of action, where appropriate. The summary and interpretations should lead logically to your recommendations.

- The summary accurately reflects the body of the report.
- The overall interpretation is consistent with the findings in the summary.
- The recommendations are consistent with the purpose of the report, the evidence presented, and the interpretations given.

As you read the following conclusion, evaluate how effectively it provides a clear and consistent perspective on the whole document.

CONCLUSION
Summary and Overall Interpretation of Findings
Electromagnetic fields exist wherever electricity flows; the stronger the current, the higher the EMF intensity. While no "safe" EMF level has been identified, long-term exposure to intensities greater than 2.5 milligauss is considered dangerous. Although home appliances can generate high EMFs during use, power lines can generate constant EMFs, typically at 2 to 3 milligauss in buildings within 150 feet. Our elementary school is less than 100 feet from a high-voltage power line corridor.

Notable epidemiologic studies implicate EMFs in increased rates of medical disorders such as cancer, miscarriage, stress, depression, and learning disabilities—all directly related to intensity and duration of exposure. Laboratory studies show that EMFs cause the kinds of cellular, metabolic, and behavioral changes that could produce these disorders.

An overall judgment about what the findings mean

Though still controversial and inconclusive, the various findings are strikingly similar and they underscore the need for more research and for risk avoidance, especially as far as children are concerned. Especially striking is the 1997 discovery of a direct biological link between EMF exposure and DNA damage.

Concerned citizens nationwide are beginning to prevail over resistant school and health officials and utility companies in reducing EMF risk to schoolchildren. And even though our local power company is taking reasonable risk-avoidance steps, our community can do more to learn about the issues and diminish potential risk.

Recommendations

Given the limitations of what we know, drastic and enormously expensive actions (such as burying all the town's power lines, increasing the height of utility towers, or using metal shields to deflect EMFs) seem inadvisable (Moulder). Moreover, these might turn out to be the wrong actions. However, the following inexpensive steps can be taken immediately to address EMF risk:

A reasonable and realistic course of action

- A version of this report should be distributed to all Bocaville residents.
- Our school board should hire a qualified contractor to take milligauss readings throughout the elementary school, to determine the extent of the problem, and to suggest reasonable corrective measures.
- Our Town Council should meet with County Electric Company representatives to explore options and costs for rerouting or burying the segment of the power lines near the school.
- A town meeting should then be held to answer citizens' questions and to solicit opinions.
- A committee (consisting of at least one physician, one engineer, and other experts) should be appointed to review emerging research as it relates to our school and town.

A closing call to action

As we await conclusive answers, we need to learn all we can about the EMF issue, and to do all we can to diminish this potentially significant health risk.

The Works Cited section for the preceding report appears on pages 592 and 594. Simoneau uses MLA documentation style.

Supplements

Submit your completed report with these supporting documents, in this order:

- title page
- letter of transmittal
- table of contents

- list of tables and figures
- abstract
- report text (introduction, body, conclusion)
- glossary (as needed)
- appendixes (as needed)
- Works Cited page (or alphabetical or numbered list of references)

A SAMPLE SITUATION REQUIRING AN ANALYTICAL REPORT

The report in Figure 25.2 on pages 542–49, patterned after the model outline, combines a feasibility analysis with a comparative analysis.

The situation: Richard Larkin, author of the following report, has a work-study job fifteen hours weekly in his school's placement office. His boss, John Fitton (placement director), likes to keep up with trends in various fields. Larkin, an engineering major, has developed an interest in technical marketing and sales. Needing a report topic for his writing course, Larkin offers to analyze the feasibility of a technical marketing and sales career, both for his own decision making and for technical and science graduates in general. Fitton accepts Larkin's offer, looking forward to having the final report in his reference file for use by students choosing careers. Larkin wants his report to be useful in three ways: (1) to satisfy a course requirement, (2) to help him in choosing his own career, and (3) to help other students with their career choices.

With his topic approved, Larkin begins gathering his primary data, using interviews, letters of inquiry, telephone inquiries, and lecture notes. He supplements these primary sources with articles in recent publications. He will document his findings in APA (author-date) style.

As a guide for designing his final report, Larkin completes the following audience and use profile (based on the worksheet on page 29).

Audience and Use Profile for a Formal Report

Audience Identity and Needs

My primary audience consists of John Fitton, Placement Director, and the students who will refer to my report. The secondary audience is my writing instructor.

Because he is familiar with the marketing field, Fitton will need very little background to understand my report. Many student readers, however, will have questions like these:

- What, exactly, is technical marketing and sales?
- What are the requirements for this career?
- What are the pros and cons of this career?

- Could this be the right career for me?
- How do I enter the field?

Attitude and Personality
Readers likely to be affected by this document are students making career choices. I expect readers' attitudes will vary:

- Some readers should have a good deal of interest, especially those seeking a people-oriented career.
- Others might be only casually interested as they investigate a range of possible careers.
- Some readers might be skeptical about something written by a fellow student instead of by some expert. To connect with all these people, I need to persuade them that my conclusions are based on reliable information and careful reasoning.

Expectations about the Document
All readers expect me to spell things out, but to be concise. Visuals will help compress and emphasize material throughout.

Essential information will include an expanded definition of technical marketing and sales, the skills and attitudes needed for success, the career's advantages and drawbacks, and a description of various paths for entering the career.

This report combines feasibility and comparative analysis, so I'll want to structure the report to reveal a clear line of reasoning: in the feasibility section, reasons for and reasons against; in the comparison section, a block structure and a table that compares the four entry paths point by point. The report will close with recommendations based solidly on my conclusions.

For various readers who might not wish to read the entire report, I will include an informative abstract.

This report's front matter (title page and so on) and end matter are shown and discussed in Chapter 16, pages 324–34.

Feasibility Analysis of a Career in Technical Marketing

INTRODUCTION

The escalating cutbacks in aerospace, defense-related, and other goods-producing industries have narrowed career opportunities for many of today's science and engineering graduates. A study by the Massachusetts Institute of Technology, for example, found that leading industries hired 80 to 90 percent fewer engineers in the mid-1990s than in the mid-1980s (Solomon, 1996). This trend is expected to continue—if not worsen—in the foreseeable future.

With the notable exception of computer engineering, employment in all engineering specialties will grow at rates ranging from average to far below average to near static through the year 2006. In some specialties (e.g., mining and petroleum engineering), employment will actually decline ("Occupational Employment," 1998, p. 16).

Given such bleak employment prospects, recent graduates might consider alternative careers in which they could apply their technical training. One especially attractive alternative is in marketing and selling of technology products or services.

Customer orientation is an ever-growing part of today's business and manufacturing climate. Beginning in the 1980s, U.S. industry ceased to be "manufacturing driven" (where customers would buy whatever products were available). Technology companies became "service driven" by the demand for customized, efficiently serviced products (Basta, 1984, p. 84).

In the product-oriented industries of the 1970s, technical marketing accounted for only 39 percent of top management backgrounds, but that number nearly doubled by 1995. Also, 1998 employment listings for recent graduates showed countless major companies offering positions in technical marketing and sales (College Placement Council, 1998, p. 402) .

Undergraduates interested in this career need answers to these basic questions:

- *Is this the right career for me?*
- *If so, how do I enter the field?*

To help answer these questions, this report analyzes information gathered from professionals as well as from the literature.

After defining *technical marketing,* the following analysis examines the field's employment outlook, required skills and personal qualities, career benefits and drawbacks, and various entry options.

Figure 25.2 An Analytical Report

COLLECTED DATA

Key Factors in a Technical Marketing Career
Anyone considering technical marketing needs to assess whether this career fits his or her interests, abilities, and aspirations.

THE TECHNICAL MARKETING PROCESS. Although the terms *marketing* and *sales* are often used interchangeably, technical marketing traditionally has involved far more than sales work. The process itself (identifying, reaching, and selling to customers) entails six major activities (Cornelius & Lewis, 1983, p. 44):

1. *Market research:* gathering information about the size and character of the target market for a product or service.
2. *Product development and management:* producing the goods to fill a specific market need.
3. *Cost determination and pricing:* measuring every expense in the production, distribution, advertising, and sales of the product, to determine its price.
4. *Advertising and promotion:* developing and implementing all strategies for reaching customers.
5. *Product distribution:* coordinating all elements of a technical product or service, from its conception through its final delivery to the customer.
6. *Sales and technical support:* creating and maintaining customer accounts, and servicing and upgrading products.

Fully engaged in all these activities, the technical marketing professional gains a detailed understanding of the industry, the product, and the customer's needs (Figure 1).

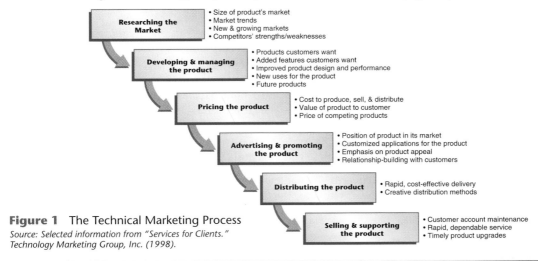

Figure 1   The Technical Marketing Process
Source: Selected information from "Services for Clients."
Technology Marketing Group, Inc. (1998).

Figure 25.2   An Analytical Report   *Continued*

EMPLOYMENT OUTLOOK. For graduates with the right combination of technical and personal qualifications, the employment outlook for technical marketing appears excellent. While engineering jobs will increase at barely one half the average growth rate for jobs requiring a Bachelor's degree, marketing jobs will exceed the average growth rate (Figure 2).

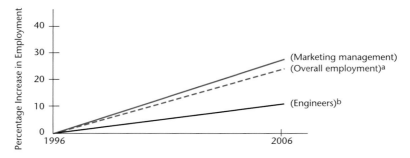

Figure 2 The Employment Outlook for Technical Marketing
[a]Jobs requiring a Bachelor's degree.
[b]Excluding outlying rates for engineering specialties at extreme ends of the spectrum (computer engineers: +109%; mining and petroleum engineers: –14%).
Source: Data from U.S. Department of Labor. (1998, Winter). Occupational Outlook Quarterly, *11–16.*

Although highly competitive, these marketing positions call for the very kinds of technical, analytical, and problem-solving skills that engineers can offer (Solomon, 1996)—especially in an automated environment.

TECHNICAL SKILLS REQUIRED. Computer networks, interactive media, and multimedia will increasingly influence the way products are advertised and sold. Also, marketing representatives increasingly work from a "virtual" office. Using laptop computers, fax networks, and personal digital assistants, representatives in the field have real-time access to electronic catalogs of product lines, multimedia presentations, pricing for customized products, inventory data, product distribution, and customized sales contacts (Tolland, 1999).

With their rich background in computer, technical, and problem-solving skills, engineering graduates are ideally suited for (a) working in automated environments, and (b) implementing and troubleshooting these complex and often sensitive electronic systems.

OTHER SKILLS AND QUALITIES REQUIRED. In marketing and sales, not even the most sophisticated information can substitute for the "human factor": the ability to connect with customers on a person-to-person level (Young, 1995, p. 95). One senior sales engineer

Figure 25.2 An Analytical Report   *Continued*

praises the efficiency of her automated sales system, but thinks that automation will "get in the way" of direct customer contact. Other technical marketing professionals express similar views about the continued importance of human interaction (94).

Besides a strong technical background, marketing requires a generous blend of those traits summarized in Figure 3.

Figure 3 Requirements for a Technical Marketing Career

Motivation is essential for marketing work. Professionals must be energetic and able to function with minimal supervision. Career counselor Phil Hawkins describes the ideal candidates as people who can plan and program their own tasks, who can manage their time, and who have no fear of hard work (personal interview, February 11, 1999). Leadership potential, as demonstrated by extracurricular activities, is an asset.

Motivation alone provides no guarantee of success. Marketing professionals are paid to communicate the virtues of their products or services. This career therefore requires skill in communication, both written and oral. Documents for readers outside the organization include advertising copy, product descriptions, sales proposals, sales letters, and user manuals and online help. In-house writing includes recommendation reports, feasibility studies, progress reports, memos, and email correspondence (U.S. Department of Labor, 1997, p. 8).

Skilled oral presentation is vital to any sales effort, as Phil Hawkins points out. Technical marketing professionals need to speak confidently and persuasively—to represent their products and services in the best possible light (personal interview, February 11, 1999). Sales presentations often involve public speaking at conventions, trade shows, and other similar forums.

Figure 25.2 An Analytical Report   *Continued*

Beyond motivation and communication skills, interpersonal skills are the ultimate requirement for success in marketing (Solomon, 1996). Consumers are more likely to buy a product or service when they like the person selling it. Marketing professionals are extroverted, friendly, and diplomatic; they can motivate people without alienating them.

ADVANTAGES OF THE CAREER. As shown in Figure 2, technical marketing offers diverse experience in every phase of a company's operation, from a product's design to its sales and service. Such broad exposure provides excellent preparation for countless upper-management positions.

In fact, sales engineers with solid experience often open their own businesses as "manufacturers' agents" representing a variety of companies. These agents represent products for companies who have no marketing staff of their own. In effect their own bosses, manufacturers' agents are free to choose, from among many offers, the products they wish to represent (Tolland, 1999).

Another career benefit is the attractive salary. Marketing professionals typically receive a base pay plus commissions. According to John Turnbow, managing recruiter of National Electric's technical marketing program, new marketing engineers for NE average over $60,000 in first-year wages. Moreover, salaries often reach six figures—sometimes higher than executive salaries (personal communication, April 5, 1999).

Technical marketing is especially attractive for its geographic and job mobility. Companies nationwide seek recent graduates, but especially in the Southeast and on the east and west coasts ("Electronic sales," 1996, pp. 1134–37). In addition, the types of interpersonal and communication skills that marketing professionals develop are highly portable. This is especially important in our current, rapidly shifting economy, in which job security is disappearing in the face of more and more temporary positions (Jones, 1997, p. 51).

DRAWBACKS OF THE CAREER. Technical marketing is by no means a career for every engineer. Sales engineer Roger Cayer cautions that personnel might spend most of their time traveling to meet potential customers. Success requires hard work over long hours, evenings, and occasional weekends. Above all, the job is stressful because of constant pressure to meet sales quotas (phone interview, February 8, 1999). Anyone considering this career should be able to work and thrive in a highly competitive environment.

A Comparison of Entry Options

Engineers and other technical graduates enter technical marketing through one of four options. Some join small companies and learn their trade directly on the job. Others join companies that offer formal training programs. Some begin by getting experience in their

Figure 25.2   An Analytical Report   *Continued*

technical specialty. Others earn a graduate degree beforehand. These options are compared below.

OPTION 1: ENTRY-LEVEL MARKETING WITH ON-THE-JOB TRAINING. Smaller manufacturers offer marketing positions in which people learn on the job. Elaine Carto, president of ABCO Electronics, believes small companies offer a unique opportunity; entry-level salespersons learn about all facets of an organization, and have a good possibility for rapid advancement (personal interview, February 10, 1999). Career counselor Phil Hawkins says, "It's all a matter of whether you prefer to be a big fish in a small pond or a small fish in a big pond" (personal interview, February 11, 1999).

Entry-level marketing offers immediate income and a chance for early promotion. A disadvantage, however, might be the loss of any technical edge one might have acquired in college.

OPTION 2: A MARKETING AND SALES TRAINING PROGRAM. Formal training programs offer the most popular entry into sales and marketing. Large to mid-size companies typically offer two formats: (a) a product-specific program, focused on a particular product or product line, or (b) a rotational program, in which trainees learn about an array of products and work in the various positions outlined in Figure 2. Programs last from weeks to months.

Former trainees Roger Cayer, of Allied Products, and Bill Collins, of Intrex, speak of the diversity and satisfaction such programs offer: specifically, solid preparation in all phases of marketing, diverse interaction with company personnel, and broad knowledge of various product lines (phone interviews, February 8, 1999).

Like direct entry, this option offers the advantage of immediate income and early promotion. With no chance to practice in their technical specialty, however, trainees might eventually find their technical expertise compromised.

OPTION 3: PRIOR EXPERIENCE IN ONE'S TECHNICAL SPECIALTY. Instead of directly entering marketing, some candidates first gain experience in their specialty. This option combines direct exposure to the workplace with the chance to sharpen technical skills in practical applications. In addition, some companies, such as Roger Cayer's, will offer marketing and sales positions to outstanding staff engineers, as a step toward upper management (phone interview, February 8, 1999).

Although this option delays a candidate's entry into technical marketing, industry experts consider direct workplace and technical experience key assets for career growth in any field. Also, work experience becomes an asset for applicants to top MBA programs (Shelley, 1997, pp. 30–31).

Figure 25.2 An Analytical Report *Continued*

OPTION 4: GRADUATE PROGRAM. Instead of direct entry, some people choose to pursue an MS degree in their specialty or an MBA. According to engineering professor Mary McClane, MS degrees are usually unnecessary for technical marketing unless the particular products are highly complex (personal interview, April 2, 1999).

In general, jobseekers with an MBA have a distinct competitive advantage. More significantly, new MBAs with a technical bachelor's degree and one to two years of experience command salaries from 10 to 30 percent higher than MBAs who lack work experience and a technical bachelor's degree. In fact, no more than 3 percent of job candidates offer a "techno-MBA" specialty, making this unique group highly desirable to employers (Shelley, 1997, p. 30).

A motivated student might combine graduate degrees. Dora Anson, president of Susimo Cosmic Systems, sees the MS/MBA combination as ideal preparation for technical marketing (1999).

One disadvantage of a full-time graduate program is lost salary, compounded by school expenses. These costs must be weighed against the prospect of promotion and monetary rewards later in one's career.

AN OVERALL COMPARISON BY RELATIVE ADVANTAGE. Table 1 compares the four entry options on the basis of three criteria: immediate income, rate of advancement, and long-term potential.

Table 1   Relative Advantages Among Four Technical-Marketing Entry Options

Option	Relative Advantages		
	Early, immediate income	Greatest advancement in marketing	Long-term potential
Entry level, no experience	yes	yes	no
Training program	yes	yes	no
Practical experience	yes	no	yes
Graduate program	no	no	yes

Figure 25.2   An Analytical Report   *Continued*

CONCLUSION

Summary of Findings

Technical marketing and sales involves identifying, reaching, and selling the customer a product or service. Besides a solid technical background, the field requires motivation, communication skills, and interpersonal skills. This career offers job diversity and excellent income potential, balanced against hard work and relentless pressure to perform.

College graduates interested in this field confront four entry options: (1) direct entry with on-the-job training, (2) a formal training program, (3) prior experience in a technical specialty, and (4) graduate programs. Each option has benefits and drawbacks based on immediacy of income, rate of advancement, and long-term potential.

Interpretation of Findings

For graduates with a strong technical background and the right skills and motivation, technical marketing offers attractive career prospects. Anyone contemplating this field, however, needs to be able to enjoy customer contact and thrive in a highly competitive environment.

Those who decide that technical marketing is for them can choose among the various entry options:

- For hands-on experience, direct entry is the logical option.
- For sophisticated sales training, a formal program with a large company is best.
- For sharpening technical skills, prior work in one's specialty is invaluable.
- If immediate income is not vital, graduate school is an attractive option.

Recommendations

If your interests and abilities match the requirements, consider these suggestions:

1. To get a firsthand view, seek the advice and opinions of people in the field.
2. Before settling on an entry option, consider all its advantages and disadvantages and decide whether this option best coincides with your career goals. (Of course, you can always combine options during your professional life.)
3. When making any career decision, consider career counselor Phil Hawkins' advice: "Listen to your brain and your heart " (personal interview, February 11, 1999). Choose an option or options that offer not only professional advancement but also personal satisfaction.

REFERENCES

[The complete list of references is shown and discussed in Appendix A, page 604.]

Figure 25.2   An Analytical Report   *Continued*

Usability Checklist FOR ANALYTICAL REPORTS

For evaluating your research methods and reasoning, refer also to the checklist on page 170. (Numbers in parentheses refer to the first page of discussion.)

Content

- ☐ Does the report grow from a clear statement of purpose? (524)
- ☐ Is the report's length adequate and appropriate for the subject? (522)
- ☐ Are all limitations of the analysis clearly acknowledged? (529)
- ☐ Are visuals used whenever possible to aid communication? (526)
- ☐ Are all data accurate? (525)
- ☐ Are all data unbiased? (147)
- ☐ Are all data complete? (525)
- ☐ Are all data full interpreted? (150)
- ☐ Is the documentation adequate, correct, and consistent? (578)
- ☐ Are the conclusions logically derived from accurate interpretation? (526)
- ☐ Do the recommendations constitute an appropriate response to the question or problem? (526)

Arrangement

- ☐ Is there a distinct introduction, body, and conclusion? (529)
- ☐ Are headings appropriate and adequate? (316)
- ☐ Are there enough transitions between related ideas? (Appendix C)
- ☐ Is the report accompanied by all needed front matter? (539)
- ☐ Is the report accompanied by all needed end matter? (540)

Style and Page Design

- ☐ Is the level of technicality appropriate for the stated audience? (24)
- ☐ Is the writing style throughout clear, concise, and fluent? (215)
- ☐ Is the language convincing and precise? (235)
- ☐ Is the report written in grammatical English? (Appendix C)
- ☐ Is the page design inviting and accessible? (305)

EXERCISE

Prepare an analytical report, using these guidelines:

a. Choose a subject for analysis from the list at the end of this exercise, from your major, or from a subject of interest.

b. Identify the problem or question so that you will know exactly what you are looking for.

c. Restate the main question as a declarative sentence in your statement of purpose.

d. Identify an audience—other than your instructor—who will use your information for a specific purpose.

e. Hold a private brainstorming session to generate major topics and subtopics.

f. Use the topics to make an outline based on the model outline in this chapter. Divide as far as necessary to identify all points of discussion.

g. Make a tentative list of all sources (primary and secondary) that you will investigate. Verify that adequate sources are available.

h. Write your instructor a proposal memo, describing the problem or question and your plan for analysis. Attach a tentative bibliography.

i. Use your working outline as a guide to research and observation. Evaluate sources and evidence, and interpret all evidence fully. Modify your outline as needed.

j. Submit a progress report to your instructor describing work completed, problems encountered, and work remaining.

k. Compose an audience and use profile. (Use the sample on pages 540 and 541 as a model, along with the profile worksheet on page 29.)

l. Write the report for your stated audience. Work from a clear statement of purpose, and be sure that your reasoning is shown clearly. Verify that your evidence, conclusions, and recommendations are consistent. Be especially careful that your recommendations observe the critical-thinking guidelines in Figure 25.1.

m. After writing your first draft, make any needed changes in the outline and revise your report according to the revision checklist. Include all necessary supplements.

n. Exchange reports with a classmate for further suggestions for revision.

o. Prepare an oral report of your findings for the class as a whole.

Here are some possible subjects for analysis:

- long-term effects of a vegetarian diet
- causes of student disinterest in campus activities
- student transportation problem to and from your college
- comparison of two or more brands of equipment
- noise pollution from nearby airport traffic
- effects of toxic chemical dumping in your community
- adequacy of veterans' benefits
- causes/effects of acid rain in your area
- influence of a civic center or stadium on your community
- best microcomputer to buy for a specific need
- qualities employers seek in a job candidate

- feasibility of opening a specific business
- best location for a new business
- causes of the high dropout rate in your college
- pros and cons of condominium ownership for you
- feasibility of moving to a certain area of the country
- job opportunities in your field
- effects of budget cuts on public higher education in your state
- effect of population increase on your local water supply
- feasibility of biological pest control as an alternative to pesticides
- feasibility of large-scale desalination of sea water as a source of fresh water
- effective water conservation measures that can be used in your area
- effects of legalizing gambling in your state
- causes of low morale in the company where you work part-time
- effects of thermal pollution from a local power plant on marine life
- feasibility of converting your home to solar heating
- adequacy of police protection in your town
- best energy-efficient, low-cost housing design for your area
- measures for improving productivity in your place of employment
- reasons for the success of a specific business in your area
- feasibility of operating a campus food co-op
- advisability of home birth (as opposed to hospital delivery)
- adequacy of the evacuation plan for your area in a nuclear emergency, hurricane, or other disaster
- feasibility and cost of improving security in the campus dorms
- advisability of pursuing a graduate degree in your field, instead of entering the work force with a bachelor's degree
- major causes of the wage gap between women and men in a particular field

COLLABORATIVE PROJECTS

1. Divide into small groups. Choose a subject for group analysis—preferably, a campus issue—and partition the topic by group brainstorming. Next, select major topics from your list and classify as many items as possible under each major topic. Finally, draw up a working outline that could be used as an analytical report on this subject.

2. Prepare a questionnaire based on your work above, and administer it to members of your campus community. List the findings of your questionnaire and your conclusions in clear and logical form. (Review pages 135–40, on questionnaires and surveys.)

Oral
Presentations

Oral presentations vary in style, range, complexity, and formality. They may include convention speeches, reports at national meetings, reports via teleconferencing networks, technical briefings for colleagues, and speeches to community groups. These talks may be designed to inform (to describe new government safety requirements), to persuade (to induce company officers to vote a pay raise), or to do both. The higher your status, on the job or in the community, the more you will give oral presentations.

AVOIDING PRESENTATION PITFALLS

An oral presentation is only the tip of a pyramid built from many earlier labors. But such presentations often serve as the concrete measure of your overall job performance. In short, your audience's only basis for judgment may be the brief moments during which you stand before them.

The podium or lectern can be a lonely place. In fact, people usually consider speaking before a group to be the most intimidating communication task. In the words of two experts, "most persons in most presentational settings do not perform well" (Goodall and Waagen 14–15). Despite the fact that they can help make or break a person's career, oral presentations are often boring, confusing, unconvincing, or too long. Many are poorly delivered, with the presenter losing her or his place, fumbling through notes, apologizing for forgetting something, or generally seeming unprofessional. Table 26.1 lists some of the pitfalls.

Given such difficulties, how can any presenter display skill and confidence? By proceeding systematically through careful analysis, planning, and preparation.

PLANNING YOUR PRESENTATION

A successful presentation involves more than just getting up and talking. We have all sat through enough lectures and presentations during our student careers to know how to separate the excellent from the awful. The successful presenter knows how to forge a relationship with the listeners, how to establish rapport and persuade listeners their time has been well spent.

Work from an Explicit Purpose Statement

Formulate, on paper, a statement of purpose in two or three sentences. Why, exactly, are you speaking on this subject? Who are your listeners? What do you want the listeners to think, know, or do? (A solid purpose statement can also serve as the introduction to your presentation.)

Assume, for example, that you represent an environmental engineering firm that has completed a study of groundwater quality in your area. The organization that sponsored your study has asked you to give an oral version of your written report, titled "Pollution Threats to Local Groundwater," at a town meeting

After careful thought, you settle on this purpose statement:

TABLE 26.1

Common Pitfalls in Oral Presentations

Speaker ● ● ●	Visuals ✶ ✶ ✶	Setting ■ ■ ■
● makes no eye contact	✶ are nonexistent	■ is too noisy
● seems like a robot	✶ are hard to see	■ is too hot or cold
● hides behind the lectern	✶ are hard to interpret	■ is too large or small
● speaks too softly/loudly	✶ are out of sequence	■ is too bright for visuals
● sways, fidgets, paces	✶ are shown too rapidly	■ is too dark for notes
● rambles or loses her/his place	✶ are shown too slowly	■ has equipment missing
● never gets to the point	✶ have typos/errors	■ has broken equipment
● fumbles with notes or visuals	✶ are word-filled	
● has too much material		

Purpose statement | *Purpose:* By informing Cape Cod residents about the dangers to the Cape's freshwater supply posed by rapid population growth, this report is intended to increase local interest in the problem.

Now you are prepared to focus on the listeners and the speaking situation.

Analyze Your Listeners

Assess your listeners' needs, knowledge, concerns, level of involvement, and possible objections (Goodall and Waagen 16).

Many audiences include people with varied technical backgrounds. Unless you have a good idea of each person's background, speak to a general, heterogeneous audience, as in a classroom of mixed majors.

QUESTIONS FOR ANALYZING YOUR LISTENERS

- ◆ Who are my listeners (strangers, peers, superiors, clients)?
- ◆ What is their attitude toward me or the topic (hostile, indifferent, needy, friendly)?
- ◆ Why are they here (they want to be here, they are forced to be here, they are curious)?
- ◆ What kind of presentation do they expect (brief, informal; long, detailed; lecture)?
- ◆ What do these listeners already know (nothing, a little, a lot)?

- ◆ What do they need or want to know (overview, bottom line, nitty gritty)?
- ◆ How large is their stake in this topic (about layoffs, new policies, pay raises)?
- ◆ Do I want to motivate, mollify, inform, instruct, or warn my listeners?
- ◆ What are their biggest concerns or objections about this topic?
- ◆ What do I want them to think, know, or do?

Analyze Your Speaking Situation

The more you can discover about the circumstances, the setting, and the constraints for your presentation, the more deliberately you will be able to prepare.

QUESTIONS FOR ANALYZING YOUR SPEAKING SITUATION

◆ How much time will I have to speak?
◆ Will other people be speaking before or after me?
◆ How formal or informal is the setting?
◆ How large is the audience?

◆ How large is the room?
◆ How bright and adjustable is the lighting?
◆ What equipment is available?
◆ How much time do I have to prepare?

Later parts of this chapter explain how to incorporate your answers to the preceding questions in your preparation.

Select an Appropriate Delivery Method

Your presentation's effectiveness will depend largely on *how* it connects with listeners. Different types of delivery create different connections.

THE MEMORIZED DELIVERY. A memorized delivery seldom connects with listeners because the speaker is too busy reciting the lines and trying to remember everything. This type of delivery takes a long time to prepare, offers no chance for revision mid-presentation, and spells disaster if you lose your place. Use this type of delivery only when absolutely necessary.

THE IMPROMPTU DELIVERY. An impromptu (off-the-cuff) delivery can be a natural way of connecting with listeners—but only when you really know your material, feel comfortable with your listeners, and are in an informal speaking situation (group brainstorming, or a response to a question: "Tell us about your team's progress on the automation project"). Avoid impromptu deliveries for complex information—no matter how well you know the material.

Too many things can go wrong in an unplanned, spontaneous presentation: You might say something offensive or irrelevant; you might seem disorganized; you might not make sense. If you anticipate being called on, never assume "It's all in my head." Get your plan down on paper. If you have little warning, at least jot a few notes about what you want to say.

THE SCRIPTED DELIVERY. For a complex technical presentation, a conference paper, or a formal speech, you may want to read your material verbatim from a prepared script. Scripted presentations work well if you have many details to present, are talkative, or have a strict time limit (e.g., at a conference), or if this audience makes you nervous. Consider a scripted delivery when you want the content, organization, and style of your presentation to be as near perfect as possible.

Although a scripted delivery helps you control your material, it offers little chance for audience interaction and it can be boring.

If you *do* plan to read aloud, allow ample preparation time. Leave plenty of white space between lines and paragraphs. Rehearse until you are able to glance up from the script periodically without losing your place. Plan on roughly two minutes per double-spaced page.

THE EXTEMPORANEOUS DELIVERY.   An extemporaneous delivery is carefully planned, practiced, and based on notes that keep you on track. In this natural way of addressing an audience, you glance at your material and speak in a conversational style. Extemporaneous delivery is based on key ideas in sentence or topic outline form, rather than fully developed paragraphs to be read or memorized.

The dangers in extemporaneous delivery are that you might lose track of your material, forget something important, say something unclearly, or exceed your time limit. Careful preparation is the key.

Table 26.2 summarizes the various uses and drawbacks of the most common types of delivery. These need not be fixed, exclusive categories. In many instances, some combination of methods can be effective. For example, in an orientation for new employees, you might prefer the flexibility of an extemporaneous format but also read a brief passage aloud from time to time (e.g., excerpts from the company's formal code of ethics).

TABLE 26.2

A Comparison of Oral Presentation Methods

Delivery Method	*Main Uses*	•Main Drawbacks•
IMPROMPTU (inventing as you speak)	*in-house meetings *small, intimate groups *simple topics	•offers no chance to prepare •speaker might ramble •speaker might lose track
SCRIPTED (reading verbatim from a written work)	*formal speeches *large, unfamiliar groups *highly complex topics *strict time limit *cross-cultural audiences *highly nervous speaker	•takes a long time to prepare •speaker can't move around •limits human contact •can appear stiff and unnatural •might bore listeners •makes working with visuals difficult
EXTEMPORANEOUS (speaking from an outline of key points)	*face-to-face presentations *medium-sized, familiar groups *moderately complex topics *somewhat flexible time limit *visually based presentations	•speaker might lose track •speaker might leave something out •speaker might get tongue-tied •speaker might exceed time limit •speaker might fumble with notes, visuals, or equipment

PREPARING YOUR PRESENTATION

To stay in control and build confidence, plan the presentation systematically. We will assume here that your presentation is extemporaneous.

Research Your Topic

Do your homework. Be prepared to support each assertion, opinion, conclusion, and recommendation with evidence and reason. Check your facts for accuracy. Your audience expects to hear a knowledgeable speaker. Don't disappoint them.

Begin gathering material well ahead of time. Use summarizing techniques from Chapter 11 to identify and organize major points.

If your preparation is simply a spoken version of a written report, you need far less preparation. Simply expand your outline for the written report into a sentence outline.

Aim for Simplicity and Conciseness

Keep your presentation short and simple. Boil the material down to a few main points. Listeners' normal attention span is about twenty minutes. After that, they begin tuning out. Time yourself in practice sessions and trim as needed. (If the material requires a lengthy presentation, plan a short break, with refreshments if possible, about halfway through.)

Anticipate Audience Questions

Consider those parts of your presentation that listeners might question or challenge. You might need to clarify or justify information that is new, controversial, disappointing, or surprising.

Outline Your Presentation

Review Chapter 12 for organizing and outlining strategies. Each sentence in the following presentation outline is a topic sentence for a paragraph that a well-prepared speaker can develop in detail.

Presentation outline

POLLUTION THREATS TO LOCAL GROUNDWATER

ARNOLD BORTHWICK

I. Introduction to the Problem
  A. Do you know what you are drinking when you turn on the tap and fill a glass?
  B. The quality of our water is good, but not guaranteed to last forever.
  C. Cape Cod's rapid population growth poses a serious threat to our freshwater supply. (Visual #1)
  D. Measurable pollution in some town water supplies has already occurred. (Visual #2)
  E. What are the major causes and consequences of this problem and what can we do about it? (Visual #3)

II. Description of the Aquifer

A. The groundwater is collected and held in an aquifer.

    1. This porous rock formation forms a broad, continuous arch beneath the entire Cape. (Visual #4)

    2. The lighter freshwater flows on top of the heavier salt water.

B. This type of natural storage facility, combined with rapid population growth, creates potential for disaster.

III. Hazards from Sewage and Landfills

A. With increasing population, sewage and solid waste from landfill dumps increasingly invade the aquifer.

B. The Cape's sandy soil promotes rapid seepage of wastes into the groundwater.

C. As wastes flow naturally toward the sea, they can invade the drawing radii of town wells. (Visual #5)

IV. Hazards from Saltwater Intrusion

A. Increased population also causes overdraw on some town wells, resulting in saltwater intrusion. (Visual #6)

B. Salt and calcium used in snow removal add to the problem by seeping into the aquifer from surface runoff.

V. Long-Term Environmental and Economic Consequences

A. The environmental effects of continuing pollution of our water table will be far-reaching. (Visual #7)

    1. Drinking water will have to be piped in more than 100 miles from Quabbin Reservoir.

    2. The Cape's beautiful freshwater ponds will be unfit for swimming.

    3. Aquatic and aviary marsh life will be threatened.

    4. The Cape's sensitive ecology might well be damaged beyond repair.

B. Such environmental damage would, in turn, spell economic disaster for Cape Cod's major industry—tourism.

VI. Conclusion and Recommendations

A. This problem is becoming more real then theoretical.

B. The conclusion is obvious: If the Cape is to survive ecologically and financially, we must take immediate action to preserve our *only* water supply.

C. These recommendations offer a starting point for action. (Visual #8)

    1. Restrict population density in all Cape towns by creating larger building lot requirements.

    2. Keep strict watch on proposed high-density apartment and condominium projects.

    3. Create a committee in each town to educate residents about water conservation.

    4. Prohibit salt, calcium, and other additives in sand spread on snow-covered roads.

    5. Explore alternatives to landfills for solid waste disposal.

D. Given its potential effects on our quality of life, such a crucial issue deserves the active involvement of every Cape resident.

Before practicing the delivery, transfer your presentation outline to 3" × 5" note-cards (one side only, each card numbered), which you can hold in one hand and shuffle as needed. Or insert the outline pages in a looseleaf binder for easy flipping. Type or print clearly, leaving enough white space to locate material at a glance.

Plan Your Visuals

Visuals increase listeners' interest, focus, understanding, and memory. Select visuals that will clarify and enhance your talk—without making you fade into the background.

DECIDE WHERE VISUALS WILL WORK BEST.   Use visuals to emphasize a point and whenever *showing* would be more effective than just *telling*.

DECIDE WHICH VISUALS WILL WORK BEST.   Will you need numerical or prose tables, graphs, charts, graphic illustrations, computer graphics? How complex should these visuals be? Should they impress or simply inform? Use the visual planning sheet in Chapter 14 to guide your decisions.

DECIDE HOW MANY VISUALS ARE APPROPRIATE.   Prefer an array of lean and simple visuals that present material in digestible amounts to one or two over-stuffed visuals that people end up staring at endlessly.

CREATE A STORYBOARD.   A presentation storyboard is a double-column format in which your discussion is outlined in the left column, aligned with the specific supporting visuals in the right column (Figure 26.1).

DECIDE WHICH VISUALS ARE ACHIEVABLE.   Fit each visual to the situation. The visuals you select will depend on the room, the equipment, and the production resources available.

Fit each visual to the situation

How large is the room and how is it arranged? Some visuals work well in small rooms, but not large ones, and vice versa. How well can the room be darkened? Which lights can be left on? Can the lighting be adjusted selectively? What size should visuals be, to be seen clearly by the whole room? (A smaller, intimate room is usually better than a room that is too big.)

What hardware is available (slide projector, opaque projector, overhead projector, film projector, videotape player, terminal with large-screen monitor)? What graphics programs are available? Which is best for your purpose and listeners? How far in advance does this equipment have to be requested?

What resources are available for producing the visuals? Can drawings, charts, graphs, or maps be created as needed? Can transparencies (for overhead projection) be made or slides collected? Can handouts be typed and reproduced? Can multimedia displays be created?

Pollution Threats to Local Groundwater

1.0   Introduce the Problem

    1.1   Do you know what you are drinking when you turn on the tap and fill the glass?

    1.2   The quality of our water is good, but not guaranteed to last forever.

    1.3   Cape Cod's rapid population growth poses a serious threat to our freshwater supply.   *(transparency: a line graph showing twenty-year population growth)*

    1.4   Measurable pollution in some town water supplies has already occurred.   *(transparency: two side-by-side tables showing twenty-year increases in nitrate and chloride concentrations in three town wells)*

    1.5   What are the major causes and consequences of this problem and what can we do about it?   *(poster: a multicolored list that previews my five subtopics)*

2.0   Describe the Aquifer

    2.1   The groundwater is collected and held in an aquifer.

        2.1.1.   This porous rock formation forms a broad, continuous arch beneath the entire Cape.   *(transparency: a cutaway view of the aquifer's geology)*

Figure 26.1   A Partial Storyboard

SELECT YOUR MEDIA.   Fit the medium to the situation. Which medium or combination is best for the topic, setting, and listeners? How fancy do listeners expect this to be? Which media are appropriate for this occasion?

Fit the medium to the situation

- For a weekly meeting with colleagues in your department, scribbling on a blank transparency, chalkboard, or dry-erase markerboard might suffice.
- For interacting with listeners, you might use a chalkboard to record audience responses to your questions.
- For immediate orientation, you might begin with a poster listing key visuals/ideas/themes to which you will refer repeatedly.
- For helping listeners take notes, absorb technical data, or remember complex material, you might distribute a presentation outline as a preview or provide handouts.
- For displaying and discussing written samples listeners bring in, you might use an opaque projector.
- For a presentation to investors, clients, or upper management, you might require polished and professionally prepared visuals, including computer graphics, such as an electronic slide presentation using Powerpoint™ software.

Chalkboard

Uses
- simple, on-the-spot visuals
- recording audience responses
- informal settings
- small, well-lighted rooms

Tips
- copy long material in advance
- write legibly
- make it visible to everyone
- use washable markers on markerboard
- speak to the listeners—not to the board

Poster

Uses
- overviews, previews, emphasis
- recurring themes
- formal or informal settings
- small, well-lighted rooms

Tips
- use 20" x 30" posterboard (or larger)
- use intense, washable colors
- keep each poster simple and uncrowded
- arrange/display posters in advance
- make each visible to the whole room
- point to what you are discussing

Flip Chart

Uses
- a sequence of visuals
- back-and-forth movement
- formal or informal settings
- small, well-lighted rooms

Tips
- use an easel pad and easel
- use intense, washable colors
- work from a story board
- check your sequence beforehand
- point to what you are discussing

Handouts

Uses
- present complex material
- help listeners follow along
- help listeners take notes
- help listeners remember

Tips
- staple or bind the packet
- number the pages
- try saving for the end
- if you must distribute up front, ask listeners to await instructions before reading the material

Figure 26.2 Selecting Media for Visual Presentations

Opaque Projection

Uses • direct display of paper documents
• samples listeners bring in
• pages from books, reports
• informal settings
• very small and dark rooms
• when spontaneity is more important than image quality

Tips • allow projector to warm up
• don't shut off projector until you're done
• use a light source for viewing your notes
• use a laser or telescoping pointer
• place listeners close enough to see clearly

Slide Projection

Uses • professional-quality visuals
• formal setting
• small or large dark rooms

Tips • work from a storyboard
• check the slide sequence beforehand
• use a light source for viewing your notes
• use a laser or telescoping pointer

Overhead Projection

Uses • on-the-spot or prepared visuals
• overlaid visuals[1]
• formal or informal settings
• small- or medium-sized rooms
• rooms needing to remain lighted

Tips • use cardboard mounting frames for your acetate transparencies
• write discussion notes on each frame
• check your sequence beforehand
• turn projector off when not using it
• face the audience—not the screen
• point directly on the transparency
• use erasable color markers to highlight items as you discuss them

Computer Projection

Uses • sophisticated charts, graphs, maps
• multimedia presentations
• formal settings
• small, dark rooms

Tips • take lots of time to prepare/practice
• work from a storyboard
• check the whole system beforehand
• have a default plan in case something goes wrong

1. The overlay technique begins with one transparency showing a basic image, over which additional transparencies (with coordinated images, colors, labels, and so on) are added to produce an increasingly complex image.

Figure 26.2 Selecting Media for Visual Presentations *Continued*

Figure 26.2 presents the various common media in approximate order of availability and ease of preparation.

Prepare Your Visuals

As you prepare visuals, focus on economy, clarity, and simplicity.

BE SELECTIVE.   Use a visual only when it truly serves a purpose. Use restraint in choosing what to highlight with visuals. Try not to begin or end the presentation with a visual. At those times, listeners' attention should be focused on the presenter instead of the visual.

MAKE VISUALS EASY TO READ AND UNDERSTAND.   Think of each visual as an image that flashes before your listeners. They will not have the luxury of studying the visual at leisure. Listeners need to know at a glance what they are looking at and what it means. Following are guidelines for achieving readability.

HOW TO DESIGN READABLE VISUALS

- Make visuals large enough to be read anywhere in the room.
- Don't cram too many words, ideas, designs, or typestyles into a single visual.
- Keep wording and images simple.
- Boil your message down to the fewest words possible.

- Break information into small chunks.
- Summarize with key words, phrases, or short sentences.
- Use 18–24 point sans serif type (White, *Great Pages* 80).

In addition to being able to *read* the visual, listeners need to *understand* it. Following are suggestions for achieving clarity.

HOW TO DESIGN UNDERSTANDABLE VISUALS

- Display only one point per visual—unless previewing or reviewing (White, *Great Pages* 79).
- Give each visual a title that announces the topic.
- Use color sparingly, to highlight key words, facts, or the bottom line.

- Use the brightest color for what is most important (White, *Great Pages* 78–79).
- Label each part of a diagram or illustration.
- Proofread each visual carefully.

When your material is extremely detailed or complex, distribute handouts to each listener.

LOOK FOR ALTERNATIVES TO WORD-FILLED VISUALS.   Instead of just presenting overhead versions of printed pages, explore the full *visual* possibilities of your media. For example, anyone who tries to write a verbal equivalent of the visual message in Figure 26.3 will soon appreciate the power of images in relation to words alone. Whenever possible, use drawings, graphs, charts, photographs, and other visual representation discussed in Chapter 14.

USE ALL AVAILABLE TECHNOLOGY.   Using desktop publishing systems and presentation software such as *Powerpoint™* or *Inspiration™*, you can create professional-quality visuals and display technical concepts. Using hypertext and multimedia systems, you can create dynamic presentations that appeal to the listener's multiple senses. Using an automatic, remote-controlled transparency feeder and a pencil-sized laser pointer, you can deliver a smooth and elegant presentation. These are just a few of the possibilities inherent in the technology.

CHECK THE ROOM AND SETTING BEFOREHAND.   Make sure you have enough space, electrical outlets, and tables for your equipment. If you will be addressing a

Figure 26.3
Images More
Powerful Than
Words
Source: "Distribution of the World's Water," as appeared in WWF Atlas of the Environment by Geoffrey Lean and Don Hinrichsen. Copyright © 1994 Banson Marketing Ltd. Reprinted with permission.

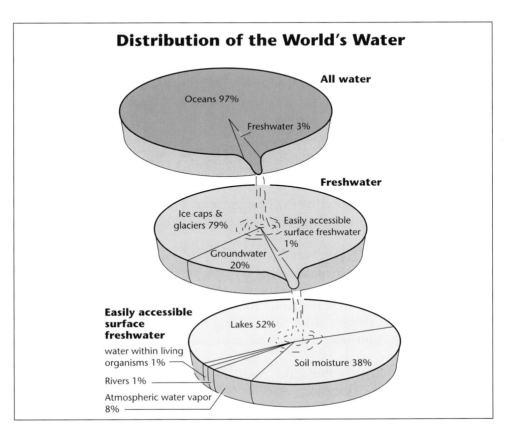

large audience by microphone and plan to point to features on your visuals, be sure the microphone is movable. Pay careful attention to lighting, especially for chalkboards, flipcharts, and posters. Don't forget a pointer if you need one.

Rehearse Your Delivery

Hold ample practice sessions to learn the geography of your report. Try to rehearse at least once in front of friends, or use a full-length mirror and a tape recorder. Assess your delivery from listeners' comments or from your taped voice (which will sound high to you). Use the evaluation sheet on page 572 as a guide.

If at all possible, rehearse using the actual equipment (overhead projector and so on) in the actual setting, to ensure that you have all you need and that everything works.

DELIVERING YOUR PRESENTATION

You have planned and prepared carefully. Now consider the following simple steps to make your actual presentation enjoyable instead of terrifying.

Cultivate the Human Landscape

A successful presentation involves relationship building between presenter and audience.

GET TO KNOW YOUR AUDIENCE.   Try to meet some audience members before your presentation. We all feel more comfortable with people we know. Don't be afraid to smile.

DISPLAY ENTHUSIASM AND CONFIDENCE.   Nobody likes a speaker who seems half dead. Clean up verbal tics ("er," "ah," "uuh"). Overcome your shyness; research indicates that shy people are seen as less credible, trustworthy, likable, attractive, and knowledgeable.

BE REASONABLE.   Don't make your point at someone else's expense. If your topic is controversial (layoffs, policy changes, downsizing), decide how to speak candidly and persuasively with the least chance of offending anyone. For example, in your presentation about groundwater pollution, you don't want to attack the developers, since the building trade is a major producer of jobs, second only to tourism, on Cape Cod. Avoid personal attacks.

DON'T PREACH.   Speak like a person talking—not someone giving a sermon or the Gettysburg Address. Use *we, you, your, our,* to establish commonality with the audience. Avoid jokes or wisecracks.

Keep Your Listeners Oriented

Help your listeners to focus their attention and organize their understanding. Give them a map, some guidance, and highlights.

INTRODUCE YOUR TOPIC CLEARLY.   Open with a preview of your discussion:

A presentation preview

> Today, I want to discuss A, B, and C.

Use visuals to highlight your main points and reveal your organization. For example, you might outline main points on a poster, a chalkboard, or a transparency, or hand out a presentation outline.

FOCUS ON LISTENERS' CONCERNS.   Say something to establish immediate common ground. Show how your presentation has meaning for them, personally.

An appeal to listener's concerns

> Do you know what you will be drinking when you turn on the tap?

Listeners who have a definite stake in the issues will be far more attentive and receptive.

PROVIDE EXPLICIT TRANSITIONS.   Alert your listeners whenever you are preparing to switch gears:

Explicit transitions

> For my next point. . . .
>
> Turning now to my second point. . . .
>
> The third point I want to emphasize. . . .

Repeat key points or terms to keep them fresh in listeners' minds.

GIVE CONCRETE EXAMPLES.   Good examples are informative and persuasive.

A concrete example

> Overdraw from town wells in Maloket and Tanford (two of our most rapidly growing towns) has resulted in measurable salt infusion at a yearly rate of 0.1 mg per liter.

Use examples that focus on listener concerns.

REVIEW AND INTERPRET.   Last things are best remembered. Help listeners remember the main points:

A review of main points

> To summarize the dangers to our groundwater. . . .

Also, be clear about what this material means. Be emphatic about what listeners should be doing, thinking, or feeling:

An emphatic
conclusion

| The conclusion is obvious: If the Cape is to survive, we must. . . .

Try to conclude with a forceful answer to this implied question from each listener: "What does this all mean to me personally?"

Manage Your Visuals

Presenting visuals effectively is a matter of good timing and careful management.

PREPARE EVERYTHING BEFOREHAND.   If you plan to draw on a chalkboard or poster, do the drawings beforehand (in multicolors). Otherwise, listeners will be sitting idly while you draw away.

Prepare handouts if you want listeners to remember or study certain material. Distribute these *after* your talk. (You want the audience to be looking at and listening to you, instead of reading the handout.) Distribute handouts before or during the talk only if you want listeners to take notes—or if your equipment breaks down. When you do distribute handouts beforehand, ask listeners to await your instructions before they turn to a particular page.

Use transparency mounting frames for easy handling (the white cardboard frames allow you to number the transparencies and prepare notes for yourself on the frame).

Increasingly available are automatic transparency feeders with a remote control, enabling your transparency presentation to work like a slide show. This device attaches easily to your overhead projector and allows you to reveal each point line by line.

ARRANGE EVERYTHING BEFOREHAND.   Make sure you organize your media materials and the physical layout beforehand, to avoid fumbling during the presentation. Check your visual sequence against your storyboard.

FOLLOW A FEW SIMPLE GUIDELINES.   Make your visuals part of a seamless presentation. Avoid listener distraction, confusion, and frustration by observing the following suggestions:

HOW TO PRESENT VISUALS

- Try not to begin with a visual.
- Try not to display a visual until you are ready to discuss it.
- Tell viewers what they should be looking for in the visual.
- Point to what is important.
- Stand aside when discussing a visual, so everyone can see it.
- Don't turn your back on the audience.
- After discussing the visual, remove it promptly.
- Switch off equipment that is not in use.
- Try not to end with a visual.

Manage Your Presentation Style

Think about how you are moving, how you are speaking, and where you are looking. These are all elements of your personal style.

USE NATURAL MOVEMENTS AND REASONABLE POSTURES.   Move and gesture as you normally would in conversation, and maintain reasonable postures. Avoid foot shuffling, pencil tapping, swaying, slumping, or fidgeting.

ADJUST VOLUME, PRONUNCIATION, AND RATE.   With a microphone, don't speak too loudly. Without one, don't speak too softly. Be sure you can be heard clearly without shattering eardrums. Ask your audience about the sound and speed of your delivery after a few sentences.

Nervousness causes speakers to gallop along and mispronounce words. Slow down and pronounce clearly. Usually, a rate that seems a bit slow to you will be just right for listeners.

MAINTAIN EYE CONTACT.   Look directly into listeners' eyes. With a small audience, eye contact is one of your best connectors. As you speak, establish eye contact with as many listeners as possible. With a large group, maintain eye contact with those in the first rows. Establish eye contact immediately—before you even begin to speak—by looking around.

Manage Your Speaking Situation

Do everything you can to keep things running smoothly.

BE RESPONSIVE TO LISTENER FEEDBACK.   Assess listener feedback continually and make adjustments as needed. If you are laboring through a long list of facts or figures and people begin to doze or fidget, you might summarize. Likewise, if frowns, raised eyebrows, or questioning looks indicate confusion, skepticism, or indignation, you can backtrack with a specific example or explanation. By tuning in to your audience's reactions, you can avoid leaving them confused, hostile, or simply bored.

STICK TO YOUR PLAN.   Say what you came to say, then summarize and close—politely and on time. Don't punctuate your speech with digressions that pop into your head. Unless a specific anecdote was part of your original plan to clarify a point or increase interest, avoid excursions. We often tend to be more interested in what we have to say than our listeners are! Don't exceed your time limit.

LEAVE LISTENERS WITH SOMETHING TO REMEMBER.   Before ending, take a moment to summarize the major points and reemphasize anything of special importance. Are listeners supposed to remember something, have a different attitude, take a specific action? Let them know! As you conclude, thank your listeners.

ALLOW TIME FOR QUESTIONS AND ANSWERS.   At the very beginning, tell your listeners that a question-and-answer period will follow. Use the following suggestions for managing listener questions diplomatically and efficiently.

HOW TO MANAGE LISTENER QUESTIONS

- Announce a specific time limit, to avoid prolonged debates.
- Listen carefully to each question.
- If you can't understand a question, ask that it be rephrased.
- Repeat every question, to ensure that everyone hears it.
- Be brief in your answers.
- If you need extra time, arrange for it after the presentation.
- Don't be defensive about a hostile or skeptical reaction.

- If anyone attempts lengthy debate, offer to continue *after* the presentation instead of getting into a contest.
- If you can't answer a question, say so and move on.
- End the session with, "We have time for one more question," or some such signal.
- Don't be defensive about a hostile or skeptical reaction.

In Brief ORAL PRESENTATIONS FOR CROSS-CULTURAL AUDIENCES

Imagine that you've been assigned to represent your company at an international conference or before international clients (e.g., of passenger aircraft or mainframe computers). As you plan and prepare your presentation, remain sensitive to various cultural expectations.

For example, some cultures might be offended by a presentation that gets right to the point without first observing formalities of politeness, well wishes, and the like.

Certain communication styles are welcomed in some cultures, but considered offensive in others. In southern Europe and the Middle East, people expect direct and prolonged eye contact as a way of showing honesty and respect. In Southeast Asia, this may be taken as a sign of aggression or disrespect (Gesteland 24). A sampling of the questions to consider:

- *Should I smile a lot or look serious? (Hulbert, Overcoming" 42)*
- *Should I rely on expressive gestures and facial expressions?*

- *How loudly or softly, rapidly or slowly should I speak?*
- *Should I come out from behind the podium and approach the audience or keep my distance?*
- *Should I get right to the point or take plenty of time to lead into and discuss the matter thoroughly?*
- *Should I focus only on the key facts or on all the details and various interpretations?*
- *Should I be assertive in offering interpretations and conclusions, or should I allow listeners to reach their own conclusions?*
- *Which types of visuals and which media might or might not work?*
- *Should I invite questions from this audience, or would this be offensive?*

To account for language differences, prepare a handout of your entire script for distribution after the presentation, along with a copy of your visuals. This way, your audience will be able to study your material at their leisure.

EXERCISES

1. In a memo to your instructor, identify and discuss the kinds of oral reporting duties you expect to encounter in your career.
2. Design an oral presentation for your class. (Base it on a written report.) Make a sentence outline and a storyboard that includes at least three visuals Practice with a tape recorder or a friend. Use the peer evaluation sheet on page 572 to assess your delivery.
3. Observe a lecture or speech, and evaluate it according to the peer evaluation sheet. Write a memo to your instructor (without naming the speaker), identifying strong and weak areas and suggesting improvements.
4. In an oral presentation to the class, present your findings, conclusions, and recommendations from the analytical report assignment in Chapter 25.

Peer Evaluation Sheet for Oral Presentations

Presentation Evaluation for (name/topic) _____

Comments

Content
- ☐ Began with a clear purpose.   _____
- ☐ Showed command of the material.   _____
- ☐ Supported assertions with evidence.   _____
- ☐ Used adequate and appropriate visuals.   _____
- ☐ Used material suited to this audience's
     needs, knowledge, concerns, and interests. _____
- ☐ Acknowledged opposing views.   _____
- ☐ Gave a sufficient amount of information.   _____

Organization
- ☐ Presented a clear line of reasoning.   _____
- ☐ Used transitions effectively.   _____
- ☐ Avoided needless digressions.   _____
- ☐ Summarized before concluding.   _____
- ☐ Was clear about what the listeners
     should think or do.   _____

Style
- ☐ Seemed confident, relaxed, and likable.   _____
- ☐ Seemed in control of the speaking situation. _____
- ☐ Showed appropriate enthusiasm.   _____
- ☐ Pronounced, enunciated, and spoke well.   _____
- ☐ Used appropriate gestures, tone,
     volume, and delivery rate.   _____
- ☐ Had good posture and eye contact.   _____
- ☐ Answered questions concisely and convincingly. _____

Overall professionalism: Superior _____ **Acceptable** _____ **Needs work** _____

Evaluator's signature: _____

Recording and Documenting Research Findings

TAKING NOTES

Many researchers take notes on a laptop computer, using electronic file programs or database management software that allows notes to be filed, shuffled, and retrieved by author, title, topic, date, or key words. You can also take notes in a single word-processing file, then use the "find" command to locate notes quickly. Whether you use a computer or notecards, your notes should be easy to organize and reorganize.

GUIDELINES

for Recording
Research
Findings

Record each
bibliographic
citation exactly as it
will appear in your
final report

1. *Make a separate bibliography listing for each work you consult.* Record a complete entry for each work (Figure A.1), using the citation format that will appear in your document. (See pages 581–90 for sample entries.) Record the information accurately so that you won't have to relocate a source at the last minute.

> Pinsky, Mark A. *The EMF book: What You Should Know About Electromagnetic Fields, Electromagnetic Radiation, and Your Health.* New York: Warner, 1995.

Figure A.1   Recording a Bibliographic Citation

When searching online, you can often print out the full bibliographic record for each work or save it to disk, thereby ensuring an accurate citation.

2. *Skim the work to locate relevant material.* Look over the table of contents and the index. Check the introduction for an overview or thesis. Look for informative headings.

3. *Go back and decide what to record.* Use a separate entry for each item.

4. *Be selective.* Don't copy or paraphrase every word. (See the guidelines for summarizing on page 176.)

5. *Record the item as a quotation or paraphrase.* When quoting others directly, be sure to record words and punctuation accurately. When restating material in your own words, preserve the original meaning and emphasis.

QUOTING THE WORK OF OTHERS

You must place quotation marks around all exact wording you borrow, whether the words were written, spoken (as in an interview or presentation), or whether they appeared in electronic form. Even a single borrowed sentence or phrase, or a single word used in a special way, needs quotation marks, with the exact source properly cited. These sources include people with whom you collaborate.

Plagiarism is often unintentional

If your notes don't identify quoted material accurately, you might forget to credit the source. Even when this omission is unintentional, you face the charge of *plagiarism* (misrepresenting as your own the words or ideas of someone else). Possible consequences of plagiarism include expulsion from school, loss of a job, and a lawsuit.

GUIDELINES

for Quoting the Work of Others

Expressions that warrant direct quotation

1. *Use a direct quotation only when absolutely necessary.* Sometimes a direct quotation is the only way to do justice to the author's own words—as in these instances:

   > "Writing is a way to end up thinking something you couldn't have started out thinking" (Elbow 15).

   > Think of the topic sentence as "the one sentence you would keep if you could keep only one" (USAF Academy 11).

   Consider quoting directly for these purposes:

Reasons for quoting directly

   ♦ to preserve special phrasing or emphasis
   ♦ to preserve precise meaning
   ♦ to preserve the original line of reasoning
   ♦ to preserve a striking or colorful example
   ♦ to convey the authority of expert opinion
   ♦ to convey the original's voice, sincerity, or emotional intensity

2. *Ensure accuracy.* Copy the selection word for word; record the exact page numbers; and double-check that you haven't altered the original expression in any way (Figure A.2).

Place quotation marks around all directly quoted material

> Pinsky, Mark A. pp. 29–30.
>
> "Neither electromagnetic fields nor electromagnetic radiation cause cancer per se, most researchers agree. What they may do is promote cancer. Cancer is a multistage process that requires an "initiator" that makes a cell or group of cells abnormal. Everyone has cancerous cells in his or her body. Cancer—the disease as we think of it—occurs when these cancerous cells grow uncontrollably."

Figure A.2 Recording a Quotation

3. *Keep the quotation as brief as possible.* For conciseness and emphasis, use *ellipses:* Use three spaced periods (. . .) to indicate each omission within a single sentence. Use four periods (. . . .) to indicate each omission that includes the end of a sentence or sections of multiple sentences.

Ellipsis within and between sentences

> Use three . . . periods to indicate each omission within a single sentence. . . .
> [and] four . . . to indicate. . . .

The elliptical passage must be grammatical and must not distort the original meaning. (For additional guidelines, see page 645.)

4. *Use square brackets to insert your own clarifying comments or transitions.* To distinguish your words from those of your source, place them inside of brackets:

Brackets setting off the writer's words within a quotation

> "Job stress [in aircraft ground control] can lead to disaster."

Note also the bracketed transition in example 3, above.

5. *Embed quoted material in your sentences clearly and grammatically.* Introduced integrated quotations with phrases such as "Jones argues that," or "Gomez concludes that." More importantly, use a transitional phrase to show the relationship between the quoted idea and the sentence that precedes it:

An introduction that unifies a quotation with the discussion

> One investigation of age discrimination at select Fortune 500 companies found that "middle managers over age 45 are an endangered species" (Jablonski 69).

Your integrated sentence should be grammatical:

Quoted material integrated grammatically with the writer's words

> "The present farming crisis," Marx argues, "is a direct result of rampant land speculation" (41).

(For additional guidelines, see pages 644–45.)

6. *Quote passages four lines or longer in block form.* Avoid relying on long quotations except in these instances:

Reasons for quoting a long passage

   ◆ to provide an extended example, definition, or analogy (see page 8)
   ◆ to analyze or discuss an idea or concept (see page 525)

Double-space a block quotation and indent the entire block ten spaces. Do not indent the first line of the passage, but do indent first lines of subsequent paragraphs three spaces. Do not use quotation marks.

7. *Introduce the quotation and discuss its significance.*

An introduction to quoted material

> Here is a corporate executive's description of some audiences you can expect to address:

8. *Cite the source of each quoted passage.*

Research writing is a process of independent thinking in which you work with the ideas of others in order to reach your own conclusions; unless the author's exact wording is essential, try to paraphrase, instead of quoting, borrowed material.

PARAPHRASING THE WORK OF OTHERS

Paraphrasing means more than changing or shuffling a few words; it means restating the original idea in your own words—sometimes in a clearer, more direct, and emphatic way—and giving full credit to the source.

To borrow or adapt someone else's ideas or reasoning without properly documenting the source is plagiarism. To offer as a paraphrase an original passage that is only slightly altered—even when you document the source—is also plagiarism. Equally unethical is offering a paraphrase, although documented, that distorts the original meaning.

GUIDELINES
for
Paraphrasing

1. *Refer to the author early in the paraphrase, to indicate the beginning of the borrowed passage.*
2. *Retain key words from the original, to preserve its meaning.*
3. *Restructure and combine original sentences for emphasis and fluency.*
4. *Delete needless words from the original, for conciseness.*
5. *Use your own words and phrases to clarify the author's ideas.*
6. *Cite (in parentheses) the exact source, to mark the end of the borrowed passage and to give full credit.*
7. *Be sure to preserve the author's original intent (Weinstein 3).*

Figure A.3 shows an entry paraphrased from Figure A.2. Paraphrased material is not enclosed within quotation marks, but it is documented to acknowledge your debt to the source.

Signal the beginning of the paraphrase by citing the author, and the end by citing the source

Pinsky, Mark A.

Pinsky explains that electromagnetic waves probably do not directly cause cancer. However, they might contribute to the uncontrollable growth of those cancer cells normally present—but controlled—in the human body (29–30).

Figure A.3   Recording a Paraphrase

WHAT YOU SHOULD DOCUMENT

Document any insight, assertion, fact, finding, interpretation, judgment or other "appropriated material that readers might otherwise mistake for your own" (Gibaldi and Achtert 155)—whether the material appears in published form or not. Specifically, you must document:

Sources that
require
documentation

- any source from which you use exact wording
- any source from which you adapt material in your own words
- any visual illustration: charts, graphs, drawings, or the like (see Chapter 14 for documenting visuals)

How to document
a confidential
source

In some instances, you might have reason to preserve the anonymity of unpublished sources: for example, to allow people to respond candidly without fear of reprisal (as with employee criticism of the company), or to protect their privacy (as with certain material from email inquiries or electronic newsgroups). You must still document the fact that you are not the originator of this material by providing a general acknowledgment in the text ("A number of employees expressed frustration with . . . ") along with a general citation in your list of references or works cited ("Interviews with Polex employees, May 1999").

Common
knowledge need
not be documented

You don't need to document anything considered *common knowledge:* material that appears repeatedly in general sources. In medicine, for instance, it has become common knowledge that foods containing animal fat contribute to higher blood cholesterol levels. So in a report on fatty diets and heart disease, you would probably not need to document that well-known fact. But you would document information about how the fat/cholesterol connection was discovered, what subsequent studies have found (say, the role of saturated versus unsaturated fats), and any information for which some other person could claim specific credit. If the borrowed material can be found in only one specific source, not in multiple sources, document it. When in doubt, document the source.

HOW YOU SHOULD DOCUMENT

Cite borrowed material twice: at the exact place you use that material, and at the end of your document. Documentation practices vary widely, but all systems work almost identically: a brief reference in the text names the source and refers readers to the complete citation, which allows readers to retrieve the source.

Many disciplines, institutions, and organizations publish their own style guides or documentation manuals. Here are a few:

Style guides from
various disciplines

Geographical Research and Writing
Style Manual for Engineering Authors and Editors
IBM Style Manual
NASA Publications Manual

This chapter illustrates citations and entries for three styles widely used for documenting sources in their respective disciplines:

- Modern Language Association (MLA) style, for the humanities
- American Psychological Association (APA) style, for the social sciences
- Council of Biology Editors (CBE) style, for the natural and applied sciences

Unless your audience has its own preference, any of these three styles can be adapted to most research writing. Use one style consistently throughout the document.

MLA DOCUMENTATION STYLE

Use this alternative to footnotes and bibliographies

Traditional MLA documentation of sources used superscript numbers (like this:[1]) in the text, followed by full references at the bottom of the page (footnotes) or at the end of the document (endnotes) and, finally, by a bibliography. But a more current form of documentation appears in the *MLA Style Manual and Guide to Scholarly Publishing,* 2nd ed., New York: Modern Language Association, 1998. Footnotes or endnotes are now used only to comment on material in the text or on sources or to suggest additional sources.

Cite a source briefly in text and fully at the end

In current MLA style, in-text parenthetical references briefly identify the source(s). Full documentation then appears in a Works Cited section at the end of the document.

A parenthetical reference usually includes the author's surname and the exact page number(s) of the borrowed material:

Parenthetical reference in the text

> One notable study indicates an elevated risk of leukemia for children exposed to certain types of electromagnetic fields (Bowman et al. 59).

Readers seeking the complete citation for Bowman can refer easily to the Works Cited section, listed alphabetically by author:

Full citation at document's end

> Bowman, J. D., et al. "Hypothesis: The Risk of Childhood Leukemia Is Related to Combinations of Power-Frequency and Static Magnetic Fields." *Bioelectromagnetics* 16.1 (1995): 48–59.

This complete citation includes page numbers for the entire article.

MLA Parenthetical References

For clear and informative parenthetical references, observe these rules:

How to cite briefly in text

- If your discussion names the author, do not repeat the name in your parenthetical reference; simply give the page number(s):

Citing page
numbers only

> Bowman et al. explain how their recent study indicates an elevated risk of leukemia for children exposed to certain types of electromagnetic fields (59).

♦ If you cite two or more works in a single parenthetical reference, separate the citations with semicolons:

Three works in a
single reference

> (Jones 32; Leduc 41; Gomez 293–94)

♦ If you cite two or more authors with same surnames, include the first initial in your parenthetical reference to each author:

Two authors with
identical surnames

> (R. Jones 32)
>
> (S. Jones 14–15)

♦ If you cite two or more works by the same author, include the first significant word from each work's title, or a shortened version:

Two works by one
author

> (Lamont, *Biophysics* 100–01)
>
> (Lamont, *Diagnostic Tests* 81)

♦ If the work is by an institutional or corporate author or if it is unsigned (that is, the author is unknown), use only the first few words of the institutional name or the work's title in your parenthetical reference:

Institutional,
corporate, or
anonymous author

> (American Medical Assn. 2)
>
> ("Distribution Systems" 18)

To avoid distracting the reader, keep each parenthetical reference brief. The easiest way to keep parenthetical references brief is to name the source in your discussion and place only the page number(s) in parentheses.

Where to place a
parenthetical
reference

For a paraphrase, place the parenthetical reference *before* the closing punctuation mark. For a quotation that runs into the text, place the reference *between* the final quotation mark and the closing punctuation mark. For a quotation set off (indented) from the text, place the reference two spaces *after* the closing punctuation mark.

MLA Works Cited Entries

How to space and
indent entries

The Works Cited list includes each source that you have paraphrased or quoted in your document. In preparing the list, type the first line of each entry flush with the left margin. Indent the second and subsequent lines five spaces. Double-space within and between each entry. Use one character space after any period, comma, or colon. Double-space within and between each entry.

How to city fully at
the end

Following are examples of complete citations as they would appear in the Works Cited section of your document. Shown after each citation is the corresponding parenthetical reference as it would appear in the text.

INDEX TO SAMPLE MLA WORKS CITED ENTRIES

Books

1. Book, single author
2. Book, two or three authors
3. Book, four or more authors
4. Book, anonymous author
5. Multiple books, same author
6. Book, one or more editors
7. Book, indirect source
8. Anthology selection or book chapter

Periodicals

9. Article, magazine
10. Article, journal with new pagination each issue
11. Article, journal with continuous pagination
12. Article, newspaper

Other Sources

13. Encyclopedia, dictionary, alphabetic reference
14. Report
15. Conference presentation

16. Interview, personally conducted
17. Interview, published
18. Letter, unpublished
19. Questionnaire
20. Brochure or pamphlet
21. Lecture
22. Government document
23. Document with corporate authorship
24. Map or other visual
25. Unpublished dissertation, report, or miscellaneous items

Electronic Sources

26. Online database
27. Computer software
28. CD-ROM
29. Listserv
30. Usenet
31. Email
32. Web site
33. Article in Online Periodical
34. Real-Time Communication

What to include in an MLA citation for a book

MLA WORKS CITED ENTRIES FOR BOOKS. Any citation for a book should contain the following information: author, title, editor or translator, edition, volume number, and facts about publication (city, publisher, date).

1. Book, Single Author—MLA

Kerzin-Fontana, Jane B. *Technology Management: A Handbook.* 3rd ed. Delmar, NY: American Management Assn., 1999.

Parenthetical reference: (Kerzin-Fontana 3–4)

Identify the state of publication by U.S. Postal Service abbreviations. If the city of publication is well known (Boston, Chicago), omit the state abbreviations. If several cities are listed on the title page, give only the first. For Canada, include the province abbreviation after the city. For all other countries include an abbreviation of the country name.

2. Book, Two or Three Authors—MLA

Aronson, Linda, Roger Katz, and Candide Moustafa. *Toxic Waste Disposal Methods.* New Haven: Yale UP, 2000.

Parenthetical reference: (Aronson, Katz, and Moustafa 121–23)

Shorten publisher's names, as in "Simon" for Simon & Schuster, "GPO" for Government Printing Office, or "Yale UP" for Yale University Press. For page numbers with more than two digits, give only the final digits for the second number.

3. Book, Four or More Authors—MLA

Santos, Ruth J., et al. *Environmental Crises in Developing Countries.* New York: Harper, 1998.

Parenthetical reference: (Santos et al. 9)

"Et al." is the abbreviated form of the Latin "et alia," meaning "and others."

4. Book, Anonymous Author—MLA

Structured Programming. Boston: Meredith, 1999.

Parenthetical reference: (*Structured* 67)

5. Multiple Books, Same Author—MLA

Chang, John W. *Biophysics.* Boston: Little, 1999.

---. *Diagnostic Techniques.* New York: Radon, 1994.

Parenthetical references: (Chang, *Biophysics* 123–26), (Chang, *Diagnostic* 87)

When citing more than one work by the same author, do not repeat the author's name; type three hyphens followed by a period. List the works alphabetically by title.

6. Book, One or More Editors—MLA

Morris, A. J., and Louise B. Pardin-Walker, eds. *Handbook of New Information Technology.* New York: Harper, 1996.

Parenthetical reference: (Morris and Pardin-Walker 34)

For more than three editors, name only the first, followed by "et al."

7. Book, Indirect Source—MLA

> Kline, Thomas. *Automated Systems.* Boston: Rhodes, 1992.
>
> Stubbs, John. *White-Collar Productivity.* Miami: Harris, 1999.

Parenthetical reference: (qtd. in Stubbs 116)

When your source (as in Stubbs, above) has quoted or cited another source, list each source in its appropriate alphabetical place on your Works Cited page. Use the name of the original source (here, Kline) in your text and precede your parenthetical reference with "qtd. in," or "cited in" for a paraphrase.

8. Anthology Selection or Book Chapter—MLA

> Bowman, Joel P. "Electronic Conferencing." *Communication and Technology: Today and Tomorrow.* Ed. Al Williams. Denton, TX: Assn. for Business Communication, 1994. 123–42.

Parenthetical reference: (Bowman 129)

The page numbers in the complete citation are for the selection cited from the anthology.

MLA Works Cited Entries for Periodicals.   A citation for an article should give this information (as available): author, article title, periodical title, volume or number (or both), date (day, month, year), and page numbers for the entire article—not just the pages cited. List the information in this order, as in the following examples.

What to include in an MLA citation for a periodical

9. Article, Magazine—MLA

> DesMarteau, Kathleen. "Study Links Sewing Machine Use to Alzheimer's Disease." *Bobbin* Oct. 1994: 36–38.

Parenthetical reference: (DesMarteau 36)

No punctuation separates the magazine title and date. Nor is the abbreviation "p." or "pp." used to designate page numbers.

If no author is given, list all other information:

> "Distribution Systems for the New Decade." *Power Technology Magazine* 18 Oct. 2000: 18+.

Parenthetical reference: ("Distribution Systems" 18)

This article begins on page 18 and continues on page 21. When an article does not appear on consecutive pages, give only the number of the first page, followed immediately by a plus sign. A three-letter abbreviation denotes any month spelled with five or more letters.

10. Article, Journal with New Pagination Each Issue—MLA

> Thackman-White, Joan R. "Computer-Assisted Research." *American Librarian* 51.1
>   (1999): 3–9.

Parenthetical reference: (Thackman-White 4–5)

Because each issue for a given year will have page numbers beginning with "1," readers need the issue number. The "51" denotes the volume number; "1" denotes the issue number. Omit "The" or "A" or any other introductory article from a journal or magazine title.

11. Article, Journal with Continuous Pagination—MLA

> Barnstead, Marion H. "The Writing Crisis." *Journal of Writing Theory* 12 (1998):
>   415–33.

Parenthetical reference: (Barnstead 415–16)

When page numbers continue from one issue to the next for the full year, readers won't need the issue number, because no other issue in that year repeats these same page numbers. (Include the issue number, however, if you think it will help readers retrieve the article.) The "12" denotes the volume number.

12. Article, Newspaper—MLA

> Baranski, Vida H. "Errors in Technology Assessment." *Boston Times* 15 Jan. 1999,
>   evening ed., sec. 2: 3.

Parenthetical reference: (Baranski 3)

When a daily newspaper has more than one edition, cite the edition after the date. Omit any introductory article in the newspaper's name (not *The Boston Times*). If no author is given, list all other information. If the newspaper's name does not include the city of publication, insert it, using brackets: *Sippican Sentinel* [Marion, MA].

What to include in
MLA citations for
other sources

MLA WORKS CITED ENTRIES FOR OTHER KINDS OF MATERIALS.   Miscellaneous sources range from unsigned encyclopedia entries to conference presentations to government publications. A full citation should give this information (as available): author, title, city, publisher, date, and page numbers.

13. Encyclopedia, Dictionary, Other Alphabetical Reference—MLA

> "Communication." *The Business Reference Book,* 1998 ed.

Parenthetical reference: ("Communication")

Begin a signed entry with the author's name. For any work arranged alphabetically, omit page numbers in the citation and the parenthetical reference. For a well-known reference book, include only an edition (if stated) and a date. For other reference books, give the full publication information.

14. Report—MLA

Electrical Power Research Institute (EPRI). *Epidemiologic Studies of Electric Utility Employees.* (Report No. RP2964.5). Palo Alto, CA: EPRI, Nov. 1994.

Parenthetical reference: (Electrical Power Research Institute [EPRI] 27)

If no author is given, begin with the organization that sponsored the report.

For any report or other document with group authorship, as above, include the group's abbreviated name in your first parenthetical reference, and then use only that abbreviation in any subsequent reference.

15. Conference Presentation—MLA

Smith, Abelard A. "Radon Concentrations in Molded Concrete." *First British Symposium in Environmental Engineering. London, 11–13 Oct. 1998.* Ed. Anne Hodkins. London: Harrison, 1999. 106–21.

Parenthetical reference: (Smith 109)

The above example shows a presentation that has been included in the published proceedings of a conference. For an unpublished presentation, include the presenter's name, the title of the presentation, and the conference title, location, and date, but do not underline or italicize the conference information.

16. Interview, Personally Conducted—MLA

Nasser, Gamel. Chief Engineer for Northern Electric. Personal interview. Rangeley, ME. 2 Apr. 1999.

Parenthetical reference: (Nasser)

17. Interview, Published—MLA

Lescault, James. "The Future of Graphics." *Executive Views of Automation.* Ed. Karen Prell. Miami: Haber, 2000. 216–31.

Parenthetical reference: (Lescault 218)

The interviewee's name is placed in the entry's author slot.

18. Letter, Unpublished—MLA

Rogers, Leonard. Letter to the author. 15 May 1998.

Parenthetical reference: (Rogers)

19. Questionnaire—MLA

> Taylor, Lynne. Questionnaire sent to 612 Massachusetts business executives. 14 Feb. 2000.

Parenthetical reference: (Taylor)

20. Brochure or Pamphlet—MLA

> *Investment Strategies for the 21st Century.* San Francisco: Blount Economics Assn., 1999.

Parenthetical reference: (*Investment*)

If the work is signed, begin with its author.

21. Lecture—MLA

> Dumont, R. A. "Managing Natural Gas." Lecture. University of Massachusetts at Dartmouth, 15 Jan. 1998.

Parenthetical reference: (Dumont)

If the lecture title is not known, write Address, Lecture, or Reading but do not use quotation marks. Include the sponsor and the location if available.

22. Government Document—MLA

> Virginia. Highway Dept. *Standards for Bridge Maintenance.* Richmond: Virginia Highway Dept., 1997.

Parenthetical reference: (Virginia Highway Dept. 49)

If the author is unknown (as above), list the information in this order: name of the government, name of the issuing agency, document title, place, publisher, and date.

For any congressional document, identify the house of Congress (Senate or House of Representatives) before the title, and the number and session of Congress after the title:

> United States Cong. House. Armed Services Committee. *Funding for the Military Academies.* 105th Congress, 2nd sess. Washington: GPO, 1998.

Parenthetical reference: (U.S. Cong. 41)

"GPO" is the abbreviation for the U.S. Government Printing Office.

For an entry from the *Congressional Record,* give only date and pages:

| Cong. Rec. 10 Marc. 1999: 2178–92.

Parenthetical reference: (Cong. Rec. 2184)

23. Document with Corporate Authorship—MLA

| Hermitage Foundation. *Global Warming Scenarios for the Year 2030.* Washington:
    Natl. Res. Council, 2000.

Parenthetical reference: (Hermitage Foun. 123)

24. Map or Other Visual—MLA

| *Deaths Caused by Breast Cancer, by County.* Map. *Scientific American* Oct. 1995: 32D.

Parenthetical reference: (Deaths Caused)

If the creator of the visual is listed, give that name first. Identify the type of visual (Map, Graph, Table, Diagram) immediately following the title.

25. Unpublished Dissertation, Report, or Miscellaneous Items—MLA

| Author (if known), "Title," Sponsoring organization or publisher, date, page
    numbers.

For any work that has group authorship (corporation, committee, task force), cite the name of the group or agency in place of the author's name.

What to include in an MLA citation for an electronic source

MLA Works Cited Entries for Electronic Sources. Citation for an electronic source with a printed equivalent should begin with that publication information (see relevant sections above). But whether or not a printed equivalent exists, any citation should enable readers to retrieve the material electronically.

The Modern Language Association recommends these general conventions:

Publication dates: For sources taken from the Internet, include the date the source was posted to the Internet or last updated or revised as well as the date you accessed the source.

Uniform Resource Locators: Include a full and accurate URL (electronic address) for any source taken from the Internet (with access mode identifier—*http, ftp, gopher, or telnet*). Enclose URLs in angle brackets (< >). When a URL continues from one line to the next, break it only after a slash. Do not add a hyphen.

Page numbering: Include page or paragraph numbers when given by the source.

26. Online Database—MLA

> Sahl, J. D. "Power Lines, Viruses, and Childhood Leukemia." *Cancer Causes Control* 6.1
>   (Jan. 1995): 83. *MEDLINE.* Online. DIALOG. 7 Nov. 1995.

Parenthetical reference: (Sahl 83)

For entries with a printed equivalent, begin with publication information, then the database title (underlined or italicized), the "Online" designation to indicate the medium, and the service provider (or URL or email address) and the date of access. The access date is important because frequent updatings of databases can produce different versions of the material.

For entries with no printed equivalent, give the title and date of the work in quotation marks, followed by the electronic source information:

> Argent, Roger R. "An Analysis of International Exchange Rates for 1999." *Accu-Data.*
>   Online. Dow Jones News Retrieval. 10 Jan. 2000.

Parenthetical reference: (Argent 4)

If the author is not known, begin with the work's title.

27. Computer Software—MLA

> *Virtual Collaboration.* Diskette. New York: Harper, 1994.

Parenthetical reference: (*Virtual*)

Begin with the author's name, if known.

28. CD-ROM—MLA

> Cavanaugh, Herbert A. "EMF Study: Good News and Bad News."
>   *Electrical World* Feb. 1995: 8. *ABI/INFORM.* CD-ROM. Proquest. Sept. 1995.

Parenthetical reference: (Cavanaugh 8)

If the material also is available in print, begin with the information about the printed source, followed by the electronic source information: name of the database (underlined), CD-ROM designation, vendor name, and electronic publication date. If the material has no printed equivalent, list its author (if known) and title (in quotation marks), followed by the electronic source information.

How to cite an
abstract

If you are citing an abstract of the complete work, insert "Abstract," followed by a period, immediately after the work's page number(s)—"8" in the previous entry.

For CD-ROM reference works and other material not routinely updated, give the title of the work, followed by the CD-ROM designation, place, electronic publisher, and date:

| *Time Almanac.* CD-ROM. Washington: Compact, 1994.

Parenthetical reference: (Time Almanac 74)

Begin with the author's name, if known.

29. Listserv—MLA

| Korsten, a. "Major Update of the WWWVL Migration and Ethnic Relations." 7 Apr.
1998. Online posting. ERCOMER News. 8 Apr. 1998
<http://www.ercomer.org/ archive/ercomer-news/0002.html>.

Parenthetical reference: (Korsten)

Begin with the author's name (if known), followed by the title of the work (in quotation marks), publication date, the Online posting designation, name of discussion group, date of access, and the URL. The parenthetical reference includes no page number because none is given in an online posting.

30. Usenet—MLA

| Dorsey, Michael. "Enviromentalism or Racism." 25 Mar. 1998. Online posting. 1 Apr.
1998 <news:alt.org.sierra-club>.

Parenthetical reference: (Dorsey)

31. Email—MLA

| Wallin, John Luther. "Frog Reveries." email to the author. 12 Oct. 1999.

Parenthetical reference: (Wallin)

Cite personal email as you would printed correspondence. If the document has a subject line or title, enclose it in quotation marks.

For publicly posted email (say, a newsgroup or discussion list) include the address and date of access.

32. Web Site—MLA

| Dumont, R. A. "An Online Course in Technical Writing." 10 Dec. 1999. UMASS
Dartmouth Online. 6 Jan. 2000
<http://www.umassd.edu/englishdepartment.html>.

Parenthetical reference: (Dumont 7–9)

Begin with the author's name (if known), followed by title of the work (in quotation marks), posting date, name of Web site, date of access, and Web address (in angle brackets). Note that a Web address that continues from one line to the next is broken only after the slash(es). No hyphen is added.

33. Article in an Online Periodical—MLA

> Jones, Francine L. "The Effects of NAFTA on Labor Union Membership." *Cambridge Business Review* 2.3 (1999): 47–64. 4 Apr. 2000 <http://www.mun.ca/cambrbusrev/1999vol2/jones2.html>.

Parenthetical reference: (Jones 44–45)

Information about the printed version is followed by the date of access to the Web site and the electronic address.

34. Real-Time Communication—MLA

Synchronous communication occurs in a "real time" forum and includes MUDs (multiuser dungeons), MOOs (MUD object-oriented software), IRC (Internet relay chat), and FTPs (file transfer protocols). The message typed in by the sender appears instantly on the screen of the recipient, as in a personal interview.

> Mendez, Michael R. Online debate. "Solar power versus Fossil Fuel Power." 3 Apr. 1998. CollegeTownMoo. 3 Apr. 1998 <telnet://next.cs.bvc.edu.777>.

Parenthetical reference: (Mendez)

Begin with the name of the communicator(s) and indicate the type of communication (personal interview, online debate, and so on), topic title, posting date, name of forum, access date, and electronic address.

MLA Sample Works Cited Page

Place your Works Cited section on a separate page at the end of the document. Arrange entries alphabetically by author's surname. When the author is unknown, list the title alphabetically according to its first word (excluding introductory articles). For a title that begins with a digit ("5," "6," etc.), alphabetize the entry as if the digit were spelled out.

The list of works cited in Figure A.4 accompanies the report on electromagnetic fields, pages 530–39. In the left margin, colored numbers refer to the elements discussed on the page facing Figure A.4. Bracketed labels identify different types of sources cited.

APA DOCUMENTATION STYLE

One popular alternative to MLA style appears in the *Publication Manual of the American Psychological Association,* 4th ed., Washington: American Psychological Association, 1994. APA style is useful when writers wish to emphasize the publication dates of their references. A parenthetical reference in the text briefly identifies the source, date, and page number(s):

Reference cited in the text

> In a recent study, mice continuously exposed to an electromagnetic field tended to die earlier than mice in the control group (de Jager & de Bruyn, 1994, p. 224).

The full citation then appears in the alphabetical listing of "References," at the report's end:

Full citation at document's end

> de Jager, L., & de Bruyn, L. (1994). Long-term effects of a 50 Hz electric field on the life-expectancy of mice. *Review of Environmental Health, 10* (3–4), 221–224.

Because it emphasizes the date, APA style (or some similar author-date style) is preferred in the sciences and social sciences, where information quickly becomes outdated.

APA Parenthetical References

How APA and MLA parenthetical references differ

APA's parenthetical references differ from MLA's (pages 579–80) as follows: the APA citation includes the publication date; a comma separates each item in the reference; and "p." or "pp." precedes the page number (which is optional in the APA system). When a subsequent reference to a work follows closely after the initial reference, the date need not be included. Here are specific guidelines:

♦ If your discussion names the author, do not repeat the name in your parenthetical reference; simply give the date and page number:

Author named in the text

> Researchers de Jager and de Bruyn explain that experimental mice exposed to an electromagnetic field tended to die earlier than mice in the control group (1994, p. 224).

When two authors of a work are named in the text, their names are connected by "and," but in a parenthetical reference, their names are connected by an ampersand, "&."

♦ If you cite two or more works in a single reference, list the authors in alphabetical order and separate the citations with semicolons:

Two or more works in a single reference

> (Jones, 1994; Gomez, 1992; Leduc, 1996)

♦ If you cite a work with three to five authors, try to name them in your text, to avoid an excessively long parenthetical reference, as on page 595.

Works Cited

Beardsley, Tim. "Say That Again?" *Scientific American* Dec. 1997: 20.

Broad, William J. "Cancer Fear Is Unfounded, Physicists Say." *New York Times* 14 May
    1995: A19.   *[newspaper article]*

Brodeur, Paul, "Annals of Radiation: The Cancer at Slater School." *New Yorker* 7 Dec.
    1992: 86+.   *[magazine article]*

Cavanaugh, Herbert A. "EMF Study: Good News and Bad News." *Electrical World*
    Feb. 1995: 8. *ABI/INFORM.* CD-ROM. Proquest. Sept. 1998.
                                        [trade magazine article from CD-ROM database]

Dana, Amy, and Tom Turner. "Currents of Controversy." *Amicus Journal* Summer
    1993: 29–32.   *[alternative press]*

de Jager, L., and L. de Bruyn. "Long-Term Effects of a 50 HZ Electric Field on the Life-
    Expectancy of Mice." *Review of Environmental Health* 10.3 (1994): 221–24.
    MEDLINE. Online. DIALOG. 8 Mar. 1999.   *[journal article from online database]*

DesMarteau, Kathleen. "Study Links Sewing Machine Use to Alzheimer's Disease."
    Bobbin Oct. 1994: 36–38. *ABI/INFORM.* CD-ROM. Proquest. Aug. 1998.

"Electrophobia: Overcoming Fears of EMFs." *University of California Wellness Letter*
    Nov. 1994: 1.   *[newsletter]*

Goodman, E. M., B. Greenebaum, and M. T. Marron. "Effects of Electromagnetic
    Fields on Molecules and Cells." *International Review of Cytology* 158 (1995):
    279–338. *MEDLINE.* Online. DIALOG. 8 Mar. 1999.

Halloran-Barney, Marianne B. Energy Service Advisor for County Electric. Email to the
    author. 3 Apr. 1999.   *[email inquiry]*

Jauchem, J. "Alleged Health Effects of Electromagnetic Fields: Misconceptions in the
    Scientific Literature." *Journal of Microwave Power and Electromagnetic Energy*
    26.4 (1991): 189–95.   *[journal article from print source]*

Kirkpatrick, David. "Can Power Lines Give You Cancer?" *Fortune* 31 Dec. 1990: 80–85.

Lee, J. M., Jr., et al. *Electrical and Biological Effects of Transmission Lines: A Review.* U.S.
    Dept. of Energy. NTIS no. PC A06/MF A01. Washington: GPO, 1989.
                                                                  [government report]

Figure A.4   A List of Works Cited (MLA Style)

Discussion of Figure A.4

1. Center the Works Cited title at the top of the page. Use one-inch margins. Double-space the entries, and order them alphabetically. Number works cited pages consecutively with text pages.

2. Indent five spaces for the second and subsequent lines of an entry.

3. Place quotation marks around article titles. Underline or italicize periodical or book titles. Capitalize the first letter of key words in all titles (also articles, prepositions, and conjunctions, but only if they come first or last). When an article skips pages in a publication, give only the first page number followed by a plus sign.

4. Do not cite a magazine's volume number, even if it is given.

5. For a CD-ROM database that is updated often (such as *Proquest*), conclude your citation with the date of electronic publication.

6. For additional perspective beyond "establishment" viewpoints, examine "alternative" publications (such as the *Amicus Journal* and *In These Times,* in this list).

7. In citing an online database, include the date you accessed the source.

8. Use a period and one space to separate a citation's three major items (author, title, publication data). Skip one space after a comma or colon. Use no punctuation to separate magazine title and date.

9. Alphabetize hyphenated surnames according to the name that appears first.

10. Include the issue number for a journal with new pagination in each issue. For page numbers of more than two digits, give only the final digits in the second number.

11. For government reports, name the sponsoring agency and include all available information for retrieving the document. Use the first author's name and "et al." for works with four or more authors or editors.

Maugh, Thomas H. "Studies Link EMF Exposure to Higher Risks of Alzheimer's."
　　Los Angeles Times 31 July 1994: A3.

Mevissen, M., M. Keitzmann, and W. Loscher. "In Vivo Exposure of Rats to a Weak
　　Alternating Magnetic Field Increases Ornithine Decarboxylase Activity in the
　　Mammary Gland by a Similar Extent as the Carcinogen DMBA." *Cancer Letter*
　　90.2 (1995): 207–14. *MEDLINE.* Online. DIALOG. 8 Mar. 1999.

Miller, M. A., et al. "Variation in Cancer Risk Estimates for Exposure to Powerline
　　Frequency Electromagnetic Fields: A Meta-analysis Comparing EMF Measurement
　　Methods." *Risk Analysis* 15.2 (1995): 281–87. *MEDLINE.* Online. DIALOG.
　　8 Mar. 1999.

Miltane, John. Chief Engineer for County Electric. Personal interview. 5 Apr. 1999.

Monmonier, Mark. *Cartographies of Danger: Mapping Hazards in America.* Chicago:
　　U of Chicago P, 1997: 190.　　　　　　　　　　　　　　　*[book–one author]*

Moore, Taylor. "EMF Health Risks: The Story in Brief." *EPRI Journal* Mar./Apr. 1995: 7–17.

Moulder, John. "Power Lines and Cancer." 6 Oct. 1998. Online posting. 10 Mar. 1999
　　<news:powerlines.cancer.FAQ:USENET>.　　　*[World Wide Web newsgroup]*

Palfreman, John. "Apocalypse Not." *Technology Review* 24 April 1996: 24–33.

Pinsky, Mark A. *The EMF Book: What You Should Know about Electromagnetic Fields,
　　Electromagnetic Radiation, and Your Health.* New York: Warner, 1995.

Raloff, Janet. "Electromagnetic Fields May Trigger Enzymes." *Science News* 153.8
　　(1998): 119.

---. "EMFs' Biological Influences." *Science News* 153.2 (1998): 29–31.

---. "Jury Is Still Out on EMFs and Cancer." *Science News* 154.25 (1998): 127.

Stix, Gary. "Are Power Lines a Dead Issue?" *Scientific American* Mar. 1998: 33–34.

Taubes, Gary. "Fields of Fear." *Atlantic Monthly* Nov. 1994: 94–108. Online. U of
　　Virginia Electronic Text Center. Internet. 15 Mar. 1999
　　<http:/etext. libvirginia.edu/english.html>.　　　　　　　　*[Web site]*

United States Environmental Protection Agency. *EMF in Your Environment.* Washington:
　　GPO, 1992.

White, Peter. "Bad Vibes." *In These Times* 28 June 1993: 14–17.

Figure A.4　A List of Works Cited (MLA Style)　*Continued*

Discussion of Figure A.4 *Continued*

12. Use three-letter abbreviations for months with five or more letters.

13. Because an online conference source such as a listserv or newsgroup provides no page numbers, you can eliminate the in-text parenthetical reference by referring directly to that source in your discussion ("Dr. Jones of Harvard points out that").

14. When the privacy of the electronic source is not an issue (e.g., a library versus an email correspondent), include the electronic address in your entry.

15. Shorten publisher's names (as in "Simon" for Simon & Schuster; "Knopf" for Alfred A. Knopf, Inc.; "GPO" for Government Printing Office; or "Yale UP" for Yale University Press).

A work with three to five authors

Franks, Oblesky, Ryan, Jablar, and Perkins (1993) studied the role of electromagnetic fields in tumor formation.

In any subsequent references to this work, name only the first author, followed by "et al." (Latin abbreviation for "and others").

♦ If you cite two or more works by the same author published in the same year, assign a different letter to each work:

Two or more works by the same author in the same year

(Lamont 1990a, p. 135)

(Lamont 1990b, pp. 67–68)

Other examples of parenthetical references appear with their corresponding entries in the following discussion of the list of references.

APA Reference List Entries

The APA reference list includes each source that you have cited in your document. In preparing the list of references, type the first line of each entry flush with the left margin. Indent the second and subsequent lines five spaces. Skip one character space after any period, comma, or colon. Double-space within and between each entry.

How to space and indent entries

Following are examples of complete citations as they would appear in the References section of your document. Shown immediately below each entry is its corresponding parenthetical reference as it would appear in the text. Note the capitalization, abbreviation, spacing, and punctuation in the sample entries.

INDEX TO SAMPLE ENTRIES FOR APA REFERENCES

Books

1. Book, single author
2. Book, two to five authors
3. Book, six or more authors
4. Book, anonymous author
5. Multiple books, same author
6. Book, one to five editors
7. Book, indirect source
8. Anthology selection or book chapter

Periodicals

9. Article, magazine
10. Article, journal with new pagination each issue
11. Article, journal with continuous pagination
12. Article, newspaper

Other Sources

13. Encyclopedia, dictionary, alphabetic reference

14. Report
15. Conference presentation
16. Interview, personally conducted
17. Interview, published
18. Personal correspondence
19. Brochure or pamphlet
20. Lecture
21. Government document
22. Miscellaneous items

Electronic Sources

23. Online database abstract
24. Online database article
25. Computer software or software manual
26. CD-ROM abstract
27. CD-ROM reference work
28. Electronic bulletin board, discussion list, email
29. Web site

What to include in an APA citation for a book

APA ENTRIES FOR BOOKS.   Any citation for a book should contain all applicable information in the following order: author, date, title, editor or translator, edition, volume number, and facts about publication (city and publisher).

I. Book, Single Author—APA

Kerzin-Fontana, J. B. (1999). *Technology management: A handbook* (3rd ed.). Delmar, NY: American Management Association.

Parenthetical reference: (Kerzin-Fontana, 1999, pp. 3–4)

Use only initials for an author's first and middle name. Capitalize only the first words of a book's title and subtitle and any proper names. Identify a later edition in parentheses between the title and the period.

2. Book, Two to Five Authors—APA

Aronson, L., Katz, R., & Moustafa, C. (2000). *Toxic waste disposal methods.* New Haven: Yale University Press.

Parenthetical reference: (Aronson, Katz, & Moustafa, 2000)

Use an ampersand (&) before the name of the final author listed in an entry. As an alternative parenthetical reference, name the authors in your text and include date (and page numbers, if appropriate) in parentheses.

3. Book, Six or More Authors—APA

Fogle, S. T., et al. (1998). *Hyperspace technology.* Boston: Little, Brown.

Parenthetical reference: (Fogle et al., 1998, p. 34)

"Et al." is the Latin abbreviation for "et alia," meaning "and others."

4. Book, Anonymous Author—APA

Structured programming. (1995). Boston: Meredith Press.

Parenthetical reference: (Structured Programming, 1995, p. 67)

In your list of references, place an anonymous work alphabetically by the first key word (not *The, A,* or *An*) in its title. In your parenthetical reference, capitalize all key words in a book, article, or journal title.

5. Multiple Books, Same Author—APA

Chang, J. W. (1997a). *Biophysics.* Boston: Little, Brown.

Chang, J. W. (1997b). *MindQuest.* Chicago: John Pressler.

Parenthetical references: (Chang, 1997a)

(Chang, 1997b)

Two or more works by the same author not published in the same year are distinguished by their respective dates alone, without the added letter.

6. Book, One to Five Editors—APA

Morris, A. J., & Pardin-Walker, L. B. (Eds.). (1999). *Handbook of new information technology.* New York: HarperCollins.

Parenthetical reference: (Morris & Pardin-Walker, 1999, p. 79)

For more than five editors, name only the first, followed by "et al."

7. Book, Indirect Source—APA

Stubbs, J. (1998). *White-collar productivity.* Miami: Harris.

Parenthetical reference: (cited in Stubbs, 1998, p. 47)

When your source (as in Stubbs, above) has cited another source, list only this second source in the References section, but name the original source in the text: "Kline's study (cited in Stubbs, 1998, p. 47) supports this conclusion."

8. Anthology Selection or Book Chapter—APA

Bowman, J. (1994). Electronic conferencing. In A. Williams (Ed.), *Communication and technology: Today and tomorrow* (pp. 123–142). Denton, TX: Association for Business Communication.

Parenthetical reference: (Bowman, 1994, p. 126)

The page numbers in the complete reference are for the selection cited from the anthology.

APA Entries for Periodicals.   A citation for an article should give this information (as available), in order: author, publication date, article title (without quotation marks), volume or number (or both), and page numbers for the entire article—not just the page(s) cited.

9. Article, Magazine—APA

DesMarteau, K. (1994, October). Study links sewing machine use to Alzheimer's disease. *Bobbin, 36,* 36–38.

Parenthetical reference: (DesMarteau, 1994, p. 36)

If no author is given, provide all other information. Capitalize the first word in an article's title and subtitle, and any proper nouns. Capitalize all key words in a periodical title. Underline or italicize the periodical title, volume number, and commas (as above).

10. Article, Journal with New Pagination for Each Issue—APA

Thackman-White, J. R. (1999). Computer-assisted research. *American Library Journal, 51*(1), 3–9.

Parenthetical reference: (Thackman-White, 1999, pp. 4–5)

Because each issue for a given year has page numbers that begin at "1," readers need the issue number (in this instance, "1"). The "51" denotes the volume number, which is underlined or italicized.

11. Article, Journal with Continuous Pagination—APA

Barnstead, M. H. (1999). "The Writing Crisis." *Journal of Writing Theory 12*, 415–433.

Parenthetical reference: (Barnstead, 1999, pp. 415–416)

The "12" denotes the volume number. When page numbers continue from issue to issue for the full year, readers won't need the issue number, because no other issue in that year repeats these same page numbers. (You can still include the issue number if you think it will help readers retrieve the article more easily.)

12. Article, Newspaper—APA

Baranski, V. H. (1999, January 15). Errors in technology assessment. *The Boston Times*, p. B3.

Parenthetical reference: (Baranski, 1999, p. B3)

In addition to the year of publication, include the month and day. If the newspaper's name begins with "The," include it in your citation. Include "p." or "pp." before page numbers. For an article on nonconsecutive pages, list each page, separated by a comma.

What to include in an APA citation for a miscellaneous source

APA Entries for Other Sources. Miscellaneous sources range from unsigned encyclopedia entries to conference presentations to government documents. A full citation should give this information (as available): author, publication date, work title (and report or series number), page numbers (if applicable), city, and publisher.

13. Encyclopedia, Dictionary, Alphabetic Reference—APA

Communication. (1998). In *The business reference book*. Boston: Business Resources Press.

Parenthetical reference: ("Communication," 1998)

For an entry that is signed, begin with the author's name and publication date.

14. Report—APA

Electrical Power Research Institute. (1994). *Epidemiologic studies of electric utility employees* (Report No. RP2964.5). Palo Alto, CA: Author.

Parenthetical reference: (Electrical Power Research Institute [EPRI], 1994, p. 12)

If authors are named, list them first, followed by the publication date. When citing a group author, as above, include the group's abbreviated name in your first parenthetical reference, and use only that abbreviation in any subsequent reference. When the agency (or organization) and publisher are the same, list "Author" in the publisher's slot.

15. Conference Presentation—APA

> Smith, A. A. (1999). Radon concentrations in molded concrete. In A. Hodkins (Ed.), *First British Symposium on Environmental Engineering* (pp. 106–121). London: Harrison Press, 2000.

Parenthetical reference: (Smith, 1999, p. 109)

In parentheses is the date of the presentation. The name of the symposium is a proper name, and so is capitalized. Following the publisher's name is the date of publication.

For an unpublished presentation, include the presenter's name, year and month, title of the presentation (underlined or italicized), and all available information about the conference or meeting: "Symposium held at. . . ." Do not underline or italicize this last information.

16. Interview, Personally Conducted—APA

Parenthetical reference: (G. Nasser, personal interview, April 2, 1999)

This material is considered a nonrecoverable source, and so is cited in the text only, as a parenthetical reference. If you name the respondent in text, do not repeat the name in the citation.

17. Interview, Published—APA

> Jable, C. K. (1998). The future of graphics [Interview with James Lescault]. In K. Prell (Ed.), *Executive Views of Automation* (pp. 216–231). Miami: Haber Press, 1999.

Parenthetical reference: (Jable, 2000, pp. 218–223)

Begin with the name of the interviewer, followed by the interview date and title (if available), the designation (in brackets), and the publication information, including the date.

18. Personal Correspondence—APA

Parenthetical reference: (L. Rogers, personal correspondence, May 15, 1998)

This material is considered nonrecoverable data, and so is cited in the text only, as a parenthetical reference. If you name the correspondent in text, do not repeat the name in the citation.

19. Brochure or Pamphlet—APA

This material follows the citation format for a book entry (page 596). After the title of the work, include the designation "Brochure" in brackets.

20. Lecture—APA

> Dumont, R. A. (1998, January 15). *Managing natural gas.* Lecture presented at the University of Massachusetts at Dartmouth.

Parenthetical reference: (Dumont, 1998)

If you name the lecturer in text, do not repeat the name in the citation.

21. Government Document—APA

> Virginia Highway Department. (1997). *Standards for bridge maintenance.* Richmond: Author.

Parenthetical reference: (Virginia Highway Department, 1997, p. 49)

If the author is unknown, present the information in this order: name of the issuing agency, publication date, document title, place, and publisher. When the issuing agency is both author and publisher, list "Author" in the publisher's slot.

For any congressional document, identify the house of Congress (Senate or House of Representatives) before the date.

> U. S. House Armed Services Committee. (1998). *Funding for the military academies.* Washington, DC: U.S. Government Printing Office.

Parenthetical reference: (U.S. House, 1998, p. 41)

22. Miscellaneous Items (unpublished manuscripts, dissertations, and so on)—APA

> Author (if known). (Date of publication). *Title of work.* Sponsoring organization or publisher.

For any work that has group authorship (corporation, committee, and so on), cite the name of the group or agency in place of the author's name.

What to include in an APA citation for an electronic source

APA Entries for Electronic Sources.   APA documentation standards for electronic sources continue to be refined and defined. A sampling of currently preferred formats is presented below. Any citation for electronic media should allow readers to identify the original source (printed or electronic) and provide an electronic path for retrieving the material.

Begin with the publication information for the printed equivalent. Then, in brackets, name the electronic source ([Online], [CD-ROM], [Computer software]), the protocol[2] (Bitnet, Dialog, FTP, Telnet), and any other items that define

---

[2]A *protocol* is a body of standards that ensures compatibility among the different products designed to work together on a particular network.

a clear path (service provider, database title, access code, retrieval number, or site address).

23. Online Database Abstract—APA

> Sahl, J. D. (1995). Power lines, viruses, and childhood leukemia [Online]. *Cancer Causes Control, 6* (1), 83. Abstract from: DIALOG File: *MEDLINE* Item: 93–04881

Parenthetical reference: (Sahl, 1995)

Note the absence of closing punctuation in items 23, 24, 26, and 29. Any punctuation added to the availability statement could interfere with retrieval.

24. Online Database Article—APA

> Alley, R. A. (1999, January). Ergonomic influences on worker satisfaction [29 paragraphs]. *Industrial Psychology* [Online serial], *5*(11). Available FTP: Hostname: publisher.com Directory: pub/journals/ industrial.psychology/1999

Parenthetical reference: (Alley, 1995)

Give the length of the article [in paragraphs], after its title.

25. Computer Software or Software Manual—APA

> Virtual collaboration [Computer software]. (1994). New York: HarperCollins.

Parenthetical reference: (Virtual, 1994)

For citing a manual, replace the "Computer software" designation in brackets with "Software manual."

26. CD-ROM Abstract—APA

> Cavanaugh, H. (1995). An EMF study: Good news and bad news [CD-ROM]. *Electrical World, 209*(2), 8. Abstract from: Proquest File: ABI/Inform: 978032

Parenthetical reference: (Cavanaugh, 1995)

The "8" in the entry above denotes the page number of this one-page article.

27. CD-ROM Reference Work—APA

> *Time almanac.* (1994). Washington: Compact, 1994.

Parenthetical reference: (*Time almanac,* 1994)

If the work on CD-ROM has a printed equivalent, APA currently prefers that it be cited in its printed form. As more works appear in electronic form, this convention may be revised.

28. Electronic Bulletin Board, Discussion List, Email—APA

Parenthetical reference: Fred Flynn (personal communication, May 10, 1999) provided these statistics.

This material is considered personal communication in APA style. Instead of being included in the list of references, it is cited directly in the text. According to APA's current standards, material from discussion lists and electronic bulletin boards has limited research value because it does not undergo the kind of review and verification process used in scholarly publications.

29. Web Site—APA

Dumont, R. A. (1999). *An Online course in composition* [Online Web site]. Available WWW.http://www.umassd.edu/englishdepartment.html

If the Web address continues from one line to the next, divide it only after the slash(es).

APA Sample Reference List

APA's References section is an alphabetical listing (by author) equivalent to MLA's Works Cited section. Like Works Cited, the reference list includes only those works actually cited. (A bibliography would usually include background works or works consulted as well.) Unlike MLA style, APA style calls for only "recoverable" sources to appear in the reference list. Therefore, personal interviews, email messages, and other unpublished materials are cited in the text only.

The list of references in Figure A.5 accompanies the report on technical marketing, pages 542–49. In the left margin, colored numbers denote elements discussed on the page facing Figure A.5. Bracketed labels on the right identify different types of sources.

CBE NUMERICAL DOCUMENTATION STYLE

In the numerical system of documentation preferred by the Council of Biology Editors, each work is assigned a number the first time it is cited. This same number then is used for any subsequent reference to that work. Numerical documentation is often used in the physical sciences (astronomy, chemistry, geology, physics) and the applied sciences (mathematics, medicine, engineering, computer science).

Preferred documentation styles for particular disciplines are defined in style manuals such as these:

◆ American Chemical Society, *The ACS Style Guide for Authors and Editors* (1985)
◆ American Institute of Physics. *AIP Style Manual* (1990)

1   **References**

2   Anson, D. (1999, March 12). *Engineering graduates and the job market.* Lecture
         presented at the University of Massachusetts at Dartmouth.   *[lecture]*

3   Basta, N. (1984, September). Take a good look at sales engineering. *Graduating
         Engineer, 32,* 84–87.   *[journal article]*

4   College Placement Council. (1998). *CPC Annual* (42nd ed.). Bethlehem, PA: Author.
                                    [reference book—author/organization as publisher]

5   Cornelius, H., & Lewis, W. (1983). *Career guide for sales and marketing* (2nd ed.). New
         York: Monarch Press.   *[book with two authors]*

6   Electronic sales positions. (1996). *The national job bank.* Holbrook, MA: Bob Adams,
         Inc.   *[directory entry—no author named]*

         Jones, B. (1997, December). Giving women the business. *Harper's Magazine, 296*
         (1772), 47–58.

7   Occupational employment. (1997–98, Winter) *Occupational Outlook Quarterly, 41*(4),
         6–24.   *[govt. periodical—no author named]*

         Shelley, K. J. (1997, Fall). A portrait of the M.B.A. *Occupational Outlook Quarterly,
         41*(3), 26–33.   *[govt. periodical—author named]*

         Solomon, S. D. (1996, January). An engineer goes to Wall Street [10 pages].
         Technology Review [Online serial], 99(1). Available WWW:
         http://www.web.mit.edu/techreview/www/   *[online article]*

8   Technology Marketing Group, Inc. (1998). *Services for clients* [Online Web site].
         Available WWW:http://www.technology-marketing.com   *[Web site]*

9   Tolland, M. (1999, April). *Alternate careers in marketing.* Presentation at Electro '99
         Conference in Boston.   *[unpublished conference presentation]*

         U.S. Department of Labor. (1997) *Tomorrow's jobs.* Washington, DC: Author.
                                                [govt. publication—no author named]

         Young, J. (1995, August). Can computers really boost sales? *Forbes ASAP,* 84–101.
                                                [magazine article—no vol. or issue number]

Figure A.5   A List of References (APA Style)

Discussion of Figure A.5

1. Center the References title at the top of the page. Use one-inch margins. Number reference pages consecutively with text pages. Include only recoverable data (material that readers could retrieve for themselves); cite personal interviews, unpublished lectures, electronic discussion lists, and email and other personal correspondence parenthetically in the text only. See also item 9 in this list.

2. Double-space entries and order them alphabetically by author's last name (excluding *A*, *An*, or *The*). List initials only for authors' first and middle names. Write out names of all months. In student papers, indent the second and subsequent lines of an entry five spaces. In papers submitted for publication in an APA journal, the *first* line is indented instead.

3. Do not enclose article titles in quotation marks. Underline or italicize periodical titles. Capitalize the first word in article or book titles and subtitles, and any proper nouns. Capitalize all key words in magazine or journal titles.

4. Identify the edition of a book in parentheses. If the author is also the publisher, use the word "Author" after the place of publication. Otherwise, write out the publisher's name in full.

5. For more than one author or editor, use ampersands instead of spelling out "and."

6. Use the first key word in the title to alphabetize works whose author is not named.

7. Use italics or a continuous underline for a journal article's title, volume number, and the comma. Give the issue number in parentheses only if each issue begins on page 1. Do not include "p." or "pp." before journal page numbers (only before page numbers from a newspaper).

8. Omit punctuation from the end of an electronic address.

9. Treat an unpublished conference presentation as a recoverable source; include it in your list of references instead of only citing it parenthetically in your text.

- ◆ American Mathematical Society. *A Manual for Authors of Mathematical Papers* (1990)
- ◆ American Medical Association. *Manual of Style* (1989)

One widely consulted guide for numerical documentation is *Scientific Style and Format: The CBE Manual for Authors, Editors, and Publishers,* 6th ed., 1994, from the Council of Biology Editors.

CBE Numbered Citations

In one version of CBE style, a citation in the text appears as a superscript number immediately following the source to which it refers:

Numbered citations in the text

> A recent study[1] indicates an elevated leukemia risk among children exposed to certain types of electromagnetic fields. Related studies[2–3] tend to confirm the EMF/cancer hypothesis.

When referring to two or more sources in a single note (as in "[2–3]" above) separate the numbers by a hyphen if they are in sequence and by commas but no space if they are out of sequence: ("[2,6,9]").

The full citation for each source then appears in the numerical listing of references at the end of the document.

REFERENCES

Full citations at document's end

1. Bowman JD, et al. Hypothesis: the risk of childhood leukemia is related to combinations of power-frequency and static magnetic fields. Bioelectromagnetics 1995; 16(1): 48–59.

2. Feychting M, Ahlbom A. Electromagnetic fields and childhood cancer: meta-analysis. Cancer Causes Control 1995 May; 6(3): 275–277.

CBE Reference List Entries

CBE's References section lists each source in the numerical order in which it was first cited. In preparing the list, which should be double-spaced, begin each entry on a new line. Type the number flush with the left margin, followed by a period and a space. Align subsequent lines directly under the first word of line one.

Following are examples of complete citations as they would appear in the Reference section for your document.

CBE ENTRIES FOR BOOKS.   Any citation for a book should contain all available information in the following order: number assigned to the entry, author or editor, work title (and edition), facts about publication (place, publisher, date), and number of pages. Note the capitalization, abbreviation, spacing, and punctuation in the sample entries.

1. Book, Single Author—CBE

1. Kerzin-Fontana JB. Technology management: a handbook. 3rd ed. Delmar, NY: American Management Assn.; 1999. 356p.

INDEX TO SAMPLE CBE ENTRIES

1. Book, single author
2. Book, multiple authors
3. Book, anonymous author
4. Book, one or more editors
5. Anthology selection or book chapter
6. Article, magazine

7. Article, journal with new pagination each issue
8. Article, journal with continuous pagination
9. Article, newspaper
10. Article, online source

---

2. Book, Multiple Authors—CBE

2. Aronson L, Katz R, Moustafa, C. Toxic waste disposal methods. New Haven: Yale Univ. Pr.; 2000. 316p.

3. Book, Anonymous Author—CBE

3. [Anonymous]. Structured programming. Boston: Meredith Pr.; 1995. 267p.

4. Book, One or More Editors—CBE

4. Morris AJ, Pardin-Walker LB, editors. Handbook of new information technology. New York: Harper; 1999. 345p.

5. Anthology Selection or Book Chapter—CBE

5. Bowman JP. Electronic conferencing. In: Williams A, editor. Communication and technology: today and tomorrow. Denton, TX: Assn. for Business Communication; 1994. p. 123–42.

CBE Entries for Periodicals. Any citation for an article should contain all available information in the following order: number assigned to the entry, author, article title, periodical title, date (year, month), volume and issue number, and inclusive page numbers for the article. Note the capitalization, abbreviation, spacing, and punctuation in the sample entries.

6. Article, Magazine—CBE

6. DesMarteau K. Study links sewing machine use to Alzheimer's disease. Bobbin 1994 Oct: 36–38.

7. Article, Journal with New Pagination Each Issue—CBE

7. Thackman-White JR. Computer-assisted research. American Library Jour 1999; 51(1): 3–9.

8. Article, Journal with Continuous Pagination—CBE

8. Barnstead MH. The writing crisis. Jour of Writing Theory 1999; 12: 415–433.

9. Article, Newspaper—CBE

9. Baranski VH. Errors in technology assessment. Boston Times 1999 Jan 15; Sect B: 33 (col 2).

10. Article, Online Source—CBE

10. Alley RA. Ergonomic influences on worker satisfaction. Industrial Psychology [serial online] 1998 Jan; 5(11). Available from: ftp. pub/journals/ industrial psychology/1995 via the INTERNET. Accessed 1999 Feb 10.

For more detailed guidelines on CBE style, consult the CBE manual.

The Problem-Solving
Process Illustrated

CRITICAL THINKING IN THE WRITING PROCESS

A SAMPLE WRITING SITUATION

YOUR OWN WRITING SITUATION

Every writing situation requires deliberate decisions for *working with the information* and for *planning, drafting,* and *revising* the document. Some of these decisions are illustrated in Figure B.1. Each writer approaches the process through a sequence of decisions that works best for *that* person. No stage of decisions is complete until *all* stages are complete. This looping structure of the writing process is illustrated in Figure B.2.

CRITICAL THINKING IN THE WRITING PROCESS

The writing process is a *critical thinking* process: the writer makes a series of deliberate decisions in response to a situation. The actual "writing" (putting words on the page) is only a small part of the overall process—probably the least significant part.

In this section, we will follow one working writer through an important writing situation; we will see how he solves his unique information, persuasion, and ethics problems and how he collaborates to design a useful and efficient document.

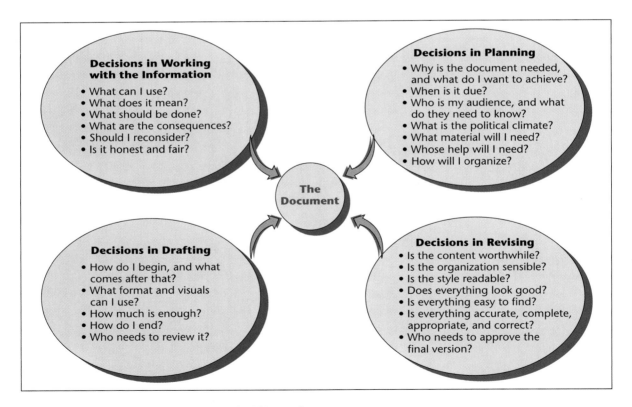

Figure B.1   Typical Decisions during the Writing Process

Figure B.2
A Flowchart for
the Writing
Process

Decisions in the
writing process are
recursive: No one
stage of decisions is
complete until *all*
stages are
complete.

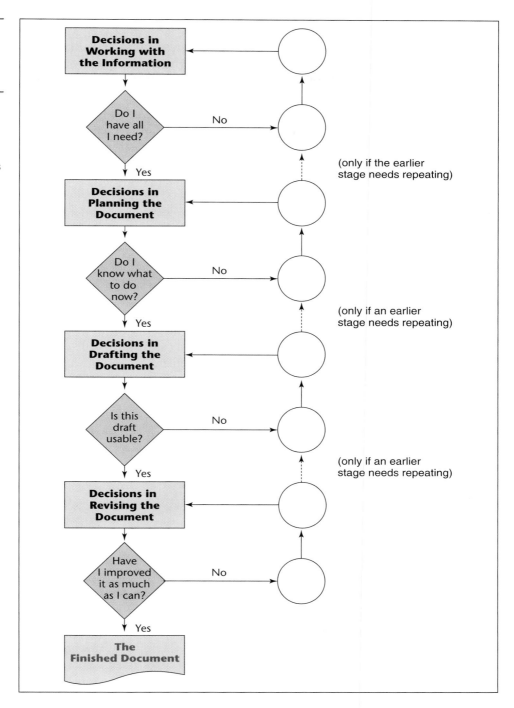

A SAMPLE WRITING SITUATION

The company is Microbyte, maker of portable microcomputers. The writer is Glenn Tarullo (BS, Management: Minor: Computer Science). Glenn has been on the job three months as Assistant Training Manager for Microbyte's Marketing and Customer Service Division.

For three years, Glenn's boss, Marvin Long, has periodically offered a training program for new managers. Long's program combines an introduction to the company with instruction in management skills (time management, motivation, communication). Long seems satisfied with his two-week program but has asked Glenn to evaluate it and write a report as part of a company move to upgrade training procedures.

Glenn knows his report will be read by Long's boss, George Hopkins (Assistant Vice President, Personnel), and Charlotte Black (Vice President, Marketing, the person who devised the upgrading plan). Copies will go to other division heads, to the division's chief executive, and to Long's personnel file.

Glenn spends two weeks (Monday, October 3, to Friday, October 14) sitting in and taking notes on Long's classes. On October 14, the trainees evaluate the program. After reading these evaluations and reviewing his notes, Glenn concludes that the program was successful but could stand improvement. How can he be candid without harming or offending anyone (instructors, his boss, or guest speakers)? Figure B.3 depicts Glenn's problem.

Glenn is scheduled to present his report in conference with Long, Black, and Hopkins on Wednesday, October 19. Right after the final class (1 P.M., Friday, the 14th), Glenn begins work on his report.

Figure B.3
Glenn's Fourfold
Problem

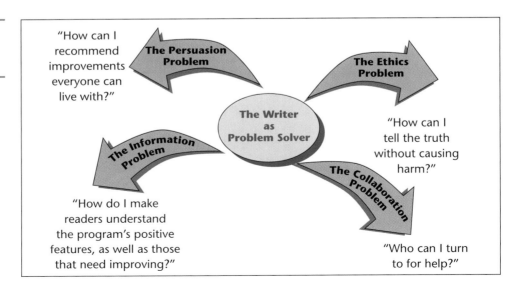

Working with the Information

Glenn spends half of Friday afternoon fretting over the details of his situation, the readers and other people involved, the political realities, constraints, and consequences. (He knows no love is lost between Long and Black, and he wants to steer clear of their ongoing conflict.) By 3 P.M., Glenn hasn't written a word. Desperate, he decides to write whatever comes to mind:

Glenn's first draft

Although the October Management Training Session was deemed quite successful, several problems have emerged that require our immediate attention.

- Too many of the instructors had poor presentation skills. A few never arrived on time. One didn't stick to the topic but rambled incessantly. Jones and Wells seemed poorly prepared. Instructors in general seemed to lack any clear objectives. Also, because too few visual aids were used, many presentations seemed colorless and apparently bored the trainees.
- The trainees (all new people) were not at all cognizant of how the company was organized or functioned, so the majority of them often couldn't relate to what the speakers were talking about.
- It is my impression that this was a weak session due to the fact that there were insufficient members (only five trainees). Such a small class makes the session a waste of time and money. For instance, Lester Beck, Senior Vice President of Personnel, came down to spend over one hour addressing only a handful of trainees. Another factor is that with fewer trainees in a class, less dialogue occurs, with people tending to just sit and get talked at.
- Last but not least, executive speakers generally skirted the real issues, saying nothing about what it was really like to work here. They never really explained how to survive politically (e.g., never criticize your superior; never complain about the hard work or long hours; never tell anyone what you *really* think; never observe how few women are in executive or managerial positions, or how disorganized things seem to be). New employees shouldn't have to learn these things the hard way.

In the final analysis, if these problems can be addressed immediately, it is my opinion we can look forward in the future to effectuating management training sessions of even higher quality than those we now have.

Glenn completes this draft at 5:10 P.M. Displeased with the results but not sure how to improve the piece, he asks an experienced colleague for advice and feedback. Blair Cordasco, a senior project manager, has collaborated with Glenn on several earlier projects. Cordasco agrees to study Glenn's draft over the weekend. Because of this document's sensitive nature, they decide to work on it face-to-face instead of transacting via email.

At 8:05 Monday morning, Cordasco reviews the document with Glenn. First, she points out obvious style problems: wordiness ("due to the fact that"), jargon ("effectuating"), triteness ("in the final analysis"), implied bias ("weak presentation," "skirted"), among others. Can you identify other style problems in Glenn's draft?

Cordasco points out other problems. The piece is disorganized, and even though Glenn is being honest, he isn't being particularly fair. The emphasis is too critical (making Glenn's boss look bad to his superiors), and the views are too subjective (no one is interested in hearing Glenn gripe about the company's political problems). Moreover, the report lacks persuasive force because it contains little useful advice for solving the problems he identifies. The tone is bossy and judgmental. Glenn is in no position to make this kind of *power connection* (p. 38). In this form, the report will only alienate people and harm Glenn's career. He needs to be more fair, diplomatic, and reasonable.

Planning the Document

Glenn realizes he needs to begin by focusing on his writing situation. His audience and use analysis goes like this:[1]

> I'd better decide *exactly* what my primary reader wants.
>
> Long requested the report, but only because Black developed the scheme for division-wide improvements. So I really have two primary readers: my boss and the big boss.
>
> My major question here: Am I including enough detail for all the bosses? The answer to this question will require answers to more specific questions:

Anticipated readers' questions	What are we doing right, and how can we do it better?
	What are we doing wrong, and does it cost us money?
	Have we left anything out, and does it matter?
	How, specifically, can we improve the program, and how will those improvements help the company?

> Because all readers have participated in these sessions (as trainees, instructors, or guest speakers), they don't need background explanations.
>
> I should begin with the *positive* features of the last session. Then I can discuss the problems and make recommendations. Maybe I can eliminate the bossy and judgmental tone by *suggesting improvements* instead of *criticizing weaknesses*. Also, I could be more persuasive by describing the *benefits* of my suggestions.

Glenn realizes that if he wants successful future programs, he can't afford to alienate anyone. After all, he wants to be seen as a loyal member of the company, yet preserve his self-esteem and demonstrate he is capable of making objective recommendations.

> Now, I have a clear enough sense of what to do.

---

[1]Throughout this section, Glenn's analysis will address *all* the areas illustrated in the audience and use profile sheet (page 29).

**Statement of
purpose**

The purpose of my document is to provide my supervisor and interested executives with an evaluation of the workshop by describing its strengths, suggesting improvements, and explaining the benefits of these changes.

From this plan, I should be able to revise my first draft, but that first draft lacks important details. I should brainstorm to get *all* the details (including the *positive* ones) I want to include.

GLENN'S BRAINSTORMING LIST. Glenn's first draft touched on several topics: incorporating them into his brainstorming, he came up with the following list.

1. better-prepared instructors and more visuals
2. on-the-job orientation *before* the training session
3. more members in training sessions
4. executive speakers should spell out qualities needed for success
5. beneficial emphasis on interpersonal communication
6. need follow-up evaluation (in six months?)
7. four types of training evaluations:
    a. trainees' reactions
    b. testing of classroom learning
    c. transference of skills to the job
    d. effect of training on the organization (high sales, more promotions, better-written reports)
8. videotaping and critiquing of trainee speeches worked well
9. acknowledge the positive features of the session
10. ongoing improvement ensures quality training
11. division of class topics into two areas was a good idea
12. additional trainees would increase classroom dialogue
13. the more trainees in a session, the less time and money wasted
14. instructors shouldn't drift from the topic
15. on-the-job training to give a broad view of the division
16. clear course objectives, to increase audience interest and to measure the program's success
17. Marvin Long has done a great job with these sessions over the years

By 9:05 A.M., the office is hectic. Glenn puts his list aside to spend the day on work that has been piling up. Not until 4 P.M. does he return to his report.

Now what? I should delete whatever my audience already knows or doesn't need, or whatever seems unfair or insincere: 7 can go (this audience needs no lecture in training theory); 14 is too negative and critical—besides, the same idea is stated more positively in 4; 17 is obvious brown-nosing, and I'm in no position to make such grand judgments.

Maybe I can unscramble this list by arranging items within categories (strengths, suggested changes, and benefits) from my statement of purpose.

GLENN'S BRAINSTORMING LIST REARRANGED.   Notice here how Glenn discovers additional *content* (see italic type) while he's deciding about *organization*.

Strengths of the Workshop
- division of class topics into two areas was useful
- emphasis on interpersonal communication
- videotaping of trainees' oral reports, followed by critiques

Well, that's one category done. Maybe I should combine *suggested changes* with *benefits,* since I'll want to cover them together in the report.

Suggested Changes/Benefits
- more members per session would increase dialogue and use resources more efficiently
- varied on-the-job experiences before the training sessions would give each member a broad view of the marketing division
- executive speakers should spell out qualities required for success and *future sessions should cover professional behavior, to provide trainees with a clear guide*
- follow-up evaluation in six months *by both supervisors and trainees would reveal the effectiveness of this training and suggest future improvements*
- clear course objectives and more visual aids would increase *instructor efficiency* and audience interest

Now that he has a fairly sensible arrangement, Glenn can get this list into report form, even though he will probably think of more material to add as he works. Since this is *internal* correspondence, he uses a memo format.

Drafting the Document
Glenn produces a usable draft—one containing just about everything he wants to cover. (Sentences are numbered for later reference.)

A later draft

[1]In my opinion, the Management Training Session for the month of October was somewhat successful. [2]This success was evidenced when most participants rated their training as "very good." [3]But improvements are still needed.

[4]First and foremost, a number of innovative aspects in this October session proved especially useful. [5]Class topics were divided into two distinct areas. [6]These topics created a general-to-specific focus. [7]An emphasis on interpersonal communication skills was the most dramatic innovation. [8]This helped class members develop a better attitude toward things in general. [9]Videotaping of trainees' oral reports, followed by critiques, helped clarify strengths and weaknesses.

[10]A detailed summary of the trainees' evaluations is attached. [11]Based on these and on my past observations, I have several suggestions.

- [12]All management training sessions should have a minimum of ten to fifteen members. [13]This would better utilize the larger number of managers involved and the time expended in the implementation of the training. [14]The quality of class interaction with the speakers would also be improved with a larger group.

- [15]There should be several brief on-the-job training experiences in different sales and service areas. [16]These should be developed prior to the training session. [17]This would provide each member with a broad view of the duties and responsibilities in all areas of the marketing division.
- [18]Executive speakers should take a few minutes to spell out the personal and professional qualities essential for success with our company. [19]This would provide trainees with a concrete guide to both general company and individual supervisors' expectations. [20]Additionally, by the next training session we should develop a presentation dealing with appropriate attitudes, manners, and behavior in the business environment.
- [21]Do a six-month follow-up. [22]Get feedback from supervisors as well as trainees. [23]Ask for any new recommendations. [24]This would provide a clear assessment of the long-range impact of this training on an individual's job performance.
- [25]We need to demand clearer course objectives. [26]Instructors should be required to use more visual aids and improve their course structure based on these objectives. [27]This would increase instructor quality and audience interest.
- [28]These changes are bound to help. [29]Please contact me if you have further questions.

Although now developed and organized, this version still is some way from the finished document. Glenn has to make further decisions about his style, content, arrangement, audience, and purpose.

Blair Cordasco offers to review the piece once again and to work with Glenn on a thorough edit.

Revising the Document

At 8:15 Tuesday morning, Cordasco and Glenn begin a sentence-by-sentence revision for worthwhile content, sensible organization, and readable style. Their discussion goes something like this:

Sentence 1 begins with a needless qualifier, has a redundant phrase, and sounds insulting ("somewhat successful"). Sentence 2 should be in the passive voice, to emphasize the training—not the participants. Also, 1 and 2 are choppy and repetitious, and should be combined.

Original	In my opinion, the Management Training Session for the month of October was somewhat successful. This success was evidenced when most participants rated their training as "very good." (28 words)
Revised	The October Management Training Session was successful, with training rated "very good" by most participants. (15 words)[2]

---

[2]Notice throughout how careful revision sharpens the writer's meaning while cutting needless words.

Sentence 3 is too blunt. An orienting sentence should forecast content diplomatically. This statement can be candid without being so negative.

Original   But improvements are still needed.

Revised   A few changes—beyond the recent innovations—should result in even greater training efficiency.

In sentence 4, "First and foremost" is trite, "aspects" is a clutter word, and word order needs changing to improve the emphasis (on innovations) and to lead into the examples.

Original   First and foremost, a number of innovative aspects in this October session proved especially useful.

Revised   Several program innovations were especially useful in this session.

Sentences 5 and 6 need combining, and content in 5 needs beefing up: *name* the two areas. If 7 is labeled the "most dramatic innovation," it ought to come last, for emphasis. In 8, "things in general" is too indefinite. And 7 and 8 need combining. Sentence 9 seems okay. All three examples would be more readable in a *list,* with parallel phrasing (maybe starting each with an "ing" phrase to signal the "action" verb, as in 9).

Original   Class topics were divided into two distinct areas. These topics created a general-to-specific focus. An emphasis on interpersonal communication skills was the most dramatic innovation. This helped class members develop a better attitude toward things in general. Videotaping of trainees' oral reports, followed by critiques, helped clarify strengths and weaknesses.

Revised
- Dividing class topics into two areas created a general-to-specific focus: *the first week's coverage of company structure and functions created a context for the second week's coverage of management skills.*[3]
- Videotaping and critiquing trainees' oral reports clarified strengths and weaknesses.
- Emphasizing interpersonal communication skills (*listening, showing empathy, and reading nonverbal feedback*) *generated enthusiasm and a sense of ease about the group, their training, and the company.*

As a design consideration, a couple of clear headings ("Workshop Strengths," "Suggested Changes/Benefits") would segment the text, improving readability and appearance.

---

[3]This revision has more words, but also much more concrete and specific detail (in italic type). Completeness of information always takes priority over word count.

In collaboration with his colleague, Glenn continues this editing and revising process throughout. Wednesday morning, after much revising and proofreading, Glenn prints out the final draft, shown in Figure B.4.

Glenn's final report is both informative and persuasive. But this document did not appear magically. Glenn made deliberate decisions about purpose, audience, content, organization, and style. He sought advice and feedback on every aspect of the document. Most importantly, he *spent time revising*.[4]

YOUR OWN WRITING SITUATION

Writers work in different ways. Some begin by brainstorming. Some begin with an outline. Others hate outlining and simply write and rewrite. Some write a quick draft before thinking through their writing situation. Introductions and titles are often written last. Whether you write alone or collaborate in preparing a document, whether you are receiving feedback or providing it, no one step in the process is complete until *the whole* is complete. Notice, for instance, how Glenn sharpens his content *and* style while he organizes. Every document you write will require *all* these decisions, but you will rarely make them in the same sequence from day to day.

No matter what the sequence, *revision* is a fact of life. It is the one *constant* in the writing process. When you've finished a draft, you have in a sense only begun. Sometimes you will have more time to compose than Glenn did, sometimes much less. Whenever your deadline allows, leave time to revise.

---

[4]A special thanks to Glenn Tarullo for his perseverance. I made his task doubly difficult by having him explain each of his decisions during this writing process.

:::. MICRO**BYTE**

October 19, 20XX

To: Marvin Long
From: Glenn Tarullo
Subject: October Management Training Program: Evaluation
         and Recommendations

The October Management Training Session was successful, with training rated as "very good" by most participants. A few changes, beyond the recent innovations, should result in even greater training efficiency.

Workshop Strengths
Especially useful in this session were several program innovations:

—Dividing class topics into two areas created a general-to-specific focus: The first week's coverage of company structure and functions created a context for the second week's coverage of management skills.

—Videotaping and critiquing trainees' oral reports clarified their speaking strengths and weaknesses.

—Emphasizing interpersonal communication skills (listening, showing empathy, and reading nonverbal feedback) created a sense of ease about the group, the training, and the company.

Innovations like these ensure high-quality training. And future sessions could provide other innovative ideas.

Suggested Changes/Benefits
Based on the trainees' evaluation of the October session (summary attached) and my observations, I recommend these additional changes:

—We should develop several brief (one-day) on-the-job rotations in different sales and service areas before the training session. These rotations would give each member a real-life view of duties and responsibilities throughout the company.

—All training sessions should have at least ten to fifteen members. Larger classes would make more efficient use of resources and improve class-speaker interaction.

Margin annotations:

Begins on a positive note, and cites evidence

States his claim

Gives clear examples of "innovations"

Cites the bases for his recommendations

Figure B.4   Glenn's Final Draft

Long, Oct. 19, 20XX, page 2

Supports each
recommendation
with convincing
reasons

—We should ask instructors to follow a standard format (based on definite course objectives) for their presentations, and to use visuals liberally. These enhancements would ensure the greatest possible instructor efficiency and audience interest.

—Executive speakers should spell out personal and professional traits that are essential to success in our company. Such advice would give trainees a concrete guide to both general company and individual supervisor expectations. Also, by the next training session, we should assemble a presentation dealing with appropriate attitudes, manners, and behavior in the business environment.

—We should do a six-month follow-up of trainees (with feedback from supervisors as well as ex-trainees) to gain long-term insights, to measure the influence of this training on job performance, and to help design advanced training.

Closes by appealing
to shared goals
(efficiency and
profit)

Inexpensive and easy to implement, these changes should produce more efficient training.

Copies: B. Hull, C. Black, G. Hopkins, J. Capilona, P. Maxwell, R. Sanders, L. Hunter

Figure B.4 Glenn's Final Draft *Continued*

Editing for Grammar, Usage, and Mechanics

COMMON SENTENCE ERRORS
EFFECTIVE PUNCTUATION
TRANSITIONS
EFFECTIVE MECHANICS

The rear endsheet displays editing and revision symbols and corresponding page references. When your instructor marks a symbol on your paper, turn to the appropriate section for explanations and examples.

COMMON SENTENCE ERRORS

The following common sentence errors are easy to repair.

frag

Sentence Fragment

A sentence expresses a logically complete idea. Any complete idea must contain a subject and a verb and must not depend on another complete idea to make sense. Your sentence might contain several complete ideas, but it must contain at least one!

> [Incomplete idea]   [complete idea]   [complete idea]
> Although Mary was injured, she grabbed the line, and she saved the boat.

Omitting some essential element (the subject, the verb, or another complete idea), leaves only a piece of a sentence—a *fragment.*

> Grabbed the line. [*a fragment because it lacks a subject*]
>
> Although Mary was injured. [*a fragment because—although it contains a subject and a verb—it needs to be joined with a complete idea to make sense*]
>
> Sam an electronics technician.

This last statement leaves the reader asking, "What about Sam the electronics technician?" The verb—the word that makes action happen—is missing. Adding a verb changes this fragment to a complete sentence.

Simple verb	Sam **is** an electronics technician.
Verb plus adverb	Sam, an electronics technician, **works hard.**
Dependent clause, verb, and subjective complement	**Although he is well paid,** Sam, an electronics technician, **is not happy.**

Do not, however, mistake the following statement—which seems to contain a verb—for a complete sentence:

> Sam being an electronics technician.

Such "-ing" forms do not function as verbs unless accompanied by other verbs such as **is, was,** and **will be.**

| **Sam,** being an electronics technician, **was responsible for checking the circuitry.**

Likewise, the "*to* + verb" form (infinitive) is not a verb.

Fragment   To become an electronics technician.

Complete   To become an electronics technician, **Sam had to complete a two-year apprenticeship.**

Sometimes we inadvertently create fragments by adding certain words (**because, since, it, although, while, unless, until, when, where,** and others) to an already complete sentence.

| **Although** Sam is an electronics technician.

Such words subordinate the words that follow them; that is, they make the statement dependent on an additional idea, which must itself have a subject and a verb and be a complete sentence. (See also "Subordination"—pages 629–30.) We can complete the subordinate statement by adding an independent clause.

| Although Sam is an electronics technician, **he hopes to become an electrical engineer.**

Note: Because the incomplete idea (dependent clause) depends on the complete idea (independent clause) for its meaning, you need only a *pause* (symbolized by a comma), not a *break* (symbolized by a semicolon). Here are some fragments from students' writing. Each is repaired in two ways. Can you think of other ways to make these statements complete?

Fragment   She spent her first week on the job as a researcher. **Selecting and compiling technical information from digests and journals.**

Revised   She spent her first week on the job as a researcher, selecting and compiling technical information from digests and journals.

In her first week on the job as a researcher, she selected and compiled technical information from digests and journals.

Fragment   **Because the operator was careless.** The new computer was damaged.

Revised   Because the operator was careless, the new computer was damaged.

The operator was careless; as a result, the new computer was damaged.

Fragment   **When each spool is in place.** Advance your film.

Revised   When each spool is in place, advance your film.

Be sure that each spool is in place before advancing your film.

Acceptable Fragments

A fragmented sentence is acceptable in commands or exclamations because the subject ("you") is understood.

Acceptable fragments	Slow down.
	Give me a hand.
	Look out!

Also, questions and answers are sometimes expressed as incomplete sentences.

Acceptable fragments	How? By investing wisely.
	When? At three o'clock.
	Who? Bill.

In general, however, avoid fragments unless you have good reason to use one for special tone or emphasis.

Comma Splice

In a comma splice, two complete ideas (independent clauses) that should be *separated* by a period or a semicolon are incorrectly *joined* by a comma:

| Sarah did a great job, she was promoted.

You can choose among several possibilities for repair:

1. Substitute a period followed by a capital letter:

| Sarah did a great job. She was promoted.

2. Substitute a semicolon to signal a relationship between the two items:

| Sarah did a great job; she was promoted.

3. Use a semicolon with a connecting adverb (a transitional word):

| Sarah did a great job; **consequently,** she was promoted.

4. Use a subordinating word to make the less important clause incomplete and thus dependent on the other:

| **Because** Sarah did a great job, she was promoted.

5. Add a connecting word after the comma:

| Sarah did a great job, **and** she was promoted.

The following revisions show that your choice of construction will depend on the exact meaning or tone you wish to convey:

Comma splice	This is a fairly new product, therefore, some people don't trust it.
Revised	This is a fairly new product. Some people don't trust it.
	This is a fairly new product; therefore, some people don't trust it.
	Because this is a fairly new product, some people don't trust it.
Comma splice	Ms. Gomez was a strict supervisor, she was well liked by her employees.
Revised	Ms. Gomez was a strict supervisor. She was well liked by her employees.
	Ms. Gomez was a strict supervisor; **however,** she was well liked by her employees.
	Although Ms. Gomez was a strict supervisor, she was well liked by her employees.
	Ms. Gomez was a strict supervisor, **but** she was well liked by her employees.

ro   Run-On Sentence

The run-on sentence, a cousin to the comma splice, crams too many ideas together without needed breaks or pauses.

Run-on	The hourglass is more accurate than the waterclock for the water in a waterclock must always be at the same temperature in order to flow with the same speed since water evaporates it must be replenished at regular intervals thus not being as effective in measuring time as the hourglass.
Revised	The hourglass is more accurate than the waterclock because water in a waterclock must always be at the same temperature to flow at the same speed. Also, water evaporates and must be replenished at regular intervals. These temperature and volume problems make the waterclock less effective than the hourglass in measuring time.

agr sv   Faulty Agreement—Subject and Verb

The subject should agree in number with the verb. Faulty agreement seldom occurs in short sentences, where subject and verb are not far apart: "Jack eat too much" instead of "Jack eats too much." But when the subject is separated from the verb by other words, we sometimes lose track of the subject-verb relationship.

Faulty	The lion's **share** of diesels **are** sold in Europe.

Although **diesels** is closest to the verb, the subject is **share,** a singular subject that needs a singular verb.

Revised   The lion's **share** of diesels **is** sold in Europe.

Agreement errors are easy to correct once subject and verb are identified.

Faulty   There **is** an estimated 29,000 **women** living in our city.
Revised   There **are** an estimated 29,000 **women** living in our city.

Faulty   A **system** of lines **extend** horizontally to form a grid.
Revised   A **system** of lines **extends** horizontally to form a grid.

A second problem with subject-verb agreement occurs with indefinite subject pronouns such as **each, everyone, anybody,** and **somebody.** They usually take a singular verb.

Faulty   **Each** of the crew members **were** injured during the storm.
Revised   **Each** of the crew members **was** injured during the storm.

Faulty   **Everyone** in the group **have** practiced long hours.
Revised   **Everyone** in the group **has** practiced long hours.

Collective nouns such as **herd, family, union, group, army, team, committee,** and **board** can call for a singular or plural verb, depending on your intended meaning. When denoting the group as a whole, use a singular verb.

Correct   The **committee meets** weekly to discuss new business.
                 The editorial **board** of this magazine **has** high standards.

To denote individual members of the group, use a plural verb.

Correct   The **committee disagree** on whether to hire Jim.
                 The editorial **board are** all published authors.

When two subjects are joined by **either . . . or** or **neither . . . nor,** the verb is singular if both subjects are singular and plural if both subjects are plural. If one subject is plural and one is singular, the verb agrees with the subject that is closer to the verb.

Correct   Neither **John** nor **Bill works** regularly.
                 Either **apples** or **oranges are** good vitamin sources.
                 Either Felix or his **friends are** crazy.
                 Neither the boys nor their **father likes** the home team.

If, on the other hand, two subjects (singular, plural, or mixed) are joined by **both . . . and,** the verb will be plural.

> Correct   Both Joe and Bill **are** resigning.

A single **and** between subjects makes for a plural subject.

`agr p` Faulty Agreement—Pronoun and Referent

A pronoun must refer to a specific noun (its referent or antecedent), with which it must agree in gender and number.

> Correct   Jane lost **her** book.
> The **students** complained that **they** had been treated unfairly.

When an indefinite pronoun such as **each, everyone, anybody, someone,** or **none** serves as the pronoun referent, the pronoun is singular.

> Correct   **Anyone** can get **his** degree from that college.
> **Anyone** can get **his** or **her** degree from that college.
> **Each** candidate described **her** plans in detail.

`coord` Faulty Coordination

Give equal emphasis to ideas of equal importance by joining them within simple or compound sentences, with coordinating conjunctions: **and, but, or, nor, for, so,** and **yet.**

> This course is difficult **but** worthwhile.
> My horse is old **and** gray.
> We must decide to support **or** reject the dean's proposal.

But do not confound your meaning by excessively coordinating.

> Excessive coordination   The climax in jogging comes after a few miles **and** I can no longer feel stride after stride **and** it seems as if I am floating **and** jogging becomes almost a reflex **and** my arms **and** legs continue to move **and** my mind no longer has to control their actions.
>
> Revised   The climax in jogging comes after a few miles when I can no longer feel stride after stride. By then I am jogging almost by reflex, nearly floating, my arms and legs still moving, my mind no longer having to control their actions.

Notice how the meaning becomes clear when the less important ideas (**nearly floating, arms and legs still moving, my mind no longer having**) are shown as dependent on, rather than equal to, the most important idea (**jogging almost by reflex**)—the idea that contains the lesser ones.

Avoid coordinating ideas that cannot be sensibly connected:

Faulty	John had a drinking problem **and** he dropped out of school.
Revised	John's drinking problem depressed him so much that he couldn't study, so he quit school.
Faulty	I was late for work **and** wrecked my car.
Revised	Late for work, I backed out of the driveway too quickly, hit a truck, and wrecked my car.

Instead of *try and,* use *try to.*

Faulty	I will try and help you.
Revised	I will try to help you.

sub

Faulty Subordination

Proper subordination shows that a less important idea is dependent on a more important idea. By using subordination, you can combine simple sentences into complex sentences and emphasize the most important idea. Consider these complete ideas:

| Joe studies hard. He has severe math anxiety.

Because these ideas are expressed as simple sentences, they appear to be coordinate (equal in importance). But if you wanted to indicate your opinion of Joe's chances of succeeding in math, you would need a third sentence: **His disability will probably prevent him from succeeding,** or **His willpower will help him succeed.** To communicate the intended meaning concisely, combine ideas and subordinate the one that deserves less emphasis.

| Despite his severe math anxiety *(subordinate idea),* Joe studies hard *(independent idea).*

This first version suggests that Joe will succeed. Below, subordination is used to suggest the opposite meaning:

| Despite his diligent studying *(subordinate idea),* Joe has severe math anxiety *(independent idea).*

A dependent (or subordinate) clause in a sentence is signalled by a subordinating conjunction: **because, so, if, unless, after, until, since, while, as,** and **although,** among others. Be sure to place the idea you want emphasized in the independent clause. Do not write

| Although Mary is receiving excellent medical treatment, she is seriously ill.

if you mean to suggest that Mary has a good chance of recovering.

Do not coordinate when you should subordinate:

> **Weak**   Television viewers can relate to an athlete they idolize and they feel obliged to buy the product endorsed by their hero.

Of the two ideas in the sentence above, one is the cause, the other the effect. Emphasize this relationship through subordination:

> **Revised**   Because television viewers can relate to an athlete they idolize, they feel obliged to buy the product endorsed by their hero.

When combining several ideas within a sentence, decide which is most important, and subordinate the other ideas to it—do not merely coordinate:

> **Faulty**   This employee is often late for work, and he writes illogical reports, and he is a poor manager, and he should be fired.
>
> **Revised**   Because this employee is often late for work, writes illogical reports, and has poor management skills, **he should be fired.** (*The last clause is independent.*)

Do not overstuff sentences by excessive subordination:

> **Overstuffed**   This job, which I took when I graduated from college, while I waited for a better one to come along, which is boring, where I've gained no useful experience, makes me anxious to quit.
>
> **Revised**   Upon college graduation, I took this job while waiting for a better one to come along. Because I find it boring and have gained no useful experience, I am eager to quit.

ca

Faulty Pronoun Case

A pronoun's case (nominative, objective, or possessive) is determined by its role in the sentence: as subject, object, or indicator of possession.

If the pronoun serves as the subject of a sentence (**I, we, you, she, he, it, they, who**), its case is *nominative*.

> **She** completed her graduate program in record time.
>
> **Who** broke the chair?

When a pronoun follows a version of the verb **to be** (a linking verb), it explains (complements) the subject, so its case is nominative.

> The killer was **she.**
>
> The professor who perfected our new distillation process is **he.**

If the pronoun serves as the object of a verb or a preposition (**me, us, you, her, him, it, them, whom**), its case is *objective.*

Object of the verb	The employees gave **her** a parting gift.
Object of the preposition	To **whom** do you wish to complain?

If a pronoun indicates possession (**my, mine, our, ours, your, yours, his, her, hers, its, their, whose**), its case is *possessive.*

> The brown briefcase is **mine.**
>
> **Her** offer was accepted.
>
> **Whose** opinion do you value most?

Here are some frequent errors in pronoun case:

Faulty	**Whom** is responsible to **who**? [*The subject should be nominative and the object should be objective.*]
Revised	**Who** is responsible to **whom**?
Faulty	The debate was between Marsha and **I**. [*As object of the preposition, the pronoun should be objective.*]
Revised	The debate was between Marsha and **me**.
Faulty	**Us** students are accountable for our decisions. [*The pronoun accompanies the subject, "students," and thus should be nominative.*]
Revised	**We** students are accountable for our decisions.
Faulty	A group of **we** students will fly to California. [*The pronoun accompanies the object of the preposition, "students," and thus should be objective.*]
Revised	A group of **us** students will fly to California.

Deleting the accompanying noun from the two latter examples reveals the correct pronoun case ("We . . . are accountable . . . "; "A group of us . . . will fly . . . ").

mod Faulty Modification

A sentence's word order (syntax) helps determine its effectiveness and meaning. Words or groups of words are modified by adjectives, adverbs, phrases, or clauses. Modifiers explain, define, or add detail to other words or ideas. Prepositional phrases, for example, usually define or limit adjacent words:

> the foundation **with the cracked wall**
>
> the repair job **on the old Ford**
>
> the journey **to the moon**

So do phrases with "-ing" verb forms:

> the student **painting the portrait**
>
> **Opening the door,** we entered quietly.

Phrases with "to + verb" form limit:

> **To succeed,** one must work hard.

Some clauses also limit:

> the person **who came to dinner**

Problems with word order occur when a modifying phrase begins a sentence and has no word to modify.

Dangling modifier	**Dialing the phone,** the cat ran out the open door.

dgl

The cat obviously did not dial the phone, but because the modifier **Dialing the phone** has no word to modify, the word order suggests that the noun beginning the main clause (*cat*) names the one who dialed the phone. Without any word to join itself to, the modifier *dangles*. By inserting a subject, we can repair this absurd message.

Correct	As **Joe** dialed the phone, the cat ran out the open door.

A dangling modifier can also obscure your meaning.

Dangling modifier	**After completing the student financial aid application form,** the Financial Aid Office will forward it to the appropriate state agency.

Who completes the form—the student or the financial aid office?

Here are some other dangling modifiers that make the message confusing, inaccurate, or downright absurd:

Dangling modifier	**While walking,** a cold chill ran through my body.
Correct	While **I** walked, a cold chill ran through my body.
Dangling modifier	Impurities have entered our bodies **by eating chemically processed foods.**
Correct	Impurities have entered our bodies by **our** eating chemically processed foods.

Dangling modifier	**By planting different varieties of crops,** the pests were unable to adapt.
Correct	By planting different varieties of crops, **farmers** prevented the pests from adapting.

The order of adjectives and adverbs in a sentence is as important as the order of modifying phrases and clauses. Notice how changing word order affects the meaning of these sentences:

I **often** remind myself of the need to balance my checkbook.

I remind myself of the need to balance my checkbook **often.**

Be sure that modifiers and the words they modify follow an order that reflects your meaning.

Misplaced modifier	Joe typed another memo on our computer **that was useless.** (*Was the typewriter or the memo useless?*)
Correct	Joe typed another useless memo on our computer.
	or
	Joe typed another memo on our useless computer.
Misplaced modifier	He read a report on the use of nonchemical pesticides **in our conference room.** (*Are the pesticides to be used in the conference room?*)
Correct	In our conference room he read a report on the use of non-chemical pesticides.
Misplaced modifier	She volunteered **immediately** to deliver the radioactive shipment. (*Volunteering immediately, or delivering immediately?*)
Correct	She immediately volunteered to deliver . . .
	or
	She volunteered to deliver immediately . . .

par

Faulty Parallelism

To reflect relationships among items of equal importance, express them in identical grammatical form:

Correct	We here highly resolve . . . that government **of the people, by the people, for the people** shall not perish from the earth.

The statement above describes the government with three modifiers of equal importance. Because the first modifier is a prepositional phrase, the others must be also. Otherwise, the message would be garbled, like this:

Faulty	We here highly resolve . . . that government **of the people, which the people created and maintain, serving the people** shall not perish from the earth.

If you begin the series with a noun, use nouns throughout the series; likewise for adjectives, adverbs, and specific types of clauses and phrases.

Faulty	The new apprentice is **enthusiastic, skilled,** and **you can depend on her.**
Correct	The new apprentice is **enthusiastic, skilled,** and **dependable.** (*all subjective complements*)
Faulty	In his new job, he felt **lonely** and **without a friend.**
Correct	In his new job, he felt **lonely** and **friendless.** (*both adjectives*)
Faulty	She plans **to study** all this month and **on scoring well** in her licensing examination.
Correct	She plans **to study** all this month and **to score well** in her licensing examination. (*both infinitive phrases*)
Faulty	She **sleeps** well and **jogs** daily, **as well as eating** high-protein foods.
Correct	She **sleeps** well, **jogs** daily, and **eats** high-protein foods. (*all verbs*)

To improve coherence in long sentences, repeat words that introduce parallel expressions:

Faulty	Before buying this property, you should decide whether you will settle down and raise a family, travel for a few years, or pursue a graduate degree.
Correct	Before buying this property, you should decide whether **to settle** down and raise a family, **to travel** for a few years, or **to pursue** a graduate degree.

shift	**Sentence Shifts**

Shifts in point of view result in incoherence. If you begin a sentence or paragraph with one subject or person, do not shift to another.

Shift in person	When **one** finishes such a great book, **you** will have a sense of achievement.
Revised	When **you** finish such a great book, **you** will have a sense of achievement.
Shift in number	**One** should sift the flour before **they** make the pie.
Revised	**One** should sift the flour before **one** makes the pie. (*Or better: Sift the flour before making the pie.*)

Do not begin a sentence in the active voice and then shift to the passive voice.

> **Shift in voice**   **He** delivered the plans for the apartment complex, and the building site **was also inspected by him.**
>
> **Revised**   He **delivered** the plans for the apartment complex and also **inspected** the building site.

Do not shift tenses without good reason.

> **Shift in tense**   She **delivered** the blueprints, **inspected** the foundation, **wrote** her report, and **takes** the afternoon off.
>
> **Revised**   She **delivered** the blueprints, **inspected** the foundation, **wrote** her report, and **took** the afternoon off.

Do not shift from one verb mood to another (as from imperative to indicative mood in a set of instructions).

> **Shift in mood**   **Unscrew** the valve and then steel wool **should be used** to clean the fittings.
>
> **Revised**   **Unscrew** the valve and then **use** steel wool to clean the fittings.

pct   EFFECTIVE PUNCTUATION

Punctuation marks are like road signs and traffic signals. They govern reading speed and provide clues for navigation through a network of ideas.

Let's review the four used most often. These marks can be ranked in order of relative strength.

1. *Period.* A period signals a complete stop at the end of an independent idea (independent clause). The first word in the idea following the period begins with a capital letter.

| Jack is a fat cat. His friends urge him to diet.

2. *Semicolon.* A semicolon signals a brief stop after an independent idea but does not end the sentence; instead, it announces that the forthcoming independent idea is **closely related** to the preceding idea.

| Jack is a fat cat; he eats too much.

3. *Colon.* A colon usually follows an independent idea and, like the semicolon, signals a brief stop but does not end the sentence. The colon and semicolon, however, are never interchangeable. A colon symbolizes "explanation to follow." Information after the colon (which need not be an independent idea) explains or clarifies the idea expressed before the colon.

Jack is a fat cat: he weighs forty pounds. (*The information after the colon answers "How fat?"*)

or

Jack is a fat cat: forty pounds worth! [*The second clause is not independent.*]

4. *Comma.* The weakest of these four marks, a comma signals only a pause within or between ideas in the sentence. A comma often indicates that the word, phrase, or clause set off from the independent idea cannot stand alone but must rely on the independent idea for its meaning.

Jack, a fat cat, is a jolly fellow.
Although he diets often, Jack is a fat cat.

A comma is used between two independent clauses only if accompanied by a coordinating conjunction (**and, but, or, nor, yet**).

Comma splice	Jack is a party animal, he is loved everywhere.
Correct	Jack is a party animal, **and** he is loved everywhere.

End Punctuation
Like a red traffic light, the three marks of end punctuation—period, question mark, and exclamation point—signal a complete stop.

./   **PERIOD.**   A period ends a sentence and is the final mark in some abbreviations.

Ms.   Assn.   Inc.

Periods serve as decimal points in numbers.

$15.95
21.4%

?/   **QUESTION MARK.**   A question mark follows a direct question.

Where is the essay that was due today?

Do not use a question mark to end an indirect question.

Faulty	Professor Grim asked if all students had completed the essay?
Revised	Professor Grim asked if all students had completed the essay.
	or
	Professor Grim asked, "Did all students complete the essay?"

EXCLAMATION POINT. Use an exclamation point only when expression of strong feeling is appropriate.

> **Appropriate**   Oh, no!
>   Pay up!

Semicolon

Like a blinking red traffic light at an intersection, a semicolon signals a brief but definite stop.

SEMICOLONS SEPARATING INDEPENDENT CLAUSES. Semicolons separate independent clauses (logically complete ideas) whose contents are closely related and are not connected by a coordinating conjunction.

> The project was finally completed; we had done a good week's work.

The semicolon can replace the conjunction-comma combination that joins two independent ideas.

> The project was finally completed, and we were elated.
> The project was finally completed; we were elated.

The second version emphasizes the sense of elation.

SEMICOLONS USED WITH ADVERBS AS CONJUNCTIONS AND OTHER TRANSITIONAL EXPRESSIONS. Semicolons must accompany conjunctive adverbs like **besides, otherwise, still, however, furthermore, moreover, consequently, therefore, on the other hand, in contrast,** or **in fact.**

> The job is filled; however, we will keep your résumé on file.
> Your background is impressive; in fact, it is the best among our applicants.

SEMICOLONS SEPARATING ITEMS IN A SERIES. When items in a series contain internal commas, semicolons provide clear separation between items.

> I am applying for summer jobs in Santa Fe, New Mexico; Albany, New York; Montgomery, Alabama; and Moscow, Idaho.
> Members of the survey crew were Juan Jimenez, a geologist; Hector Lightfoot, a surveyor; and Mary Shelley, a graduate student.

Colon

Like a flare in the road, a colon signals you to stop and then proceed, paying attention to the situation ahead. Usually a colon follows an introductory statement that requires a follow-up explanation.

We need this equipment immediately: a voltmeter, a portable generator, and three pairs of insulated gloves.

She is an ideal colleague: honest, reliable, and competent.

Except for salutations in formal correspondence (e.g., Dear Ms. Jones:) colons follow independent (logically and grammatically complete) statements.

> **Faulty** My plans include: finishing college, traveling for two years, and settling down in Sante Fe.

No punctuation should follow "include."

Colons can introduce quotations.

The supervisor's message was clear enough: "You're fired."

A colon can replace a semicolon between two related, complete statements when the second one explains or amplifies the first.

Pam's reason for accepting the lowest-paying job offer was simple: she had always wanted to live in the Northwest.

Comma

The comma is the most frequently used—and abused—punctuation mark. It works like a blinking yellow traffic light, for which you slow down briefly without stopping. Never use a comma to signal a *break* between independent ideas.

COMMA AS A PAUSE BETWEEN COMPLETE IDEAS.   In a compound sentence in which a coordinating conjunction (**and, or, nor, for, but**) connects equal (independent) statements, a comma usually precedes the conjunction.

This is an excellent course, **but** the work is difficult.

COMMA AS A PAUSE BETWEEN AN INCOMPLETE AND A COMPLETE IDEA.   A comma is usually placed between a complete and an incomplete statement in a complex sentence when the incomplete statement comes first.

Because he is a fat cat, Jack diets often.

When he eats too much, Jack gains weight.

When the order is reversed (complete idea followed by incomplete), the comma is usually omitted.

Jack diets often **because he is a fat cat.**

Jack gains weight **when he eats too much.**

Reading a sentence aloud should tell you whether to pause (and use a comma).

COMMAS SEPARATING ITEMS (WORDS, PHRASES, OR CLAUSES) IN A SERIES.
Use commas after items in a series, including the next-to-last item.

> Helen, Joe, Marsha, and John are joining us on the term project.
>
> He works hard **at home, on the job,** and even **during his vacation.**
>
> The new employee complained **that the hours were long, that the pay was low, that the work was boring, and that the supervisor was paranoid.**

Use no commas if **or** or **and** appears between all items in a series.

> She is willing to study in San Francisco or Seattle or even in Anchorage.

COMMA SETTING OFF INTRODUCTORY PHRASES.
Infinitive, prepositional, or verbal phrases introducing a sentence are usually set off by commas, as are interjections.

Infinitive phrase	**To be or not to be,** that is the question.
Prepositional phrase	**In Rome,** do as the Romans do.
Participial phrase	**Being fat,** Jack was slow at catching mice. **Moving quickly,** the army surrounded the enemy.
Interjection	**Oh, is** that the verdict?

COMMAS SETTING OFF NONRESTRICTIVE ELEMENTS.
A *restrictive* phrase or clause modifies or defines the subject in such a way that deleting the modifier would change the meaning of the sentence.

> All students **who have work experience** will receive preference.

Without **who have work experience,** which *restricts* the subject by limiting the category **students,** the meaning would be entirely different. All students will receive preference.

Because this phrase is essential to the sentence's meaning, it is *not* set off by commas.

A *nonrestrictive* phrase or clause could be deleted without changing the sentence's meaning and *is* set off by commas.

> Our new manager, **who has only six weeks' experience,** is highly competent.

Modifier deleted	Our new manager is highly competent.

| This house, **riddled with carpenter ants,** is falling apart.

Modifier
deleted   This house is falling apart.

COMMAS SETTING OFF PARENTHETICAL ELEMENTS.   Items that interrupt the flow of a sentence (such as **of course, as a result, as I recall,** and **however**) are called parenthetical and are enclosed by commas. They may denote emphasis, afterthought, clarification, or transition.

Emphasis   This deluxe model, **of course,** is more expensive.
Afterthought   Your essay, **by the way,** was excellent.
Clarification The loss of my job was, **in a way,** a blessing.
Transition   Our warranty, **however,** does not cover tire damage.

Direct address is parenthetical.

| Listen, **my children,** and you shall hear. . . .

A parenthetical expression at the beginning or the end of a sentence is set off by a comma.

Naturally, we will expect a full guarantee.

My friends, I think we have a problem.

You've done a good job, **Jim.**

Yes, you may use my name in your advertisement.

COMMAS SETTING OFF QUOTED MATERIAL.   Quoted items within a sentence are set off by commas.

| The customer said, **"I'll take it,"** as soon as he laid eyes on our new model.

COMMAS SETTING OFF APPOSITIVES.   An appositive, a word or words explaining a noun and placed immediately after it, is set off by commas when the appositive is nonrestrictive. (See page 639.)

Martha Jones, **our new president,** is overhauling all personnel policies.

Alpha waves, **the most prominent of the brain waves,** are typically recorded in a waking subject whose eyes are closed.

Please make all checks payable to Sam Sawbuck, **school treasurer.**

COMMAS USED IN COMMON PRACTICE.   Commas set off the day of the month from the year, in a date.

| May 10, 1989

Commas set off numbers in three-digit intervals.

> 11,215
> 6,463,657

They also set off street, city, and state in an address.

> Mail the bill to J. B. Smith, 18 Sea Street, Albany, Iowa 01642.

When the address is written vertically, however, the commas that would otherwise occur at the end of each address line are omitted.

> J. B. Smith
> 18 Sea Street
> Albany, Iowa 01642

Commas set off an address or date in a sentence.

> Room 3C, Margate Complex, is my summer address.
> June 15, 1987, is my graduation date.

Commas set off degrees and titles from proper nouns.

> Roger P. Cayer, M.D.
> Gordon Browne, Jr.
> Sandra Mello, Ph.D.

COMMAS USED INCORRECTLY.   Avoid needless or inappropriate commas. Read a sentence aloud to identify inappropriate pauses.

Faulty	The instructor told me, that I was late. [*separates the indirect from the direct object*]
	The most universal symptom of the suicide impulse, is depression. [*separates the subject from its verb*]
	This has been a long, difficult, semester. [*second comma separates the final adjective from its noun*]
	John, Bill, and Sally, are joining us on the trip home. [*third comma separates the final subject from its verb*]
	An employee, who expects rapid promotion, must quickly prove his or her worth. [*separates a modifier that should be restrictive*]

I spoke by phone with John, and Marsha. [*separates two nouns linked by a coordinating conjunction*]

The room was, 18 feet long. [*separates the linking verb from the subjective complement*]

We painted the room, red. [*separates the object from its complement*]

| ap/ |

Apostrophe

Apostrophes indicate the possessive, a contraction, and the plural of numbers, letters, and figures.

APOSTROPHE INDICATING THE POSSESSIVE.   At the end of a singular word, or of a plural word that does not end in *s*, add an apostrophe plus *s* to indicate the possessive. Single-syllable nouns that end in *s* take the apostrophe before an added *s*.

> The **people's** candidate won.
> The chainsaw was **Emma's**.
> The **women's** locker room burned.
> I borrowed **Chris's** book.

Do not add *s* to words that already end in *s* and have more than one syllable; add an apostrophe only.

> **Aristophanes'** death

Do not use an apostrophe to indicate the possessive form of either singular or plural pronouns.

> The book was **hers**.
> **Ours** is the best school in the county.
> The fault was **theirs**.

At the end of a plural word that ends in *s*, add an apostrophe only.

> the **cows'** water supply
> the **Jacksons'** wine cellar

At the end of a compound noun, add an apostrophe plus *s*.

| my **father-in-law's** false teeth

At the end of the last word in nouns of joint possession, add an apostrophe plus *s* if both own one item.

| **Joe and Sam's** lakefront cottage

Add an apostrophe plus **s** to both nouns if each owns specific items.

| **Joe's** and **Sam's** passports

APOSTROPHE INDICATING A CONTRACTION.   An apostrophe shows that you have omitted one or more letters in a phrase that is usually a combination of a pronoun and a verb.

| I'm   they're
| he's   you'd
| you're   who's

Don't confuse **they're** with **their** or **there.**

Faulty	there books
	their now leaving
	living their
Correct	their books
	they're now leaving
	living there

Remember the distinction this way:

| Their friend knows they're there.

It's means "it is." **Its** is the possessive.

| It's watching its reflection in the pond.

Who's means "who is." **Whose** indicates the possessive.

| Who's interrupting whose work?

Other contractions are formed from the verb and the negative.

isn't	can't
don't	haven't
won't	wasn't

Apostrophe Indicating the Plural of Numbers, Letters, and Figures

The **6**'s on this new printer look like smudged **G**'s, **9**'s are illegible, and the **%**'s are unclear.

"/"   Quotation Marks

Quotation marks set off the exact words borrowed from another speaker or writer. The period or comma at the end is placed within the quotation marks.

<table>
<tr>
<td>Periods and commas belong within quotation marks</td>
<td>"Hurry up," Jack whispered.

Jack told Felicia, "I'm depressed."</td>
</tr>
</table>

The colon or semicolon is always placed outside quotation marks.

<table>
<tr>
<td>Colons and semicolons belong outside quotation marks</td>
<td>Our student handbook clearly defines "core requirements"; however, it does not list all the courses that fulfill the requirement.</td>
</tr>
</table>

When a question mark or exclamation point is part of a quotation, it belongs within the quotation marks, replacing the comma or period.

<table>
<tr>
<td>Some punctuation belongs within quotation marks</td>
<td>"Help!" he screamed.

Marsha asked John, "Can't we agree about anything?"</td>
</tr>
</table>

But if the question mark or exclamation point pertains to the attitude of the person quoting instead of the person being quoted, it is placed outside the quotation mark.

<table>
<tr>
<td>Some punctuation belongs outside quotation marks</td>
<td>Why did Boris wink and whisper, "It's a big secret"?</td>
</tr>
</table>

Use quotation marks around titles of articles, paintings, book chapters, and poems.

Certain titles belong within quotation marks   The enclosed article, "The Job Market for College Graduates," should provide some helpful insights.

But titles of books, journals, or newspapers should be underlined or italicized. Finally, use quotation marks (with restraint) to indicate irony.

Quotation marks to indicate irony   She is some "friend"!

Ellipses

. . ./

Three dots in a row (. . .) indicate you have omitted material from a quotation. If the omitted words come at the end of the original sentence, a fourth dot indicates the period. (Also see page 576.)

> "Three dots . . . indicate . . . omitted . . . material. . . . A fourth dot indicates the period. . . .

Italics

ital

In typing or longhand writing, indicate italics by underlining. On a word processor, use italic print for titles of books, periodicals, films, newspapers, and plays; for the names of ships; for foreign words or scientific names; sparingly, for emphasizing a word; and for indicating the special use of a word.

> The *Oxford English Dictionary* is a handy reference tool.
>
> The *Lusitania* sank rapidly.
>
> She reads the *Boston Globe* often.
>
> My only advice is *caveat emptor.*
>
> *Bacillus anthracis* is a highly virulent organism.
>
> *Do not* inhale these fumes under any circumstances!
>
> Our contract defines a *work-study student* as one who works a minimum of twenty hours weekly.

Parentheses

()/

As a general rule, use commas to set off most parenthetical elements, dashes to give some emphasis to the material that is set off, and parentheses to enclose material that defines or explains the statement that precedes it.

> An anaerobic **(airless)** environment must be maintained for the cultivation of this organism.
>
> The cost of running our college has increased by fifteen percent in one year **(see Appendix A for full cost breakdown).**
>
> This new calculator **(made by Ilco Corporation)** is perfect for science students.

Material in parentheses, like all other parenthetical material discussed earlier, can be deleted without harming the logical and grammatical structure of the sentence.

Brackets

Brackets in a quotation set off material that was not in the original quotation but is needed for clarification, such as an antecedent (or referent) for a pronoun.

> "She [Amy] was the outstanding candidate for the scholarship."

Brackets can enclose information taken from some other location within the context of the quotation.

> "It was in early spring [April 2, to be exact] that the tornado hit."

Use **sic** ("thus," or "so") to indicate an error that appears in the original source.

> The assistant's comment was clear: "He don't [sic] want any."

Dashes

Dashes can be effective—if they are not overused. Parentheses de-emphasize the enclosed material; dashes emphasize it.

> Have a good vacation—but watch out for sandfleas.
> Mary—a true friend—spent hours helping me rehearse.

TRANSITIONS

Transitional expressions **announce** relations between ideas. Words or phrases such as **for example, meanwhile, however,** and **moreover** work like bridges between thoughts. Each has a definite meaning—even without a specific context—as shown below.

Transition	Relation
X; meanwhile, Y	X and Y are occurring at the same time.
X; however, Y	Y is in contrast or exception to X.
X; moreover, Y	Y is in addition to X.
X; thus, Y	Y is a result of X.

Here is a paragraph in which these transitions are used to clarify the writer's line of thinking (emphasis added):

USING TRANSITIONS TO BRIDGE IDEAS

Psychological and social problems of aging are too often aggravated by the final humiliation: poverty. One out of every three older Americans lives near or below the

COMMON TRANSITIONS AND THE RELATIONS THEY INDICATE

◆ **An addition:** *moreover, in addition, and, also*

> I am majoring in naval architecture; *also,* I spent three years crewing on a racing yawl.

◆ **Results:** *thus, hence, therefore, accordingly, thereupon, as a result, and so, as a consequence*

> Mary enjoyed all her courses; *therefore,* she worked especially hard last semester.

◆ **An example or illustration:** *for instance, to illustrate, namely, specifically*

> Competition for part-time jobs is fierce; *for example,* 80 students applied for the clerk's job at Sears.

◆ **An explanation:** *in other words, simply stated, in fact*

> Louise had a terrible semester; *in fact,* she flunked three courses.

◆ **A summary or conclusion:** *in closing, to conclude, to summarize, in brief, in summary, to*

sum up, all in all, on the whole, in retrospect, in conclusion

> Our credit is destroyed, our bank account is overdrawn, and our debts are piling up; *in short,* we are bankrupt.

◆ **Time:** *first, next, second, then, meanwhile, at length, later, now, the next day, in the meantime, in turn, subsequently*

> Mow the ball field this morning; *then,* clean the dugouts.

◆ **A comparison:** *likewise, in the same way, in comparison*

> Our reservoir is drying up because of the drought; *similarly,* water supplies in neighboring towns are dangerously low.

◆ **A contrast or alternative:** *however, nevertheless, yet, still, in contrast, otherwise, but, on the other hand, to the contrary, notwithstanding, conversely*

> Felix worked hard; *however,* his grades remained poor.

> poverty level. *Meanwhile,* only one out of every nine younger adults lives in poverty. The American public assumes that Social Security and Medicare provide adequate support for the aged. These benefits alone, *however,* are rarely enough to raise an older person's living standards above the poverty level. *Moreover,* older people are the only group living in poverty whose population has recently increased rather than decreased. More and more of our aging citizens *thus* confront the prospect of living with less and less.

NOTE   *Transitional expressions should be a limited option for achieving coherence. Use them sparingly, and only when a relationship is not already clear.*

Note, too, that whole sentences can serve as transitions between paragraphs and a whole paragraph can serve as a transition between sections of writing. Assume, for instance, that you work as a marketing intern for a stereo manufacturer. You have just completed a section of a memo on the advantages of the new AKS amplifier and are now moving to a section on selling the idea to consumers. This next paragraph might link the two sections:

A TRANSITIONAL PARAGRAPH

Because the AKS amplifier increases bass range by fifteen percent, it should be installed as a standard item in all our stereo speakers. Tooling and installation adjustments, however, will add roughly $50 to the list price of each model. We must, therefore, explain the cartridge's long-range advantages to consumers. Let's consider ways of explaining these advantages.

Notice that this transitional paragraph *contains* transitional expressions as well.

EFFECTIVE MECHANICS

Correctness in abbreviation, hyphenation, capitalization, use of numbers, and spelling demonstrates your attention to detail.

ab

Abbreviations

Avoid abbreviations in formal writing or in situations that might confuse your reader. When in doubt, write the word out.

Abbreviate some words and titles when they precede or immediately follow a proper name, but not military, religious, or political titles.

Correct	Mr. Jones
	Dr. Jekyll
	Raymond Dumont, Jr.
	Reverend Ormsby
	President Clinton

Abbreviate time designations only when they are used with actual times.

Correct	400 B.C.
	5:15 a.m.
Faulty	Plato lived sometime in the BC period.
	She arrived in the a.m.

Abbreviate a unit of measurement only when it appears often in your report and is written out in full on first use. Use only abbreviations you are sure readers will understand. Abbreviate items in a visual aid only if you need to save space. Most dictionaries provide an alphabetical list of other abbreviations. For abbreviations in documentation of research sources, see pages 579–90.

-/

Hyphen

Hyphens divide words at line breaks and join two or more words used as a single adjective if they precede the noun (but not if they follow it):

com-puter

the rough-hewn wood

COMMON ABBREVIATIONS FOR UNITS OF MEASUREMENT

AC	alternating current	kw	kilowatt
amp	ampere	kwh	kilowatt hour
Å	angstrom	l	liter
az	azimuth	lat	latitude
bbl	barrel	lb	pound
BTU	British Thermal Unit	lin	linear
C	Celsius	long	longitude
cal	calorie	log	logarithm
cc	cubic centimeter	m	meter
circ	circumference	max	maximum
cm	centimeter	mg	milligram
CPS	cycles per second	min	minute
cu ft	cubic foot	ml	milliliter
dB	decibel	mm	millimeter
DC	direct current	mo	month
dm	decimeter	mph	miles per hour
doz	dozen	oct	octane
DP	dewpoint	oz	ounce
F	Fahrenheit	psf	pounds per square foot
F	farad	psi	pounds per square inch
fbm	foot board measure	qt	quart
fl oz	fluid ounce	r	roentgen
FM	frequency modulation	rpm	revolutions per minute
freq	frequency	sec	second
ft	foot	sp gr	specific gravity
ft lb	foot pound	sq	square
gal	gallon	t	ton
GPM	gallons per minute	temp	temperature
gr	gram	tol	tolerance
hp	horsepower	ts	tensile strength
hr	hour	V	volt
in	inch	VA	volt ampere
IU	international unit	W	watt
J	joule	wk	week
ke	kinetic energy	WL	wavelength
kg	kilogram	yd	yard
km	kilometer	yr	yearl

the all-too-human error

The wood was rough hewn.

The error was all too human.

Some other commonly hyphenated words:

◆ Most words that begin with the prefix self-. (Check your dictionary.)

COMMON ABBREVIATIONS FOR REFERENCE IN MANUSCRIPTS

anon.	anonymous	fig.	figure
app.	appendix	i.e.	that is
b.	born	illus.	illustrated
©	copyright	jour.	journal
c., ca.	about (c. 1988)	l., ll.	line(s)
cf.	compare	ms., mss.	manuscript(s)
ch.	chapter	no.	number
col.	column	p., pp.	page(s)
d.	died	pt., pts.	part(s)
ed.	editor	rev.	revised or review
e.g.	for example	sec.	section
esp.	especially	sic.	thus, so (to cite an
et al.	and others		error in the quotation)
etc.	and so on	trans.	translation
ex.	example	vol.	volume
f. or ff.	the following page or pages		

> self-reliance
>
> self-discipline

◆ Combinations that might be ambiguous.

> re-creation [*a new creation*]
>
> recreation [*leisure activity*]

◆ Words that begin with *ex* only if *ex* means "past."

> ex-faculty member
>
> excommunicate

◆ All fractions, along with ratios that are used as adjectives and that precede the noun (but not those that follow it), and compound numbers from twenty-one through ninety-nine.

> a **two-thirds** majority
>
> In a **four-to-one** vote, the student senate defeated the proposal.
>
> The proposal was voted down **four to one.**
>
> **Thirty-eight** windows were broken.

---

cap

Capitalization

Capitalize the first words of all sentences as well as titles of people, books, and chapters; languages; days of the week; the months; holidays; names of organizations or groups; races and nationalities; historical events; important documents; and names of structures or vehicles. In titles of books, films, and the like, capitalize the first word and all those following except articles or prepositions.

Items that are capitalized	Joe Schmoe	Russian
	A Tale of Two Cities	Labor Day
	Protestant	Dupont Chemical Company
	Wednesday	Senator Barbara Boxer
	the *Queen Mary*	France
	the Statue of Liberty	The War of 1812

Do not capitalize the seasons (**spring, winter**) or general groups (**the younger generation, the leisure class**).

Capitalize adjectives that are derived from proper nouns.

Chaucerian English

Capitalize titles preceding a proper noun (but not those following).

State Senator Marsha Smith

Marsha Smith, state senator

Capitalize words such as **street, road, corporation,** and **college** only when they accompany a proper noun.

Bob Jones University

High Street

The Rand Corporation

Capitalize **north, south, east,** and **west** when they denote specific locations, not when they are simply directions.

the South

the Northwest

Turn east at the next set of lights.

Use of Numbers

Numbers expressed in one or two words can be written out or written as numerals. Use numerals to express larger numbers, decimals, fractions, precise technical figures, or any other exact measurements.

543	2,800,357
$3\frac{1}{4}$	15 pounds of pressure
50 kilowatts	4,000 rpm

Use numerals for dates, census figures, addresses, page numbers, exact units of measurement, percentages, times with a.m. or p.m. designations, and monetary and mileage figures.

page 14	1:15 p.m.
18.4 pounds	9 feet
12 gallons	$15

Do not begin a sentence with a numeral. If your figure needs more than two words, revise your word order.

> Six hundred students applied for the 102 available jobs.
>
> The 102 available jobs brought 780 applicants.

Do not use numerals to express approximate figures, time not designated as a.m. or p.m., or streets named by numbers less than 100.

> about seven hundred fifty
>
> four fifteen
>
> 108 East Forty-second Street

In contracts and other documents in which precision is vital, a number can be stated both in numerals and in words:

> The tenant agrees to pay a rental fee of three hundred seventy-five dollars ($375.00) monthly.

sp

Spelling

Take the time to use your dictionary for all writing assignments. When you read, note the spelling of words that give you trouble. Compile a list of troublesome words.

Works Cited

Abelman, Arthur F. "Legal Issues in Scholarly Publishing." *MLA Style Manual.* 2nd ed. New York: Modern Language Association, 1998: 30–57.

Adams, Gerald R., and Jay D. Schvaneveldt. *Understanding Research Methods.* New York: Longman, 1985.

Adler, Jerry. "For Humans, Evolution Ain't What It Used to Be." *Newsweek* 29 Sept. 1997: 17.

The Aldus Guide to Basic Design. Aldus Corporation, 1988.

Alleman, James E., and Brooke T. Mossman. "Asbestos Revisited." *Scientific American* July 1997: 70–75.

American Psychological Association. *Publication Manual of the American Psychological Association.* 4th ed. Washington: Author, 1994.

"And the Winner of the Dubious-Study-of-Year Award Is . . . ." *University of California at Berkeley Wellness Letter* 14.6 (1998): 1+.

"Any Alternative?" *The Economist* 1 Nov. 1997: 83–84.

"Are We in the Middle of a Cancer Epidemic?" *University of California at Berkeley Wellness Letter* 10.9 (1994): 4–5.

Armstrong, William H. "Learning to Listen." *American Educator* Winter 1997–98: 24+.

Baker, Russ. "Surfer's Paradise." *Inc.* Nov. 1997: 57+.

Ball, Charles. "Figuring the Risks of Closer Runways." *Technology Review* Aug./Sept. 1996: 12–13.

Barbour, Ian. *Ethics in an Age of Technology.* New York: Harper, 1993.

Barfield, Woodrow, Mark Haselkorn, and Catherine Weatbrook. "Information Retrieval with a Printed User's Manual and with Online Hypercard Help." *Technical Communication* 37.1 (1990): 22–27.

Barnes, Shaleen. "Evaluating Sources Checklist." Information Literacy Project. 10 June 1997. Online Posting. 23 June 1998 <http://www.2lib.umassd.edu/library2/INFOLIT/prop.html>.

Barnett, Arnold. "How Numbers Can Trick You." *Technology Review* Oct. 1994: 38–45.

Barnum, Carol, and Robert Fisher. "Engineering Technologists as Writers: Results of a Survey." *Technical Communication* 31.2 (1984): 9–11.

Bashein, Barbara J., and M. Lynne Markus. "A Credibility Equation for IT Specialists." *Sloan Management Review* 38.4 (Summer 1997): 35–44.

Baumann, K. E., et al. "Three Mass Media Campaigns to Prevent Adolescent Cigarette Smoking." *Preventive Medicine* 17 (1988): 510–30.

Baumeister, Roy F. "Should Schools Try to Boost Self-Esteem?" *American Educator* Summer 1996: 14+.

Bazerman, Max H., Kimberly P. Morgan, and George F. Loewenstein. "The Impossibility of Auditor Independence." *Sloan Management Review* 38.4 (Summer 1997): 89–94.

Beamer, Linda. "Learning Intercultural Communication Competence." *Journal of Business Communication* 29.3 (1992): 285–303.

Bedford, Marilyn S., and F. Cole Stearns. "The Technical Writer's Responsibility for Safety." *IEEE Transactions on Professional Communication* 30.3 (1987): 127–32.

Begley, Sharon. "Bad Days on the Lily Pad." *Newsweek* 13 July 1998: 67.

———. "Is Science Censored?" *Newsweek* 14 Sept. 1992: 63.

———. "Odds on the Greenhouse." *Newsweek* 1 Dec. 1997: 72.

Belkin, Lisa. "How Can We Save the Next Victim?" *The New York Times Magazine* 15 June 1997: 28+.

Benson, Phillipa J. "Visual Design Consideration in Technical Publications." *Technical Communication* 32.4 (1985): 35–39.

Bernstein, Mark. "Deeply Intertwingled Hypertext: The Navigation Problem Reconsidered." *Technical Communication* 38.1 (1991): 41–47.

Berry, Stephen R. "Scientific Information in the Electronic Era." *Professional Ethics Report* [American Association for the Advancement of Science] X.2 (Spring 1997): 1+.

Bjerklie, David. "E-Mail: The Boss Is Watching." *Technology Review* 14 Apr. 1993: 14–15.

Black, Ken. *Business Statistics: Contemporary Decision Making.* St. Paul, MN: West, 1994.

Blaser, Martin J. "The Bacteria behind Ulcers." *Scientific American* Feb. 1996: 140+.

Blinder, Alan S., and Richard E. Quandt. "The Computer and the Economy." *The Atlantic Monthly* Dec. 1997: 26–32.

Blum, Deborah. "Investigative Science Journalism." *Field Guide for Science Writers.* Eds. Deborah Blum and Mary Knudson. New York: Oxford 1997. 86–93.

Bogert, Judith, and David Butt. "Opportunities Lost, Challenges Met: Understanding and Applying Group Dynamics in Writing Projects." *Bulletin of the Association for Business Communication* 53.2 (1990): 51–53.

Boiarsky, Carolyn. "Using Usability Testing to Teach Reader Response." *Technical Communication* 39.1 (1992): 100–02.

Bosley, Deborah. "International Graphics: A Search for Neutral Territory." *INTERCOM* Aug./Sept. 1996: 4–7.

Boucher, Norman. "Back to the Everglades." *Technology Review* Aug./Sept. 1995: 24–35.

Boyd, Ruth-Anne. "Plain Language: Making It Work." *INTERCOM* Nov. 1997: 16–18.

Branscum, Deborah. "bigbrother@the.office.com." *Newsweek* 27 Apr. 1998: 78.

Brimelow, Peter. "Income Gap." *Forbes* 27 July 1998: 51.

Broad, William J. "NASA Budget Cuts Raise Concerns over Safety of Shuttle. *The New York Times* 8 Mar. 1994, sec. B: 5+.

Brody, Herb. "Clinking onto Webzines." *Technology Review* May/June 1997: 40–47.

Brownell, Judi, and Michael Fitzgerald. "Teaching Ethics in Business Communication: The Effective/Ethical Balancing Scale." *Bulletin of the Association for Business Communication* 55.3 (1992): 15–18.

Bruhn, Mark J. "E-Mail's Conversational Value." *Business Communication Quarterly* 58.3 (1995): 43–44.

Bryan, John. "Down the Slippery Slope: Ethics and the Technical Writer as Marketer." *Technical Communication Quarterly* 1.1 (1992): 73–88.

Burger, Katrina. "Righteousness Pays." *Forbes* 22 Sept. 1997: 200–01.

Burghardt. M. David. *Introduction to the Engineering Profession.* New York: Harper, 1991.

Burnett, Rebecca E. "Substantive Conflict in a Cooperative Context: A Way to Improve the Collaborative Planning of Workplace Documents." *Technical Communication* 38.4 (1991): 532–39.

Busiel, Christopher, and Tom Maeglin. *Researching Online.* New York: Addison, 1998.

Byrd, Patricia, and Joy M. Reid. *Grammar in the Composition Classroom.* Boston: Heinle, 1998.

Caher, John M. "Technical Documentation and Legal Liability." *Journal of Technical Writing and Communication* 25.1 (1995): 5–10.

Carliner, Saul. "Demonstrating Effectiveness and Value: A Process for Evaluating Technical Communication Products and Services." *Technical Communication* 44.3 (1997): 252–65.

Caswell-Coward, Nancy. "Cross-Cultural Communication: Is It Greek to You?" *Technical Communication* 39.2 (1992): 264–66.

Chauncey, C. "The Art of Typography in the Information Age." *Technology Review* Feb./Mar. (1986): 26+.

Christians, C. G., et al. *Media Ethics: Cases and Moral Reasoning.* 2nd ed. White Plains, NY: Longman, 1978.

Clark, Gregory. "Ethics in Technical Communication: A Rhetorical Perspective." *IEEE Transactions on Professional Communication* 30.3 (1987): 190–95.

Clark, Thomas. "Teaching Students How to Write to Avoid Legal Liability." *Business Communication Quarterly* 60-3(1997): 71–77.

Clement, David E. "Human Factors, Instructions, and Warnings, and Product Liability." *IEEE Transactions on Professional Communication* 30.3 (1987): 149–56.

Cochran, Jeffrey K., et al. "Guidelines for Evaluating Graphical Designs." *Technical Communication* 36.1 (1989): 25–32.

Coe, Marlana. *Human Factors for Technical Communicators.* New York: Wiley, 1996.

———. "Writing for Other Cultures: Ten Problem Areas." *INTERCOM* Jan. 1997: 17–19.

Cohn, Victor. "Coping with Statistics." *A Field Guide for Science Writers.* Eds. Deborah Blum and Mary Knudson. New York: Oxford, 1997. 102–09.

Cole-Gomolski. "Users Loathe to Share Their Know-How." *Computerworld* 17 Nov. 1997: 6.

Communication Concepts, Inc. "Electronic Media Poses New Copyright Issues." *Writing Concepts* ©. Reprinted in *INTERCOM* Nov. 1995: 13+.

"Consequences of Whistle Blowing in Scientific Misconduct Reported." *Professional Ethics Report* [American Association for the Advancement of Science] IX.4 (Winter 1996): 2.

Consumer Product Safety Commission. *Fact Sheet No. 65.* Washington: GPO, 1989.

Cooper, Lyn O. "Listening Competency in the Workplace: A Model for Training." *Business Communication Quarterly* 60.4 (Dec. 1997): 75–84.

"Copyright Protection and Fair Use of Printed Information" *Addison Wesley Longman Author's Guide.* New York: Longman, 1998.

Cortese, Amy. "Automatic Web Downloads—without the Overload." *Business Week* 24 Nov. 1997: 152.

Cotton, Robert, ed. *The New Guide to Graphic Design.* Secaucus, NJ: Chartwell, 1990.

Council of Biology Editors. *Scientific Style and Format: The CBE Manual for Authors, Editors, and Publishers.* 6th ed. Chicago: Cambridge UP, 1994.

Croal, N'Gai. "Want a Job? Get Online." *Newsweek* 9 June 1997: 81–82.

Cronin, Mary J. "Knowing How Employees Use the Intranet Is Good Business." *Fortune* 21 July 1997: 103.

———. "Using the Web to Push Key Data to Decision Makers." *Fortune* 29 Sept. 1997: 254.

Cross, Mary. "Aristotle and Business Writing: Why We Need to Teach Persuasion." *Bulletin of the Association for Business Communication* 54.1 (1991): 3–6.

Crossen, Cynthia. *Tainted Truth: The Manipulation of Fact in America.* New York: Simon, 1994.

Curry, Jerome. "Tapping the Internet's Job Search Resources." *Business Communication Quarterly* 61.2 (1998): 100–06.

D'Aprix, Roger. "Related Thoughts." *Journal of Employee Communication Management* Nov./Dec. 1997: 66–70.

Daugherty, Shannon. "The Usability Evaluation: A Discount Approach to Usability Testing." *INTERCOM* Dec. 1997: 16–20.

Davenport, Thomas H. *Information Ecology.* New York: Oxford, 1997.

Debs, Mary Beth, "Collaborative Writing in Industry." In *Technical Writing: Theory and Practice.* Ed. Bertie E. Fearing and W. Keats Sparrow. New York: Modern Language Assn., 1989: 33–42.

———. "Recent Research on Collaborative Writing in Industry." *Technical Communication* 38.4 (1991): 476–85.

December, John. "An Information Development Methodology for the World Wide Web." *Technical Communication* 43.3 (1996): 369–75.

Desmond, Edward W. "How Your Data May Soon Seek You Out." *Fortune* 8 Sept. 1997: 149–50.

Detjen, Jim. "Environmental Writing." *A Field Guide for Science Writers.* Eds. Deborah Blum and Mary Knudson. New York: Oxford, 1997. 173–79.

Dillard, James P., Denise H. Solomon, and Jennifer A. Samp. "Framing Social Reality: The Relevance of Relational Judgments." *Communication Research* 23.6 (1996): 703–22.

Dombrowski, Paul M. "Challenger and the Social Contingency of Meaning: Two Lessons for the Technical Communication Classroom." *Technical Communication Quarterly* 1.3 (1992): 73–86.

Dorner, Deitrich. *The Logic of Failure.* Reading, MA: Addison, 1996.

Dowd, Charles. "Conducting an Effective Journalistic Interview." *INTERCOM* May 1996: 12–14.

Doyle, Rodger. "Amphibians and Risk." *Scientific American* Aug. 1998: 27.

Dragga, Sam, and Gwendolyn Gong. *Editing: The Design of Rhetoric.* Amityville, NY: Baywood, 1989.

Dulude Jennifer. "The Web Marketing Handbook." Thesis. University of Massachusetts Dartmouth, 1997.

Dumont, R. A. "Writing, Research, and Computing." *Critical Thinking: An SMU Dialogue, 1988.* No. Dartmouth MA, Southeastern Mass. Univ., 1988.

Dumont, R. A., and J. M. Lannon, *Business Communications.* 3rd ed. Glenview, IL: Scott, 1990.

Dyson, Esther. "Intellectual Value." *Wired* July 1995: 134+.

"Earthquake Hazard Analysis for Nuclear Power Plants." *Energy and Technology Review* June 1984: 8.

Easton, Thomas, and Stephan Herrara. "J&J's Dirty Little Secret." *Forbes* 12 Jan. 1998: 42–44.

Elbow, Peter. *Writing without Teachers.* New York: Oxford, 1973.

"Electronic Mentors." *The Futurist* May 1992: 56.

Elias, Stephen. *Patent, Copyright, and Trademark.* Berkeley: Nolo Press, 1997.

Elliot, Joel. "Evaluating Web Sites: Questions to Ask." 18 Feb. 1997. Online Posting. List for Multimedia and New Technologies in Humanities Teaching. 9 Mar. 1997 <http://www.learnnc.org/documents/webeval.html>.

Evans, James. "Legal Briefs." *Internet World* Feb. 1998: 22.

———. "Whose Web Site Is It Anyway?" *Internet World* Sept. 1997: 46+.

Extejt, Marian M. "Teaching Students to Correspond Effectively Electronically." *Business Communication Quarterly* 61.2 (1998): 57–67.

Fackelmann, Kathleen. "Science Safari in Cyberspace." *Science News* 152.50 (1997): 397–98.

Facts and Figures about Cancer. Boston: Dana-Farber Cancer Institute, 1995.

"Fair Use." 10 June 1993. Online. 2 pp. United States Copyright Office, Library of Congress. 19 June 1998 <http://www.loc.gov/copyright>.

"Fat Chance." *University of California at Berkeley Wellness Letter* 14.7 (1998): 2–3.

Fawcett, Heather. "*The New Oxford Dictionary* Project." *Technical Communication* 40.3 (1993): 379–82.

Felker, Daniel B., et al. *Guidelines for Document Designers.* Washington: American Institutes for Research, 1981.

Fineman, Howard, "The Power of Talk." *Newsweek* 8 Feb. 1993: 24–28.

Finkelstein, Leo, Jr. "The Social Implications of Computer Technology for the Technical Writer." *Technical Communication* 38.4 (1991): 466–73.

Fisher, Anne. "Can I Stop Gay Bashing?" *Fortune* 7 July 1997: 205–06.

———. "My Team Leader Is a Plagiarist." *Fortune* 27 Oct. 1997: 291–92.

Florman, Samuel, quoting Donald D. Rikard. "Toward Liberal Learning for Engineers." *Technology Review* Mar. 1986: 18–25.

Foster, Edward. "Why Users Beef about Documentation. *INTERCOM* Nov. 1998: 10.

Fox, Justin, "A Startling Notion—The Whole Truth." *Fortune* 24 Nov. 1997: 303.

Franke, Earnest A. "The Value of the Retrievable Technical Memorandum System to the Engineering Company." *IEEE Transactions on Professional Communication* 32.1 (Mar. 1989): 12–16.

Freundlich, Naomi. "When the Cure May Make You Sicker." *Business Week* 16 Mar. 1998: 14.

"From Mir to Mars." PBS Online. 11 Nov. 1998 <http://www.PBS.org>.

Fugate, Alice E. "Mastering Search Tools for the Internet." *INTERCOM* Jan. 1998: 40–41.

Gallagher, Leigh. "Isn't That Special?" *Forbes* 9 Mar. 1998: 39.

Gannon, Joseph P. "From GUI Guru to Web Weaver: Making the Transition." *INTERCOM* Dec. 1997: 21–25.

Garfield, Eugene, "What Scientific Journals Can Tell Us about Scientific Journals." *IEEE Transactions on Professional Communication* 16.4 (1973): 200–02.

Garner, Rochelle. "IS Newbies: Eager, Motivated, Clueless." *Computerworld* 1 Dec. 1997: 85–86.

Garten, Jeffrey E. "Globalism Doesn't Have to Be Cruel." *Business Week* 9 Feb. 1998: 26.

Gartiganis. Arthur. "Lasers." *Occupational Outlook Quarterly* Winter 1984: 22–26.

Gerstner, John. "Print Is Obsolete, but It Won't Go Away." *Journal of Employee Communication Management* Nov./Dec. 1997: 42–47.

Gesteland, Richard R. "Cross-Cultural Compromises." *Sky* May 1993: 20+.

Gibaldi, Joseph. *MLA Handbook for Writers of Research Papers.* 4th ed. New York: Modern Language Assn., 1995.

Gibaldi, Joseph, and Walter S. Achtert. *MLA Handbook for Writers of Research Papers.* 3rd ed. New York: Modern Lanuage Assn., 1988.

Gibbs, W. Wayt. "The Price of Silence." *Scientific American* Nov. 1996: 15–16.

Gilbert, Nick, "1-800-ETHIC." *Financial World* 16 Aug. 1994: 20+.

Gilsdorf, Jeanette W. "Executives' and Academics' Perception of the Need for Instruction in Written Persuasion." *Journal of Business Communication* 23.4 (1986): 55–68.

———. "Write Me Your Best Case for . . ." *Bulletin of the Association for Business Communication* 54.1 (1991): 7–12.

Girill, T. R. "Technical Communication and Art. *Technical Communication* 31.2 (1984): 35.

———. "Technical Communication and Ethics." *Technical Communication* 34.3 (1987): 178–79.

———. "Technical Communication and Law" *Technical Communication* 32.3 (1985): 37.

Glassman, James K. "Dihydrogen Monoxide: It's a Killer." *Daily Hampshire Gazette* 22 Oct. 1997: 6.

Glidden, H. K. *Reports, Technical Writing and Specifications.* New York: McGraw, 1964.

Golen, Steven, et al. "How to Teach Ethics in a Basic Business Communications Class." *Journal of Business Communication* 22.1 (1985): 75–84.

Goodall, H. Lloyd, Jr., and Christopher L. Waagen. *The Persuasive Presentation.* New York: Harper, 1986.

Goodman, Danny. *Living at Light Speed.* New York: Random, 1994.

Goodman, Ellen. "Fear of Taxes Trumps Risk of Cancer." *Daily Hampshire Gazette* 24 June 1998: 6.

Goubil-Gambrell, Patricia. "Designing Effective Internet Assignments in Introductory Technical Communication Courses." *IEEE Transactions on Professional Communication* 39.4 (1996): 224–31.

Gouran, Dennis S., et al. "A Critical Analysis of Factors Related to Decisional Processes Involved in the Challenger Disaster." *Central States Speech Journal* 37.3 (1986): 119–35.

Grant, Linda. "Where Did the Snap, Crackle, & Pop Go?" *Fortune* 4 Aug. 1997: 223+.

Grassian, Esther. "Thinking Critically about World Wide Web Resources." 20 Aug. 1997. UCLA College Library. 25 Oct. 1997 <http://www.library.ucla.edu/libraries/college/instruct/critical.htm>.

Graybill, Nina. "Freelancers and the Information Highway." *AWP Chronicle* Sept. 1997: 27.

Gribbons, William M. "Organization by Design: Some Implications for Structuring Information." *Journal of Technical Writing and Communication* 22.1 (1992): 57–74.

Grice, Roger A. "Document Development in Industry." In *Technical Writing: Theory and Practice.* Ed. Bertie E. Fearing and W. Keats Sparrow. New York: Modern Lanuage Assn., 1989. 27–32.

————. "Focus on Usability: Shazam!" *Technical Communication* 42.1 (1995): 131–33.

Grice Roger A., and Lenore S. Ridgway. "Presenting Technical Information in Hypermedia Format: Benefits and Pitfalls." *Technical Communication Quarterly* 4.1 (1995): 35–46.

Gross, Neil. "Between a Rock and a Hard Place." *Business Week* 20 Apr. 1998: 134+.

Grossman, Wendy M. "Downloading as a Crime." *Scientific American* Mar. 1998: 37.

Hafner, Kate, "Have Your Agent Call My Agent." *Newsweek* 27 Feb. 1995: 76–77.

Hahn, Laura, "How Communicators Can 'Coach' the Workforce into Excellence." *Journal of Employee Communication Management* Nov./Dec., 1997: 10–21.

Hall, Judith G. "Medicine on the Web: Finding the Wheat, Leaving the Chaff." *Technology Review* Mar./Apr. 1998: 60–61.

Halpern, Jean W. "An Electronic Odyssey." In *Writing in Nonacademic Settings.* Eds. Dixie Goswami and Lee Odell. New York: Guilford, 1985. 157–201.

Hamblen, Matt. "Volvo Taps AT&T for Global Net." *Computerworld* 1 Dec. 1997: 51+.

Hammett, Paula. "Evaluating Web Resources." 29 Mar. 1997. Ruben Salazar Library, Sonoma State University. 26 Oct. 1997 <http://libweb.sonoma.edu/resources/eval.html>.

Harcourt, Jules. "Teaching the Legal Aspects of Business Communication." *Bulletin of the Association for Business Communication* 53.3 (1990): 63–64.

Harris, Richard F. "Toxics and Risk Reporting." *A Field Guide for Science Writers.* Eds. Deborah Blum and Mary Knudson. New York: Oxford, 1997. 166–72.

Harris, Robert. "Evaluating Internet Research Sources." 17 Nov. 1997. Online. Internet. 23 June 1998 <http://www.sccu.edu/faculty/R_Harris/evalu8it.htm>.

Harrison, Bennett. "Don't Blame Technology This Time." *Technology Review* July 1997: 62.

Hart, Geoff. "Accentuate the Negative: Obtaining Effectve Reviews through Focused Questions." *Technical Communication* 44.1 (1997): 52–57.

Hartley, James. *Designing Instructional Text.* 2nd ed. London: Kogan Page, 1985.

Haskin, David. "The Extranet Team Play." *Internet World* Aug. 1997: 57–60.

————. "Meetings without Walls." *Internet World* Oct. 1997: 53–60.

————. "A Push in the Right Direction." *Internet World* Sept. 1997: 75+.

Hauser, Gerald. *Introduction to Rhetorical Theory.* New York: Harper, 1986.

Hayakawa, S. I. *Language in Thought and Action,* 3rd ed. New York: Harcourt, 1972.

Hays, Robert. "Political Realities in Reader/Situation Analysis." *Technical Communication* 31.1 (1984): 16–20.

Hein, Robert G. "Culture and Communication." *Technical Communication* 38.1 (1991): 125–26.

Hill-Duin, Ann. "Terms and Tools: A Theory and Research-Based Approach to Collaborative Writing." *Bulletin of the Association for Business Communication* 53.2 (1990): 45–50.

Hodges, Mark. "Is Web Business Good Business?" *Technology Review* Aug./Sept. 1997: 22+.

Hoger, Elizabeth, James J. Cappel, and Mark A. Myerscough. "Navigating the Web with a Typology of Corporate Uses." *Business Communication Quarterly* 61.2 (1998): 39–47.

Hogge, Robert. Unpublished review of *Technical Writing.* 2nd ed.

Holler, Paul F. "The Challenge of Writing for Multimedia." *INTERCOM* July/Aug. 1995: 25.

Holloway, Marguerite. "Sounding Out Science." *Scientific American* Oct. 1996: 106–13.

Hollwitz, John C., and Donna Pawlowski. "The Development of an Ethical Integrity Interview for Pre-Employment Screening." *The Journal of Business Communication* 34.2 (1997): 203–19.

Holyoak, K. J. "Symbolic Connectionism." *Toward Third-Generation Theories of Expertise: Prospects and Limits.* Eds. K. A. Ericsson and J. Smith. New York: Cambridge UP, 1991. 331–35.

Hopkins-Tanne, Janice. "Writing Science for Magazines." *A Field Guide for Science Writers.* Eds. Deborah Blum and Mary Knudson. New York: Oxford, 1997. 17–26.

Horgan, John. "Multicultural Studies." *Scientific American* Nov. 1996: 24+.

Horton, William. "Is Hypertext the Best Way to Document Your Product?" *Technical Communication* 38.1 (1991): 20–30.

————. "Mix Media, Not Metaphors." *Technical Communication* 41.4 (1994): 781–83.

Howard, Tharon. "Property Issue in E-Mail Research." *Bulletin of the Association for Business Communication* 56.2 (1993): 40–41.

Huff, Darrell. *How to Lie with Statistics.* New York: Norton, 1954.

Hugenberg, Lawrence W., Renee M. LaCivita, and Andra M. Lubanovic. "International Business and Training: Preparing for the Global Economy." *The Journal of Business Communication* 33.2 (1996): 205–22.

Hulbert, Jack E. "Developing Collaborative Insights and Skills." *Bulletin of the Association for Business Communication* 57.2 (1994): 53–56.

————. "Overcoming Intercultural Communication Barriers." *Bulletin of the Association for Business Communication* 57.2 (1994): 41–44.

Humphreys, Donald S. "Making Your Hypertext Interface Usable." *Technical Communication* 40.4 (1993): 754–61.

Hunt, Kevin. "Establishing a Presence on the World Wide Web: A Rhetorical Approach." *Technical Communication* 46.4 (1996): 376–87.

Hutheesing, Nikhil. "What Are You Doing on That Porn Site?" *Forbes* 3 Nov. 1997: 368–69.

IBM Corporation. "IBM Solutions." [Advertisement] 1997.

"In Brief." *Scientific American* Oct. 1997: 28.

"International Copyright." 3 Feb. 1996. Online. United States Copyright Office, Library of Congress. 19 June 1998 <http://www.loc.gov/copyright>.

Isaacs, Arlene B. "Tact Can Seal a Global Deal." *The New York Times* 26 July 1998, sec. B: 43.

James-Catalano, P. "Fight for Privacy." *Internet World* Jan. 1997: 32+.

Jameson, Daphne A. "Using a Simulation to Teach Intercultural Communication in Business Communication Courses." *Bulletin of the Association for Business Communication* 56.1 (1993): 3–11.

Janis, Irving L. *Victims of Groupthink: A Psychological Study of Foreign Policy Decisions and Fiascos.* Boston: Houghton, 1972.

Johannesen, Richard L. *Ethics in Human Communication.* 2nd ed. Prospect Heights, IL: Waveland, 1983.

Jones, Barbara. "Giving Women the Business." *Harper's Magazine* Dec. 1997: 47–58.

Journet, Debra. Unpublished review of *Technical Writing.* 3rd ed.

Kane, Kate. "Can You Perform under Pressure?" *Fast Company* Oct./Nov. 1997: 54+.

Karaim, Reed. "The Invasion of Privacy." *Civilization* Oct./Nov. 1996: 70–77.

Kawasaki, Guy. "Get Your Facts Here." *Forbes* 23 Mar. 1998: 156.

————. "The Rules of E-Mail." *MACWORLD* Oct. 1995: 286.

Kelly-Reardon, Kathleen. *They Don't Get It Do They?: Communication in the Workplace—Closing the Gap between Women and Men.* Boston: Little, 1995.

Kelman, Herbert C. "Compliance, Identification, and Internalization: Three Processes of Attitude Change." *Journal of Conflict Resolution* 2 (1958): 51–60.

Keyes, Elizabeth. "Typography, Color, and Information Structure." *Technical Communication* 40.4 (1993): 638–54.

Kiely, Thomas. "The Idea Makers." *Technology Review* Jan. 1993: 33–40.

Kim, Heather. "Diversity among Asian American High School Students." 14 Mar. 1997. Online. Abstract. 10 June 1998 <http://www.ets.org>.

Kipnis, David, and Stuart Schmidt. "The Language of Persuasion." *Psychology Today* Apr. 1985: 40–46. Rpt. in Raymond S. Ross, *Understanding Persuasion.* 3rd ed. Englewood Cliffs: Prentice, 1990.

Kirsh, Lawrence. "Take It from the Top." *MACWORLD* Apr. 1986: 112–15.

Kleimann, Susan D. "The Complexity of Workplace Review." *Technical Communication* 38.4 (1991): 520–26.

Kohl, John R., et al. "The Impact of Language and Culture on Technical Communication in Japan." *Technical Communication* 40.1 (1993): 62–72.

Koretz, Gene. "Income Swings Can Be Deadly. *Business Week* 2 Mar. 1998: 32.

Kotulak, Ronald. "Reporting on Biology of Behavior." *A Field Guide for Science Writers.* Eds. Deborah Blum and Mary Knudson. New York: Oxford, 1997. 142–51.

Kraft. Stephanie. "Whistleblower Bill's Holiday Adventures." *The Valley Advocate* [Northhampton, MA] 6 Jan. 1994: 5–6.

Krause, Tim. "Preparing an Online Résumé." *Business Communication Quarterly* 60.1 (1997): 159–61.

Kremers, Marshall. "Teaching Ethical Thinking in a Technical Writing Course." *IEEE Transactions on Professional Communication* 32.2 (1989): 58–61.

Lambert, Steve. *Presentation Graphics on the Apple® Macintosh.* Bellevue, WA: Microsoft, 1984.

Lang, Thomas A., and Michelle Secic. *How to Report Statistics in Medicine.* Philadelphia: American College of Physicians, 1997.

Larson, Charles U. *Persuasion: Perception and Responsibility.* 7th ed. Belmont, CA: Wadsworth: 1995.

Lavin, Michael R. *Business Information: How to Find It, How to Use It.* 2nd ed. Phoenix, AZ: Oryx, 1992.

Lederman, Douglas. "Colleges Report Rise in Violent Crime," *Chronicle of Higher Education* 3 Feb. 1995, sec. A: 5+.

Lee, Susan. "Death by Charcoal?" *Forbes* 25 Aug. 1997: 280.

Leki, Ilona. "The Technical Editor and the Non-native Speaker of English." *Technical Communication* 37.2 (1990): 148–52.

Lemonick, Michael. "The Evils of Milk?" *Time* 15 June 1998:85.

Lenzer, Robert, and Carrie Shook. "Whose Rolodex Is It, Anyway?" *Forbes* 23 Feb. 1998: 100–04.

Lewis, Howard L. "Penetrating the Riddle of Heart Attack." *Technology Review* Aug./Sept. 1997: 39–44.

Lewis, Kate Bohner. "Maybe Don't Take Two Aspirin." *Forbes* 23 May 1994: 222–23.

Lewis, Philip L., and N. L. Reinsch. "The Ethics of Business Communication." Proceedings of the American Business Communication Conference. Champaign, IL., 1981. in *Technical Communication and Ethics.* Eds. John R. Brockman and Fern Rook. Washington: Soc. for Technical Communication, 1989: 29–44.

Littlejohn, Stephen W., and David M. Jabusch. *Persuasive Transactions.* Glenview, IL: Scott, 1987.

Machlis, Sharon. "Surfing into a New Career as Webmaster." *Computerworld* 1 Dec. 1997: 45+.

MacKenzie, Nancy, Unpublished review of *Technical Writing.* 5th ed.

Mackin, John. "Surmounting the Barrier between Japanese and English Technical Documents." *Technical Communication* 36.4 (1989): 346–51.

Maeglin, Thomas. Unpublished review of *Technical Writing,* 7th ed.

Manning, Michael. "Hazard Communication 101." *INTERCOM* June 1998: 12–15.

Martin, James A. "A Road Map to Graphics Web Sites." *MACWORLD* Oct. 1995: 135–36.

Martin, Jeanette S., and Lillian H. Chaney. "Determination of Content for a Collegiate Course in Intercultural Business Communication by Three Delphi Panels." *Journal of Business Communication* 29.3 (1992): 267–83.

Martin, Justin. "Changing Jobs? Try the Net." *Fortune* 2 Mar. 1998: 205+.

———. "So, You Want to Work for the Best . . ." *Fortune* 12 Jan. 1998: 77–78.

Matson, Eric. "The Seven Sins of Deadly Meetings." *Fast Company* Oct./Nov. 1997: 27–31.

———. "(Search) Engines." *Fast Company.* Oct./Nov. 1997: 249–52.

Max, Robert R. "Wording It Correctly." *Training and Development Journal* Mar. 1985: 5–6.

McDonald, Kim A. "Covering Physics." *A Field Guide for Science Writers.* Eds. Deborah Blum and Mary Knudson. New York: Oxford, 1997. 188–95.

———. "Some Physicists Criticize Research Purporting to Show Links between Low-Level Electromagnetic Fields and Cancer." *Chronicle of Higher Education* 3 May 1991, sec. A: 5+.

McGuire, Gene. "Shared Minds: A Model of Collaboration." *Technical Communication* 39.3 (1992): 467–68.

McNair, Catherine. "New Technologies and Your Résumé." *INTERCOM* June 1997: 12–14.

McWilliams, Gary, and Marcia Stepanik. "Taming the Info Monster." *Business Week* 22 June 1998: 170+.

Melymuka, Kathleen. "Not Another #$!&!$ Survey!!" *Computerworld* 24 Nov. 1997: 82.

Menz, Mary. "Clip Art Comes of Age." *INTERCOM* May 1997: 4–8.

Meyer, Benjamin D. "The ABCs of New-Look Publications." *Technical Communication* 33.1 (1986): 13–20.

Meyerson, Moe. "Grand Illusions." *Inc. Tech* 2 (1997): 35–36.

Microsoft Word User's Guide: Word Processing Program for the Macintosh, Version 5.0. Redmond, WA: Microsoft Corporation, 1992.

Miller, Julie. "Trade Journals. "*A Handbook for Science Writers.* Eds. Deborah Blum and Mary Knudson. New York: Oxford, 1997. 27–30.

Mirel, Barbara, Susan Feinberg, and Leif Allmendinger. "Designing Manuals for Active Learning Styles." *Technical Communication* 38.1 (1991): 75–87.

Mirsky, Steve. "Wonderful Town." *Scientific American* July 1996: 29.

"Misconduct Scandal Shakes German Science." *Professional Ethics Report* [American Assoc. for the Advancement of Science] X.3 (Summer 1997): 2.

Mokhiber, Russell. "Crime in the Suites." *Greenpeace* May 1989: 14–16.

Monastersky, Richard. "Courting Reliable Science." *Science News* 153.16 (1998): 249–51.

———. "Do Clouds Provide a Greenhouse Thermostat?" *Science News* 142 (1992): 69.

Monmonier. Mark. *Cartographies of Danger: Mapping Hazards in America.* Chicago: U of Chicago P, 1997.

Morgan, Meg. "Patterns of Composing: Connections between Classroom and Workplace Collaborations." *Technical Communication* 38.4 (1991): 540–42.

Morgenson, Gretchen. "Would Uncle Sam Lie to You?" *Worth* Nov. 1994: 53+.

Morse, June. "Hypertext—What Can We Expect?" *INTERCOM* Feb. 1992: 6–7.

Munger, David. Unpublished review of *Technical Writing,* 8th ed.

Munter, Mary. "Meeting Technology: From Low-Tech to High-Tech." *Business Communication Quarterly* 61.2 (1998): 80–87.

Murphy, Kate. "Separating Ballyhoo from Breakthrough." *Business Week* 13 July 1998: 143.

Nakache, Patricia. "Is It Time to Start Bragging about Yourself?" *Fortune* 27 Oct. 1997: 287–88.

Nantz, Karen S., and Cynthia L. Drexel. "Incorporating Electronic Mail with the Business Communication Course." *Business Communication Quarterly* 58.3 (1995): 45–51.

Neilsen, Jakob. "Be Succinct! (Writing for the Web)." 15 Mar. 1997. Alertbox. 8 Aug. 1998 <http://www.useit.com/alertbox/9719a.html>.

———. "Global Web: Driving the International Network Economy." Apr. 1998. Alertbox. 8 Aug. 1998 <http://www.useit.com/alertbox/9710a.html>.

———. "How Users Read on the Web." Oct. 1997. Alertbox. 8 Aug. 1998 <http://www.useit.com/alertbox/9710a.html>.

———. "International Web Usability." Aug. 1996. Alertbox. 8 Aug. 1998 <http://www.useit.com/alertbox/9710a.html>.

———. "Inverted Pyramids in Cyberspace." June 1996. Alertbox. 8 Aug. 1998 <http://www.useit.com/alertbox/9710a.html>.

Nelson, Sandra J., and Douglas C. Smith. "Maximizing Cohesion and Minimizing Conflict in Collaborative Writing Groups. *Bulletin of the Association for Business Communication* 53.2 (1990): 59–62.

Nickels-Shirk, Henrietta. "'Hyper' Rhetoric: Reflections on Teaching Hypertext." *Technical Writing Teacher* 18.3 (1991): 189–200.

"No Need to Phone." Graph. *Newsweek* 26 Jan. 1998: 15.

Nordenberg, Tamar. "Direct to You: TV Drug Ads That Make Sense." *FDA Consumer* Jan./Feb. 1998: 7–10.

Nydell Margaret K. *Understanding Arabs: A Guide for Westerners.* New York: Logan, 1987.

"Olive Oil and Breast Cancer: How Strong a Connection?" *University of California at Berkeley Wellness Letter* 11.7 (1995): 1–2.

"On Line." *Chronicle of Higher Education* 21 Sept. 1992, sec. A: 29.

Ornatowski, Cezar M. "Between Efficiency and Politics: Rhetoric and Ethics in Technical Writing." *Technical Communication Quarterly* 1.1 (1992): 91–103.

Outing, Steve. "Does Your Site Contribute to Data Smog?" 28 May 1997. *Editor and Publisher Interactive.* 8 Aug. 1998 <http://www.mediainfo.com/ephome/news/newsshtm/stop/st052897.htm>.

Pace, Roger C. "Technical Communication, Group Differentiation, and the Decision to Launch the Space Shuttle Challenger." *Journal of Technical Writing and Communication* 18.3 (1988): 207–20.

Parrish, Deborah. "The Scientific Misconduct Definition and Falsification of Credentials." *Professional Ethics Report* [American Assoc. for the Advancement of Science] IX.4 (1996): 1+.

Parsons, Gerald M. Review of *Technical Writing,* 6th ed. *Journal of Technical Writing and Communication* 25.3 (1995): 322–24.

Pearce, C. Glenn, Johnson, Iris W., and Randolph T. Barker. "Enhancing the Student Listening Skills and Environment." *Business Communication Quarterly* 58.4 (Dec. 1995): 28–33.

Pender, Kathleen. "Dear Computer, I Need a Job." *Worth* Mar. 1995: 120–21.

"People, Performance, Profits." *Forbes* 20 Oct. 1997: 57.

Perelman, Lewis, J. "How Hypermation Leaps the Learning Curve." *Forbes ASAP* 25 Oct. 1993: 78+.

Perloff, Richard M. *The Dynamics of Persuasion.* Hillsdale, NJ: Erlbaum, 1993.

Peterson, Ivars. "Web Searches Fall Short." *Science News* 153.18 (1998): 286.

Petroski, Henry. *Invention by Design.* Cambridge, MA: Harvard UP, 1996.

Peyser, Marc, and Steve Rhodes. "When E-Mail is Oops-Mail." *Newsweek* 16 Oct. 1995: 82.

Pinelli, Thomas E., et al. "A Survey of Typography, Graphic Design, and Physcial Media in Technical Reports." *Technical Communication* 33.2 (1986): 75–80.

Plumb, Carolyn, and Jan H. Spyridakis. "Survey Research in Technical Communication: Designing and Administering Questionnaires." *Technical Communication* 39.4 (1992): 625–38.

Pool, Robert. "When Failure Is Not an Option." *Technology Review* July 1997: 38–45.

Porter, James E. "Truth in Technical Advertising: A Case Study." *IEEE Transactions on Professional Communication* 30.3 (1987): 182–89.

Powell, Corey S. "Science in Court." *Scientific American* October 1997: 32+.

Presidential Commission. *Report to the President on the Space Shuttle Challenger Accident.* Vol I. Washington: GPO, 1986.

Pugliano, Fiore. Unpublished review of *Technical Writing.* 5th ed.

Rao, Srikumar. "Diaper-Beer Syndrome." *Forbes* 9 Apr. 1998: 128.

Redish, Janice C., and David A. Schell. "Writing and Testing Instructions for Usability." In *Technical Writing: Theory and Practice*. Eds. Bertie E. Fearing and W. Keats Sparrow. New York: Modern Language Assn., 1987: 61–71.

Redish, Janice C., et al. "Making Information Accessible to Readers." In *Writing in Nonacademic Settings*. Eds. Lee Odell and Dixie Goswami: New York: Guilford, 1985.

Reichard, Kevin. "Web-Site Watchdogs." *Internet World* Dec. 1997: 106+.

Reiss, Tammy. "You've Got Mail, and Mail, and Mail . . ." *Business Week* 20 July 1998: 6.

Rensberger, Boyce. "Covering Science for Newspapers." *A Field Guide for Science Writers*. Eds. Deborah Blum and Mary Knudson. New York: Oxford, 1997. 7–16.

Research Triangle Institute. *Consequences of Whistleblowing for the Whistleblower in Misconduct in Science Cases*. (Report prepared for the Office of Research Integrity.) Washington: ORI, 1995.

Richards, Thomas O., and Ralph A. Richards, "Technical Writing." Paper presented at the University of Michigan, 11 July 1941. Published by the Society for Technical Communication.

Rifkin, William, and Brian Martin. "Negotiating Expert Status: Who Gets Taken Seriously." *IEEE Technology and Society Magazine* Spring 1997: 30–39.

Riney, Larry A. *Technical Writing for Industry*. Englewood Cliffs: Prentice, 1989.

Ritzenthaler, Gary, and David H. Ostroff. "The Web and Corporate Communication: Potentials and Pitfalls." *IEEE Transactions on Professional Communication* 39.1 (1996): 16–20.

Robart, Kay. "Submitting Résumés via E-Mail." *INTERCOM* July/Aug. 1998: 13–14.

Robinson, Edward A. "Beware—Job Seekers Have No Secrets." *Fortune* 29 Dec. 1997: 285.

Rokeach, Milton. *The Nature of Human Values*. New York: Free, 1973.

Ross, Philip E. "Enjoy It While It Lasts." *Forbes* 27 July 1998: 206.

———. "Lies, Damned Lies, and Medical Statistics." *Forbes* 14 Aug. 1995: 130–35.

Ross, Raymond S. *Understanding Persuasion*. 3rd ed. Englewood Cliffs: Prentice, 1990.

Ross-Flanigan, Nancy. "The Virtues (and Vices) of Virtual Collaboration." *Technology Review* Mar./Apr. 1998: 50–59.

Rottenberg, Annette, T. *Elements of Argument*. 3rd ed. New York: St. Martin's, 1991.

Rowland, D. *Japanese Business Etiquette: A Practical Guide to Success with the Japanese*. New York: Warner: 1985.

Rowland Robert C. "The Relationship between the Public and the Technical Spheres of Argument: A Case Study of the Challenger Seven Disaster." *Central States Speech Journal* 37.3 (1986): 136–46.

Rubens, Philip M. "Reading and Employing Technical Information in Hypertext." *Technical Communication* 38.1 (1991): 36–40.

———. "Reinventing the Wheel?: Ethics for Technical Communicators." *Journal of Technical Writing and Communication* 11.4 (1981): 329–39.

Ruggiero, Vincent R. *The Art of Thinking*. 3rd ed. New York: Harper, 1991.

Ruggiero, Vincent R. *The Art of Thinking*. 5th ed. New York: Addison, 1998.

Ruhs, Michael A. "Usability Testing: A Definition Analyzed." *Boston Broadside* [Newsletter of the Soc. for Technical Communication] May/June 1992: 8+.

Samuelson, Robert J. "The Endless Paper Chase." *Newsweek* 1 Dec. 1997: 53.

———. "Merchants of Mediocrity." *Newsweek* 1 Aug. 1994: 44.

Schafer, Sarah. "Is Your Data Safe?" *Inc.* Feb. 1997: 93–97.

Schenk, Margaret T., and James K. Webster. *Engineering Information Resources*. New York: Decker, 1984.

Schrage, Michael. "Time for Face Time." *Fast Company* Oct./Nov. 1997: 232.

Scott, James C., and Diana J. Green. "British Perspectives on Organizing Bad-News Letters: Organizational Patterns Used by Major U.K. Companies." *Bulletin of the Association for Business Communication* 55.1 (1992): 17–19.

Seglin, Jeffrey L. "Would You Lie to Save Your Company?" *Inc.* July 1998: 53+.

Selber, Stuart. Unpublished review of *Technical Writing*, 7th ed. 1998.

Seligman, Dan. "Gender Mender." *Forbes* 6 Apr. 1998: 72+.

Selzer, Jack. "Composing Processes for Technical Discourse." In *Technical Writing: Theory and Practice.* Eds. Bertie E. Fearing and W. Keats Sparrow. New York: Modern Language Assn., 1989: 43–50.

Senge, Peter M. "The Leader's New Work: Building Learning Organizations." *Sloan Managment Review* 32.1 (Fall 1990): 1–17.

Shenk, David. "Data Smog: Surviving the Information Glut." *Technology Review* May/June 1997: 18–26.

Sherblom, John C., Claire F. Sullivan, and Elizabeth C. Sherblom, "The What, the Whom, and the Hows of Survey Research," *Bulletin of the Association for Business Communication* 56:12 (1993): 58–64.

Sherif, Muzapher, et al. *Attitude and Attitude Change: The Social Judgment-Involvement Approach.* Philadelphia: Saunders, 1965.

Sharzenski, Emily. "Tips for Creating ASCII and HTML Resumes." *INTERCOM* June/July 1996: 17–18.

Sklaroff, Sara, and Michael Ash. "American Pie Charts." *Civilization* April/May 1997: 84–85.

Snyder, Joel. "Finding It on Your Own." *Internet World* June 1995: 89–90.

Sowell, Thomas. "Magic Numbers." *Forbes* 20 Oct. 1997: 120.

Spencer, SueAnn. "Use Self-Help to Improve Document Usability." *Technical Communication* 46.1 (1996): 73–77.

Spragins, Ellyn E. "The Numbers Racket." *Newsweek* 5 May 1997: 77.

Spyridakis, Jan H. "Conducting Research in Technical Communication: The Application of True Experimental Design." *Technical Communication* 39.4 (1992): 607–24.

Spyridakis, Jan H., and Michael J. Wenger. "Writing for Human Performance: Relating Reading Research to Document Design." *Technical Communication* 39.2 (1992): 202–15.

Stanton, Mike. "Fiber Optics." *Occupational Outlook Quarterly* Winter 1984: 27–30.

Stedman, Craig. "Data Mining for Fool's Gold." *Computerworld* 1 Dec. 1997: 1+.

———. "Extranet Brokers Access to Funds Data." *Computerworld* 20 Oct. 1997: 49+.

Steinberg, Stephen. "Travels on the Net." *Technology Review* July 1994: 20–31.

Stemmer, John. "Citing Internet Sources." 4 Mar. 1997. Online posting. Political Science Research and Teaching List. 22 April 1997 <polpsrt@h—met.msu.edu>.

Stepanek, Marcia. "When in Bejing, Mum's the Word." *Business Week* 13 July 1998: 4.

———. "When the Devil Is in the E-Mails." *Business Week* 8 June 1998.

Stevenson, Richard W. "Workers Who Turn in Bosses Use Law to Seek Big Rewards." *New York Times* 10 July 1989, sec. A: 7.

Stix, Gary. "Plant Matters: How Do You Regulate an Herb?" *Scientific American* Feb. 1998: 30+.

Stone, Peter H. "Forecast Cloudy: The Limits of Global Warming Models." *Technology Review* Feb./Mar. 1992: 32–40.

Stonecipher, Harry. *Editorial and Persuasive Writing.* New York: Hastings, 1979.

Sturges, David L. "Internationalizing the Business Communication Curriculum." *Bulletin of the Association for Business Communication* 55.1 (1992): 30–39.

Taubes, Gary. "Telling Time by the Second Hand." *Technology Review* May/June 1998: 76–78.

Taylor, John R. *Introduction to Error Analysis.* 2nd ed. Sausalito, CA; University Science Books, 1997.

Teague, John H. "Marketing on the World Wide Web." *Technical Communication* 42.2 (1995): 236–42.

Templeton, Brad. "10 Big Myths about Copyright Explained." 29 Nov. 1994. Online posting. 6 May 1995 <http://www.law/copyright/FAQ/myths/part1>

"The Safest Car May Be a Truck." *Fortune* 21 July 1997: 72.

Thrush, Emily A. "Bridging the Gap: Technical Communication in an Intercultural and Multicultural Society." *Technical Communication Quarterly* 2.3 (1993): 271–83.

Trafford, Abigail. "Critical Coverage of Public Health and Government." *A Field Guide for Science Writers.* Eds. Deborah Blum and Mary Knudson. New York: Oxford, 1997. 131–41.

Turner, John R. "Online Use Raises New Ethical Issues." *INTERCOM* Sept. 1995: 5+.

Unger, Stephen H. *Controlling Technology: Ethics and the Responsible Engineer.* New York: Holt, 1982.

U.S. Air Force Academy. *Executive Writing Course.* Washington: GPO, 1981.

U.S. Department of Commerce. *Statistical Abstract of the United States.* Washington: GPO, 1994, 1997.

U.S. Department of Labor. *Tips for Finding the Right Job.* Washington: GPO, 1993.

———. *Tomorrow's Jobs.* Washington: GPO 1995.

U.S. General Services Administration. *Your Rights to Federal Records.* Washington: GPO, 1995.

"Using Icons as Communication." *Simply Stated* [Newsletter of the Document Design Center, American Institutes for Research] 75 (Sept./Oct. 1987): 1+.

van der Meij, Hans, and John M. Carroll. "Principles and Heuristics for Designing Minimalist Instruction." *Technical Communication* 42.2 (1995): 243–61.

Van Pelt, William. Unpublished review of *Technical Writing.* 3rd ed.

Varner, Iris I., and Carson H. Varner. "Legal Issues in Business Communications." *Journal of the American Association for Business Communication* 46.3 (1983): 31–40.

Vaughan, David K. "Abstracts and Summaries: Some Clarifying Distinctions." *Technical Writing Teacher* 18.2 (1991): 132–41.

Velotta, Christopher. "How to Design and Implement a Questionnaire." *Technical Communication* 38.3 (1991): 387–92.

Victor, David A. *International Business Communication.* New York: Harper, 1992.

"Vital Signs." *Internet World* Jan. 1998: 18.

"Vitamin C under Attack." *University of California, Berkeley Wellness Letter* 14.10 (1998): 1.

Walker, Janice R. "MLA-Style Citations of Electronic Sources." April 1995. Online posting. The Alliance for Computers and Writing. 10 Jan. 1996 <http://prairie-island.ttu.edu/acw.html>.

"Walking to Health." *Harvard Men's Health Watch* 2.12 (1998): 3–4.

Wallace, Bob. "Restaurant Franchiser Puts Intranet on Menu." *Computerworld* 10 Nov. 1997: 12.

Wallich, Paul. "Not So Blind, After All." *Scientific American* May 1996: 20+.

Walter, Charles, and Thomas F. Marsteller. "Liability for the Dissemination of Defective Information." *IEEE Transactions on Professional Communication* 30.3 (1987): 164–67.

Wandycz, Katarzyna. "Damn Yankees." *Forbes* 10 March, 1997: 22–23.

Warshaw, Michael. "Have You Been House-Trained?" *Fast Company* Oct. 1998: 46+.

Weingarten, Tara. "The All-Day, All-Night, Global, No-Trouble Job Search." *Newsweek* 6 Apr. 1998: 17.

Weinstein, Edith K. Unpublished review of *Technical Writing.* 5th ed.

Weiss, Edmond H. *How to Write a Usable User Manual.* Philadelphia: ISI, 1985.

"Wellness Facts." *University of California, Berkeley Wellness Letter* 14.10 (1998): 1.

Weymouth, L. C. "Establishing Quality Standards and Trade Regulations for Technical Writing in World Trade." *Technical Communication* 37.2 (1990): 143–47.

White, Jan. *Color for the Electronic Age.* New York: Watson-Guptill, 1990.

———. *Editing by Design.* 2nd ed. New York: Bowker, 1982.

———. *Great Pages.* El Segundo, CA: Serif, 1990.

———. *Visual Design for the Electronic Age.* New York: Watson-Guptill, 1988.

Wicclair, Mark R., and David K. Farkas. "Ethical Reasoning in Technical Communication: A Practical Framework." *Technical Communication* 31.2 (1984): 15–19.

Wickens, Christopher D. *Engineering Psychology and Human Performance.* 2nd ed. New York: Harper, 1992.

Wiggins, Richard. "The Word Electric." *Internet World* Sept. 1995: 31–34.

Wight, Eleanor, "How Creativity Turns Facts into Usable Information." *Technical Communication* 32.1 (1985): 9–12.

Wilkinson, Theresa A. "Defining Content for a Web Site." *INTERCOM* June 1998: 33–34.

Williams, Robert I. "Playing with Format, Style, and Reader Assumptions." *Technical Communication* 30.3 (1983): 11–13.

Winsor, D. A. "Communication Failures Contributing to the Challenger Accident: An Example for Technical Communicators." *IEEE Transactions on Professional Communication* 31.3 (1988): 101–07.

Wojahn, Patricia G. "Computer-Mediated Communication: The Great Equalizer between Men and Women?" *Technical Communication* 41.4 (1994): 747–51.

Woodhouse, E. J., and Dean Nieusma. "When Expert Advice Works and When It Does Not." *IEEE Technology and Society Magazine* Spring 1997: 23–29.

Wurman, Richard Saul. *Information Anxiety.* New York: Doubleday, 1989.

Yoos, George. "A Revision of the Concept of Ethical Appeal." *Philosophy and Rhetoric* 12.4 (1979): 41–58.

Young, Patrick. "Writing Articles for Science Journals." *A Field Guide for Science Writers.* Eds. Deborah Blum and Mary Knudson. New York: Oxford, 1997. 110–16.

Zinsser, William. *On Writing Well.* New York: Harper, 1980.

Index

EDITING AND REVISION SYMBOLS

Symbol	Problem	Page*	Symbol	Problem	Page*
ab	wrong abbreviation	648	ital	italics	645
agr p	error in pronoun agreement	628	( ) /	parentheses	645
agr sv	error in subject-verb agreement	626	. /	period	635
amb	ambiguous phrasing	216	? /	question mark	636
av	active voice needed	220	" / "	quotation	644
bias	biased language	248	; /	semicolon	635
ca	wrong pronoun case	630	pref	needless preface	227
cap	capital letter needed	650	prep	needless preposition	228
cl	word adds clutter	230	pv	passive voice needed	223
comb	sentences need to be combined	232	qual	needless qualifier	231
cont	faulty contraction	643	red	redundant phrase	225
coord	faulty coordination	628	rep	needless repetition	225
cs	comma splice	625	ref	faulty reference	216
dgl	dangling modifier	218	ro	run-on sentence	626
euph	euphemism	239	sexist	sexist usage	249
frag	sentence fragment	623	shift	sentence shift	634
jarg	needless jargon	237	short	short sent. needed	234
mod	misplaced modifier	218	simple	simpler word needed	235
neg	negative phrasing	230	spec	specific word needed	241
nom	nominalization	228	sub	faulty subordination	629
offen	offensive usage	249	tel	telegraphic writing	217
os	overstuffed sentence	224	th	faulty sent. opener	226
over	overstatement	240	tone	inappropriate tone	244
par	faulty parallelism	633	trans	transition needed	646
pct	faulty punctuation	635	trite	overused expression	239
ap/	apostrophe	642	ts	faulty topic sentence	206
[ ] /	brackets	646	var	sent. variety needed	233
: /	colon	637	w	too many words	225
, /	comma	638	wo	faulty word order	219
– – /	dash	646	wv	weak verb	227
. . . /	ellipses	645	ww	wrong word	240
! /	exclamation point	637	#	wrong use of numbers	651
– /	hyphen	649	¶	new paragraph needed	205
			¶ coh	para. lacks coherence	207
			¶ lgth	para. too long or short	208
			¶ un	paragraph lacks unity	206

*Numbers refer to the first page of major discussion in the text.

150 Best Jobs™ for Your Skills

Part of JIST's Best Jobs™ Series

Michael Farr
America's Career Expert **and Laurence Shatkin, Ph.D.**

Foreword by Linda Kobylarz, President, Kobylarz & Associates Career Consultants

Also in JIST's *Best Jobs* Series

- *Best Jobs for the 21st Century*
- *200 Best Jobs for College Graduates*
- *300 Best Jobs Without a Four-Year Degree*
- *250 Best Jobs Through Apprenticeships*
- *50 Best Jobs for Your Personality*

- *40 Best Fields for Your Career*
- *225 Best Jobs for Baby Boomers*
- *250 Best-Paying Jobs*
- *175 Best Jobs Not Behind a Desk*

JIST Works
America's Career Publisher

150 Best Jobs for Your Skills

© 2008 by JIST Publishing, Inc.

Published by JIST Works, an imprint of JIST Publishing, Inc.
8902 Otis Avenue
Indianapolis, IN 46216-1033

Phone: 1-800-648-JIST   Fax: 1-800-JIST-FAX
E-mail: info@jist.com   Web site: www.jist.com

Some Other Books by the Authors

Michael Farr

The Quick Resume & Cover Letter Book
Getting the Job You Really Want
The Very Quick Job Search
Overnight Career Choice

Laurence Shatkin

90-Minute College Major Matcher

Quantity discounts are available for JIST products. Have future editions of JIST books automatically delivered to you on publication through our convenient standing order program. Please call 1-800-648-JIST or visit www.jist.com for a free catalog and more information.

Visit www.jist.com for information on JIST, free job search information, book excerpts, and ordering information on our many products.

Acquisitions Editor: Susan Pines
Development Editor: Stephanie Koutek
Cover and Interior Designer: Aleata Howard
Cover Illustrations: Image Zoo
Interior Layout: Aleata Howard
Proofreaders: Linda Seifert, Jeanne Clark
Indexer: Cheryl Lenser

Printed in the United States of America

12 11 10 09 08 07   9 8 7 6 5 4 3 2 1

Library of Congress Cataloging-in-Publication Data

Farr, J. Michael.
    150 best jobs for your skills / Michael Farr and Laurence Shatkin.
        p. cm. -- (JIST's best jobs series)
    Includes index.
    ISBN 978-1-59357-417-8 (alk. paper)
1. Vocational guidance--United States. 2. Vocational interests--United States. 3. Occupations--United States. 4. Job hunting--United
    States. I. Shatkin, Laurence. II. Title. III. Title: One hundred and fifty best jobs for your skills.
    HF5382.5.U5F3634 2007
    331.7020973--dc22
                                    2007011236

We have been careful to provide accurate information throughout this book, but it is possible that errors and omissions have been introduced. Please consider this in making any career plans or other important decisions. Trust your own judgment above all else and in all things. Results provided from the skills assessment are part of a whole-person approach to the assessment process. They provide useful information that individuals can use to identify their strengths and to identify training needs and occupations that they may wish to explore further. The skills assessment should never be used by businesses or organizations to screen applicants for jobs or for programs of education or training.

ISBN 978-1-59357-417-8